Advances in

INORGANIC CHEMISTRY

Volume 57

Advances in
INORGANIC CHEMISTRY
Including Bioinorganic Studies

EDITED BY

Rudi van Eldik

Institute for Inorganic Chemistry
University of Erlangen-Nürnberg
91058 Erlangen
Germany

Ivano Bertini

CERM and Department of Chemistry
University of Florence
I-50019 Sesto Fiorentino
Italy

VOLUME 57: Relaxometry of Water-Metal Ion Interactions

ELSEVIER
ACADEMIC PRESS

Amsterdam Boston Heidelberg London New York Oxford
Paris San Diego San Francisco Singapore Sydney Tokyo

ELSEVIER B.V.
Radarweg 29
P.O. Box 211, 1000 AE
Amsterdam, The Netherlands

ELSEVIER Inc.
525 B Street, Suite 1900
San Diego, CA 92101-4495
USA

ELSEVIER Ltd
The Boulevard, Langford Lane
Kidlington, Oxford OX5 1GB
UK

ELSEVIER Ltd
84 Theobalds Road
London WC1X 8RR
UK

First edition 2005

Library of Congress Cataloging in Publication Data
A catalog record is available from the Library of Congress.

British Library Cataloguing in Publication Data
A catalogue record is available from the British Library.

ISBN: 0-12-023657-5
ISSN: 0898-8838

♾ The paper used in this publication meets the requirements of ANSI/NISO Z39.48-1992 (Permanence of Paper).
Printed in The Netherlands.

CONTENTS

Introduction: General Theory of Nuclear Relaxation

DANIEL CANET

NMR Relaxation in Solution of Paramagnetic Complexes: Recent Theoretical Progress for $S \geq 1$

JOZEF KOWALEWSKI, DANUTA KRUK and GIACOMO PARIGI

^{1}H NMRD Profiles of Paramagnetic Complexes and Metalloproteins

IVANO BERTINI, CLAUDIO LUCHINAT and GIACOMO PARIGI

Gd(III)-Based Contrast Agents for MRI

SILVIO AIME, MAURO BOTTA and ENZO TERRENO

Relaxation by Metal-containing Nanosystems

R. N. MULLER, L. VANDER ELST, A. ROCH, J. A. PETERS, E. CSAJBOK, P. GILLIS and Y. GOSSUIN

Magnetic Relaxation Dispersion in Porous and Dynamically Heterogeneous Materials

JEAN-PIERRE KORB and ROBERT G. BRYANT

Water and Proton Exchange Processes on Metal Ions

LOTHAR HELM, GAËLLE M. NICOLLE and ANDRÉ E. MERBACH

Nuclear Magnetic Relaxation Studies on Actinide Ions and Models of Actinide Complexes

JEAN F. DESREUX

Technical Aspects of Fast Field Cycling

GIANNI FERRANTE and STANISLAV SYKORA

PREFACE

This thematic volume focuses on “Relaxometry of water-metal ion interactions”. Water interacts with metal ions in a variety of contexts: from aqueous solutions of inorganic salts to enzymatic catalysis. The investigation of water-metal ion interactions is conveniently performed through water ^{1}H NMR at different magnetic fields. The technique is referred to as relaxometry. The most interesting magnetic fields are between 0.01 and 100 MHz, where proton relaxation undergoes maximal variations. ^{2}H and ^{17}O NMR are complementary and meaningful and, indeed, occasionally such measurements are also performed.

I have invited Professor Ivano Bertini (University of Florence, Italy), who is an expert in this field, to be the co-editor of this volume. Professor Bertini is a well-known inorganic chemist that has significantly contributed, both experimentally and theoretically, to the understanding of the relations between water ^{1}H relaxation and unpaired electron relaxation. On its turn, electron relaxation determines the high resolution NMR behavior of paramagnetic substances, a field in which he is quite active.

The present volume includes nine contributions. In Chapter 1, Daniel Canet presents a general introduction to the topic and focuses on general theory and nuclear relaxation. This is followed by a chapter by Jozef Kowalewski, Danuta Kruk and Giacomo Parigi on the theoretical approaches to NMR relaxation in solution of paramagnetic metal complexes. In Chapter 3, Ivano Bertini, Claudio Luchinat and Giacomo Parigi treat ^{1}H nuclear magnetic resonance dispersion profiles of paramagnetic complexes and metalloproteins and discuss the dynamic and energetic parameters obtained from the measurements. Gd(III)-based contrast agents for magnetic resonance imaging is reviewed by Silvio Aime, Mauro Botta and Enzo Terreno in Chapter 4. Relaxation by metal containing nano systems is covered by Muller, Vander Elst, Roch, Peters, Csajbok, Gillis and Gossuin in Chapter 5. The theme of magnetic relaxation dispersion in porous and dynamically heterogeneous materials is covered in Chapter 6 by Jean-Pierre Korb and Robert Bryant. The following chapter covers water and proton exchange processes on metal ions and is composed by Lothar Helm, Gaelle Nicolle and André Merbach. Nuclear magnetic relaxation studies on actinide ions and models of actinide complexes are covered by Jean Desreux in Chapter 8. The final chapter of this thematic issue is devoted to technical aspects of fast field cycling and is presented by Gianni Ferrante and Stanislav Sykora, from the Stelar company that has commercialized such equipment. The first commercial prototype was produced within a European Union funded project and now a much more sensitive instrument is being commercialized again within another European Union funded project.

The mentioned authors are gratefully acknowledged for their efficient collaboration and remarkable efforts made to meet all set deadlines for the production of this volume.

I thoroughly believe that these contributions cover important advances in inorganic and bioinorganic chemistry, since information on the dynamics of water and on the molecules/ions to which water is bound is obtained. Sometimes, structural and energetic parameters are also obtained. These are precious and general data which are uniquely obtained by this approach. Applications in MRI are common. I trust that the inorganic and bioinorganic chemistry community will benefit from this thematic volume.

Rudi van Eldik
University of Erlangen-Nürnberg
Germany
September 2004

ERRATUM

"Potential applications for the use of lanthanide complexes as luminescent biolabels" by Graham R. Moston, Jean S. Flemming and Sally Brooker

Advances in Inorganic Chemistry, Vol. 55 (2004) pp. 361–442.

Due to a printing error, details are missing from the structures depicted in Figs. 29 and 31. The Figures are reprinted here in their entirety.

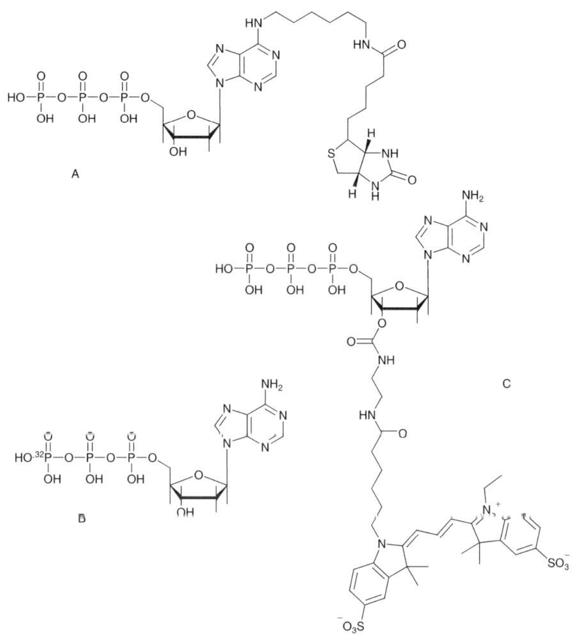

Fig. 29. Structure of a variety of labeled nucleotides. A: biotin-labeled UTP; B: Gamma labeled ^{32}P-ATP; C: The ribose-modified fluorescent ATP analogue, Cy3-EDA-ATP, which was synthesized by conjugating Cy3 reactive dye with 2′(3′)-*O*-[*N*-(2-aminoethyl)carbamoyl] ATP (EDA-ATP).

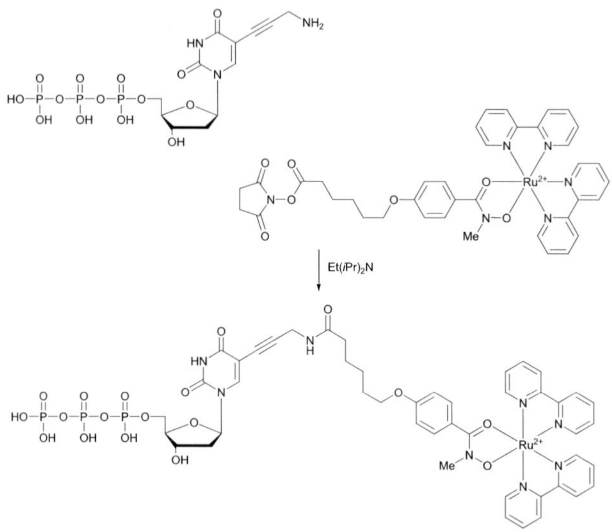

Fig. 31. Connection of 5-(3″-aminopropynyl)-2′-deoxyuridine-5′-triphosphate to the succinimide active ester arm of a ruthenium polypyridyl complex.

INTRODUCTION: GENERAL THEORY OF NUCLEAR RELAXATION

DANIEL CANET

Méthodologie RMN, Université Henri Poincaré – Nancy I, Faculté des Sciences, B.P. 239, 54506 Vandoeuvre-les-Nancy (Cedex), France

I. What Is Nuclear Relaxation?

Relaxation is a general term for phenomena that bring a system back toward its equilibrium state. We can think of a pendulum, the oscillations of which are damped by friction forces. In Nuclear Magnetic Resonance (NMR), it is the restoration of the nuclear magnetization to its equilibrium configuration after it has been perturbed. Let us recall that an ensemble of nuclear spins, in the presence of a static magnetic field B_0, leads to a macroscopic magnetization aligned with B_0 (this component is usually called longitudinal, while components in a plane perpendicular to B_0 are usually called transverse). Obtaining an NMR spectrum implies to take the nuclear magnetization into a transverse configuration. This is one possible perturbation, but, as we see later, there are many others. It can, however, be stated at the onset that longitudinal and transverse components of magnetization behave independently, the former recovering to the equilibrium magnetization according to a longitudinal relaxation time T_1 (also called spin-lattice relaxation time), the latter disappearing according to a transverse relaxation

ADVANCES IN INORGANIC CHEMISTRY
VOLUME 57 ISSN 0898-8838 / DOI 10.1016/S0898-8838(05)57001-6

time T_2 (also called spin–spin relaxation time). However, in the case of the usual liquids, T_1 and T_2 are mostly equal and lie in the 100 ms–30 s range. Conversely, in solids, T_2 may be quite short – some tens of microseconds – whereas T_1 may become exceedingly long – several hundred of seconds. Although, the consideration of only two relaxation times is a simplistic view, it may help one to understand the basics of the somewhat complicated aspects of spin relaxation. Historically, as reported by Slichter in one of the most famous books on NMR (*1*), the very first attempt to detect a nuclear resonance was hampered by the choice of a solid material with a too long longitudinal relaxation time. In general, regardless of the detection mode of the NMR signal, the line-width at half height $\Delta \upsilon$ is, in principle, controlled by the transverse relaxation time T_2; indeed $\Delta \upsilon = 1/(\pi T_2)$ and this tells us that short T_2s will be responsible for a lack of resolution and may require some numerical procedures in order to separate overlapping peaks. In fact, the effective transverse relaxation time (denoted by T_2^*) is generally shorter than T_2 as it comprises contains the effects of B_0inhomogeneity (this latter contribution can be denoted by $1/T_2^{'}$ so that $1/T_2^* = 1/T_2 + 1/T_2^{'}$). In addition to the static magnetic field B_0, one needs an oscillating magnetic field, also called radio-frequency (rf) field (generally denoted as B_1) to induce transitions. NMR experiments are performed by rf pulse methods which yield the so-called free induction decay (fid) or interferogram. As Fourier transform of the fid leads to the conventional spectrum, this form of spectroscopy is usually referred to as Fourier Transform NMR (FTNMR). One of its major interests lies in the ability to improve efficiently the signal-to-noise (S/N) ratio (*2*). Indeed, it is well known that NMR is characterized by a poor sensitivity, compared to most other spectroscopy methods, and is often necessary to repeat the experiment in order to add coherently the relevant fids. Again, this accumulation procedure is governed by T_1 and some care has to be exercised in order to obtain quantitative results. Some more subtle effects, for instance cross-relaxation, which manifests itself through the nuclear Overhauser effect (nOe), must be considered for thoroughly analyzing an NMR spectrum. This is discussed later in the case of multispin systems.

The Relaxation parameters are of great interest as they arise from the various interactions to which nuclear spins are subjected with the most common being the classical dipolar interaction (between the magnetic dipoles associated with nuclear spins). In order to be active, these interactions must be time dependent, for instance *via* molecular motions. Therefore relaxation parameters not only encompass structural information (through the interactions themselves), but also dynamical information (through this time modulation). This latter information can be about (i) overall motions, translational or rotational, governed essentially by the molecular volume and the viscosity of the medium in which are embedded the molecules under investigation, (ii) local motions such as internal rotations around C–C bonds, or segmental motions in molecules or aggregates of appreciable size, (iii) exchange between two different molecules or between two distinct sites within the same molecule. Before discussing in detail the various relaxation mechanisms, we consider the experimental methods which yield relaxation parameters with accuracy and reliability.

A. Longitudinal Relaxation

Any experimental method aiming at the measurement of a dynamical parameter invariably starts by a perturbation which moves the system out of its equilibrium state; this is followed by an evolution period whose duration is generally denoted by τ, which is such that the dynamical parameters of interest manifest themselves. The ultimate stage consists of reading the actual state of the system. Generally, this measurement is performed for different durations of the evolution period so as to properly evaluate the dynamical parameters. Regarding longitudinal relaxation, it seems obvious that the first two stages (initial perturbation and evolution) involve only longitudinal magnetization. However, the read stage necessarily implies transverse components which are the only ones giving rise to a detectable signal. In other words, after evolution of longitudinal magnetization, it must be converted into transverse magnetization. The efficiency of the whole process will be optimal, thus warranting the greatest accuracy for the determination of the considered relaxation parameter, provided that the initial perturbation is maximum. For longitudinal relaxation, this corresponds to a *complete inversion* by means of a 180° (or π) rf pulse, in such a way that magnetization *remains longitudinal* even during the evolution period. Subsequent to the evolution period, the ongoing value of the longitudinal magnetization is read with a 90° pulse which converts it into directly detectable transverse magnetization . This is the well-known *inversion recovery* (*3*) experiment which is schematized in Fig. 1.

At this point, it may be instructive to recall briefly how a rf pulse works. Let us denote by ν_r the transmitter frequency and by ν_0 the resonance frequency corresponding to a given peak in the NMR spectrum. It is convenient to define a frame rotating at the frequency ν_r around the z axis which is the direction of the static magnetic field B_0. In this rotating frame, it can be shown that the radiofrequency field B_1 acts as if it is stationary, say along x. Furthermore, if the condition $(\gamma B_1/2\pi) \gg |\nu_r - \nu_0|$ is fulfilled (the gyromagnetic ratio, γ is a constant characteristic of the considered isotope), the sole magnetic field which is able to act on nuclear magnetization is B_1, so that a precessional motion around B_1 occurs at the angular velocity γB_1 (in the same way that transverse magnetization precesses at an angular velocity γB_0 – the so-called Larmor frequency – in the presence of a magnetic field B_0). Thus starting from magnetization along z, a rf field applied for a time τ_{90}, such that $\gamma B_1 \tau_{90} = \pi/2$, takes the magnetization onto the y axis whereas if it is applied for twice this duration, it produces a complete inversion.

In the first approach, we shall limit ourselves to the hypothesis where nuclear magnetization conforms to the Bloch equations. Concerning the

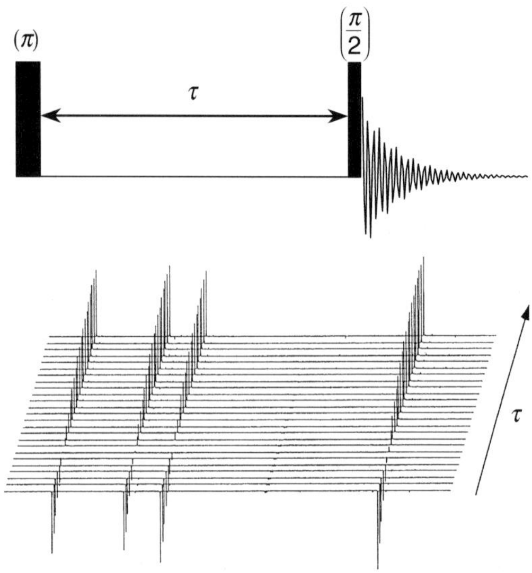

FIG. 1. Top: Scheme of an inversion recovery experiment yielding the longitudinal relaxation time (inversion is achieved by mean of the (π) radiofrequency (rf) pulse, schematized by a filled vertical rectangle). Free induction decays (fid; represented by a damped sine function) resulting from the (π/2) read pulse are subjected to a Fourier transform and lead to a series of spectra corresponding to the different τ values (evolution period). Spectra are generally displayed with a shift between two consecutive values of τ. The analysis of the amplitude evaluation of each peak from $-M_0$ to M_0 provides an accurate evaluation of T_1. Bottom: the example concerns carbon-13 T_1 of *trans*-crotonaldehyde with the following values (from left to right): 20.5 s, 19.8 s, 23.3 s, and 19.3 s.

longitudinal component, one has:

$$\frac{dM_z}{dt} = -\frac{M_z - M_0}{T_1} \tag{1}$$

M_0 is the equilibrium magnetization.

This first-order differential equation can be easily solved to yield

$$M_z(\tau) = M_0 + [M_z(0) - M_0]\exp(-\tau/T_1). \tag{2}$$

We can verify that the right-hand side of (2) is equal to $M_z(0)$ for $\tau = 0$ and to M_0 for $\tau \to \infty$. For the ideal case of a perfect inverting (180°) pulse, $M_z(0) = -M_0$ and (2) reduces to

$$M_z(\tau) = M_0[1 - 2\exp(-\tau/T_1)], \tag{3}$$

or, in a logarithmic form

$$\ln\left[\frac{M_0 - M_z(\tau)}{2M_0}\right] = -\frac{\tau}{T_1}. \tag{4}$$

This latter form is especially convenient for a quick determination of T_1, which is the inverse of the slope of the linear representation of $\ln\left[\frac{M_0 - M_z(\tau)}{2M_0}\right]$ vs. τ. M_0 can be measured by means of a single read-pulse or for a time τ of the order of $5T_1$ (which ensures a return to equilibrium for greater than 99% of the magnetization).

It turns out that, in the case of an imperfect inverting pulse, the factor of 2 in equation (3) must be substituted by an unknown factor K (< 2); it is thus recommended to turn to a non-linear fit of $M_z(\tau) = M_0[1 - K\exp(-\tau/T_1)]$, where the three quantities M_0, K, and T_1 have to be refined, starting for example from values deduced from (4). It must be stressed that the T_1 measurement, as described above, does not require the system to have returned to equilibrium between two experiments with different τ values, or between two consecutive scans if accumulation is necessary for improving the S/N ratio (*4*). If the repetition time T is smaller than $5T_1$, the factor K of the above equation depends on the ratio T/T_1 here and, just as before, has to be adjusted for each resonance.

An interesting alternative to the inversion recovery experiment is the saturation recovery experiment (*5*) which proceeds by an initial saturation of the nuclear magnetization (total cancellation by application of long scrambling rf pulses). In that case, the initial condition is $M_z(0) = 0$ instead of $-M_0$ and the evolution is governed by $M_z(\tau) = M_0[1 - \exp(-\tau/T_1)]$. Although, the dynamic range is reduced by a factor of two, there are two distinct advantages i.e., neither a waiting time nor an adjustment of the π pulse is needed. It can be emphasized that this method becomes valuable when magnetization cannot be inverted in the case of a very short T_2 (because of the relaxation taking place during the pulse) or in the case of the so-called "radiation damping" (*6*). This occurs when the NMR signal is strong, and induces a current in the detection coil capable of producing an alternating magnetic field which counteracts the magnetization motion, thus tending to bring it back to $+z$; this phenomenon is becoming very common with high field sensitive spectrometers when dealing with a highly concentrated species or the solvent for instance (and more especially, the huge water proton signal in aqueous solutions).

The above considerations are valid as long as the NMR signal is strong enough. The subject of this volume is, in part, NMRD (Nuclear Magnetic Resonance Dispersion) profiles, that is the T_1 evolution with the applied B_0 field or rather with the Larmor frequency given by $\nu_0 = \gamma B_0/2\pi$ including frequencies in the kilohertz range (so as to probe large correlation times, as seen later). It turns out that, in such circumstances, the NMR signal becomes very weak thus precluding its direct observation and making impossible the determination of T_1 by the methods described earlier.

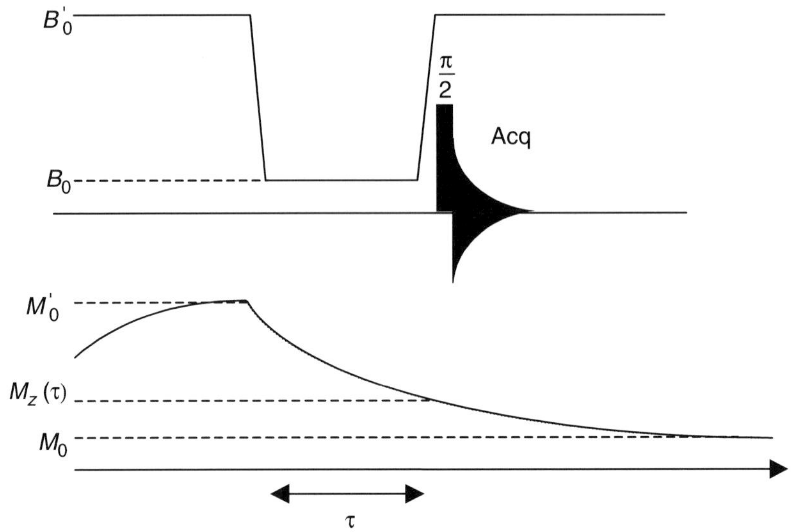

FIG. 2. The different stages of a field-cycling experiment. Spins are pre-polarized in a relatively high static magnetic field (B'_0). The relevant magnetization then decays to its equilibrium value in the field B_0 according to the longitudinal relaxation time of interest (at a frequency equal to $\gamma B_0/2\pi$). For sensitivity reasons, magnetization is read again in the B'_0 field (a $\pi/2$ pulse yielding an fid: Acq).

Alternative procedures, based on detection of the NMR signal at a higher frequency, have therefore to be worked out. The idea is to pre-polarize nuclear spins at a relatively high static magnetic field (B'_0), then to let them evolve in the (weak) B_0 field of interest and to finally "read" the state of nuclear magnetization again in the relatively high B_0, all for sensitivity reasons. Two approaches can be considered, one which uses a mechanical (pneumatic) device which moves (as quickly as possible) the sample from high to low field and *vice-versa* (*7,8*) and the other which relies on rapid modifications of the B_0 field values (*9*). The latter procedure is known as "field cycling" and will be briefly described with reference to Fig. 2 (more details can be found in Chapter 9 of this volume).

When longitudinal magnetization has reached its equilibrium value (M'_0) in the B'_0 field, the latter is suddenly lowered to the value of interest, i.e., B_0. Longitudinal magnetization then tends to decay toward its new equilibrium value (M_0) according to

$$M_z(\tau) = M'_0 \exp(-\tau/T_1) + M_0 \tag{5}$$

T_1 in (5) is precisely the quantity that has to be measured (the longitudinal relaxation time which prevails for the B_0 field). $M_z(\tau)$ is measured in the higher magnetic field in order to benefit from a larger signal (let us recall that, in a general way, the NMR signal is proportional to $\gamma^3 B^2$, where B is the static magnetic field at the measurement stage). This implies again a rapid switch, this time from B_0 toward B_0'. The method, although very efficient, is hampered by two drawbacks: (i) it is limited to the determination of T_1 (transverse relaxation times are not presently accessible by field cycling methods), (ii) it is necessarily a low-resolution experiment (B_0' generally does not exceed 1 T; furthermore, a magnet with rapid switching capabilities cannot deliver a very homogeneous magnetic field).

B. Transverse Relaxation

As stated before, the initial perturbation should be maximal with respect to the equilibrium state. Since we are dealing with transverse magnetization here, this maximal perturbation is obviously a 90° pulse. However, it can be immediately noticed that signals collected after a simple read-pulse, decay exponentially according to a time constant T_2^* which differs from the genuine T_2 by a contribution due to the static induction B_0 inhomogeneity:

$$\frac{1}{T_2^*} = \frac{1}{T_2} + (B_0 \text{ inhomogeneity}) \tag{6}$$

Experimental methods must therefore be devised for removing this contribution. For reaching this goal, an ingenious procedure was proposed by Hahn (*10*) in the early 1950s, soon after the first NMR experiments. It consists of the so-called echo sequence which involves a 180° pulse in the middle of an interval of duration 2τ. To understand how this sequence works, let us look at the events occurring after the application of the 90° pulse. Whenever nuclear magnetization is taken to the transverse plane, it precesses at the frequency $\nu_0 = \frac{\gamma B_0}{2\pi}(1-\sigma)$, where σ stands for the shielding coefficient which defines the chemical shift, that is the differentiation of various nuclei of the same isotopic species as a function of the electronic distribution by which they are surrounded (it is indeed this feature which makes NMR such a powerful technique for identifying nuclei – for instance protons – belonging to different chemical environments). A coil whose axis is in the transverse plane is therefore capable of detecting a signal (through an induced emf i.e., electromotive force) oscillating at the frequency ν_0. Now if this signal is detected with respect to a reference frequency precisely equal to ν_r (the frequency of the rotating frame, already invoked for describing the effects of a radiofrequency pulse, corresponding to the experimental arrangement used in most spectrometers), precession actually occurs at a low frequency $|\nu_0 - \nu_\gamma|$. Of course, the signal is attenuated by T_2^* and, assuming a (crude) exponential decay due to B_0 inhomogeneity (a situation generally encountered in liquid phase), one ends up with a damped sine or cosine function of the form:

$$M_0 \cos[2\pi(\nu_0 - \nu_r)t]\exp(-t/T_2^*) \tag{7}$$

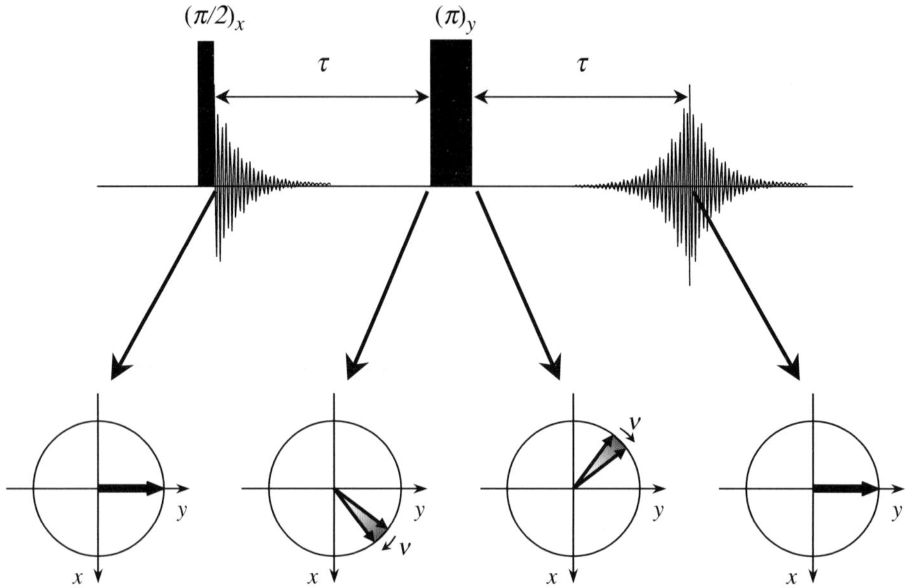

FIG. 3. The basic Hahn sequence for the measurement of the transverse relaxation time T_2. Any precession motion characterized by the frequency ν in the rotating frame is refocused. This precession may arise either from chemical shift or from B_0 inhomogeneity (symbolized by the shaded area, which has been strongly reduced for visualization purposes; owing to the fast decay of the fid, it should in fact extend to the whole circle).

Let us now suppose that the initial 90° rf pulse acts along the x axis of the rotating frame taking the magnetization to the y axis and that after a time τ, we apply a 180° pulse along the y axis as shown in Fig. 3; this has the virtue of reversing any chemical shift effect so that after an additional time τ, the magnetization lies again along the y axis and a signal reappears in the form of an echo.

This pulse sequence, $(\pi/2)_x - \tau - (\pi)_y - \tau -$ acquisition, is the basic Hahn sequence, which yields in principle the true T_2 since any precession effect is removed (in particular, precession due to B_0 inhomogeneity), leaving a transverse magnetization attenuated according to the transverse relaxation time. By reference to the Bloch equations relative to transverse magnetization

$$\frac{dM_{x,y}}{dt} = -\frac{M_{x,y}}{T_2} \tag{8}$$

one can recognize that the Fourier transform of the half-echo leads to a signal of amplitude $M_0\exp(-2\tau/T_2)$. For a set of τ values, it appears thus possible to extract an accurate value of T_2 for each line in the spectrum. Unfortunately, this analysis does not take into account translational diffusion phenomena that will be now considered.

We shall assume that the B_0 induction is not perfectly homogeneous; for simplicity and without loss of generality, hypothesize that it varies linearly across the whole sample in the X direction of the laboratory frame, so that the induction sensed by a molecule at abscissa X is of the form

$$B(X) = B_0 + g_0 X \tag{9}$$

Here g_0 stands for a *uniform gradient* of the static field. Let us recall that such a gradient can be purposely created with the aim of producing NMR images or measuring the translational diffusion coefficient (*11*). This gradient can also result from a non-existing or incomplete compensation of the genuine inhomogeneity of B_0 (imperfect *shimming;* shims are additional coils used for correcting the main field B_0 inhomogeneity). In any event, the precession frequency depends on the *location* of the considered molecule *via* the *spatial* dependence of B_0. We have just seen that any precession effect is refocused by an echo sequence whose goal is precisely to get rid of B_0 inhomogeneity. However, this feature is impaired if, during the refocusing process, molecules translate from a location X to a location X' for which the precession frequency is different from the one at location X. This should produce a further signal attenuation or an additional "defocusing" which arises from self-diffusion phenomena. The latter are characterized by a self-diffusion coefficient D which results in the following expression (*11,12*):

$$M_{x,y}(2\tau) = M_0 \exp(-2\tau/T_2) \exp\left[-\left(D\gamma^2 g_0^2/3\right)\left(2\tau^3\right)\right] \tag{10}$$

It can be recognized that the compensation of all precession effects is still accounted for in (9). It can also be stressed that translational diffusion in the presence of a gradient generates a decay depending on τ^3, whereas transverse relaxation produces a decay depending on τ. The idea of Carr and Purcell (*13*) was to minimize the effects of diffusion in a T_2 experiment. This goal is achieved by a train of 180° pulses of Fig. 4 and stems from the fact that the differential equation must be solved for each interval following a 180° pulse since, at that point, the magnetization sign is suddenly modified; consequently, new boundary conditions prevail.

Extending the analysis of the Hahn sequence to the present experiment, we obtain the amplitude of the nth echo:

$$M_{x,y}(2n\tau) = M_0 \exp(-2n\tau/T_2) \exp\left[-\left(D\gamma^2 g_0^2/3\right)\left(2n\tau^3\right)\right] \tag{11}$$

We note that for an evolution whose duration is identical to that of a Hahn sequence, the argument of the exponential relevant to diffusion has been divided by n^2. Therefore, the remedy to make negligible translational diffusion effects consists simply of increasing n, which amounts to bring the π pulses closer to each other.

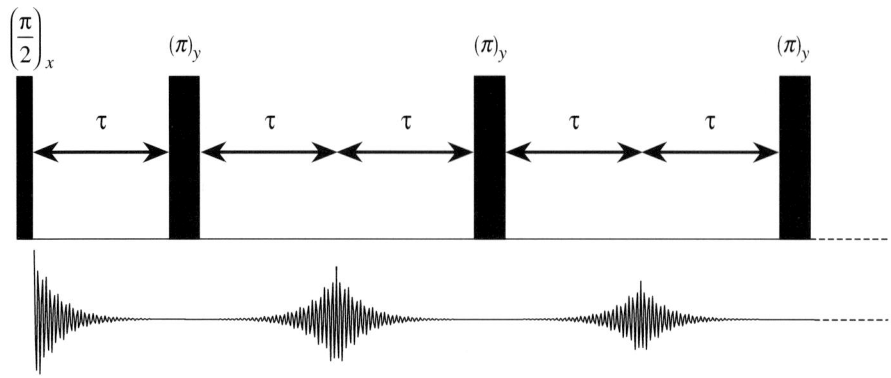

FIG. 4. The CPMG pulse sequence. An echo is formed halfway between two consecutive π pulses. The echo amplitude (or the Fourier transform of the half-echo) provides an evaluation of T_2 less affected by translational diffusion than in the simple Hahn sequence. The phase change of π pulses with respect to the initial π/2 pulse cancels the effect of (π) pulse imperfections.

The trick introduced by Meiboom and Gill (*14*) is to *dephase* all π pulses in the Carr–Purcell train by an angle of 90° with respect to the initial π/2 pulse. It is easily shown that, without this phase change, imperfections of the π pulses are cumulative, whereas with the 90° phase change, a self-compensation occurs for all echoes of even number. The CPMG (Carr–Purcell–Meiboom–Gill) experiment can be handled in two ways:

- If the spectrum involves only one resonance (or if linewidths do not allow for the separation of several resonances), a *single experiment* can be run with acquisition of the amplitude of each echo along the pulse train (for sensitivity enhancement, accumulations can be carried out). This experiment is especially valuable for determining the relative proportions of two species which differ by their transverse relaxation time, for instance the two types of water (free and bound), if exchange between these two states is sufficiently slow. For this type of measurement, a "low resolution" spectrometer (without any shim system) proves to be quite sufficient.
- If the spectrum involves several well-resolved resonances, one can proceed by Fourier transform of the half-echo (possibly with signal accumulation). Many experiments are performed by varying n so that the successive values of $2n\tau$ (see formula (11)) lead to T_2 with the required accuracy. Although of little relevance in the present context, it can be mentioned that in homonuclear coupled spin systems, the magnetization decay is affected by "*J*-modulation" (*15*). It can be recalled that, in addition to chemical shift, high resolution NMR spectra (of liquid samples) are characterized by a fine structure arising from the so-called J coupling (or indirect coupling or scalar coupling), which occurs *via* the spins of bonding electrons.

C. The Rotating Frame Relaxation Time $T_{1\rho}$ (*16*)

Most of the drawbacks mentioned above can be circumvented by a simple experiment, though somewhat instrumentally demanding. The sequence is depicted in Fig. 5. First, a standard $(\pi/2)_x$ flips the nuclear magnetization toward the y axis of the rotating frame; thereafter a radiofrequency field B_1 is applied along that direction for a duration τ. This rf field must be sufficiently strong so as to avoid off-resonance effects but not too strong so as to prevent probe deterioration. During the τ interval, magnetization should nutate around B_1; since both are collinear, magnetization is stationary along the y axis of the rotating frame. It is said to be locked, hence the terminology of "spin-lock" associated with this experiment.

Any modification of the magnetization thus arises from relaxation phenomena. The transverse magnetization spin-locked along B_1 must end up at its thermal equilibrium value, that is zero. The corresponding evolution is exponential with a time constant denoted by $T_{1\rho}$ (relaxation time in the rotating frame), very close (if not identical) to T_2. In practice, the signal is measured (and subsequently Fourier transformed) for a set of τ values, in successive experiments, and obeys the equation

$$S(\tau) = S_0 \exp(-\tau/T_{1\rho}) \tag{12}$$

where S_0 is the signal amplitude immediately after the $(\pi/2)_x$ pulse. An immediate advantage of the method is the absence of precession during the evolution period and thus, among other things, the absence of J-modulation. Another interesting feature is the dependence of $T_{1\rho}$ upon the amplitude of the spin-locking field, which constitutes a form of relaxometry capable of probing small motions in the kHz range.

D. Cross-relaxation Rates – Nuclear Overhauser Enhancement (nOe) Factors

Experimental methods outlined in the preceding section concern those systems which obey Bloch equations and which consequently exhibit

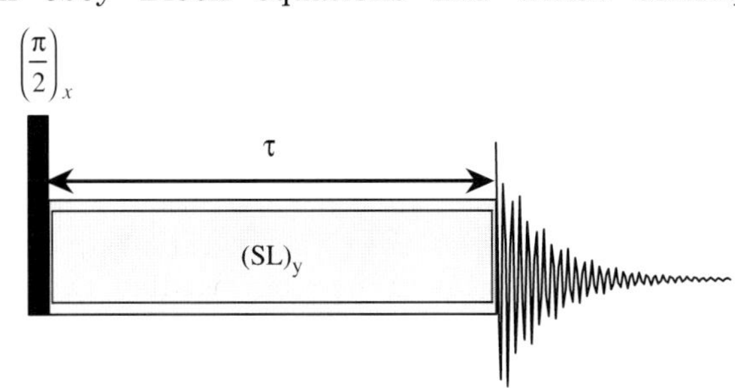

Fig. 5. Principle of a spin-lock experiment leading to the determination of the relaxation time in the rotating frame ($T_{1\rho}$). $(SL)_y$ stands for the spin-lock period which corresponds to the application of a rf field along the y axis of the rotating frame.

mono-exponential evolutions according to a time constant T_1 for longitudinal magnetization and T_2 for transverse magnetization. Although, in some instances, Bloch equations are either strictly valid or represent a good approximation, there exist situations where bi-exponentiality shows up obviously and it would be a shame not to exploit this feature. Such a situation occurs whenever two spins I and S interact by a time-dependent mechanism (they can be (i) two nuclear spins or, (ii) one nuclear spin I and one electron spin S). Such a process is dubbed *cross-relaxation* and means that any modification of the I magnetization induces a modification of the S magnetization which adds up to the specific evolution of S magnetization (however, due to a factor of 658 between electron and proton gyromagnetic ratios, perturbing I will have a negligible effect on S if it is an electron spin); the symmetrical process, between S and I, of course, holds as well. More formally, this coupling can be expressed *via* the famous Solomon equations (*17*) which are written below for the two longitudinal magnetizations I_z and S_z (as already mentioned, any coupling between longitudinal and transverse magnetizations is impossible):

$$\begin{aligned} \boldsymbol{d}I_z|\boldsymbol{d}t &= -R_1^I(I_z - I_{eq}) - \sigma(S_z - S_{eq}) \\ \boldsymbol{d}S_z|\boldsymbol{d}t &= -R_1^S(S_z - S_{eq}) - \sigma(I_z - I_{eq}) \end{aligned} \tag{13}$$

R_1^I represents the specific longitudinal relaxation rate of spin I ($R_1^I = 1/T_1^I$). σ is the *cross-relaxation* rate which reflects the coupling between the two magnetizations alluded to above; I_{eq} stands for the equilibrium magnetization. It can be seen that Solomon equations are in fact Bloch equations to which the cross-relaxation rate σ has been appended. σ may have two origins:

- The dipolar interaction between the two spins I and S, is modulated by molecular motions. This contribution is very interesting with regard to the information it contains because (as detailed later), it is proportional to r_{IS}^{-6}, where r_{IS} is the distance between I and S; this will therefore tell us if the two spins are close to each other (this is of course an essential information in terms of molecular structure or intermolecular proximity).
- A chemical exchange process (*18*) such as spins move from site A to site B and conversely from site B to site A (of course, we are only concerned here about two nuclear spins). If we denote by τ the residence time in each site (this implies identical concentrations for A and B, thus in accord, precisely with the situation considered for the two spins I and S), then $\sigma = -1/\tau = -k$. Moreover the exchange rate k must be added to each specific rate R_1^I and R_1^S.

In the following, we shall be concerned with the determination of the cross-relaxation rate σ, essentially when I and S are nuclear spins. The bi-exponential character of the evolution of I_z and S_z (see (13)) and the difficulties associated with the analysis of such evolutions have prompted the design of more direct methods based on the *initial behavior* after an appropriate perturbation. Again, we start from equation (13) and we assume that t is short enough for ensuring that a first order expansion is adequate:

$$\begin{aligned} I_z(t) &= I_z(0) + t(\boldsymbol{d}I_z/\boldsymbol{d}t)_{t=0} \\ S_z(t) &= S_z(0) + t(\boldsymbol{d}S_z/\boldsymbol{d}t)_{t=0} \end{aligned} \tag{14}$$

It is convenient, from a conceptual and practical point of view, to express equations (14) in the form of the *initial slopes* of the reduced quantities:

$$\begin{aligned} S_I &= \frac{\boldsymbol{d}}{\boldsymbol{d}t}\left(\frac{I_{eq}-I_z}{2I_{eq}}\right)_{t=0} = -R_1^I\left(\frac{I_{eq}-I_z(0)}{2I_{eq}}\right) - \sigma\left(\frac{S_{eq}-S_z(0)}{2I_{eq}}\right) \\ S_S &= \frac{\boldsymbol{d}}{\boldsymbol{d}t}\left(\frac{S_{eq}-S_z}{2S_{eq}}\right)_{t=0} = -R_1^S\left(\frac{S_{eq}-S_z(0)}{2S_{eq}}\right) - \sigma\left(\frac{I_{eq}-I_z(0)}{2S_{eq}}\right) \end{aligned} \tag{15}$$

These initial slopes are readily evaluated since we can measure the signal corresponding to I_{eq} (or S_{eq} if S is a nuclear spin) and, since the relevant instrumental factor is identical to the one which prevails in the measurement of I_{eq} (or S_{eq}). Moreover, initial conditions can be devised for determining separately one of the three relaxation parameters R_1^I, R_1^S or σ. The simplest experiment consists of *selectively* inverting one of the two magnetizations. Consider first the selective inversion of I magnetization for which the following initial conditions hold: $I_z(0) = -I_{eq}$ and $S_z(0) = S_{eq}$. This yields for the initial slopes

$$\begin{aligned} S_I(I \text{ selectively inverted}) &= -R_1^I \\ S_s(I \text{ selectively inverted}) &= -(\gamma_I/\gamma_s)\sigma \end{aligned} \tag{16}$$

The complementary experiment (selective inversion of S magnetization) leads to

$$\begin{aligned} S_I(S \text{ selectively inverted}) &= -(\gamma_s/\gamma_I)\sigma \\ S_s(S \text{ selectively inverted}) &= -R_I^s \end{aligned} \tag{17}$$

For paramagnetic complexes (S would be an electron spin), EPR lines are usually very broad (due to very short relaxation times), and, as a consequence, inverting S cannot be considered. Furthermore and as already mentioned, because of the large difference in gyromagnetic ratios (by a factor of 658), perturbing nuclear spins has negligible effect upon electron spins. Thus, in the first of equations (13), the second term involving the cross-relaxation rate can be safely ignored (see the Solomon–Bloembergen theory in Section II.A of Chapter 2).

Another popular way of measuring the cross-relaxation rate σ relies upon the saturation (by continuous irradiation) of one of the two nuclei, say S. Referring again to the first of equations (13) with $S_z \equiv 0$ (saturation of spin S)

and with $dI_z/dt = 0$ which implies that a steady state has been reached prior to the measurement (this means that the irradiation of spin S has been applied for a sufficiently long time with respect to relaxation times), we obtain a new value for the longitudinal magnetization of spin I which will be denoted as I_{st}.

$$I_{st} = I_{eq}[1 + (\gamma_S/\gamma_I)(\sigma/R_1^I) \qquad (18)$$

Hence, provided that I_{eq} is known and that R_1^I has been determined by means of an independent experiment, I_{st} provides the cross-relaxation rate σ. This enhancement is called *nuclear Overhauser effect* (nOe) (*17,19*) from Overhauser (*20*) who was the first to recognize that, by a related method, electron spin polarization could be transferred to nuclear spins (such a method can be worked out whenever EPR lines are relatively sharp; it is presently known as DNP for *Dynamic Nuclear Polarization*). This effect is usually quantified by the so-called nOe factor η

$$\eta = (I_{st} - I_{eq})I_{eq} = (\gamma_S/\gamma_I)(\sigma/R_1^I) \qquad (19)$$

For the sake of completeness, we shall now briefly describe related two-dimensional techniques (*21*) which provide in a single experiment all cross-relaxation rates. The first one, the acronym of which is NOESY (*22*) (Nuclear Overhauser Effect SpectroscopY), is generally reserved to the study of homonuclear cross-relaxation. In a general way, the analysis of a signal $S(t_1, t_2)$ (depending on two time variables) by means of a double Fourier transform leads to a two-dimensional map in the frequency domains (ν_1, ν_2), whose cross-peaks (off-diagonal peaks for which $\nu_1 = \nu_1'$ and $\nu_2 = \nu_2'$) contain generally a correlation information regarding two nuclei resonating at frequencies ν_1' and ν_2'. Concerning the NOESY sequence depicted in Fig. 6, this correlation arises from cross-relaxation and indicates either a dipolar interaction (thus a spatial proximity of the two considered nuclei) or chemical exchange between the two relevant sites.

Let us emphasize again that the advantage of this two-dimensional technique stems from the global view of all dipolar or exchange correlations existing within the molecular system under investigation. The preparation period allows for the nuclear magnetization to return to equilibrium. The evolution period provides a labeling according to the resonance frequency of each nucleus by repeating the experiment for incremented values of t_1. It is assumed that transverse components during the mixing time t_m do not affect the final result (actually this can be achieved by several means including phase cycling procedures). This amounts to consider, prior to the mixing period, exclusively longitudinal components which are off equilibrium by an amount depending on the chemical shift of each nucleus. It is therefore conceivable that the measured signal contains effects of specific relaxation and cross-relaxation, encoded in the t_1 dimension according to the chemical shift. Thus, the diagonal peaks depend on specific longitudinal relaxation rates whereas cross-peaks (symmetrically located with respect to the diagonal) depend solely on the cross-relaxation rate between the two relevant nuclei. A similar experiment exists for measuring heteronuclear cross-relaxation rates (HOESY (*23,24*) for Heteronuclear Overhauser SpectroscopY). Of course, it employs rf pulses which must act on both nuclei. This experiment has been mostly used for studying intermolecular proximity.

We turn now to the determination of *transverse* cross-relaxation rates, by contrast to the NOESY experiment which deals with *longitudinal* cross-relaxation rates. It will be shown that the latter may be positive or negative, depending on the molecular mobility, and thus may go to zero; this situation occurs at certain measuring frequencies for slowly tumbling molecules such as proteins or nucleic acids. On the contrary, transverse cross-relaxation rates are always positive. In that sense, a ROESY experiment (*25,26*) (Rotating-frame Overhauser Effect SpectroscopY) which yields transverse cross-relaxation rates can be thought to be complementary to a NOESY experiment. For theoretical reasons explained later, transverse cross-relaxation rates do not actually affect a conventional relaxation experiment using spin echoes. Rather, they become visible in a *spin-lock* experiment (see Section I.C). To maximize their effects, the two magnetizations of interest should be taken along the locking rf field in opposite directions. Hence, a possible design for a one-dimensional experiment is as follows: a selective pulse (i.e., where intensity has been adjusted to act only on the S nucleus) takes the S magnetization toward $-z$; it is followed by a non-selective pulse which brings both magnetizations along the spin-locking field which are then allowed to cross-talk during the mixing time t_m. During the mixing time t_m, magnetizations evolve according to the relaxation rates in the rotating frame $R_{1\rho}^I$ and $R_{1\rho}^S$ and also according to the transverse cross-relaxation rate denoted by σ_ρ(or $\sigma_\perp$). This is homologous to the relaxation parameters which govern the evolution of longitudinal magnetizations (R_1^I, R_1^S, and σ). We can therefore expect a similar behavior, with, however, relaxation rates of different values in conditions of slow molecular mobility. The two-dimensional counterpart can be deduced from the sequence described above by substituting the selective pulse by a non-selective ($\pi/2$) pulse followed by an evolution period as in the NOESY sequence of Fig. 6. It yields qualitatively the same type of result as a NOESY experiment: a cross-peak reveals a non-zero transverse cross-relaxation rate.

E. Cross-correlation Rates

J-splitting, when it exists, imposes the definition of new spin quantities. These quantities also evolve according to relaxation phenomena and may interfere (by relaxation) with the usual magnetization components. This latter interference stems precisely from cross-correlation rates, i.e., relaxation parameters which involve two different mechanisms, for instance the dipolar interaction *and* the so-called Chemical Shift Anisotropy (*27,28*) (csa)

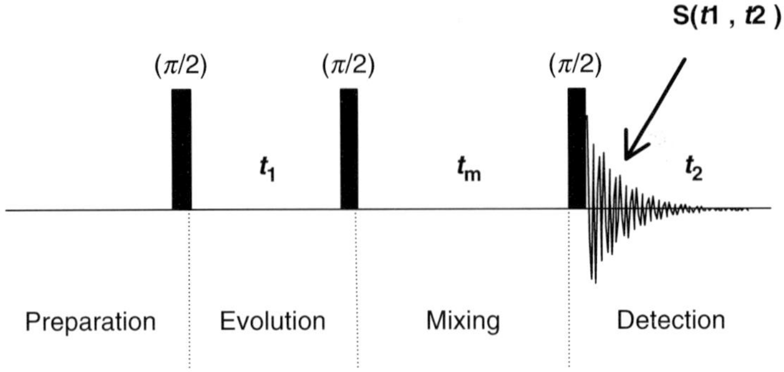

FIG. 6. Basic scheme of the NOESY sequence which provides essentially homonuclear cross-relaxation rates from a double Fourier transform of the signal $S(t_1, t_2)$.

mechanism (see Section II.B). In order to illustrate what is meant by "new spin quantities", let us consider the doublet (of splitting J) of one nucleus in a J-coupled two spin ½ system. Let us assume for simplicity that this spin is "on resonance" (that is its resonance frequency ν_0 is identical to the reference frequency ν_r which is also the frequency of the rotating frame), so that in the rotating frame, the resonance frequency is zero and the two lines of the doublet will just precess at frequency $+J/2$ and $-J/2$, respectively. As illustrated in Fig. 7, the spin state at time t can be decomposed into an "in-phase" doublet (for the components along y) and an "antiphase" doublet (for the components along x).

Using the notations of the preceding section, the in-phase doublet is obviously represented by I_y whereas it can be shown that the antiphase doublet is represented by the product operator $2I_xS_z$ (*29*). It turns out that the two quantities I_y and $2I_xS_z$ can be coupled by a csa-dipolar cross-correlation rate and, as a consequence, the two lines within the doublet possess different transverse relaxation rates (*28,30*). This is a simple example of spin relaxation by cross-correlation. In many other situations, cross-correlation effects can be visible (*31–33*). This will be discussed further in Chapter 2 (Section III.B) about interference between dipolar interaction (between two nuclear spins) and the so-called dipolar shielding anisotropy (*34*) which, in paramagnetic systems, arises from the thermally averaged magnetic moment per molecule and is formally equivalent to chemical shift anisotropy. Thus, we shall pursue with the latter turning to the longitudinal-spin order represented by $2I_zS_z$, which is the complement of I_z and S_z so as to fully describe longitudinal relaxation in a system of two spin ½ nuclei. This latter spin state can be obtained by applying, to the previously discussed antiphase doublet ($2I_xS_z$), a $(\pi/2)_y$ selective pulse which acts exclusively on the I spin. It therefore transforms I_x into I_z and ($2I_xS_z$) into ($2I_zS_z$). On the other hand, it turns out that longitudinal magnetization is coupled to the longitudinal order by csa-dipolar cross-correlation (as further discussed in Section II.D of Chapter 2), according to the extended

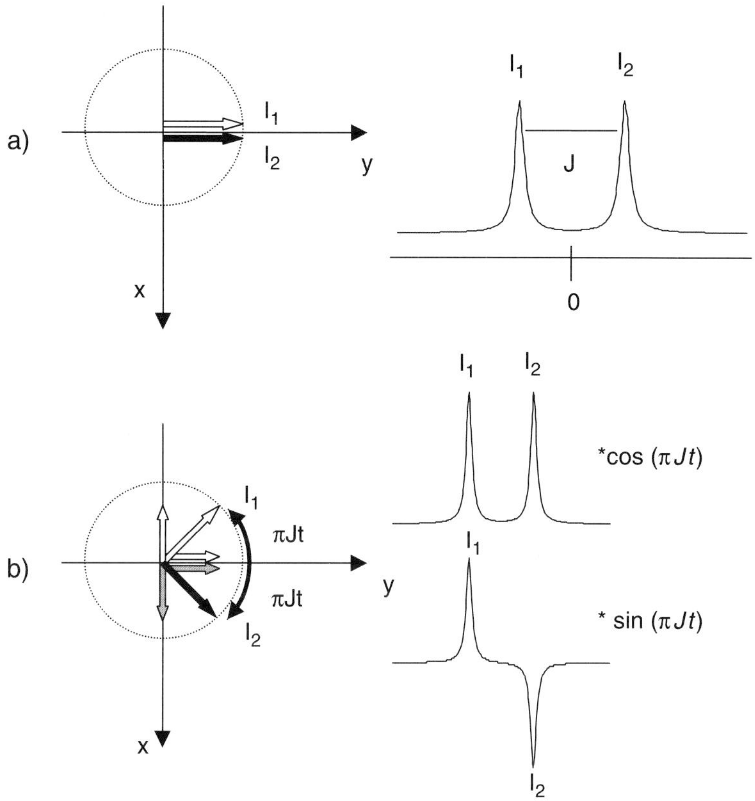

FIG. 7. Spin states of a doublet (a) after a $(\pi/2)_x$ pulse; (b) some time t later and its decomposition into an in-phase and an antiphase doublets.

Solomon equations (see, for instance, Ref. (*33*))

$$\begin{aligned} dI_z/dt &= -R_1^I(I_z - I_{eq}) - \sigma(S_z - S_{eq}) - \delta_I(2I_zS_z) \\ dS_z/dt &= -R_1^S(S_z - S_{eq}) - \sigma(I_z - I_{eq}) - \delta_S(2I_zS_z) \\ d(2I_zS_z)/dt &= -R_1^{IS}(2I_zS_z) - \delta_I(I_z - I_{eq}) - \delta_S(S_z - S_{eq}) \end{aligned} \tag{20}$$

These equations involve additional parameters with respect to the classical Solomon equations (see (13)): R_1^{IS},which is the longitudinal order specific relaxation rate, δ_I and δ_S, which are the cross-correlation rates or interference terms between the I–S dipolar interaction and chemical shift anisotropy at I and S, respectively. As a consequence, the longitudinal order can be created in the course of a standard T_1 measurement and may alter the corresponding experimental results. This feature has also become especially apparent in measurements performed on proteins with spectrometers equipped with a magnet delivering a high magnetic field B_0 (*35*).

II. Relaxation Mechanisms

One way to perturb a spin system from its equilibrium configuration or to take it back to its equilibrium state is to induce transitions. This, of course, can be done by application of a rf field or through magnetic fields governing relaxation phenomena. Consider a spin ½ system, i.e., a system involving only two energy levels. In order to induce a transition between these two states, the experimenter can apply a rf field at the Larmor frequency ν_0 such that: $h\nu_0 = \Delta E = \frac{\gamma h B_0}{2\pi}$ (in this expression, the shielding coefficient has been omitted). On the other hand, within a sample, an elementary nuclear magnetic moment μ is subjected to a local magnetic field $\boldsymbol{b}(t)$ originating from the various interactions to which this magnetic moment is subjected. Due to molecular motions these local fields are time dependent and consequently may be able to induce transitions. They must, however, mimic the action of a rf field and therefore fulfill the following conditions: (i) present some degree of coherence and (ii) be active at the frequency of the considered transition. A first global treatment, taking into account simple randomly fluctuating magnetic fields (without specifying their characteristics), will be presented. Depending on the origin of $\boldsymbol{b}(t)$, different *mechanisms* can be considered and the ones relevant to this book will be detailed thereafter. It can be borne in mind that each of them is capable of contributing to relaxation rates in an *additive* way.

A. Interaction with Randomly Fluctuating Magnetic Fields (*36*)

The magnetic field $\boldsymbol{b}(t)$ is, for instance, created by another spin (nuclear spin or the spin of an unpaired electron); in that case, it is proportional to $1/r^3$ (r: distance between the two spins). Its time dependency arises from the orientation of r and/or from the distance fluctuation. However, in this section, we shall disregard the origin of $\boldsymbol{b}(t)$ and we only rest on its general properties (assuming an isotropic medium) which arise from the random nature of molecular motions:

(i) the three components of a local magnetic field are not correlated

$$\overline{b_x(t)b_y(t)} = \overline{b_x(t)b_z(t)} = \overline{b_y(t)b_z(t)} = 0 \tag{21}$$

(the bar represents an ensemble average)

(ii) the mean value of each component is zero

$$\overline{b_x(t)} = \overline{b_y(t)} = \overline{b_z(t)} = 0 \tag{22}$$

(iii) the mean values of their squares are identical

$$\overline{b_x^2(t)} = \overline{b_y^2(t)} = \overline{b_z^2(t)} = \overline{b^2} \tag{23}$$

To understand the two latter points, one can think of a quantity whose modulus is non-zero but which can take opposite values with the same probability.

The coherence of a random field can be evaluated by its correlation function, i.e.,

$$\overline{b_x(t)b_x(0)} \neq 0 \tag{24}$$

which reflects its memory properties. Indeed, if the two quantities are totally uncorrelated (independent), one has:

$$\overline{b_x(t)b_x(0)} = \overline{b_x(t)}\ \overline{b_x(0)} = 0 \tag{25}$$

Conversely, a rf field is totally correlated because it is represented by a sine (or cosine) function and, as a consequence, its value at any time t can be predicted from its value at time zero. The efficiency of a random field at a given frequency ω can be appreciated by the Fourier transform of the above correlation function

$$J_{\mathrm{rf}}(\omega) = \int_0^\infty \overline{b_x(t)b_x(0)}\exp(-i\omega t)dt \tag{26}$$

(rf representing "random fields")
These quantities are called spectral densities and are involved in all relaxation parameters. Very often $\overline{b_x(t)b_x(0)}$ can be written as $\overline{b^2}e^{-t/\tau_c}$, where $_c$ is a *correlation time* associated with molecular reorientation (it can be considered as the time required for one radian rotation; in liquids, τ_c lies generally between 10^{-8} and 10^{-12}s). This is especially true in the case of a sphere, the reorientation of which is governed by the classical diffusion equation

$$\partial\psi(\Omega, t)/\partial t = D\Lambda\psi(\Omega, t) \tag{27}$$

Here $\psi(\Omega, t)$ is an arbitrary function of the two usual polar angles defining the sphere orientation (denoted by Ω) and of time t. Λ is the angular Laplacian while D is the rotational diffusion coefficient:

$$D = \frac{k_B T}{8\pi a^3 \eta} \tag{28}$$

(a: sphere radius; k_B: Boltzmann constant; T: absolute temperature; η: medium vicosity). For most situations encountered here, $\psi(\Omega, t)$will be the spherical harmonics Y_2^0, which is proportional to $[3\cos^2\theta(t) - 1]$, where θ is the angle between a molecular axis (the molecule of interest being a sphere) and the B_0 direction of the laboratory frame (generally, the dependence upon a single polar angle stems from the isotropy of the medium). Upon solving the diffusion equation by means of an expansion over spherical harmonics (*37*), one arrives indeed at a solution of the form e^{-t/τ_c} with $\tau_c = 1/6D$. It can be emphasized that such a result is also obtained when the correlation function is simply $\overline{\cos^2\theta(t)\cos^2\theta(0)}$. It can be remembered that, as early as 1934, Perrin (*38*) established and solved the rotational Brownian equations for an ellipsoid. The solution involves at most three different diffusion coefficients which lead to appropriate spectral densities (*39*). Returning to a simple sphere, the Fourier transform of (26) yields a Lorentzian function of the type $\frac{\tau_c}{1+\omega^2\tau_c^2}$. More precisely, we shall denote in the following as "normalized spectral density" $\tilde{J}(\omega)$, the Fourier transform (multiplied by 4π) of the correlation function of $Y_2^0(\theta) = \sqrt{5/16\pi}(3\cos^2\theta - 1)$.

$$\tilde{J}(\omega) = 4\pi \int_{-\infty}^{+\infty} \overline{Y_2^0[\theta(t)]\,Y_2^0[\theta(0)]}\exp(-i\omega t)dt \tag{29}$$

As outlined above, the normalized spectral density is especially simple when a single correlation time is involved (for instance, when the molecular reorientation can be approximated by the motion of a sphere):

$$\tilde{J}(\omega) = \frac{2\tau_c}{1 + \omega^2\tau_c^2} \tag{29A}$$

When the condition $\omega^2\tau_c^2 \ll 1$ is fulfilled (and this is the case for fast motions since the NMR measurement frequency is lower than 10^9 Hz), the spectral density becomes frequency independent. In such cases, one says that "extreme narrowing" conditions prevail.

We now intend to derive the Bloch equations in order to express T_1 and T_2 according to spectral densities at appropriate frequencies. The starting point is the evolution equation of an elementary magnetic moment μ subjected to a random field $\boldsymbol{b}$

$$\frac{d}{dt}\mu = \gamma\mu \wedge \boldsymbol{b} \tag{30}$$

One has for example μ_z

$$\frac{d}{dt}\mu_z = \gamma\left(\mu_{x'}b_{y'} - \mu_{y'}b_{x'}\right) \tag{31}$$

The prime indicates that one has switched to the so-called rotating frame in order to remove any precession effect at the angular velocity $\omega_0 = 2\pi\nu_0$, ν_0 being the resonance frequency.

$$\begin{aligned} b_{x'} &= b_x\cos(\omega_0 t) - b_y\sin(\omega_0 t) \\ b_{y'} &= b_x\sin(\omega_0 t) + b_y\cos(\omega_0 t) \end{aligned} \tag{32}$$

One can always write

$$\mu_z(t) = \mu_z(0) + \int_0^t \frac{d\mu_z}{dt'}dt' \tag{33}$$

and proceed by time-dependent perturbation methods (*40*). It can be remembered that the quantity of interest is the z component of nuclear magnetization

$$M_z(t) = \overline{\mu_z(t)} \tag{34}$$

On applying algebra (and other mathematical tricks), one arrives at

$$\frac{d}{dt}M_z = -(M_z - M_0)\gamma^2 \int_0^{\infty} \left[\overline{b_x(t)b_x(0)} + \overline{b_y(t)b_y(0)}\right]e^{-i\omega_0 t}dt \tag{35}$$

where M_0 is the equilibrium magnetization. This leads to the familiar Bloch equation pertaining to longitudinal magnetization:

$$\frac{d}{dt}M_z = -\left(\frac{1}{T_1}\right)(M_z - M_0) \tag{36}$$

From (35) and (26), we can write

$$\left(\frac{1}{T_1}\right)_{\text{rf}} = (R_1)_{\text{rf}} = 2\gamma^2 \boldsymbol{J}_{\text{rf}}(\omega_0) \tag{37}$$

(as in Eq. (26), rf stands for "random fields").

Similar calculations applied to transverse components lead to

$$\frac{d}{dt}M_{x,y} = -\left(\frac{1}{T_2}\right)M_{x,y} \tag{38}$$

with

$$\left(\frac{1}{T_2}\right)_{\text{rf}} = (R_2)_{\text{rf}} = \gamma^2[\boldsymbol{J}_{\text{rf}}(\omega_0) + \boldsymbol{J}_{\text{rf}}(0)] \tag{39}$$

Notice the presence of a spectral density *at zero frequency* in R_2 arising from $\int_0^{\infty} \overline{b_z(t)b_z(0)}dt$ (which evidently does not require to switch to the rotating frame). This zero frequency spectral density will be systematically encountered in transverse relaxation rates and, in the case of slow motions, explains why R_2 can be much larger than R_1. Moreover, owing to the form of $\boldsymbol{J}_{rf}(\omega_0)$

$$\boldsymbol{J}_{\mathrm{rf}}(\omega_0) \propto \overline{b^2}\frac{\tau_c}{1+\omega_0^2\tau_c^2} \tag{40}$$

dispersion of both R_1 and R_2 (i.e., variation with the value of the main magnetic field B_0) can be expected.

Finally, concerning the so-called spin-lattice relaxation time in the rotating frame (Section I.C), one has

$$\left(\frac{1}{T_{1\rho}}\right)_{\mathrm{rf}} = (R_{1\rho})_{\mathrm{rf}} = \gamma^2[\boldsymbol{J}_{\mathrm{rf}}(\omega_0) + \boldsymbol{J}_{\mathrm{rf}}(\omega_1)] \tag{41}$$

with $\omega_1 = \gamma B_1$, B_1 being these amplitude of the spin-locking field. It can be seen that $(R_{1\rho})_{rf}$ is very close to $(R_2)_{rf}$, $\boldsymbol{J}_{\mathrm{rf}}(0)$ being substituted by $\boldsymbol{J}_{\mathrm{rf}}(\omega_1)$. As a consequence, dispersion can arise here from B_0 and B_1.

Whenever the spin system under investigation can be approximated as a single spin 1/2, with its two energy levels, then the random field approach is appropriate (for instance, this is usually adequate for treating the intermolecular dipolar contribution). Such situations, where the interactions responsible for spin relaxation are not precisely identified, may be frequently encountered and it is interesting to look at the evolution of the two main relaxation parameters, i.e., T_1 and T_2 as a function: (i) of the correlation time (Fig. 8) and (ii) of the measurement frequency $\nu_0 = \omega_0/2\pi$ (in other words, as a function of B_0 which is nothing but NMRD; Fig. 9). Figure 8 reveals that, while T_2 decreases monotonously when the correlation time increases, T_1 first presents a similar behavior, then goes through a minimum and attains large values for very large correlation times (as found, for instance, in rigid solids). Concerning the dispersion curves shown in Fig. 9, extreme narrowing conditions ($T_1 = T_2$ independent of B_0, in the range of presently available magnetic field values) are seen to be satisfied for a short correlation time (1 ps), as expected, whereas, for a correlation time of 5 ns, T_1 and T_2 decrease when B_0 increases, yet according to two different Lorentzian functions. Of course, for both curves, the point of inflection occurs at a frequency equal to $1/\tau_c$ (here 200 MHz), demonstrating the interest of NMRD for accessing correlation times.

B. Relaxation by Chemical Shift Anisotropy (csa)

Let B_0 be the static magnetic field of the NMR experiment. The actual field sensed by the nucleus has the following components (in the laboratory frame X, Y, Z)

$$\begin{aligned} b_X &= \sigma_{XZ}B_0 \\ b_Y &= \sigma_{YZ}B_0 \\ b_Z &= (1-\sigma_{ZZ})B_0 \end{aligned} \tag{42}$$

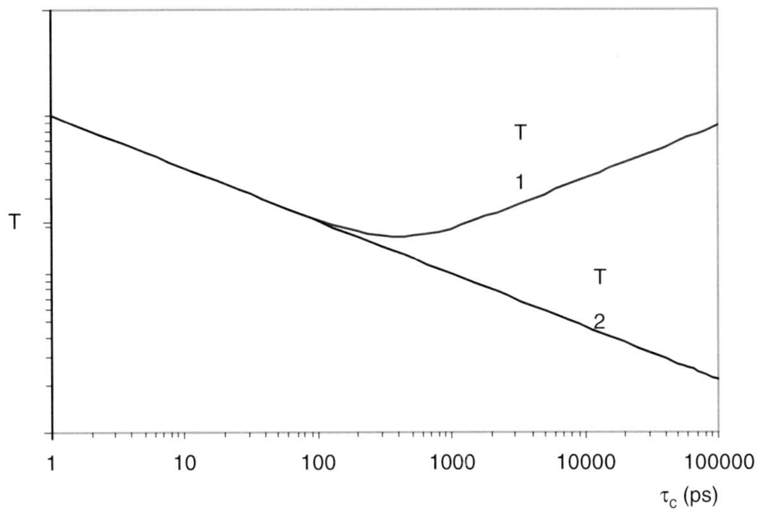

FIG. 8. Evolution of the longitudinal and transverse relaxation times (T_1 and T_2, respectively) as a function of τ_c (for a fixed measurement frequency: $\nu_0 = 400$ MHz) assuming that the considered spin is subjected to random fields whose correlation function is proportional to $e^{-t|\tau_c}$, τ_c being the correlation time. Notice the continuous decease of T_2 as τ_c increases, while T_1 goes through a minimum.

where σ_{xz}, σ_{yz}, and σ_{zz} are elements of the shielding tensor expressed in the laboratory frame. It can be recalled that these quantities are very small (of the order of 10^{-6}) and originate from the modification of B_0 caused by the electronic distribution around the nucleus under investigation. It can also be recalled that, in the liquid state due to averaging, line positions (chemical shift) depend only on the third one (element) of the tensor trace. While returning to the effective components of the static magnetic field (local field), b_z remains, of course, prominent. Nevertheless, as these three elements fluctuate in time due to molecular tumbling, the local field (random field) also does. This mechanism is therefore active as far as spin relaxation is concerned and the approach is similar to that of the previous section. However, rather than dealing with spectral densities involving random fields, it would be more informative to deal with spectral densities involving elements of the shielding tensor. One thus defines a new spectral density

$$\boldsymbol{J}_{csa}(\omega) = \int_0^\infty \overline{\sigma_{ZZ}(t)\sigma_{ZZ}(0)} \exp(-i\omega t)dt \tag{43}$$

It can be shown that, for an isotropic medium, one has *(40)*

$$\int_0^\infty \overline{\sigma_{XZ}(t)\sigma_{XZ}(0)} \exp(-i\omega t)dt = \int_0^\infty \overline{\sigma_{YZ}(t)\sigma_{YZ}(0)} \exp(-i\omega t)dt = (3/4)\boldsymbol{J}_{csa}(\omega) \tag{44}$$

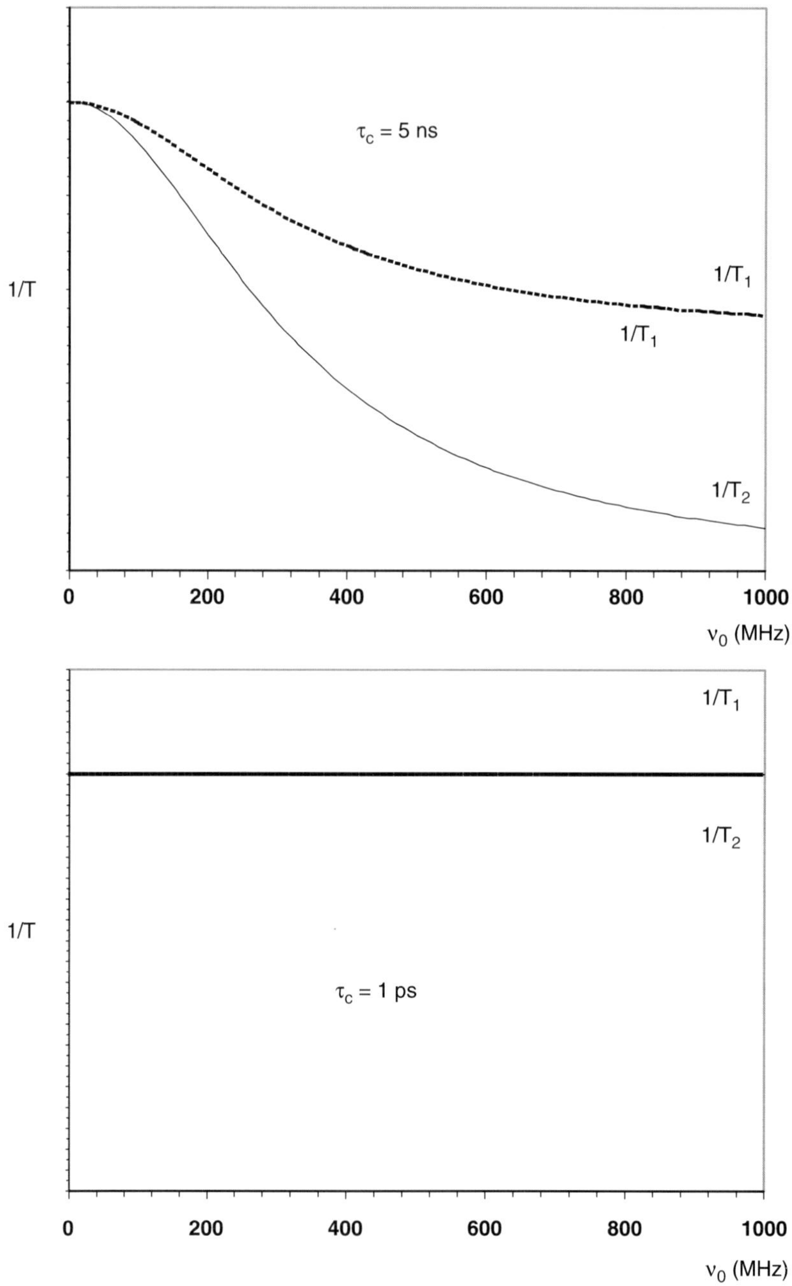

FIG. 9. Same as in figure 8, but for the evolution with the static magnetic field B_0 (or, equivalently, with the measurement frequency ν_0). For $\tau_c = 1$ ps, extreme narrowing conditions are seen to be fulfilled.

Now, from equation (35), we remember that

$$\left(\frac{1}{T_1}\right) = \gamma^2 \int_0^\infty \left(\overline{b_X(t)b_X(0)} + \overline{b_Y(t)b_Y(0)}\right)e^{-i\omega_0 t}dt \tag{45}$$

This leads to

$$(R_1)_{csa} = (3/2)\gamma^2 B_0^2 \boldsymbol{J}_{csa}(\omega_0) \tag{46}$$

Likewise

$$(R_2)_{csa} = \gamma^2 B_0^2[(3/4)\boldsymbol{J}_{csa}(\omega_0) + \boldsymbol{J}_{csa}(0)] \tag{47}$$

As immediate consequences, we can state that (i) R_1 and R_2 vary according to the square of the static magnetic field. This contribution can therefore be sorted out by experiments performed at different magnetic fields and (ii) under extreme narrowing conditions:

$$\frac{(R_2)_{csa}}{(R_1)_{csa}} = \frac{7}{6} \tag{48}$$

In order to extract some more information from the *csa* contribution to relaxation times, the next step is to switch to a molecular frame (x, y, z) where the shielding tensor is diagonal (x, y, z is called the Principal Axis System i.e., PAS). Owing to the properties reported in (44), the relevant calculations include the transformation of σ_{ZZ} into σ_{xx}, σ_{yy}, and σ_{zz} involving, for the calculation of spectral densities, the correlation function of squares of trigonometric functions such as $\overline{\cos^2\theta(t)\cos^2\theta(0)}$ (see the previous section and more importantly Eq. (29) for the definition of the normalized spectral density $\tilde{J}(\omega)$). They yield for an isotropic reorientation (the molecule is supposed to behave as a sphere)

$$\begin{aligned}(R_1)_{csa} &= \frac{1}{15}(\gamma B_0)^2(\Delta\sigma)^2\left(1 + \frac{\eta_{csa}^2}{3}\right)\tilde{J}(\omega_0)\\ (R_2)_{csa} &= \frac{1}{15}(\gamma B_0)^2(\Delta\sigma)^2\left(1 + \frac{\eta_{csa}^2}{3}\right)\left[\frac{2\tilde{J}(0)}{3} + \frac{\tilde{J}(\omega_0)}{2}\right]\end{aligned} \tag{49}$$

where $\Delta\sigma = \sigma_{zz} - (\sigma_{xx} + \sigma_{yy})/2$ is the shielding anisotropy, $\eta_{csa} = (3/2)(\sigma_{xx} - \sigma_{yy})/\Delta\sigma$ the asymmetry parameter, while the normalized spectral density contains only dynamical information. In the case of an axially symmetric shielding tensor and for any kind of molecular tumbling, one can use the above formulae with $\eta_{csa} = 0$ and evaluate the normalized spectral densities according to the actual molecular motions.

It can be seen that, in all cases, relaxation rates are directly proportional to $(\Delta\sigma)^2$. Because $\Delta\sigma$ reflects the anisotropy of the shielding tensor and because the chemical shift originates from the shielding effect, the terminology "Chemical Shift Anisotropy" is used for denoting this relaxation mechanism. Dispersion may be disconcerting because of the presence of B_0^2 (proportional to ω_0^2) in the numerator of R_1 and R_2 (Eq. (49)). Imagine that molecular reorientation is sufficiently slow so that $\omega_0^2\tau_c^2 \gg 1$ for all considered values of ω_0; from (49), it can be seen that R_1 is constant whereas R_2 increases when B_0 increases, a somewhat unusual behavior.

C. Scalar Relaxation of the Second Kind (*37*)

While the terminology "scalar relaxation of the first kind" concerns J modulation (J: spin–spin or scalar or indirect coupling constant) by exchange phenomena, the usual example of the "second kind" is a spin 1/2 nucleus (I), J coupled to a fast relaxing quadrupolar nucleus (S) with relaxation times T_1^S and T_2^S (and spin number I_S). The relevant interaction is of the form

$$J_{IS}\left(I_xS_x + I_yS_y + I_zS_z\right) \tag{50}$$

Here, J_{IS} has to be expressed in rad s^{-1} (see the unit of $J_{IS}S_x$ below).

Because of fast S relaxation (which usually prevents the observation of J splittings), $J_{IS}S_x$ can be considered as a random field γb_x acting on I. Consequently, correlation functions of the type $\overline{S_x(t)S_x(0)}$ have to be evaluated (in this case T_2^S plays the role of a correlation time). The forthcoming calculations make use of the following identity

$$\overline{S_x(0)^2} = \overline{S_y(0)^2} = \overline{S_z(0)^2} = \frac{S^2}{3} = \frac{I_S(I_S+1)}{3} \tag{51}$$

For evaluating the spectral density involving S_z, we can disregard constant terms in $S_z(t)$ and express it as $S_z(0)\exp(-t/T_1^S)$ so that

$$J_{IS}^2 \int_0^\infty \overline{S_z(t)S_z(0)}dt = J_{IS}^2 \frac{I_S(I_S+1)}{3} T_1^S \tag{52}$$

As regards S_x and S_y, precession must be accounted for. This is better carried out with the quantities S_+ and S_- defined by $S_\pm = S_x \pm iS_y$. The correlation function involved in (35) $\overline{S_x(t)S_x(0) + S_y(t)S_y(0)}$ reduces to $\overline{S_+(t)S_-(0)}$ because $\overline{S_+(t)S_-(0)} = \overline{S_-(t)S_+(0)}$ for obvious symmetry reasons. As we can write $S_+(t) = S_+(0)\exp(-t/T_2^S)\exp(i\omega_S t)$, the relevant spectral density (see (35)) is easily calculated.

$$\begin{aligned}\frac{J_{IS}^2}{2}\int_0^\infty \overline{S_+(t)S_-(0)}e^{-i\omega_I t}dt &= \frac{J_{IS}^2}{2}\int_0^\infty \overline{S_+(0)S_-(0)}e^{-t/T_2^S}e^{-i(\omega_I-\omega_S)t}dt \\ &= J_{IS}^2\frac{I_S(I_S+1)}{3}\frac{T_2^S}{1+(\omega_I-\omega_S)^2(T_2^S)^2}\end{aligned} \tag{53}$$

Finally, referring to Eqs. (35) and (39), we arrive at

$$\begin{aligned}(R_1^I)_{sc} &= \frac{2J_{IS}^2}{3}I_S(I_S+1)\frac{T_2^S}{1+(\omega_S-\omega_I)^2(T_2^S)^2} \\ (R_2^I)_{sc} &= \frac{J_{IS}^2}{3}I_S(I_S+1)\left[\frac{T_2^S}{1+(\omega_S-\omega_I)^2(T_2^S)^2}+T_1^S\right]\end{aligned} \tag{54}$$

It can be noticed that: (i) due to the importance of $(\omega_I-\omega_S)^2(T_2^S)^2$, $(R_1^I)_{SC}$ is generally negligible unless ω_s is close to ω_I, (ii) for the same reason, $(R_2^I)_{SC}\approx 4\pi^2 J_{IS}^2 T_1^S I_S(I_S+1)/3$ and is able to provide the value of the J_{IS} coupling constant.

This mechanism is identical to the one arising from the "contact interaction" between an unpaired electron and a nuclear spin (*41*). In that case, the hyperfine coupling (generally denoted by A_{SC} or A_c; and exists only if the electron density is non-zero at the considered nucleus, hence the terminology of "contact") replaces the J coupling and the earlier statement (i) may be untrue because it so happens that T_2^S becomes very short. In that case, dispersion curves provide some information about electronic relaxation. These points are discussed in detail in Section II.B of Chapter 2 and I.A.1 of Chapter 3.

D. Relaxation by Dipolar (or direct) Interaction

Here the problem is complicated by the fact that the system includes at least two spins 1/2 interacting by dipolar interaction (this is the simplest one which can be thought of) and the major difference with a spin system involving a single spin 1/2 is that one has to deal with a four-level energy diagram instead of a two-level energy diagram (Fig. 10). As a consequence, one has to consider spectral densities at frequencies corresponding to all the transitions which can be induced by local fields: ω_I and ω_S for one-quantum transitions, $(\omega_S-\omega_I)$ for the zero-quantum transition, $(\omega_S+\omega_I)$ for the double–quantum transition. It can finally be mentioned that, if I is a nuclear spin and S the spin of an unpaired electron, this relaxation mechanism is dubbed dipolar (see Sections I and II of Chapter 2).

By cosidering the local field produced by one spin on the other, we can get an idea of what the relaxation parameters should be and provide their actual expressions as given by a quantum mechanical treatment which, in principle, would be required for such a biparticle interaction. Two

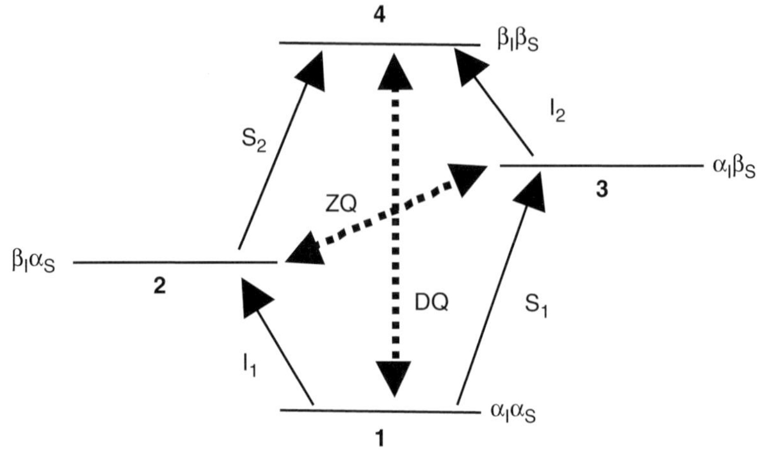

FIG. 10. Energy diagram of a *IS* spin system (two spins 1/2). α and β are the usual spin functions. One-quantum transitions (I_1, I_2, S_1, S_2, physically observable): full arrows. ZQ and DQ (dashed arrows) refer to zero-quantum and double-quantum transitions.

approaches can be considered: the one proposed by Abragam (*37*), and the other based on the Redfield equations (*42*). Both approaches are outlined in Chapter 2 (Section II). The energy corresponding to the dipolar interaction between the magnetic dipoles μ_I and μ_S associated with the considered spins (r being the distance between the two point dipoles) is given by the classical expression

$$E_{dipolar} = \frac{\mu_I.\mu_S}{r^3} - 3\frac{(\mu_I.r)(\mu_S.r)}{r^5} = \frac{\mu_I \mu_S}{r^3}(3\cos^2\theta - 1) \tag{55}$$

The last equality results from the special arrangement shown in Fig. 11 (an antiparallel configuration for the two dipoles would lead to the same result). Several important features can be deduced from (55):

- for isotropic tumbling in solution, the average value of $\cos^2\theta$ being equal to 1/3, the dipolar interaction does not lead to any splitting in the NMR spectrum.
- at the level of *each* spin system (of each molecule), the dipolar interaction depends on $1/r^3$. Consequently, the correlation function and therefore the relaxation rates will depend on $1/r^6$.
- time dependence arises from the angle θ and more precisely from $(3\cos^2\theta - 1)$, which is proportional to the spherical harmonics Y_2^0. Consequently, the normalized spectral densities defined by Eq. (29) can be used in the expressions of relaxation rates.

With

$$K_d = \frac{1}{20}\left(\frac{\mu_0}{4\pi}\right)^2\left(\frac{\gamma_I\gamma_S\hbar}{r^3}\right)^2$$

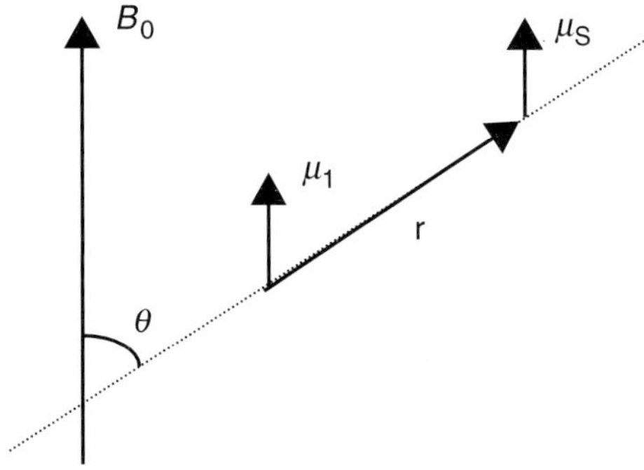

FIG. 11. Model, with the two magnetic dipoles oriented parallel to the static field B_0, used for the calculation of the dipolar interaction energy.

where μ_0 is the vacuum permittivity and $\hbar$ the Planck constant divided by 2π, one has

$$(R_1^{I,S}) = K_d[6\tilde{J}(\omega_I + \omega_S) + 3\tilde{J}(\omega_{I,S}) + \tilde{J}(\omega_S - \omega_I)] \tag{56}$$

where the double-quantum, the one-quantum and the zero-quantum transitions can be recognized, the coefficients arising from the form of the dipolar hamiltonian. Likewise, one obtains the transverse relaxation rate

$$(R_2^{I,S}) = K_d[3\tilde{J}(\omega_I + \omega_S) + 3\tilde{J}(\omega_{I,S})/2 + 3\tilde{J}(\omega_{S,I}) + \tilde{J}(\omega_S - \omega_I) + 2\tilde{J}(0)] \tag{57}$$

Concerning the spin-lattice relaxation rate in the rotating frame ($R_{1\rho}$, see Section I.C and Eq. (41)), one has to simply replace $\tilde{J}(0)$ by $\tilde{J}(\omega_1)$.

The longitudinal cross-relaxation rate (see Eq. (13)) originates solely from the terms in the dipolar Hamiltonian involving both spins, namely those terms corresponding to zero-quantum and double-quantum transitions so that

$$\sigma(= \sigma_{//}) = K_d[6\tilde{J}(\omega_I + \omega_S) - \tilde{J}(\omega_S - \omega_I)] \tag{58}$$

The transverse cross-relaxation rate does not exist due to the difference in I and S resonance frequencies. It is, however, retrieved (see Section I.D) in a spin-lock experiment (where this difference disappears):

$$\sigma_\rho^{I,S}(= \sigma_\perp) = K_d[2\tilde{J}(\omega_{I,S}) + 2J(\omega_1)] \tag{59}$$

Concerning these cross-relaxation rates, an important point is that the longitudinal cross-relaxation rate $\sigma_{//}$ can go through zero at relatively high measurement frequencies and therefore become non-observable, whereas this

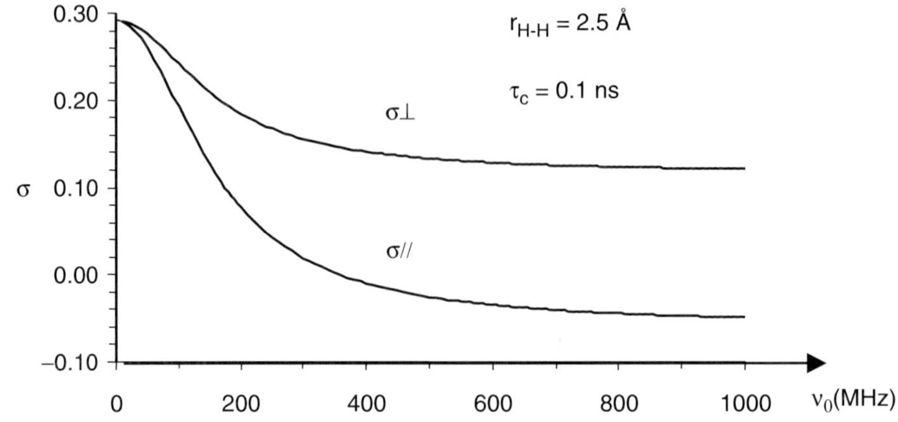

FIG. 12. Comparison of longitudinal and transverse cross-relaxation rates as a function of the measurement frequency (ν_0) for a pair of protons and a normalized spectral density of the form $2\tau_c/(1+\omega^2\tau_c^2)$ (see Eqs. (58) and (59)).

never occurs for the transverse cross-relaxation rate $\sigma_{\perp}$ (*25,26*). This feature is illustrated by the two curves of Fig. 12.

The last point is about equivalent spins (or "like spins" as the two protons of the water molecule). Referring to Solomon equations (see (13)), we can notice that, because of this equivalence, the effective longitudinal relaxation rate is obtained by adding the cross-relaxation rate to the specific longitudinal relaxation rate:

$$R_1^{like} = 3K_d[4\tilde{J}(2\omega_I) + \tilde{J}(\omega_I)] \tag{60}$$

Finally, from Eqs. (56), (57), and (60), we can appreciate that dispersion curves can be more or less complicated, depending on the involved correlation times (possibly related to rotational molecular motions). This is detailed in Chapters 2 and 3.

E. QUADRUPOLAR RELAXATION

The interaction between the quadrupole moment of a nucleus of spin greater than 1/2 (the nuclear electric charge is no longer of spherical symmetry as it is for a spin 1/2) with the electric field gradient is tensorial and the time modulation arises again from the orientation of the principal axis system with respect to the B_0 direction, as for the csa mechanism and, in this respect, similar spectral densities can be used (e.g., normalized spectral densities). However, because the spin system (i.e., the quadrupolar nucleus) possesses more than two energy levels, a quantum mechanical treatment is again required (as for dipolar relaxation) and, in general, the evolution of longitudinal and transverse magnetizations is not exponential. As far as the environment of the considered nucleus is not of spherical symmetry, the quadrupolar mechanism is overwhelming. Some useful formulas are given below.

(i) In extreme narrowing conditions, τ_c^{eff} describing the reorientation of the electric field gradient (efg) tensor z axis, longitudinal and transverse relaxations are purely exponential with

$$(R_1)_q = (R_2)_q = (3\pi^2/10)\{(2I+3)/[I^2(2I-1)]\}\chi^2\left(1+\eta_q^2\right)\tau_c^{eff} \tag{61}$$

I: spin number. $\chi = eQV_{zz}/h$: quadrupole coupling constant with e: electron charge, Q: quadrupolar momentum; V_{zz}: either the largest element of the efg tensor (which is supposed to be expressed in its principal axis system) or the element corresponding to the symmetry axis if the efg tensor is axially symmetry; $\eta_q = (V_{xx} - V_{yy})/V_{zz}$ (with $|V_{zz}| \geq |V_{yy}| \geq |V_{xx}|$) is the asymmetry parameter.

(ii) In the general case, the following approximations may be considered (*43*):

$$\begin{aligned}(R_1)_q &\approx [(2I+3)/5][8J^q(2\omega_0) + 2J^q(\omega_0)] \\ (R_2)_q &\approx [(2I+3)/5][2J^q(2\omega_0) + 5J^q(\omega_0) + 3J^q(0)]\end{aligned} \tag{62}$$

where ω_0 is the measurement frequency and $J^q(\omega)$, the spectral density associated with the quadrupolar mechanism and which can be written as a function of normalized spectral densities

$$J^q(\omega) = (3\pi^2/40)\left[\frac{1}{I^2(2I-1)}\right]\chi^2\left(1+\eta_q^2/3\right)\tilde{J}(\omega) \tag{63}$$

It can be noticed that (62) and (63) are strictly valid for a spin 1 ($^2H, ^{14}N$). Dispersion arises essentially from molecular reorientation and, as far as the water molecule is concerned, the two quadrupolar nuclei (i.e., ^2H and ^{17}O) proved to be especially useful (this will be further discussed in Section III).

III. Water NMRD in Diamagnetic Systems

The relatively low number of available nuclear species may entail some ambiguity on studies of water by spin relaxation. Those spins are due to:

- the two protons, which are however subjected to intra- and intermolecular dipolar relaxation with the obvious difficulty of separating these two contributions. Moreover, these protons are usually labile and exchange phenomena must be accounted for.
- deuterium (^2H), either at the natural abundance level or in heavy water. Although exchange is still present, its relaxation is dominated by the

quadrupolar mechanism which is exclusively intramolecular and thus easier to interpret.

- ^{17}O (it can be recalled that the abundant isotope, ^{16}O, has no spin) either at the natural abundance level (which is low) or, better, in labeled molecules. This is a quadrupolar nucleus ($I = 5/2$) with a purely intramolecular relaxation and not subjected to exchange.

As the essentials of this book are devoted to paramagnetic systems, we shall, for the sake of completeness, conclude this introductory chapter by a related discussion about diamagnetic systems and more precisely about water NMRD in protein solutions.

A. Pure Water

At ambient temperature, ^{1}H, ^{2}H and ^{17}O relaxation is in the extreme narrow range and dispersion curves are perfectly flat (see Fig. 9 bottom) precluding any correlation time determination. Furthermore as inter- and intramolecular contributions to proton relaxation cannot be easily separated and as the deuterium and ^{17}O quadrupole coupling constants are not known with sufficient accuracy, there is a real problem for determining a meaningful correlation time. This problem was solved only in the early 1980s by resorting to the ^{17}O–^{1}H cross-relaxation rate which is purely intramolecular (*44,45*). By measuring the ^{17}O enhanced ^{1}H relaxation in ^{17}O enriched water, it is possible to obtain the ^{17}O–^{1}H dipolar contribution to proton relaxation (see formula (56) in extreme narowing conditions). Thereafter, assuming an O–H bond length of 0.98 A°, one obtains the correlation time associated with the motion of the O–H bond (τ_{OH} = 1.95 ps at 25°C). Owing to a relation between ^{17}O and deuterium quadrupole coupling constants and because the effective correlation time for deuterium relaxation is identical to τ_{OH} (the two relaxation vectors have the same orientation), it became feasible, from ^{17}O relaxation, to determine a correlation time pertaining to another molecular direction (*46*). As the same correlation time is found, it can be concluded that, in pure water, molecules reorient isotropically.

The proton longitudinal relaxation rate of deoxygenated water is 0.3 s^{-1} at 25°C, with about 25% of this value being attributed to intermolecular dipolar relaxation. In that case, of course, no dispersion occurs. However, for water in equilibrium with air, due to paramagnetic molecular oxygen, the relaxation rate increases by 0.1 s^{-1} at low fields and exhibits a dispersion around 40 MHz (*47*).

B. Water in Protein Solutions (Diamagnetic Systems)

There exists an enormous amount of experimental results and the reader is referred to several reviews containing extensive data along with tentative interpretations (*48–50*). In addition to the information obtained by NOESY (see section I.D) about residue directly interacting with water by cross-relaxation (*51*), many conclusions have been drawn from the dispersion curves of water ^1H, ^2H and ^{17}O. Most of the time, longitudinal relaxation rates are considered and, in a general way, they can be expressed as (*50*)

$$R_1(\omega_0) = (1 - f_s - f_I)R_{bulk} + f_s R_s + f_I R_I(\omega_0) \tag{64}$$

ω_0 is the measurement frequency. Rapid exchange between the different fractions is assumed: the bulk, water at the protein surface (s) and interior water molecules, buried in the protein and responsible for dispersion (i). In fact, protons from the protein surface exchanging with water lead to dispersion as well and should fall into this category. Bulk and s are relevant to extreme narrowing conditions and cannot be separated unless additional data or estimations are available (for instance, an estimation of f_s from some knowledge of the protein surface). As far as quadrupolar nuclei are concerned (i.e., ^2H and ^{17}O), dispersion of R_I is relevant of Eqs. (62) and (63) (this evolves according to a Lorentzian function as in Fig. 9) and yield information about the number of water molecules inside the protein and about the protein dynamics (sensed by the buried water molecules). Two important points must be noted about Eqs. (62) and (63). First, the effective correlation time τ_C is composed of the protein rotational correlation time τ_R and of the residence time τ_W at the hydration site so that

$$1/\tau_C = 1/\tau_R + 1/\tau_W \tag{65}$$

On the other hand, the combination $4\tilde{J}(2\omega_0) + \tilde{J}(\omega_0)$ is very close to $5\tilde{J}(\sqrt{3}\omega_0)$ (*52*), explaining why a single inflection point is observed. This applies also to the dipolar relaxation of like spins (Eq. (60)).

Proton dispersion appears more complicated as the dispersion curves are generally stretched with respect to a simple Lorentzian function (Figs. 13 and 14). Although the origin of this stretching is not fully understood, dipolar relaxation with protein protons, possibly combined with exchange, can be invoked. It involves spectral densities different from those pertaining to the dipolar interaction of the two water molecule protons (see (57) and (60)) and leads to a dispersion which is no longer a single Lorentzian function (*53*), as this would be the case for relaxation arising solely from the two protons of water (like spins, mentioned earlier). Moreover, a distribution of correlation times cannot be excluded and models different from a single Lorentzian function must therefore be envisioned. This is the subject of the two subsections that follow.

B.1. The Cole–Cole Equation

The Cole and Cole function (*54*) was proposed very early for interpreting dielectric relaxation data. This empirical expression, related to the distribution of correlation times, has been used successfully in the field of dielectric relaxation and transposed by Hallenga and Koenig (*55*) for interpreting NMRD data according to the equation given below (which is given with the original notations)

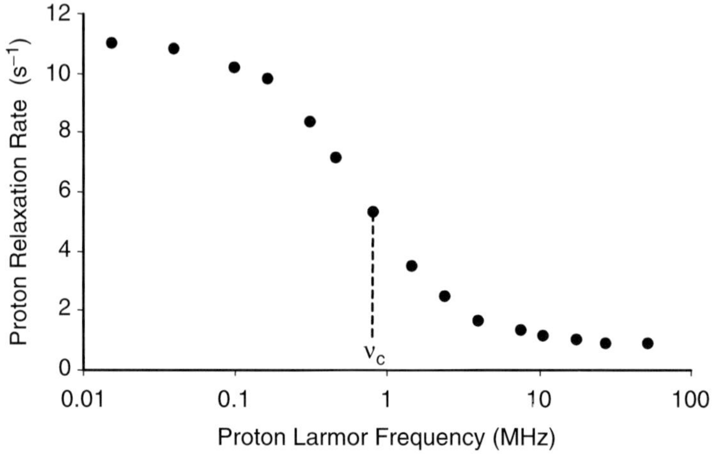

FIG. 13. Typical experimental results (adapted from ref. (*50*)) showing the dispesion curve of water protons in an aqueous protein solution. $\nu_c = \omega_c/2\pi$ is the cut-off frequency (see Eq. (66)).

$$R_1(\omega) = R_{1W} + D + A \times Re\left[\frac{1}{1 + (i\omega/\omega_c)^{\beta/2}}\right] \tag{66}$$

where *Re* stands for "real part of". R_{1W} is the bulk-water relaxation rate at the considered temperature; D has an identical meaning as R_s in Eq. (64) (relaxation rate of surface water without any dispersion in the considered frequency range); A is related to the proportion of buried water; the term inside brackets represents the distribution of correlation times *via* the exponent $\beta/2$, the central correlation time being the inverse of the frequency cut-off which corresponds to the inflection point, i.e., at half the maximum of the dispersion curve ($\tau_0 = 2\pi/\omega_c$; see Fig. 13). It can be noticed that for $\beta = 2$ the factor of A is a simple Lorenztian. A, D, τ_0, and β must be extracted from the experimental data. Although the function (66) is purely empirical, τ_0 is generally identified with the true correlation time governing the dynamics of water molecules associated with the protein whereas A would be of the form $C(0)\tau_0$ where $C(0)$ is the value at time zero of the relevant correlation function. In spite of a wide (and apparently successful) use of the Cole–Cole function since 1976 (ω_c and A can generally be correlated with the molecular weight), this type of interpretation has been toughly (and convincingly) questioned in 1998 (*53*), especially regarding the physical meaning of τ_0 and A.

B.2. The Model-Free Analysis

Halle *et al.* (*53*) have proposed to describe dispersions such as the one shown in Fig. 13 by a set of unambiguously defined parameters, which are able to fully characterize the whole curve provided that the two plateaus (at low and high frequencies) are clearly visible. Let us denote by $C(t)$ the correlation function (see (24)) which, after Fourier transformation, leads to the

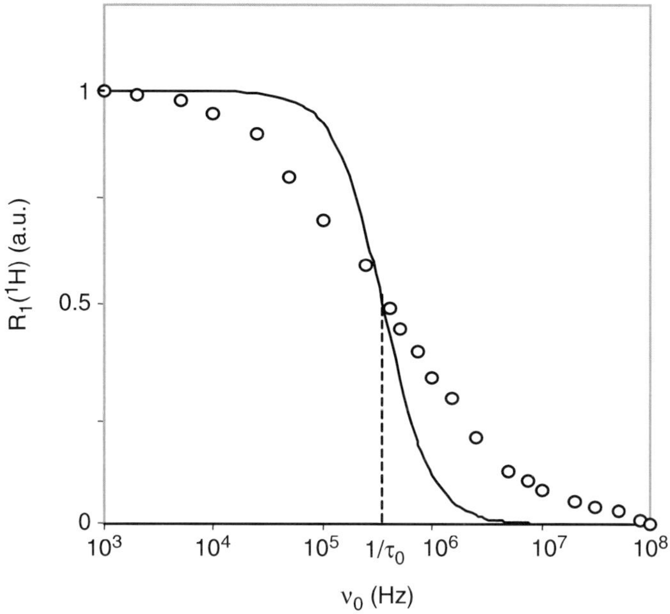

Fig. 14. Circles: typical NMRD data (water ^1H longitudinal relaxation rates in a protein aqueous solution; adapted from ref. (*55*)). The curve corresponds to a lorentzian function ($\frac{1}{1+4\pi^2\nu_0^2\tau_0^2}$) with τ_0 deduced from the half-height of the experimental data. Note the stretching of the experimental data with respect to a single lorentzian curve.

spectral density $J(\omega)$ (see (26)) which will be identified here with the experimental dispersion curve. A general property of the Fourier transform is that the integral of the function in one domain (say the frequency domain) is equal (except a known factor) to the value of the function at zero in the other domain (at time zero). This yields a first parameter which has a physical meaning, namely the correlation function at time zero (which, in terms of random fields, would be their mean square; see (24))

$$C(0) = \frac{2}{\pi}\int_0^{\infty} J(\omega)d\omega \tag{67}$$

The other quantity is a sort of average correlation time defined as

$$\langle\tau_c\rangle = \int_0^{\infty} \frac{C(t)}{C(0)}dt = \frac{J(0)}{C(0)} \tag{68}$$

It is simply obtained by dividing the value of the low-frequency plateau by $C(0)$. After having determined τ_0 as explained in the previous section (it can be reminded that the physical meaning of this parameter is dubious), the stretching can be defined by two further parameters (which do not have any obvious physical meaning)

$$\lambda_- = 2\int_0^{1/\tau_0} \frac{J_0(\omega) - J(\omega)}{\omega J(0)} d\omega$$
$$\lambda_+ = -2\int_{1/\tau_0}^{\infty} \frac{J_0(\omega) - J(\omega)}{\omega J(0)} d\omega \tag{69}$$

where J_0 is the Lorentzian homologous of J

$$J_0(\omega) = \frac{J(0)}{1 + \omega^2\tau_0^2} \tag{70}$$

Now, the numerical integration involved in Eqs. (67) and (69) may be troublesome because of the uncertainties affecting experimental data points. In order to alleviate this problem, Halle *et al.* (*53*) proposed to model the observed dispersion curve according to the following expansion

$$J(\omega) = \sum_{n=1}^{N} c_n \frac{\tau_n}{1 + \omega^2\tau_n^2} \tag{71}$$

The parameters of this expansion, as well as the number N of Lorentzian functions, are determined (from the experimental data) by a non-linear least squares fit along with statistical tests. It can be noticed that this expansion has no physical meaning but is merely a numerical device allowing for smoothing and interpolation of the experimental data. Nevertheless, this procedure proves to be statistically more significant than the Cole–Cole equation and thus to account much better for the representation of experimental data. The two physically meaningful parameters, i.e., $C(0)$ and $\langle\tau_c\rangle$, can then be easily deduced from the quantities involved in (71)

$$C(0) = \sum_n c_n$$
$$\langle\tau_c\rangle = \sum_n c_n\tau_n / \sum_n c_n \tag{72}$$

The validity of this analysis rests on the absence of any hypothesis regarding the motional model and/or the nature of the interactions. Also, it is especially reassuring to find for $\langle\tau_c\rangle$, values close to what is expected for protein reorientation (*53*).

The above features have been confirmed by a thorough study of Bertini *et al.* (*56*) who considered aqueous solutions of 14 different proteins. They interpreted their results according to

$$R_1(\omega) = \alpha + \beta[0.2J(\omega + 0.8J(2\omega)] \tag{73}$$

which is in agreement with (64) and (60). Data were perfectly fitted and it appears that $\langle\tau_c\rangle$ is somewhat different from τ_0 (Lorentzian model) and from the correlation time derived from the Cole–Cole equation. As already obseved by Halle *et al.* (*53*), they found that $\langle\tau_c\rangle$ is very close to the correlation time predicted by the Stokes equation ($\tau_R = \frac{4\pi\eta r^3}{3kT}$; η: medium viscosity; r: radius of the protein supposed to be spherical; k: Boltzmann constant; T: absolute temperature). Moreover, they noticed that β is perfectly correlated with the molecular weight, which is a strong indication that dispersion is essentially due to protein protons exchanging with water. The quantity obtained by subtracting the bulk water relaxation rate from α also shows a very good linear correlation with the molecular weight. This is interpreted as the contribution of water molecules in the vicinity of the protein surface which still have an important mobility, yet perturbed by one of the protein molecules.

References

1. Slichter, C. P. "*Principles of Magnetic Resonance*"; Harper and Row: New York, **1963**, p. 8.
2. Ernst, R. R. *Adv. Magn. Reson.* **1966**, *2*, 1–135.
3. Vold, R. L.; Waugh, J. S.; Klein, M. P.; Phelps, D. E. *J. Chem. Phys.* **1968**, *48*, 3831.
4. Canet, D.; Levy, G. C.; Peat, I. R. *J. Magn. Reson.* **1975**, *18*, 199.
5. Freeman, R.; Hill, H. D. W. *J. Chem. Phys.* **1971**, *54*, 3367.
6. Abragam, A. "*The Principles of Nuclear Magnetism*"; Clarendon Press: Oxford, **1961**, pp. 73–74.
7. Wagner, S.; Dinesen, T. R. J.; Rayner, T.; Bryant, R. G. *J. Magn. Reson.* **1999**, *140*, 172.
8. Redfield, A. G. *Magn. Reson. Chem.* **2003**, *41*, 753.
9. Noack, F. *Prog. NMR Spectrosc.* **1966**, *18*, 171.
10. Hahn, E. *Phys. Rev.* **1950**, *80*, 580.
11. Callaghan, P. J. "*Principles of Nuclear Magnetic Resonance Microscopy*"; Clarendon Press: Oxford, **1991**, Chapters 1 and 6.
12. Torrey, H. C. *Phys. Rev.* **1956**, *104*, 563.
13. Carr, H. Y.; Purcell, E. M. *Phys. Rev.* **1954**, *94*, 630.
14. Meiboom, S.; Gill, D. *Rev. Sci. Instrum.* **1958**, *29*, 688.
15. Hahn, E. L.; Maxwell, D. E. *Phys. Rev.* **1951**, *84*, 1246.
16. Look, D. C.; Lowe, I. J. *J. Chem. Phys.* **1966**, *44*, 2995.
17. Noggle, J. H.; Schirmer, R. E. "*The Nuclear Overhauser Effect. Chemical Applications*"; Academic Press: New York, **1971**, Chapter 1.
18. Delpuech, J. J. "*Timescales in NMR: Nuclear Site Exchange and Dynamics NMR,* in *Dynamics of Solutions and Fluids Mixtures by NMR*"; Ed. Delpuech, J. J.; Wiley: Chichester, **1995**, pp. 73–172.
19. Neuhaus, D.; Williamson, M. "*The Nuclear Overhauser Effect in Structural and Conformational Analysis*"; VCH Publishers: New York, **1989**.
20. Overhauser, A. W. *Phys. Rev.* **1953**, *92*, 411.
21. Ernst, R. R.; Bodenhausen, G.; Wokaun, A. "*Principle of Nuclear Magnetic Resonance in One and Two Dimensions*"; Clarendon Press:: Oxford, **1987**, Chapter 6.
22. Jeener, J.; Meier, B. H.; Bachmann, P.; Ernst, R. R. *J. Chem. Phys.* **1979**, *71*, 4546.
23. Rinaldi, P. L. *J. Am. Chem. Soc.* **1983**, *105*, 5167.
24. Yu, C.; Levy, G. C. *J. Am. Chem. Soc.* **1983**, *105*, 6994.
25. Bothner-By, A. A.; Stephens, R. L.; Lee, J.-M.; Warren, C. D.; Jeanloz, R. W. *J. Am. Chem. Soc.* **1984**, *106*, 811.
26. Bax, A.; Davis, D. G. *J. Magn. Reson.* **1985**, *63*, 207.
27. Mackor, E. L.; MacLean, C. *Progr. NMR Spectrosc.* **1967**, *3*, 129.
28. Goldman, M. *J. Magn. Reson.* **1984**, *60*, 437.

29. Sorensen, O. W.; Eich, G. W.; Levitt, M. H.; Bodenhausen, G.; Ernst, R. R. *Progr. NMR Spectrosc.* **1983**, *16*, 163.
30. Guéron, M.; Leroy, J. L.; Griffey, R. H. *J. Am. Chem. Soc.* **1983**, *105*, 7262.
31. Werbelow, L. G.; Grant, D. M. *Adv. Magn. Reson.* **1977**, *9*, 190.
32. Vold, R. L.; Vold, R. R. *Progr. NMR Spectrosc.* **1978**, *12*, 79.
33. Canet, D. *Progr. NMR Spectrosc.* **1989**, *21*, 237.
34. Bertini, I.; Kowalewski, J.; Luchinat, C.; Parigi, G. *J. Magn. Reson.* **2001**, *152*, 103.
35. Tjandra, N.; Szabo, A.; Bax, A. *J. Am. Chem. Soc.* **1996**, *118*, 6986.
36. Carrington, A.; McLachlan, A. D. "*Introduction to Magnetic Resonance*"; Chapman and Hall: London, **1967**, appendix F.
37. Abragam, A. "*The Principles of Nuclear Magnetism*"; Clarendon Press: Oxford, **1961**, Chapter 8.
38. Perrin, F. *J. Phys. Rad.* **1934**, *V*, 33.
39. Huntress, W. T. *Adv. Magn. Reson.* **1970**, *4*, 1.
40. See for instance: Canet, D. "*Nuclear Magnetic Resonance, Concepts and Methods*"; Wiley: Chichester, **1991**, Chapter 4.
41. Banci, L.; Bertini, I.; Luchinat, C. "*Nuclear and Electron Relaxation*"; VCH: Weinheim, **1991**, Chapter 5.
42. Redfield, A. G. *Adv. Magn. Reson.* **1965**, *1*, 1.
43. Halle, B.; Wennerström, H *J. Magn. Reson.* **1989**, *44*, 89.
44. Lang, E. W.; Lüdemann, H. D. *Ber. Bunsenges. Phys. Chem.* **1981**, *85*, 1016.
45. Lankhorst, D.; Schriever, J.; Leyte, J. C. *Ber. Bunsenges. Phys. Chem.* **1982**, *86*, 215.
46. van der Maarel, J. R. C.; Lankhorst, D.; de Bleiser, J.; Leyte, J. C. *Chem. Phys. Lett.* **1985**, *122*, 541.
47. Hausser, R.; Noack, F. *Z. Naturforsch.* **1965**, *20a*, 1668.
48. Koenig, S. H.; Brown, R. D. *Prog. NMR Spectrosc.* **1990**, *22*, 487.
49. Halle, B.; Denisov, V. P.; Venu, K. "*Modern Techniques in Protein NMR*"; vol. 17, Eds. Berliner, N. J.; Krishna, N. R.; Plenum: New York, **1998**.
50. Halle, B. "*Hydration Process in Biology*"; Ed. Bellisent-Funel, M.-C.; IOS Press: Amsterdam, **1999**.
51. Otting, G. *Prog. NMR Spectrosc.* **1997**, *31*, 259.
52. Koenig, S. H.; Schillinger, W. E. *J. Biol. Chem.* **1969**, *244*, 3283.
53. Halle, B.; Johannesson, H.; Venu, K. *J. Magn. Reson.* **1998**, *135*, 1.
54. Cole, K. S.; Cole, R. H. *J. Chem. Phys.* **1941**, *9*, 341.
55. Hallenga, K.; Koenig, S. H. *Biochemistry* **1976**, *15*, 4255.
56. Berini, I.; Fragai, M.; Luchinat, C.; Parigi, G. *Magn. Reson. Chem.* **2000**, *38*, 543.

NMR RELAXATION IN SOLUTION OF PARAMAGNETIC COMPLEXES: RECENT THEORETICAL PROGRESS FOR $S \geq 1$

JOZEF KOWALEWSKI[a], DANUTA KRUK[a,b] and GIACOMO PARIGI[c]

[a]Division of Physical Chemistry, Arrhenius Laboratory, Stockholm University, S-10691 Stockholm, Sweden

[b]Institute of Physics, Jagiellonian University, ul. Reymonta 4, PL-30-059 Krakow, Poland

[c]Department of Agricultural Biotechnology, University of Florence, I-50144 Florence, Italy

ADVANCES IN INORGANIC CHEMISTRY
VOLUME 57 ISSN 0898-8838 / DOI 10.1016/S0898-8838(04)57002-8

I. Introduction

Paramagnetic systems are materials with positive magnetic susceptibility, associated with unpaired electrons. The paramagnetic solutions of interest to chemists contain usually free radicals or transition metal complexes. In addition, certain species with triplet ground state, such as oxygen molecule, need to be considered sometimes. The presence of unpaired electron spins has a profound influence on NMR spectra of such solutions. The origin of the effects is found in the large value of the electronic magnetic moment, about 650 times that of the proton.

The paramagnetic species influence the NMR spectra of liquids in several ways. First, the nuclear spin relaxation rates are enhanced. The paramagnetic relaxation enhancement (PRE) is caused by a random variation of the electron spin–nuclear spin interactions, which open new pathways for longitudinal as well as transverse relaxation. The PRE is most often studied for $I = 1/2$ nuclei, but applications of oxygen-17 ($I = 5/2$) are also quite common. Second, the NMR signals may be shifted, provided the relevant electron spin–nuclear spin interaction has a non-zero average. Third, the spin–spin splittings may also be affected. Moreover, paramagnetic complexes in solution often contain exchangeable ligands, and the exchange phenomena, together with the intrinsic relaxation, shift and splitting properties, contribute to the observed lineshapes and line positions in one-dimensional as well as multi-dimensional NMR spectra. The paramagnetic effects have been used for several decades by chemists and biochemists as a source of structural, thermodynamic, and dynamic information. Applications of this type have been described in numerous books and reviews, recently by Bertini and co-workers (*1–3*). Another very active field of application of paramagnetic relaxation enhancement in recent years, is the development and use of paramagnetic materials as contrast agents in magnetic resonance imaging (MRI) (*4,5*).

The fundamental origin of the paramagnetic effects on NMR spectral parameters is the magnetic hyperfine interaction between the nuclear and electronic magnetic moments (*6*). Assuming that the orbital angular momentum is quenched (*7*), the nuclear and electronic magnetic moments are proportional to the corresponding spin operators. Employing the language of spin-Hamiltonians (*8*), the hyperfine interaction can be expressed as:

$$H_{hf} = \mathbf{I} \cdot \mathbf{A} \cdot \mathbf{S} \tag{1}$$

where $\mathbf{I}$ denotes the nuclear-spin vector operator and $\mathbf{S}$ the electron-spin operator. Throughout this paper the spin operators are assumed dimensionless and the Hamiltonians are in angular frequency units. The symbol $\mathbf{A}$ is the hyperfine coupling tensor. It is often practical to split the hyperfine interaction into two components: the dipole–dipole interaction between the nuclear magnetic moment and the electrons outside of the nucleus and the scalar interaction between the nuclear moment and the electron spin density at the site of the nucleus. The scalar interaction is also called the Fermi contact interaction, or simply contact interaction. The rotational average of the dipole–dipole interaction is sometimes denoted as a pseudocontact term.

In this review, we concentrate in the first place on the phenomenon of paramagnetic relaxation enhancement but we also mention recent theoretical developments in the neighboring fields. The first issue that needs to be clarified is the relation between macroscopic, observable properties of nuclear spins, and their microscopic counterparts. In solutions of transition metal ions or complexes, one can commonly consider a situation where the ligands carrying nuclear spins can reside in two types of environment: in the coordination sphere of the paramagnetic metal ion or in the bulk. If the ligand contains only one type of magnetic nuclei (which is the case for example for ${}^{1}H_2^{16}O$ isotopomer of water) or if we can disregard interactions between nuclear spins, each of the two sites can be characterized by nuclear spin–lattice and spin–spin relaxation times, T_1 and T_2, respectively. Assuming the paramagnetic species to be dilute, Swift and Connick (*9*) and Luz and Meiboom (*10*) derived a set of equations relating the observables of NMR experiments and the relaxation, shift and lifetime properties of nuclei in the paramagnetic complexes:

$$T_{1P}^{-1} = \frac{P_M}{\tau_M + T_{1M}} \tag{2}$$

$$T_{2P}^{-1} = \frac{P_M}{\tau_M}\left[\frac{T_{2M}^{-2} + (T_{2M}\tau_M)^{-1} + \Delta\omega_M^2}{(T_{2M}^{-1} + \tau_M^{-1})^2 + \Delta\omega_M^2}\right] \tag{3}$$

$$\Delta\omega_P = \frac{P_M \Delta\omega_M}{(\tau_M/T_{2M} + 1)^2 + \tau_M^2\Delta\omega_M^2} \tag{4}$$

T_{1P}^{-1}, T_{2P}^{-1}, and $\Delta\omega_P$ are the excess spin–lattice relaxation rate, spin–spin relaxation rate and shift measured for the ligand in solution (a difference between the quantity of interest in a paramagnetic solution and the corresponding value in a diamagnetic reference solution), while the properties with index M refer to the ligand in the paramagnetic complex. The symbol τ_M is the lifetime of the ligand in the complex and P_M is the mole-fraction of ligand nuclei in bound positions. The concept of PRE is often used synonymous with the enhancement of the spin–lattice relaxation rate. The PRE as given by Eq. (2) is commonly denoted as the "inner-sphere PRE". If only one type of paramagnetic species exists in solution, the PRE is proportional to the concentration of that species. The PRE normalized to 1 mM concentration of paramagnetic complexes is called "relaxivity".

Another limiting situation arises when the paramagnetic species interact only weakly with the molecules carrying the nuclear spins. In such a case, it is not meaningful to speak about exchange between discrete sites, but rather about free diffusion or diffusion in a potential. One then speaks about outer-sphere PRE, still referring to the enhancement of the spin–lattice relaxation rate. The outer-sphere PRE is also proportional to the concentration of paramagnetic material, in this case through the idea of "density of electron spins", and the concept of relaxivity is equally useful in this case.

The theory of PRE has been subject to a comprehensive review in the mid-1980s (*11*); different types of methods have been reviewed more recently (*12–14*). The structure of the present review is as follows. We begin in Section II with a brief discussion of the "classical" Solomon–Bloembergen theory, a high-field limit, perturbation-type theoretical model, and some of its modifications. In Section III, we proceed with the so-called Curie spin relaxation, a relaxation mechanism occurring at high magnetic fields through the thermally averaged magnetic moment associated with the electron spin. In that context, we review briefly other interesting relaxation phenomena. In Section IV, we cover the PRE theory for a more general case, when the coupling between the electron spin and classical degrees of freedom in the surroundings, in the first place molecular rotations, is so strong that the perturbation approach breaks down. Next, in Section V, we go back to the perturbation-based theories (Redfield-limit theories) for electron relaxation, but avoid some approximations inherent in the Solomon–Bloembergen approach. Spin-dynamics models, a non-analytical way of taking rotations into consideration, are covered in Section VI. Section VII deals with the outer-sphere relaxation while Section VIII investigates the relation between the electron spin relaxation and vibrational degrees of freedom. Finally, concluding remarks are provided in Section IX. This review was completed in August 2003.

II. The "Classical" Theory

A. The Solomon–Bloembergen Theory

In 1955, Solomon presented a seminal paper on dipole–dipole (DD) relaxation (*15*). The DD interaction is assumed to be of constant strength, corresponding to a constant distance r_{IS} between the two spins, I and S, acting as magnetic dipoles. The molecule-fixed vector $\mathbf{r}_{IS}$ connecting the two spins (the principal axis of the dipole–dipole interaction) changes its orientation in the laboratory frame (defined by the direction of the magnetic field) through random molecular motions. The Hamiltonian for the system is written as:

$$H = H_0 + H_1(t) \tag{5}$$

H_0 is the unperturbed Hamiltonian, describing the Zeeman interaction and $H_1(t)$ is the dipole–dipole Hamiltonian, time-dependent through variation in the orientation of the $\mathbf{r}_{IS}$ vector. The DD Hamiltonian can conveniently be expressed using scalar contraction of irreducible tensors (*16*).

$$H_1^{DD}(t) = \sum_{m=-2}^{2} (-1)^m F_m^{(2)}(t) A_{-m}^{(2)} \tag{6}$$

where $A_{-m}^{(2)}$ are spin operators and $F_m^{(2)}$ represent classical functions related to the lattice, i.e., to the classically described environment of the spins. The functions $F_m^{(2)}(t)$ are products of dipole–dipole interaction strength constant and suitably normalized spherical harmonics of the angles specifying the direction of the dipole–dipole axis with respect to the laboratory frame, in which the spin operators are defined. Because of molecular tumbling in solution, the orientation of the DD axis is a stochastic function of time. Through this, the $H_1^{DD}(t)$ and its matrix elements are also stochastic functions of time.

Using a simple kinetic model, Solomon demonstrated that the spin–lattice relaxation of the I and S spins was described by a system of coupled differential equations, with bi-exponential functions as general solutions. A single exponential relaxation for the I spin, corresponding to a well-defined T_{1I}, could only be obtained in certain limiting situations, e.g., if the other spin, S, was different from I and had an independent and highly efficient relaxation pathway. This limit is normally fulfilled if S represents an electron spin. The spin–lattice relaxation rate, T_{1I}^{-1}, for the nuclear spin, I, is in such a situation given by:

$$T_{1I}^{-1} = W_0 + 2W_{1I} + W_2 \tag{7}$$

W_0, W_{1I}, and W_2 are transition probabilities in a four-level system, *cf.* Fig. 1. The transition probabilities were calculated using time-dependent

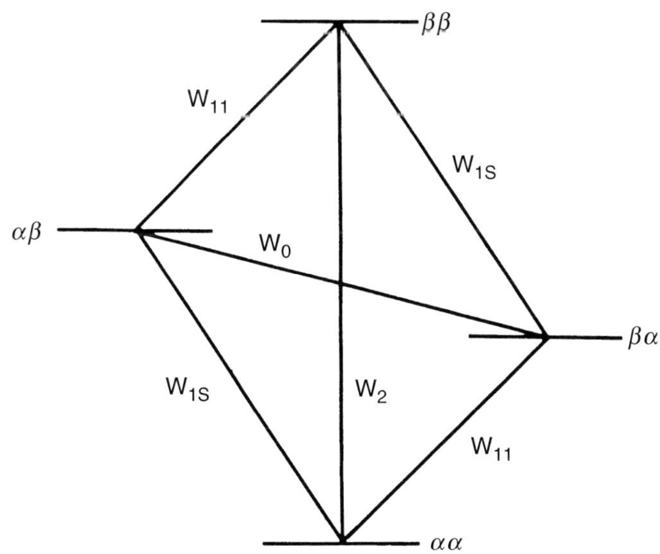

FIG. 1. Transition probabilities in a two-spin, four-level system. Reprinted from *Ann. Rep. NMR Spectrosc.*, vol. 22, Kowalewski, J., "Nuclear Spin Relaxation in Diamagnetic Fluids. Part 1. General Aspects and Inorganic Applications", pp. 307–414, Copyright 1990, with permission from Elsevier.

perturbation theory with stochastic perturbation. The result was that the transition probabilities could be expressed as:

$$W_{ij} = 2\,\mathrm{Re}\int_0^t \overline{\langle i|H_1(t)|\,j\rangle\langle\,j|H_1(t+\tau)|i\rangle}\exp(i\omega_{ij}\tau)d\tau$$

$$= 2\,\mathrm{Re}\int_0^t \overline{\langle i|H_1(0)|j\rangle\langle j|H_1(\tau)|i\rangle}\exp(i\omega_{ij}\tau)d\tau$$

$$= 2\,\mathrm{Re}\int_0^t G_{ij}(\tau)\exp(i\omega_{ij}\tau)d\tau = 2\,J_{ij}(\omega_{ij}) \qquad (8)$$

i and j denote the eigenstates of the unperturbed Zeeman Hamiltonian, ω_{ij} is the difference between the corresponding eigenvalues expressed in angular frequency units, $\langle i|H_1(t)|j\rangle$ is the matrix element of the perturbation Hamiltonian. The bar denotes ensemble average. Between the first and the second lines of Eq. (8), we assume that the perturbation is stationary (the product $\overline{\langle i\,|H_1(t)\,|\,j\rangle\langle j|H_1(t+\tau)|i\rangle}$ is independent of t). In the third line, we introduce the time-correlation function (tcf), $G_{ij}(t)$, a fundamental quantity in time-dependent statistical mechanics. Since the tcf describes the correlation of a matrix element of H_1 at one point in time with the complex conjugate of the same matrix element at another point in time, it is also denoted as autocorrelation function. In the last equality, we assume that the tcf decays rapidly compared to the upper limit of integration, t, and that the integration can consequently be extended to infinity, introducing a spectral density function $J_{ij}(\omega)$:

$$J_{ij}(\omega) = \mathrm{Re}\int_0^\infty G_{ij}(\tau)\exp(i\omega\tau)d\tau \qquad (9)$$

The last assumption is very fundamental. It results in time-independent transition probabilities and makes a "clean" theory possible. It requires that the product of the time scale of the decay time for the tcf (called the correlation time and denoted τ_c) and the strength of the perturbation (in angular frequency units) has to be much smaller than unity (*17–20*). This range is sometimes denoted as the Redfield limit or the perturbation regime.

A common assumption in the relaxation theory is that the time-correlation function decays exponentially, with the above-mentioned correlation time as the time constant (this assumption can be rigorously derived for certain limiting situations (*18*)). The spectral density function is then Lorentzian and the nuclear spin relaxation rate of Eq. (7) becomes:

$$T_{1I}^{-1} = \frac{1}{10}C_{DD}^2\left[\frac{\tau_c}{1+(\omega_S-\omega_I)^2\tau_c^2} + \frac{3\tau_c}{1+\omega_I^2\tau_c^2} + \frac{6\tau_c}{1+(\omega_S+\omega_I)^2\tau_c^2}\right] \qquad (10)$$

where C_{DD} is the dipole–dipole coupling constant, the strength of the dipole–dipole interaction, $C_{DD} = (\mu_0 \gamma_I \gamma_S \hbar / 4\pi r_{IS}^3)$, determined by the magnetogyric ratios, γ_I and γ_S, together with the interspin distance r_{IS}. Equation (10) is valid for both spins having the spin quantum number of 1/2. In Solomon's original formulation, τ_c is the rotational correlation time characterizing the rate of reorientation of the DD axis in a liquid solution. More specifically, τ_c is the time constant for the exponential decay of the tcf for the rank-two spherical harmonics of the polar angles that change with time according to the rotational diffusion equation (*18*). Now, we denote this correlation time as τ_R. In a liquid solution, τ_R depends on the molecular and hydrodynamic properties of the solution, it increases in general with increasing viscosity and molecular size. The rotational correlation time is independent of the magnetic field, but the nuclear spin relaxation rate in Eq. (10) does depend on the magnetic field, B_0, through $\omega_I = \gamma_I B_0$ and $\omega_S = \gamma_S B_0$. A plot of the field-dependence of the terms in the square parentheses in Eq. (10) is shown in Fig. 2. We can see two characteristic dispersions. Assuming $\omega_S \gg \omega_I$, the low-field dispersion occurs when $\omega_S \tau_c = 1$ (i.e., when the first and the third Lorentzian in Eq. (10) start decaying) and the high-field dispersion at $\omega_I \tau_c = 1$, when the second Lorentzian disperses. An experimental diagram corresponding to Fig. 2 and describing the field-dependence of the nuclear spin–lattice relaxation rate is called nuclear magnetic relaxation dispersion (NMRD) profile.

Solomon and Bloembergen (*21,22*) formulated a similar expression for the interaction between the two spins being scalar (*J*-coupling) rather than

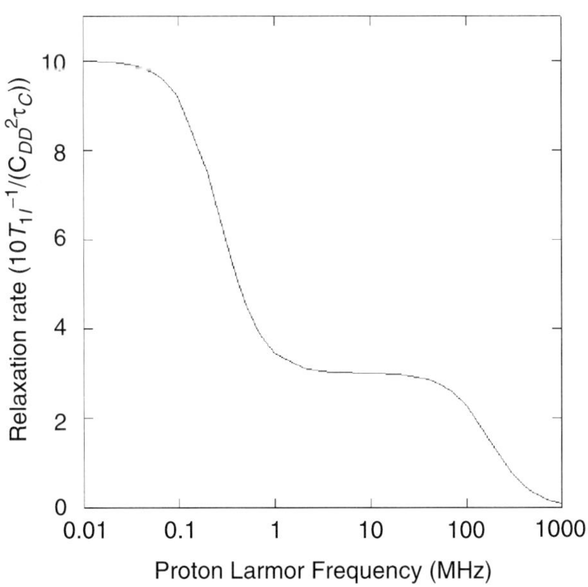

FIG. 2. Field-dependence of the PRE according to the Solomon model, Eq. (10). The rotational correlation time of 1 ns is used.

dipolar. The scalar interaction does not depend on orientation and the random modulation of the perturbation Hamiltonian, leading to relaxation of I, has its origin in rapid relaxation of the S-spin. In the terminology of Abragam (*18*), we can here talk about the relaxation of I spin as the scalar relaxation of the second kind. The scalar relaxation of the first kind refers to the situation when the scalar coupling is modulated by the chemical exchange of the S-spin between a site with J-coupling to I and a site without such a coupling (*18*).

B. THE MODIFIED SOLOMON–BLOEMBERGEN EQUATIONS AND THE SOLOMON–BLOEMBERGEN–MORGAN THEORY

In the context of paramagnetic systems, I denotes a nuclear spin and S stands for electron spin. The system of interest in this review usually contains the nuclear spin in a ligand or solvent. The nuclear spin interacts, through the dipolar and scalar part of the hyperfine interaction, with the electron spin. The reorientation of the complex modulates the DD interaction, but not the scalar interaction. The electron spin has a rapid relaxation pathway of its own, which affects both the dipolar and the scalar interaction. Concentrating for the moment on the case when the ligand exchanges between the inner-sphere of a transition metal complex and the bulk, we need in addition to consider the modulation of both parts of the hyperfine interaction by chemical exchange, characterized by the exchange lifetime τ_M. Assuming the three processes (rotation, electron spin relaxation, and exchange) being uncorrelated, and the electron spin relaxation to be described by a single spin–lattice relaxation time (T_{1e}) and a single spin–spin relaxation time (T_{2e}), the Solomon equation needs to be modified only slightly. We need to allow for the electron spin quantum number to be larger than 1/2, and to define some new correlation times:

$$\begin{aligned} \tau_{ci}^{-1} &= \tau_R^{-1} + T_{ie}^{-1} + \tau_M^{-1}; \\ \tau_{ei}^{-1} &= T_{ie}^{-1} + \tau_M^{-1}; \qquad i = 1, 2 \end{aligned} \tag{11}$$

Making this adjustment in the Solomon equation (Eq. (10)) and adding the corresponding scalar term, with the interaction strength given by the scalar coupling A_{SC} (in angular frequency units), results in the modified Solomon–Bloembergen equations (MSB-equations) for the nuclear spin–lattice relaxation rate in the first coordination sphere of a paramagnetic complex:

$$\begin{aligned} T_{1I}^{-1} = (T_{1I}^{SC})^{-1} + (T_{1I}^{DD})^{-1} &= \frac{2}{3} A_{SC}^2 S(S+1) \frac{\tau_{e2}}{1 + (\omega_S - \omega_I)^2 \tau_{e2}^2} \\ &+ \frac{2}{15} S(S+1) C_{DD}^2 \left[\frac{\tau_{c2}}{1 + (\omega_S - \omega_I)^2 \tau_{c2}^2} + \frac{3\tau_{c1}}{1 + \omega_I^2 \tau_{c1}^2} + \frac{6\tau_{c2}}{1 + (\omega_S + \omega_I)^2 \tau_{c2}^2} \right] \end{aligned} \tag{12}$$

The MSB–equations were first presented by Connick and Fiat (*23*) and by Reuben *et al.* (*24*). A formal derivation of these equations can be found, in a somewhat sketchy form, in the article by Gueron (*25*) and in a more stringent version in an article by Benetis *et al.* (*26*). Since $\omega_S \approx 658\omega_I$ if I is a proton, and even more if it represents another nuclear spin, the first and third term of the DD part of Eq. (12) can safely be combined into a "seven term", dispersing at $\omega_S\tau_{c2} = 1$ while the "three term" disperses at $\omega_I\tau_{c1} = 1$. Similar equations can also be derived for the nuclear spin–spin relaxation rate in a paramagnetic complex:

$$T_{2I}^{-1} = (T_{2I}^{SC})^{-1} + (T_{2I}^{DD})^{-1} = \frac{1}{3}A_{SC}^2 S(S+1)\left(\tau_{e1} + \frac{\tau_{e2}}{1+(\omega_S-\omega_I)^2\tau_{e2}^2}\right) + \frac{1}{15}S(S+1)C_{DD}^2\left[4\tau_{c1} + \frac{3\tau_{c1}}{1+\omega_I^2\tau_{c1}^2} + \frac{\tau_{c2}}{1+(\omega_S-\omega_I)^2\tau_{c2}^2} + \frac{6\tau_{c2}}{1+\omega_S^2\tau_{c2}^2} + \frac{6\tau_{c2}}{1+(\omega_S+\omega_I)^2\tau_{c2}^2}\right] \quad (13)$$

The relaxation rates in Eqs. (12) and (13) depend now on the magnetic field in a more complicated way. Not only are the Larmor frequencies in the denominators of the Lorentzians proportional to the field, the electron spin relaxation rates are, in principle, also field-dependent.

Bloembergen and Morgan (*27*) developed in the early 1960s a theory for electron spin relaxation, valid for $S = 1$ electron spin systems in the Redfield limit and at high magnetic field (the Zeeman limit). We shall discuss the electron spin relaxation in the Redfield limit (as well as beyond) more thoroughly later. For the moment, it suffices to say that the field-dependence of the electron spin relaxation rates in the Bloembergen–Morgan theory are expressed as:

$$T_{1e}^{-1} = \frac{1}{5}\tau_{S0}^{-1}\left[\frac{1}{1+\tau_v^2\omega_S^2} + \frac{4}{1+4\tau_v^2\omega_S^2}\right] \quad (14)$$

$$T_{2e}^{-1} = \frac{1}{10}\tau_{S0}^{-1}\left[3 + \frac{5}{1+\tau_v^2\omega_S^2} + \frac{2}{1+4\tau_v^2\omega_S^2}\right] \quad (15)$$

The symbol τ_{S0} denotes the electron spin relaxation time at zero magnetic field, where $T_1 = T_2$ and τ_v is another correlation time, associated with distortions of the paramagnetic complex caused by molecular collisions.

The Bloembergen–Morgan equations, Eqs. (14) and (15), predict that the electron spin relaxation rates should disperse at around $\omega_S\tau_v = 1$. This will make the correlation times for the dipolar and scalar interaction, τ_{ci} and τ_{ei}, respectively, in Eq. (11) dependent on the magnetic field. A combination of the modified Solomon–Bloembergen equations (12) and (13), for nuclear relaxation rates with the Bloembergen–Morgan equations for the field dependence of the correlation times yields a consistent theory, called the Solomon–Bloembergen–Morgan (SBM) theory for the field dependence of the PRE.

We wish to mention that a modified, parametrized, form of Eq. (15), for effective electron spin–spin relaxation rate in $S = 7/2$ complexes of Gd(III) and Eu(II), was used in some articles from the Merbach group (*28–30*).

Recapitulating, the SBM theory is based on two fundamental assumptions. The first one is that the electron relaxation (which is a motion in the electron spin space) is uncorrelated with molecular reorientation (which is a spatial motion influencing the dipole coupling). The second assumption is that the electron spin system is dominated by the electronic Zeeman interaction. Other interactions lead to relaxation, which can be described in terms of the longitudinal and transverse relaxation times T_{1e} and T_{2e}. This point will be elaborated on later. In this sense, one can call the modified Solomon–Bloembergen equations a Zeeman-limit theory. The validity of both the above assumptions is questionable in many cases of practical importance.

Besides these two main assumptions, the modified Solomon–Bloembergen equations, Eqs. (12) and (13), contain several additional approximations, the most important of which are the following:

(i) The electron spin is assumed to be a point-dipole centered at the metal ion. We discuss this assumption in Section II.C.
(ii) The electron g-tensor is assumed to be isotropic. The effects of deviations from this assumption were first investigated by Sternlicht (*31*) and more recently by Bertini *et al.* (*32*).
(iii) It is assumed that the reorientation of the nuclear-electron spin vector can be described by a single correlation time, τ_R. This amounts to neglecting the effects of anisotropic reorientation and internal motion. The latter issue seems to be important, in particular for polymeric systems containing several paramagnetic centers, and has been dealt with by Toth *et al.* (*33*) and Dunand *et al.* (*34,35*). [*Note added in proof*: An important contribution to the field was recently presented by Yerly *et al.* (Yerly, F.; Borel, A.; Helm, L.; Merbach, A.E. *Chem. Eur. J.* **2003**, *9*, 5468–5480).]
(iv) It is assumed that chemical exchange is uncorrelated with the remaining motions in the lattice. This assumption seems to be easy to fulfil. We shall return to the exchange effects in section VII.D.

C. Validity of the Point-Dipole Approximation

In this section, we discuss the work inquiring into the meaning of r_{IS}, the distance between the two dipoles in Eqs. (12) and (13). The simplest approximation is to assume that r_{IS} is equal to the internuclear distance between the nucleus in the ligand, the relaxation of which is being studied, and the metal ion. This amounts to the point-dipole approximation for both the nuclear and the electron spins. While such an approximation is perfectly reasonable for the nuclear spin, it is evident from a quantum chemical point of view that, due to the finite size of the electron distribution, and due to spin polarization and spin delocalization (covalency), the treatment of the electron spin as a point-dipole may be incorrect.

The first quantitatively relevant treatment of the problem was presented already in the 1970s by Waysbort and Navon (*36*). They derived an expression for the dipolar part of the T_{1I}^{-1} for an electron spin $S = 1/2$, following Solomon's original derivation but including the spatial distribution of the unpaired electron, represented by a molecular orbital, into the total wavefunction describing the eigenstates of the unperturbed Hamiltonian. The calculation was later extended to the case of a general electron spin S (*37*). A similar, but more general treatment, based on the spin density matrix formalism within the molecular orbital theory (the concept of density matrix as used in quantum chemistry should not be confused with the statistical density operator or its matrix representation, to be discussed in Section II.D and there on) has been given by Gottlieb *et al.* (*38*). Kowalewski *et al.* re-formulated the approach of Gottlieb *et al.* in a way suitable for *ab initio* calculations and used it for a series of hexaaquo transition metal ions (*39,40*) to estimate the deviations from the point-dipole approximation for both proton and oxygen-17 nuclei. In this formulation the quantity r_{IS}^{-6} in Eqs. (12) and (13) should be replaced by the inverse sixth power of an effective distance, r_{eff}, given by:

$$r_{\text{eff}}^{-6} = \frac{1}{4}\left[\langle q_{zz}\rangle^2 + \frac{1}{3}\left(\langle q_{xx}\rangle - \langle q_{yy}\rangle\right)^2\right] \tag{16}$$

where $\langle q_{\alpha\alpha}\rangle$ are diagonal elements of the spin-differential field gradient tensor, a quantity expressed in terms of spin density matrix and integrals over products of atomic orbitals, spherical harmonics and r^{-3} (*39*).

The effective distances obtained by Nordenskiöld *et al.* (*40*) are compared with the internuclear distances in Table I. Clearly, the point dipole approximation is reasonable for the hydrogen nuclei in these complexes, while substantial deviations are observed for the oxygen nuclei. The findings of these early quantum chemical studies were confirmed by Sahoo and Das (*41–43*). Wilkens *et al.* have reported DFT calculations using Eq. (16) for a 104 atom model for high-spin Fe(III) rubredoxin (*44*). Large discrepancies between the effective distances and the input distances for the calculations were found for the hyperfine-shifted nitrogen-15 resonances, as well as for proton and carbon-13 nuclei in cysteines bound to the iron center.

More approximate expressions, based on the concepts of ligand-centered dipolar relaxation, were also proposed. The review by Mispelter and co-workers (*45*) gives the expression:

$$r_{\mathit{eff}}^{-6} = r_{IS}^{-6} + a r_L^{-6}(\rho^{\pi})^2 + b r_{IS}^{-3} r_L^{-3} \rho^{\pi} \tag{17}$$

Table I

DEVIATIONS FROM THE POINT–DIPOLE IN HEXAAQUA COMPLEXES OF DIVALENT 3D METAL IONS[a]

Measurements	Metal ions Mn	Fe	Co	Ni	Cu
^{1}H Data					
r_{cryst} (pm)	290.6	290.6	278.1	275.5	270.4
r_{eff} (pm)	290.0	289.5	276.3	273.1	268.4
$r^{6}_{cryst} \cdot <1/r^{3}>^{2}_{eff^{b}}$	1.01	1.02	1.04	1.05	1.05
^{17}O Data					
r_{cryst} (pm)	222.0	222.0	209.0	206.1	201.0
r_{eff} (pm)	219.8	211.7	190.5	174.9	172.3
$r^{6}_{cryst} \cdot <1/r^{3}>^{2}_{eff^{b}}$	1.06	1.33	1.74	2.67	2.52

[a]From Ref. (*40*). M–H distances were calculated using $r_{O\text{-}H} = 95.7$ pm and HOH = 104.52°.
[b]The effective distances account for the experimental T_{1M}^{-1} by considering the electronic distribution predicted from *ab initio* MO calculations.
$r^{6}_{cryst} \cdot <1/r^{3}>^{2}_{eff^{b}}$ is the predicted enhancement coefficient of T_{1M}^{-1} with respect to a pure metal-centered dipolar relaxation mechanism.

It is assumed that most of the electron spin density resides on the metal, but that a certain small part of it, given by the quantity ρ^{π}, is delocalized to the ligand heteroatom L. The first term is the point–dipole interaction term, the second corresponds to the dipolar interaction between the nuclear spin under consideration and the spin-density on the atom L and the last term describes the cross-correlation of the two dipolar interactions (we discuss the issue of cross-correlation phenomena in more general terms in Section II.D and III.B). The quantity r_L is the effective distance from the nuclear spin to the center of the local spin density. For protons bound to L, one sets r_L to the internuclear HL distance. The coefficients a and b in Eq. (17) were also estimated by Mispelter and co-workers (*45*). The review by Bertini and co-workers advises caution in using the point-dipole approximation for protons close to the paramagnetic center (*46*), whereas the careful analysis of data for Cu(II) plastocyanin by Ma *et al.* (*47*) questions the point-dipole approximation for protons at distances longer than about 1 nm from the paramagnetic metal center. For heteronuclei, such as ^{13}C or ^{15}N, Ma *et al.* find that the ligand-centered dipolar relaxation is normally the dominant term (*47*).

D. THE REDFIELD RELAXATION THEORY

A more general formulation of relaxation theory, suitable for systems with scalar spin–spin couplings (*J*-couplings) or for systems with spin quantum numbers higher than 1/2, is known as the Wangsness, Bloch and Redfield (WBR) theory or the Redfield theory (*17*). In analogy with the Solomon–Bloembergen formulation, the Redfield theory is also based on the second-order perturbation approach, which in certain situations (not uncommon in paramagnetic systems) can be a limitation. The requirement that the product of the relevant correlation time and the strength of the perturbation (in angular frequency units) has to be much smaller than unity (*17–20*), is a necessary condition here also. Rather than dealing with the concepts of magnetizations or energy level populations, as in the Solomon formulation, the Redfield theory is given in terms of density operator σ or its matrix representation, the density matrix. Briefly, the density operator or density matrix describes the average behavior of an ensemble of quantum mechanical systems. In Redfield's original formulation (*17*) and in Slichter's classical textbook (*19*), the matrix representation of the density operator is constructed using the basis set of eigenstates, α and α', of the unperturbed Hamiltonian. The diagonal elements of the density matrix correspond to the populations, while the off-diagonal terms are called coherences. The time development of the density matrix element $\rho_{\alpha\alpha'}$ is according to the Redfield theory given by the expression:

$$\frac{d\rho_{\alpha\alpha'}}{dt} = -i\omega_{\alpha\alpha'}\rho_{\alpha\alpha'}(t) + \sum_{\beta\beta'} R_{\alpha\alpha'\beta\beta'}\left(\rho_{\beta\beta'}(t) - \rho^{0}_{\beta\beta'}\right) \tag{18}$$

The symbol $\omega_{\alpha\alpha'}$ denotes the energy difference between the two eigenstates, converted into angular frequency. The first term on the right-hand side (rhs) of Eq. (18) vanishes for the populations ($\omega_{\alpha\alpha} = 0$) and describes the precessional motion for coherences. $R_{\alpha\alpha'\beta\beta'}$ is an element of the relaxation matrix (also called relaxation supermatrix) describing various decay and transfer processes in the spin system. Under certain conditions (secular approximation), one neglects the relaxation matrix elements unless the condition $\omega_{\alpha\alpha'} = \omega_{\beta\beta'}$ is fulfilled.

Using a classical description of the environment of the spins (semi classical relaxation theory), the relaxation matrix elements are expressed as linear combinations of various spectral density functions of the form slightly more general than that of Eq. (9). The quantity $\rho^{0}_{\beta\beta'}$ (zeroes if $\beta \neq \beta'$, Boltzmann-distributed populations if $\beta = \beta'$) is not present in the simple semi-classical derivation of Eq. (18); it is inserted in an *ad hoc* manner in order to give the correct description of the equilibrium condition at finite temperature. The relaxation matrix is in principle large; for a system with n eigenstates, it has the dimension $n^2 \times n^2$. One says that the n eigenstates $|\alpha\rangle$(eigenvectors, kets) span a Hilbert space for the system at hand, while the n^2 "ket-bra" products $|\alpha\rangle\langle\alpha'|$ span the corresponding Liouville space. For a system of two spin 1/2 nuclei, $n = 4$ and $n^2 = 16$. Operators are represented in the Hilbert space as matrices. In the Liouville space, operators form the basis set. The objects operating on the basis set in the Liouville space – and represented there as matrices – are called superoperators (*48,49*). Equation (18) may be expressed in the Liouville superoperator formalism as:

$$\frac{d}{dt}|\rho(t)) = -i\hat{\hat{L}}_0|\rho(t)) - \hat{\hat{R}}\left[|\rho(t)) - |\rho^{T})\right] \tag{19}$$

The "double hat" denotes superoperators. $\hat{\hat{L}}_0$ is the unperturbed Liouville superoperator (Liouvillian) for the system. A Liouvillian is defined as a commutator with the Hamiltonian, in this case $\hat{\hat{L}}_0 = [H_0,]$, $\hat{\hat{L}}_0\rho = [H_0, \rho]$ and $\hat{\hat{L}}_0|\alpha\rangle\langle\alpha'| = \omega_{\alpha\alpha'}|\alpha\rangle\langle\alpha'|$. In the same way, $\hat{\hat{R}}$ is the relaxation superoperator. $|\rho^T)$ is the density operator at thermal equilibrium.

In an alternative formulation of the Redfield theory, one expresses the density operator by expansion in a suitable operator basis set and formulates the equation of motion directly in terms of the expectation values of the operators (*18,20,50*). Consider a system of two nuclear spins with the spin quantum number of $1/2, I$, and N, interacting with each other through the scalar J-coupling and dipolar interaction. In an isotropic liquid, the former interaction gives rise to J-split doublets, while the dipolar interaction acts as a relaxation mechanism. For the discussion of such a system, the appropriate sixteen-dimensional basis set can for example consist of the unit operator, E, the operators corresponding to the Cartesian components of the two spins, $I_x, I_y, I_z, N_x, N_y, N_z$ and the products of the components of I and the components of N (*49*). These sixteen operators span the Liouville space for our two-spin system. If we concentrate on the longitudinal relaxation (the relaxation connected to the distribution of populations), the Redfield theory predicts the relaxation to follow a set of three coupled differential equations:

$$\frac{d}{dt}\begin{pmatrix} <I_z> \\ <N_z> \\ <2I_zN_z> \end{pmatrix} = -\begin{pmatrix} \rho_I & \sigma_{IN} & \delta_{I,IN} \\ \sigma_{IN} & \rho_N & \delta_{N,IN} \\ \delta_{I,IN} & \delta_{N,IN} & \rho_{IN} \end{pmatrix}\begin{pmatrix} <I_z> - I_z^0 \\ <N_z> - N_z^0 \\ <2I_zN_z> \end{pmatrix} \tag{20}$$

The matrix on the rhs of Eq. (20) is an example of the relaxation matrix for this particular representation of the density operator. To be exact, it represents one block of a larger, block-diagonal relaxation matrix, defined in the Liouville space. The remaining part of the large matrix describes the relaxation of coherences, which can be seen as generalizations of the transverse components of the magnetization vector. In systems without degeneracies, each of the coherences decays exponentially, with its own T_2. The first 2×2 block of Eq. (20) corresponds to the Solomon case (*15*): the diagonal elements ρ of the relaxation matrix (not to be confused with density operator itself) correspond to the spin–lattice relaxation rates for the two nuclei and σ_{IN} is the cross-relaxation rate, giving rise to the phenomenon of nuclear Overhauser enhancement. The characteristic feature of the coupled spin systems is the occurrence of the expectation value of the $2I_zN_z$ operator (the two-spin order), characterized by an own decay rate ρ_{IN}, and coupled to the one-spin longitudinal operators by the terms $\delta_{I,IN}$ and $\delta_{N,IN}$.

In order to discuss the origin of these terms we need to allow the spins to have anisotropic shielding tensors. Molecular tumbling in solution makes the chemical shielding in the direction of the external magnetic field a stochastic function of time and acts therefore as a relaxation mechanism, called the chemical shielding anisotropy (CSA) mechanism. The Hamiltonian for each of the two spins, analogous to Eq. (5), contains therefore two time-dependent terms, $H_1^{DD}(t)$ and $H_1^{CSA}(t)$. When deriving the relaxation matrix elements or transition probabilities, as in Eq. (8), we now get three types of terms: two of them involve autocorrelation functions for matrix elements of $H_1^{DD}(t)$ and $H_1^{CSA}(t)$ and the third is a cross-correlation term, $\langle i|H_1^{DD}(t)|j\rangle\langle j|H_1^{CSA}(t+\tau)|i\rangle$. The relaxation matrix elements $\delta_{I,IN}$ and $\delta_{N,IN}$ arise exclusively through these DD-CSA cross-correlation or interference terms (*51*). In the case of transverse relaxation or linewidth-related phenomena, similar DD-CSA cross-correlation terms give rise to the differential line-broadening of the components of multiplets caused by scalar coupling. We shall return to the terms of this type in Section III.B.

III. The Curie-Spin Relaxation and Related Topics

A. The Curie-Spin Relaxation

Among the three dynamic processes modulating the electron-nuclear dipole–dipole interaction, summarized in a simple form by Eqs. (12) and (13) the electron spin relaxation is often the fastest, especially for large molecules with sluggish reorientation. This situation has been considered by Gueron (*25*) and by Vega and Fiat (*52*). The authors write the electron spin vector as a sum of the thermal average "Curie spin", aligned along the magnetic field, and the oscillating and zero-average remainder, **s**. The magnitude of the Curie spin is:

$$S_C = \langle S_z\rangle = g_e\mu_B S(S+1)B_0/3k_BT \tag{21}$$

where k_B is the Boltzmann constant and T temperature, while the mean square of **s** is $S(S+1)/3 - S_C^2$. The electronic g factor is denoted g_e and μ_B is the Bohr magneton. We note that $\hbar\gamma_S = g_e\mu_B$. The Curie spin is related to the magnetic susceptibility and Eq. (21) is derived under the assumption that the susceptibility is given by the Curie law (*3*). The nuclear spin interacts, through the dipole–dipole interaction with both S_C and **s**. The magnetic moment related to the Curie spin is a small fraction of the total electron spin magnetic moment, but it becomes non-negligible at sufficiently high magnetic field. At high field, we can in addition neglect the terms in Eqs. (12) and (13), which contain ω_S. Neglecting the scalar interaction terms, the relaxation rates given by Eqs. (12) and (13) become under these conditions:

$$T_{1I}^{-1} = \frac{6}{5}\left(\frac{\mu_0}{4\pi}\right)^2 \gamma_I^2 g_e^2 \mu_B^2 r_{IS}^{-6}$$
$$\left\{S_C^2\frac{\tau_D}{1+\omega_I^2\tau_D^2} + \left[\frac{1}{3}S(S+1) - S_C^2\right]\frac{\tau_{c1}}{1+\omega_I^2\tau_{c1}^2}\right\} \tag{22}$$

and

$$T_{2I}^{-1} = \frac{1}{5}\left(\frac{\mu_0}{4\pi}\right)^2 \gamma_I^2 g_e^2 \mu_B^2 r_{IS}^{-6}$$
$$\left\{S_C^2(4\tau_D + \frac{3\tau_D}{1+\omega_I^2\tau_D^2}) + \left[\frac{1}{3}S(S+1) - S_C^2\right]\left(4\tau_{c1} + \frac{3\tau_{c1}}{1+\omega_I^2\tau_{c1}^2}\right)\right\} \qquad (23)$$

The first term in both expressions corresponds to the relaxation by the Curie spin (Curie spin relaxation, CSR), while the second is very close to the dipolar parts of expressions of Eqs. (12) and (13). The Curie spin is modulated by reorientation and exchange, but not by the electron spin relaxation, and the corresponding correlation time is given by $\tau_D^{-1} = \tau_R^{-1} + \tau_M^{-1}$. If $\tau_D >> \tau_{c1}$, which is a common situation for large paramagnetic molecules with rapid electron spin relaxation, then the Curie spin terms become important and can in certain situations dominate over the modified Solomon-like terms. This is, for example, the case for aqualanthanide(III) ions except Gd(III) (*1,53*). In fact, the expressions in Eqs. (12,13,21–23) take a somewhat different form for lanthanides and actinides (*1*). Because of the very strong spin-orbit coupling in these systems, the total angular momentum should substitute the spin angular momentum. Thus, the quantum number S should be substituted by J, $\langle S_z \rangle$ should be replaced by the expectation value of the z-component of the total angular momentum, $\langle J_z \rangle$, and g_e by the Landé factor g_J. After these modifications, one obtains expressions containing factors $g_J^2\mu_B^2 J(J+1)$. These can be appropriately replaced by the square and the fourth power of the effective magnetic moments, when available (*1*).

The Curie spin relaxation is also called magnetic susceptibility relaxation. Vega and Fiat (*52*) have also considered the Curie spin relaxation originating from anisotropic magnetic susceptibility, rather than an isotropic one as assumed in Eqs. (21–23).

The magnetic susceptibility relaxation is usually more important for T_2 than for T_1. In fact, this mechanism is often dominant in determining the proton linewidth in paramagnetic proteins at high magnetic fields (*3*). Gillis and co-workers have recently developed a theory for the related case of proton linewidth in colloidal solutions of so-called superparamagnetic particles (*54,55*).

B. Paramagnetic Cross-Correlation and Interference Phenomena

Besides causing the increased linewidth (a single-spin property), the magnetic susceptibility can also give rise to more complicated multispin relaxation phenomena, if more than one nuclear spin is included. Consider again a system of two nuclear spins with the spin quantum number of 1/2, I and N, introduced in Section II.D. Along with the effects on the longitudinal relaxation mentioned there, the cross-correlation effects lead to differential line broadening of the J-split doublet components (*51*). This phenomenon has its analog in paramagnetic systems, which are related to the paramagnetic susceptibility and can be explained in the following way.

The thermally averaged electronic magnetic moment per molecule, induced by the external magnetic field, can in a general anisotropic case be expressed as:

$$\langle \mu \rangle = \frac{\chi \cdot \mathbf{B}_0}{\mu_0} \tag{24}$$

where χ is the magnetic susceptibility tensor. The Hamiltonian describing the dipolar interaction between the nuclear spin and this average magnetic moment can be written as:

$$\begin{aligned} H_\chi &= -\frac{\mu_0}{4\pi}\left[\frac{3(\gamma_I \mathbf{I}\cdot\mathbf{r})(\langle\mu\rangle\cdot\mathbf{r})}{r^5} - \frac{(\gamma_I \mathbf{I}\cdot\langle\mu\rangle)}{r^3}\right] \\ &= -\frac{1}{4\pi}\left[\frac{3(\gamma_I \mathbf{I}\cdot\mathbf{r})(\mathbf{r}\cdot\chi\cdot\mathbf{B}_0)}{r^5} - \frac{(\gamma_I \mathbf{I}\cdot\chi\cdot\mathbf{B}_0)}{r^3}\right] \end{aligned} \tag{25}$$

where the radius vector $\mathbf{r}$ is defined with respect to the paramagnetic metal ion at the origin (if more than one paramagnetic ion is present, the interaction can be expressed as a sum over terms similar to the right-hand-side of Eq. (25)). We can note that both terms in the second line of Eq. (25) contain the nuclear spin operator $\mathbf{I}$ on the left and $\mathbf{B}_0$ on the right. We can condense the term in-between the two vectors into a new tensor quantity, σ, and obtain (*3,25,56*):

$$H_\chi = -\gamma_I \mathbf{I}\cdot\sigma\cdot\mathbf{B}_0 \tag{26}$$

This formulation is identical to the anisotropic magnetic shielding term in the nuclear spin Hamiltonian. Since the origin of this shielding is the dipolar interaction with the induced magnetic moment, we call it dipolar shielding. In the principal frame of the susceptibility tensor, the dipolar shielding tensor is given by:

$$\sigma = \frac{1}{4\pi}\begin{pmatrix} (3x^2-r^2)\chi_x r^{-5} & 3xy\chi_y r^{-5} & 3xz\chi_z r^{-5} \\ 3xy\chi_x r^{-5} & (3y^2-r^2)\chi_y r^{-5} & 3yz\chi_z r^{-5} \\ 3xz\chi_x r^{-5} & 3yz\chi_y r^{-5} & (3z^2-r^2)\chi_z r^{-5} \end{pmatrix} \tag{27}$$

Consider the case where the I spin has a neighbor spin N, with which it interacts by the J-coupling as well as the dipolar coupling. Having stated the interaction with the thermally averaged electronic magnetic moment in the form of Eq. (26), we can immediately use the Redfield theory for multispin relaxation phenomena as developed for diamagnetic systems. This approach to the differential line-broadening in a system of two nuclear spins (e.g., the amide ^{1}H and ^{15}N in a paramagnetic protein), influenced by anisotropic paramagnetic susceptibility, was pursued by Bertini *et al.* (*56*), following the DD-CSA cross-correlation theory by Goldman (*51*).

The interference effects involving the susceptibility relaxation need not be treated using the CSR-CSA analogy. One can also use the fact that if more than two spins are present, several dipole–dipole interactions are possible and can interfere with each other (*50*). In fact, the multispin relaxation phenomena in paramagnetic systems can fully be explained in terms of the interference effects involving dipolar interactions, between the two nuclear spins and between one of nuclear spins and the electron spin. This argument was employed by Werbelow and Thevand (*57*), who demonstrated in 1993 that the longitudinal as well as transverse coupled spin relaxation of an *IN* nuclear spin pair is affected by a rapidly relaxing third spin S (with the spin quantum number 1/2), through an unequally-weighted superposition of two relaxed spectra associated with the $m_S = 1/2$ and $m_S = -1/2$ states of the S spin. The unequal weights are the results of the unequal population of the energy levels for the electron spin. Thus, the underlying physics is identical to the ideas of the Curie spin. At about the same time, Bertini and co-workers (*58,59*) and Qin *et al.* (*60*) used the CSR/CSA similarity and demonstrated that the interference between the proton–proton dipolar interaction and the dipolar shielding anisotropy could lead to relaxation-allowed coherence transfer and cross-peaks in two-dimensional COSY spectra of paramagnetic proteins, in full analogy with the diamagnetic DD-CSA case described by Wimperis and Bodenhausen (*61*). Using these results (*56–58*), Mäler *et al.* (*62*) studied the transfer between one-spin and two-spin longitudinal order in paramagnetic systems, fully analogous to the situation described in Eq. (20). Bertini and co-workers have also studied paramagnetic effects on zero and double-quantum relaxation in coupled proton systems (*63*).

The effects of electron spin –nuclear spin cross-correlations on differential broadening of multiplet components arising through the *J*-coupling between the amide ^{1}H (I spin) and ^{15}N (N spin) in paramagnetic proteins were also studied by Ghose and Prestegard (*64*). Assuming the isotropic paramagnetic susceptibility given by the Curie law, the expression for the difference in the linewidth of the two components of I nucleus doublet was derived, taking (after a correction of the numerical factor (*56*)) the form:

$$\Delta\nu = \frac{2}{15\pi}\left(\frac{\mu_0}{4\pi}\right)^2 \frac{B_0\gamma_I^2\gamma_N\mu_B^2 g_e^2\hbar S(S+1)}{r^3 r_{IN}^3 k_B T}\left(4\tau_c + \frac{3\tau_c}{1+\omega_I^2\tau_c^2}\right)\frac{3\cos^2\theta_{SIN}-1}{2} \tag{28}$$

The angle θ_{SIN} is that between the *IN* axis and the *IS* axis. The expression $(3\cos^2\theta-1)/2$ in Eq. (28) is characteristic for the cross-correlated relaxation effects. An analogous and somewhat more general expression for the case of anisotropic susceptibility was given by Bertini *et al.* (*56*). The cross-correlation-driven coherence transfer phenomena between nuclear spins in paramagnetic systems with anisotropic susceptibility were even earlier considered by Desvaux and Gochin (*65*).

The cross-correlated DD-CSA (or DD-CSR) spectral densities, giving rise to differential line broadening and to the order transfer phenomena summarized by Eq. (20), can in principle be complex functions. The line-broadening effects and the δ-terms in Eq. (20) are caused by real parts of the spectral densities, while the imaginary parts shift the transition frequencies slightly. The phenomenon is referred to as dynamic frequency shifts (dfs) and is very difficult to see, unless the dfs have a differential effect on multiplet components. Theory of dynamic frequency shifts was reviewed by Werbelow (*66*). For the DD/CSA cross-correlation in a two-spin (*IN*) system, he showed that the expression for the dfs for each of the NMR lines should consist of two terms, corresponding to the imaginary part of the spectral density, taken at the Larmor frequency of both spin species. The Ghose and Prestegard paper (*64*) takes up the issue of the dfs in paramagnetic systems. They show that the dfs change the positions of the *I* spin multiplet components in opposite directions and thus influences the apparent *IN* spin–spin coupling. An expression for the frequency shift of the *I* lines was given, but it is probably incorrect as it only contains the spectral density component at ω_I. Making use of the analogy between the dipolar shift anisotropy and chemical shift anisotropy, one would expect two terms, taken at different frequencies, as given by Werbelow (*66*).

The dynamic frequency shifts can also arise through cross-correlated dipolar interactions, as described by Brüschweiler (*67*). In a three-spin system with scalar (non-zero) and dipolar (zero-average) couplings, the dfs result in completely asymmetric multiplet patterns. The multiplet asymmetries can be derived as originating from three imaginary cross-correlated spectral density components, evaluated at the three Larmor frequencies characteristic for the system. If one of the three spins is an electron spin, a combined effect of the rapid relaxation and unequal spin level populations (compare Werbelow and Thevand (*57*)), results in a fairly complicated situation, currently under investigation (*68*). As the possible dfs effects can be indistinguishable from the residual dipolar couplings, resulting from the self orientation of paramagnetic proteins with anistropic susceptibility in solution, the understanding of the dfs is important from the point of view of paramagnetic constraints for the determination of protein structure (*3,69*).

IV. The General (Slow-Motion) Theory

We now come back to the simplest possible nuclear spin system, containing only one kind of nuclei *I*, hyperfine-coupled to electron spin *S*. In the Solomon–Bloembergen–Morgan theory, both spins constitute the spin system with the unperturbed Hamiltonian containing the two Zeeman interactions. The dipole–dipole interaction and the interactions leading to the electron spin relaxation constitute the perturbation, treated by means of the Redfield theory. In this section, we deal with a situation where the electron spin is allowed to be so strongly coupled to the other degrees of freedom that the Redfield treatment of the combined *IS* spin system is not possible. In Section V, we will be faced with a situation where the electron spin is in the Redfield limit, but other assumptions of the Solomon–Bloembergen–Morgan theory are not fulfilled.

A. The Redfield Limit and Beyond

Consider a general system described by the Hamiltonian of Eq. (5), where $H_1(t) \equiv H_{IL}(t)$ describes the interaction between the spin system (I) and its environment (the lattice, L). The interaction is characterized by a strength parameter ω_{IL}. When deriving the WBR (or the Redfield relaxation theory), the time-dependence of the density operator is expressed as a kind of power expansion in $H_{IL}(t)$ or ω_{IL} (*17–20*). The first (linear) term in the expansion vanishes if the ensemble average of $H_{IL}(t)$ is zero. If $\omega_{IL}\tau_c \ll 1$, where the correlation time, τ_c, describes the decay rate of the time correlation functions of $H_{IL}(t)$, the expansion is convergent and it is sufficient to retain the first non-zero term corresponding to ω_{IL}^2. This leads to the Redfield equation of motion as stated in Eq. (18) or (19). In the other limit, $\omega_{IL}\tau_c \gg 1$, the expansion tends to diverge and the time-dependence of the density operator needs to be formulated in another way.

If the spin system is allowed to contain the electron spin, the latter situation is not uncommon. The electron spins are often involved in strong anisotropic interactions, which can, in the first approximation, be considered constant in a molecule-fixed frame. As the spin-bearing molecule reorients in a liquid (with a correlation time τ_R), the anisotropic interactions vary in a stochastic way. Let us assume that the rotational average of the interaction is included in the unperturbed Hamiltonian and that we can introduce an anisotropy parameter ω_{aniso} (in angular frequency units), describing the amplitude of the coupling subject to rotational averaging. If the product of the anisotropy parameter and the rotational correlation time, $\omega_{aniso}\tau_R$, is small, then we can describe the electron spin relaxation and the electrons spin resonance (ESR) line shape using the Redfield theory. This situation arises commonly for $S = 1/2$ radicals in low-viscosity solutions, where the ω_{aniso} are related to the anisotropy of the g tensor or the hyperfine interaction. For solutions with high viscosity or for systems with $S > 1/2$, the product $\omega_{aniso}\tau_R$ may be of the order of unity or greater, and the Redfield theory does not apply. The theory of the ESR line shapes in this kind of systems has been derived by Freed and co-workers (*70–72*) and by Lynden-Bell (*73*), using the formalism based on the stochastic Liouville equation (SLE), proposed originally by Kubo (*74*).

The stochastic Liouville equation, in the form relevant for the ESR line shape calculation, can be written in a form reminiscent of the Redfield equation in the superoperator formulation, Eq. (19) (*70–73*):

$$\frac{d}{dt}|\chi(t)) = -i\hat{\hat{L}}|\chi(t)) - \hat{\hat{K}}\left[|\chi(t)) - |\chi^T)\right] \tag{29}$$

χ is the density operator, here involving not only the (electron) spin, but also the classical degrees of freedom to which the spin is strongly coupled. $\hat{\hat{L}}$ is the corresponding Liouvillian (depending on spin and the variables related to the important classical degree of freedom) and $\hat{\hat{K}}$ is a stochastic operator, describing the evolution of the classical degree of freedom under consideration. In an important class of cases, the relevant classical degree of freedom is the orientation of the molecule with respect to the laboratory frame. The superoperator $\hat{\hat{L}}$ depends in this case on the set of angles, Ω, specifying the orientation of the principal frame of the anisotropic interaction with respect to the laboratory frame and $\hat{\hat{K}}$ takes the form of the operator describing the random variation of the orientation, e.g., through rotational diffusion.

Considering the systems relevant for the PRE, we may often encounter a situation where the interaction of nuclear spins with their environment, containing the electron spin, i.e., the hyperfine interaction, is weak enough to be treated by the Redfield theory, but at the same time the Redfield theory is not applicable to the electron spin characterized by a strong anisotropic interaction. One way to treat that problem, closely related to the use of the SLE, has been proposed by Hwang and co-workers (*75,76*). Another set of models, suitable for this case, have been developed during the last twenty years by Swedish groups. The common general theoretical ground for this set of models is referred to as the slow-motion theory. The slow-motion theory was first formulated in a series of papers in the early 1980s (*26,77–79*); these early versions of the theory were reviewed some time ago (*11,13*).

The basic idea of the slow-motion theory is to treat the electron spin as a part of the lattice and limit the spin part of the problem to the nuclear spin rather than the *IS* system. The difficult part of the problem is to treat, in an appropriate way, the combined lattice, now containing the classical degrees of freedom (such as rotation in condensed matter) as well as quantized degrees of freedom (such as the electron Zeeman interaction). The Liouville superoperator formalism is very well suited for treating this type of problems.

The whole system, nuclear spin and the composite lattice, is described by the equation of motion:

$$\frac{d}{dt}\chi = i[\chi, H] = -i\hat{\hat{L}}\chi = -i\left(\hat{\hat{L}}_I + \hat{\hat{L}}_{IL} + \hat{\hat{L}}_L\right)\chi \tag{30}$$

H is the total Hamiltonian (in the angular frequency units) and $\hat{\hat{L}}$ is the total Liouvillian, divided into three parts describing the nuclear spin system ($\hat{\hat{L}}_I$), the lattice ($\hat{\hat{L}}_L$) and the coupling between the two subsystems ($\hat{\hat{L}}_{IL}$). The symbol χ is the density operator for the whole system, expressible as the direct product of the density operators for spin (ρ) and lattice (σ), $\chi = \rho \otimes \sigma$. The Liouvillian ($\hat{\hat{L}}_I$) for the spin system is the commutator with the nuclear Zeeman Hamiltonian (we thus treat the nuclear spin system as an ensemble of non-interacting spins in a magnetic field). $\hat{\hat{L}}_L$ will be defined later and $\hat{\hat{L}}_{IL}$

is the commutator with the hyperfine Hamiltonian, which we here choose to write as the scalar contraction of rank one tensors:

$$H_{IL} = \sum_{n=-1}^{1} (-1)^n I_n^1 T_{-n}^1 = H_{SC} + H_{DD} = \sum_{n=-1}^{1} (-1)^m I_n^1 (T_{-n}^{1,DD} + T_{-n}^{1,SC}) \quad (31)$$

The symbols $T_{-n}^{1,SC}$ and $T_{-n}^{1,DD}$ are rank one tensor operators; $T_{-n}^{1,SC} = A_{SC} S_{-n}^1$, where S_{-n}^1 are the components of the electron spin operator in the standard irreducible tensor form (*16,77*), and:

$$T_n^{1,DD} = C_{DD}\sqrt{30} \sum_{q\in(-1,0,1)} \begin{pmatrix} 2 & 1 & 1 \\ n-q & q & -n \end{pmatrix} S_q^{1(L)} D_{0,n-q}^2(\Omega_{ML}) \quad (32)$$

The expression of Eq. (32) implies a rank one contraction of $S^{1(L)}$ (L in parentheses in the superscript explicitly indicates that operator is given in the laboratory frame) with the rank two Wigner rotation matrices $D_{0,n-q}^2[\Omega_{ML}(t)]$, which describe the transformation from the molecule-fixed (*M*, the principal frame of the dipole–dipole interaction) to the laboratory frame through the set of Euler angles Ω_{ML}. If the scalar term is neglected, then the formulations of Eqs. (31) and (32) are fully equivalent to that of Eq. (6). The expressions given by Eqs. (31) and (32) are convenient for the case of the electron spin treated as a part of the combined lattice.

Assuming that the lattice can, on the time scale relevant for the evolution of the nuclear spin density operator, be considered to remain in thermal equilibrium, $\sigma = \sigma^T$, and applying the Redfield theory to the nuclear spin sub-system allows us to obtain the following expressions for nuclear spin–lattice and spin–spin relaxation rates:

$$T_{1I}^{-1} = 2\mathrm{Re}\big(K_{1,1}^{DD}(-\omega_I) + K_{1,1}^{SC}(-\omega_I) + K_{1,1}^{DD-SC}(-\omega_I)\big) \quad (33)$$

$$\begin{aligned} T_{2I}^{-1} &= \mathrm{Re}\big(K_{0,0}^{DD}(0) + K_{0,0}^{SC}(0) + K_{0,0}^{DD-SC}(0)\big) \\ &\quad + \mathrm{Re}\big(K_{1,1}^{DD}(-\omega_I) + K_{1,1}^{SC}(-\omega_I) + K_{1,1}^{DD-SC}(-\omega_I)\big) \end{aligned} \quad (34)$$

The symbol $\mathrm{Re}(K_{n,n}^{DD}(\omega))$ denotes the real part of the complex spectral density, corresponding to the autocorrelation of the dipolar interactions, while $\mathrm{Re}(K_{n,n}^{SC}(\omega))$ is its counterpart for the scalar interaction. The symbol $\mathrm{Re}(K_{n,n}^{DD-SC}(\omega))$ denotes the spectral density describing the cross-correlation of the two parts of the hyperfine interaction. The cross-correlation vanishes at the MSB level of the theory, but in the more complicated case of the lattice containing the electron spin, the cross term may be non-zero. A general expression for the dipolar spectral density is:

$$K_{n,n}^{DD}(-\omega) = \int_0^\infty Tr_L\Big\{T_n^{1,DD+}\Big[\exp(-i\hat{\hat{L}}_L\tau)T_n^{1,DD}\Big]\sigma^T\Big\}\exp(-i\omega\tau)d\tau \quad (35)$$

and the other spectral densities are formulated in an analogous way, *mutatis mutandis*. It can be recognized in Eq. (35) that the spectral density is the Fourier–Laplace transform of the autocorrelation function, $G^{IL}_{n,n}(\tau)$, defined as:

$$G^{DD}_{n,n}(-\tau) = Tr_L\left\{T^{1,DD+}_n\left[\exp(-i\hat{\hat{L}}_L\tau)T^{1,DD}_n\right]\sigma^T\right\} \tag{36}$$

In most of the work using the slow-motion theory (except for some of the early work (*77–79*)), the interest was concentrated on the paramagnetic enhancement of the spin–lattice relaxation and the effects of the scalar interaction were neglected. The relevant special case of Eq. (33) then becomes:

$$T^{-1}_{1I} = 2\mathrm{Re}\big(K^{DD}_{1,1}(-\omega_I)\big) \tag{37}$$

Setting Eq. (32) into (35), we obtain:

$$K^{DD}_{1,1}(\omega) = 30(C_{DD})^2 \sum_{q\in(-1,0,1)}\sum_{p\in(-1,0,1)}\begin{pmatrix} 2 & 1 & 1 \\ 1-q & q & -1\end{pmatrix}\begin{pmatrix} 2 & 1 & 1 \\ 1-p & p & -1\end{pmatrix}$$
$$\int_0^\infty Tr_L\left\{S^{1(L)^+}_q D^{2^*}_{0,1-q}(\Omega_{ML})\left[\exp(-i\hat{\hat{L}}_L\tau)S^{1(L)}_p D^2_{0,1-p}(\Omega_{ML})\right]\sigma^T\right\}$$
$$\exp(-i\omega\tau)d\tau \tag{38}$$

S^{1+}_q denotes the adjoint electron spin operator. One should notice that the expression $[\exp(-i\hat{\hat{L}}_L\tau)S^{1(L)}_p D^2_{0,1-p}(\Omega_{ML})]$ results in the S-operators and the Ω_{ML} being (implicitly) time-dependent. In order to continue any further, we need to specify the lattice and its Liouvillian.

B. The Lattice and the ZFS

As mentioned above, the electron spin system is strongly coupled to the classical degrees of freedom, in the first place the orientation of a molecule-fixed frame with respect to the lab frame, through anisotropic interactions. We concentrate here on the case of $S \geq 1$, where the main anisotropic interaction is denoted as the zero-field splitting, ZFS. In the language of spin Hamiltonians (*8,80*), the ZFS interaction is written:

$$H_{ZFS} = \mathbf{S}\cdot\mathbf{D}\cdot\mathbf{S} \tag{39}$$

D is the zero-field splitting tensor, a traceless, rank-two tensorial quantity. The ZFS tensor is a property of a molecule or a paramagnetic complex, with its origin in the mixing of the electrostatic and spin–orbit interactions (*80*). In addition, the dipole–dipole interaction between individual electron spins can contribute to the ZFS (*81*), but this contribution is believed to be unimportant for the transition metal complexes (*80,82*). The **D** tensor can be diagonalized by choosing a proper molecule-fixed Cartesian coordinate frame. One usually defines two constants given by:

$$D = D_{zz} - \frac{1}{2}(D_{xx} + D_{yy})$$
$$E = \frac{1}{2}(D_{xx} - D_{yy}) \tag{40}$$

called the axial and rhombic ZFS components. For performing calculations, it is more convenient to express the ZFS interaction using spherical tensor notation:

$$\begin{aligned} H_{ZFS} &= \sum_{n=-2}^{2} (-1)^n F_{-n}^{2(L)} S_n^{2(L)} \\ &= \sum_{n=-2}^{2} \sum_{m=-2}^{2} (-1)^n f_m^{2(P)} D_{m,-n}^2(\Omega_{PL}) S_n^{2(L)} \end{aligned} \tag{41}$$

where $F_{-n}^{2(L)}$ denotes the spherical components of the ZFS tensor in the laboratory frame, the same frame as the spin operators. In the second line of Eq. (41) it is explicitly stated that the suitable frame for the ZFS is its molecule-fixed principal frame, the P frame (with the appropriate components $f_m^{2(P)}$), while the spin operators are defined in a coordinate frame fixed in the laboratory. The spherical components $f_m^{2(P)}$ are simply related to the D and E symbols of Eq. (40): $f_0^{2(P)} = \sqrt{\frac{2}{3}}D$, $f_{\pm1}^{2(P)} = 0$, $f_{\pm2}^{2(P)} = E$. The symbol $D_{m,-n}^2(\Omega_{PL})$ is, in analogy with Eq. (32), a Wigner rotation matrix element. In the early implementations of the slow-motion theory, it was additionally assumed that the ZFS tensor is axially symmetric, $f_{\pm2}^{2(P)} = E = 0$ and that the ZFS principal frame coincides with the dipole–dipole principal frame (the M frame), which simplifies Eq. (41) and lets us use the same set of Euler angles (Ω_{ML}) in the expression for the ZFS Hamiltonian as in Eq. (32):

$$H_{ZFS} = f_0^{2(M)} \sum_{n=-2}^{2} (-1)^n D_{0,-n}^2(\Omega_{ML}) S_n^{2(L)} \tag{42}$$

The simplest possible physical picture of the lattice contains the electron Zeeman interaction, the axially symmetric ZFS (whose principal axis coincides with the dipole–dipole axis) and the molecular rotation. The corresponding Liouvillian is given by:

$$\hat{\hat{L}}_L = \hat{\hat{L}}_{S0} + \hat{\hat{L}}_{RS} + \hat{\hat{L}}_R = \hat{\hat{L}}_{S0} + \hat{\hat{L}}_{ZFS} - i\Gamma_R \tag{43}$$

$\hat{\hat{L}}_{S0}$ is the commutator with the electron spin Zeeman Hamiltonian (assuming isotropic g tensor, $H_{S0} = g\mathbf{S}\cdot\mathbf{B}_0$), $\hat{\hat{L}}_{RS} \equiv \hat{\hat{L}}_{ZFS}$ (the sub-script RS stands for coupling of the rotational and spin parts of the composite lattice) is the commutator with the ZFS Hamiltonian and $\hat{\hat{L}}_R = -i\Gamma_R$, where Γ_R is a stationary Markov operator describing the conditional probability distribution, $P(\Omega_0/\Omega, t)$, of the orientational degrees of freedom through:

$$\frac{d}{dt}P(\Omega_0/\Omega, t) = -\Gamma_R P(\Omega_0/\Omega, t) \tag{44}$$

$P(\Omega_0/\Omega, t)$ is the conditional probability of the orientation being Ω at time t, provided it was Ω_0 at time zero. The symbol $-\Gamma_R$ is the rotational diffusion operator. In the simplest possible case, Γ_R then takes the form of the Laplace operator, acting on the Euler angles (Ω_{ML}) specifying the orientation of the molecule-fixed frame with respect to the laboratory frame, multiplied with a rotational diffusion coefficient, D_R. Equation (44) then becomes identical to the isotropic rotational diffusion equation. The rotational diffusion coefficient is simply related to the rotational correlation time introduced earlier, by $\tau_R = 1/6D_R$.

The equation of motion for the density operator for the lattice can now be formulated in analogy with the SLE, Eq. (30).

$$\frac{d}{dt}\sigma = -i\hat{\hat{L}}_L\sigma = -i\left(\hat{\hat{L}}_{S0} + \hat{\hat{L}}_{ZFS} - i\Gamma_R\right)\sigma \tag{45}$$

In order to evaluate the spectral density of Eq. (35) or (38), one needs a complete basis set spanning the lattice operator space. This basis set can be obtained by taking direct products of Wigner rotation matrices, $D^L_{K,M}(\Omega)$ (which are eigenfunctions of Γ_R) and a complete set of basis operators Q^Σ_σ spanning the $(2S+1)(2S+1)$ dimensional spin space, $\{Q^\Sigma_\sigma\} \otimes \{D^L_{K,M}\}$. The mathematical problem of calculating the PRE can then be expressed as a matrix inversion problem (*26*):

$$\begin{aligned} T_{1I}^{-1} &= \frac{4}{3}(C^{DD})^2 S(S+1)\mathrm{Re}\{\mathbf{c}_1^*[i(\mathbf{L}_L + \omega_I\mathbf{1})]^{-1}\mathbf{c}_1\} \\ &= \frac{4}{3}(C^{DD})^2 S(S+1)\mathrm{Re}\{\mathbf{c}_1^*\mathbf{M}^{-1}\mathbf{c}_1\} \end{aligned} \tag{46}$$

$\mathbf{L}_L$ is the matrix representation of the lattice Liouvillian in the space of the basis operators, $\mathbf{1}$ is a unit (super)operator and $\mathbf{c}_1$ are projection vectors representing the operators $T_n^{1,DD}$ of Eq. (32) in the same space. The projection vectors contain only three non-zero elements, thus we only need a 3×3 fragment of the inverse (super)matrix $\mathbf{M}^{-1}$. The matrix $\mathbf{M}$ is sparse and, in principle, infinitely large because the set of Wigner rotation matrices is infinite. In practice, the value of the L quantum number is

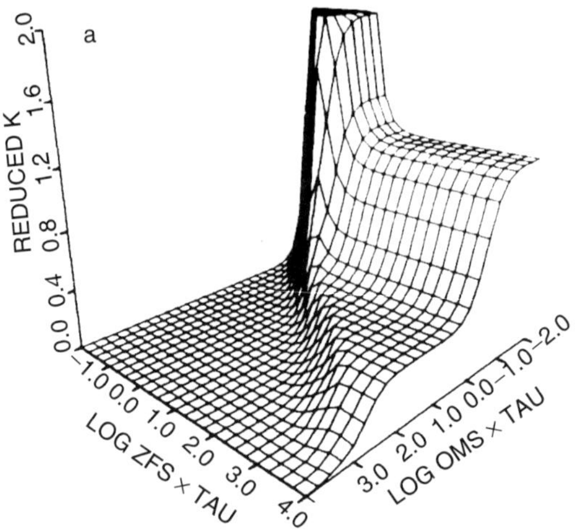

FIG. 3. Variation of the completely reduced dipole–dipole spectral density (see text) for the model of a low-symmetry complex for $S = 3/2$. Reprinted from *J. Magn. Reson.*, vol. 59, Westlund, P.O.; Wennerström, H.; Nordenskiöld, L.; Kowalewski, J.; Benetis, N., "Nuclear Spin–Lattice and Spin–Spin Relaxation in Paramagnetic Systems in the Slow-Motion Regime for Electron Spin. III. Dipole–Dipole and Scalar Spin–Spin Interaction for S = 3/2 and 5/2", pp. 91–109, Copyright 1984, with permission from Elsevier.

increased step by step until convergence of the desired accuracy is obtained. The inversion of **M** can be performed numerically using the Lanczos algorithm (*83*).

The formalism of Eqs. (35), (37), and (46) was used in the original papers presenting the slow-motion theory for PRE (*26,77–79*), where the ZFS tensor was assumed to be cylindrically symmetric. The results of calculations using this formulation are exemplified in Fig. 3, obtained for $S = 3/2$. The lattice is characterized by three parameters: the electron Larmor frequency ω_S, the ZFS parameter D (or ω_D, in the angular frequency units) and the rotational correlation time τ_R. The shape of the reduced dipolar spectral density $\mathrm{Re}K_{1,1}^{DD}(-\omega_I)$, divided by the square of the dipolar coupling constant and the rotational correlation time), is universal in the sense that it only depends on two dimensionless quantities, $\omega_S\tau_R$ and $\omega_D\tau_R$. In the range $\omega_D \ll \omega_S, \tau_R^{-1}$, the situation is very similar to the case of the SBM theory (and would in fact be identical to it for $S = 1$) and the shape of the NMRD profile (the shape of the reduced spectral density versus $\omega_S\tau_R$, OMS*TAU in the diagram, at a constant $\omega_D\tau_R$, ZFS*TAU in the diagram) changes gradually as the product $\omega_D\tau_R$ increases. Similar diagrams were reported for other types of spectral densities (K^{SC}, K^{SC-DD}) and for other values of the quantum number S (*79*). The problem with the NMRD curves calculated using this approach is that, at least for the presumably dominant DD-case, a monotonous decrease of the PRE with the increasing magnetic field is predicted, which is contrary to many experimental situations. Benetis and Kowalewski (*84*) reported similar calculations, generalizing the description of the reorientational motion to the diffusion of a symmetric top, but this did not resolve the problem.

Westlund *et al.* (*85*) developed at an early stage an alternative description of the lattice, suited for the systems where the ZFS vanishes because of high-symmetry (e.g., octahedral) environment of the transition metal ion. This picture might be applicable to, for example, hydrated nickel(II) ion. Even though a model of a highly symmetric complex (with vanishing *static* ZFS) might be reasonable when time-averaged positions of oxygen atoms are considered, instantaneous distortions of the $Ni(H_2O)_6^{2+}$ complex by collisions, as well as particular orientations of the water molecules in the first coordination sphere, create a rapidly-oscillating *transient* ZFS. This is the essence of the Bloembergen–Morgan theory for electron spin relaxation (*27*). Rubinstein and co-workers (*86*) developed the idea a little further and coined the concept of the pseudorotation model, where the time-dependence of the transient ZFS, assumed to be cylindrically symmetric, is described by a rotational diffusion equation with another diffusion coefficient, D_υ, or correlation time, τ_υ, related to the distortions (compare Eqs. (14) and (15)). One can think of the pseudorotation model as describing a rapid reorientation of the principal axis of the ZFS in a molecule-fixed frame, while the magnitude of the ZFS remains constant. Westlund *et al.* (*85*) combined the slow-motion theory with the pseudorotation model. The Liouvillian of Eq. (43) is augmented by one more term:

$$\hat{\hat{L}}_L = \hat{\hat{L}}_{S0} + \hat{\hat{L}}_{ZFS} - i\Gamma_R - i\Gamma_\upsilon \tag{47}$$

and the Liouville space basis becomes a direct product of the spin operators Q_σ^Σ and two sets of Wigner rotation matrices. Using this model, they obtained for the case of $S = 1$ (appropriate for the Ni(II)) a diagram similar to that of Fig. 3, displayed in Fig. 4. The diagram in Fig. 4 is not "universal" in the above-specified sense, as it is dependent in addition on the τ_R/τ_υ ratio. We can see in the diagram that non-monotonic NMRD profiles can be obtained at high $\omega_D\tau_R$ values. Westlund *et al.* described also, in the same paper (*85*), other possible models for the lattice, based on the concepts of classical diffusion in a harmonic potential (described by the one-dimensional Smoluchowski equation) and quantum-mechanical oscillator. The pseudorotation slow-motion model was applied successfully to interpret the NMRD profile of aqueous protons in acidic solutions of Ni(II) perchlorate (*87*). Further extensions of the pseudorotation model, within the general framework of the slow-motion theory, were proposed by Westlund *et al.* (*88*), who allowed for coupling between the complex rotation and the ZFS dynamics. In this approach, called the extended pseudorotation model, the validity of the decomposition approximation (statistical independence of rotation and electron spin relaxation) can be scrutinized for different parameter values. Svoboda *et al.* described another modification of the basic pseudorotation approach (*89*). They assumed that the transient ZFS was fully rhombic, i.e., that the energy level structure of the $S = 1$ system corresponded to three non-degenerate, equidistant levels, rather

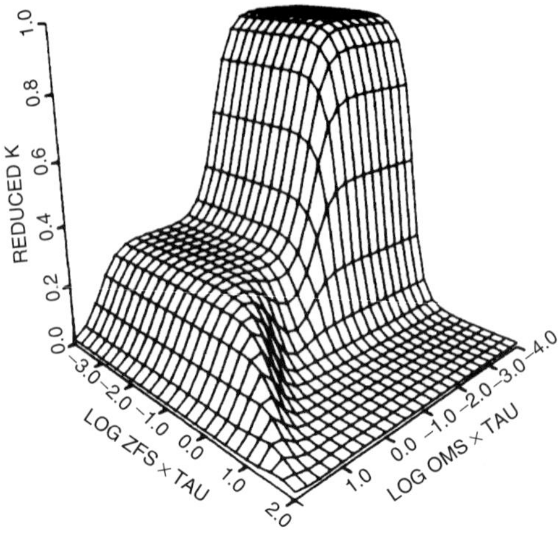

FIG. 4. Variation of the completely reduced dipole–dipole spectral density (see text) for the model of a high-symmetry complex for $S = 1$. Reproduced with permission from Westlund, P.O.; Benetis, N.; Wennerström, H. *Mol. Phys.* **1987**, *61*, 177–194. Copyright 1987 Taylor and Francis Ltd (http://www.tandf.co.uk/journals/tf/00268976.html).

than a two-level (one of them degenerate) situation of the cylindrically symmetric ZFS. The model was used to re-interpret the NMRD profile of aqueous protons in the Ni(II) solution (*87*). It should in this context be mentioned that the PRE of aqueous Ni(II) was also discussed by Friedman *et al.*, who described the motions generating the transient ZFS using a "flickering" model and approached the problem of the breach of the Redfield limit using a different method (*90*).

The first attempt to use the distortion-related ideas to complexes of inherently lower symmetry, where one cannot neglect the effects of the non-zero average (static) ZFS defining its own molecule-fixed frame, was proposed by Westlund and Larsson (*91*). They introduced the idea of both static and transient ZFS interactions present at the same time and subject to independent modulations. For the transient ZFS, they used the description based on the Smoluchowski equation allowing for the fluctuation of the magnitude, but not the orientation of the ZFS. This turned out not to cause major changes in the NMRD profiles compared to calculations containing only the rigid body rotation (*26,77–79*). Larsson *et al.* (*92*) developed further the idea of two sources of modulation of the ZFS, by introducing the anisotropic pseudo-rotation model for $S = 1$ systems. The difference compared to the work of Westlund and Larsson (*91*) is that the transient ZFS is now allowed to change the direction in the molecule-fixed frame, rather than magnitude. The ZFS Hamiltonian is formally expressed as:

$$\begin{aligned} H_{ZFS}(t) &= H_{ZFS}^{S}(t) + H_{ZFS}^{T}(t) \\ &= f_0^2 \sum_{n=-2}^{2} (-1)^n S_{-n}^2 D_{0,0}^2[\Omega_{PM}(t)] D_{0,n}^2[\Omega_{ML}(t)] \end{aligned} \tag{48}$$

while the corresponding ZFS Liouvillian, to be used in Eq. (45), becomes:

$$\hat{\hat{L}}_{ZFS} = \hat{\hat{L}}^{S}_{ZFS} + \hat{\hat{L}}^{T}_{ZFS} = [H_{ZFS}, ...] \tag{49}$$

The ZFS is assumed to be cylindrically symmetric (only the f_0^2 component is different from zero) and of constant magnitude. The static part of the H_{ZFS} is obtained by averaging the Wigner rotation matrix $D^2_{0,0}[\Omega_{PM}(t)]$ over the anisotropic distribution function, $P(\Omega_{PM})$. The principal axis of the static ZFS is, in addition, assumed to coincide with the dipole–dipole (*IS*) axis. Eq. (48) becomes equivalent to Eq. (42), with the f_0^2 component scaled by $D^2_{0,0}[\Omega_{PM}(t)]$. The transient part of the H_{ZFS} can be expressed in several ways, the simplest being (*92*):

$$H^{T}_{ZFS}(t) = H_{ZFS}(t) - H^{S}_{ZFS}(t) \tag{50}$$

The lattice dynamics in this model depends on a larger number of parameters. Except for the magnetic field, we have the magnitudes of the static and transient ZFS and the two correlation times, τ_R and τ_v. The model is very flexible and can predict NMRD profiles of widely different shapes, exemplified in Fig. 5.

Larsson *et al.* (*92*) demonstrated also that the anisotropic pseudorotation model could be used for direct fitting of experimental PRE data for aniline protons in the complex between a Ni(II) chelate (Ni(II)(dpm)$_2$) and ring-deuterated aniline in toluene-d$_8$ solution. The Ni(II) ion is in this case surrounded by four oxygen atoms and two nitrogen atoms, which should result in a sizable static ZFS, which indeed is confirmed by the fits.

C. The Effects of ZFS Orientation and Rhombicity

While the anisotropic pseudorotation model seems to capture the most relevant physics of the problem, there are still many possibilities to improve it.

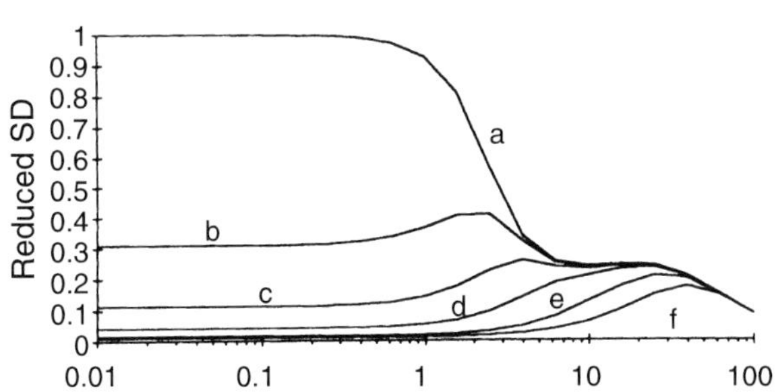

FIG. 5. NMRD profiles for an asymmetric complex calculated for different values of the transient ZFS. Reproduced with permission from Larsson, T.; Westlund, P.O.; Kowalewski, J.; Koenig, S.H. *J. Chem. Phys.* **1994**, *101*, 1116–1128. Copyright 1994 American Institute of Physics.

Two such improvements have been proposed and implemented by Nilsson and co-workers, first for $S = 1$ (*93*) and subsequently for a general S-value (*94*).

In the first one, it is no longer assumed that the static ZFS tensor is axially symmetric: the $f_{\pm 2}^{(2,M)} = E$ terms in Eq. (41) are retained. Second, the assumption that the principal axes of the static ZFS (the P frame) and of the DD interaction (the M frame) coincide is removed. This implies that the transformation between the laboratory frame and the M frame is carried out in two steps, through the P frame. It is most convenient to do this transformation explicitly for $S^{1(L)}$ in Eq. (38):

$$S_p^{1(L)} = \sum_{m\in(0,\pm 1)} \sum_{k\in(0,\pm 1)} S_m^{1(P)} D_{m,k}^1(\Omega_{PM}) D_{k,p}^1(\Omega_{ML}) \tag{51}$$

This leads to Eq. (38) taking on a correspondingly more complicated form (*93–96*). In essence, the spectral density becomes dependent on the angles θ and ϕ specifying the relative orientations of the two relevant molecule-fixed frames: the principal frame of the static ZFS and that of the dipole–dipole interaction. Disregarding the possibility of internal motions, the θ and ϕ angles are time-independent.

The rhombicity of the static ZFS tensor has an important effect on the fine structure of the energy levels. Examples of the energy level changes due to ZFS rhombicity for $S=1$ and 3/2 are shown in Fig. 6. We can note that the energy level structure changes much more significantly for the case of $S=1$ (integer spin) than for $S=3/2$ (half integer spin). The feature is more general and is related to the Kramers' degeneracy (*80*) for half-integer spin systems. The effect of the ZFS rhombicity on the PRE for different S-values is illustrated in Fig. 7. We can note that the non-zero rhombic static ZFS parameter (E_S) always reduces the PRE at low magnetic field; the effect is most dramatic for $S=1$, it is clearly pronounced for higher integer S-values and is smaller for half-integer spins.

The role of the θ angle, between the principal axes of the ZFS and DD interactions, deviating from zero is similar to the effect of non-zero rhombic

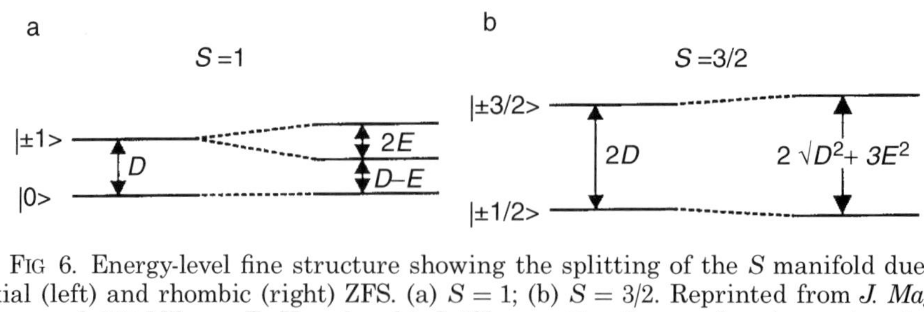

FIG 6. Energy-level fine structure showing the splitting of the S manifold due to axial (left) and rhombic (right) ZFS. (a) $S = 1$; (b) $S = 3/2$. Reprinted from *J. Magn. Reson.* vol. 146, Nilsson, T.; Kowalewski, J., "Slow-motion theory of nuclear spin relaxation in paramagnetic low-symmetry complexes: A generalization to high electron spin", pp. 345–358, Copyright 2000, with permission from Elsevier.

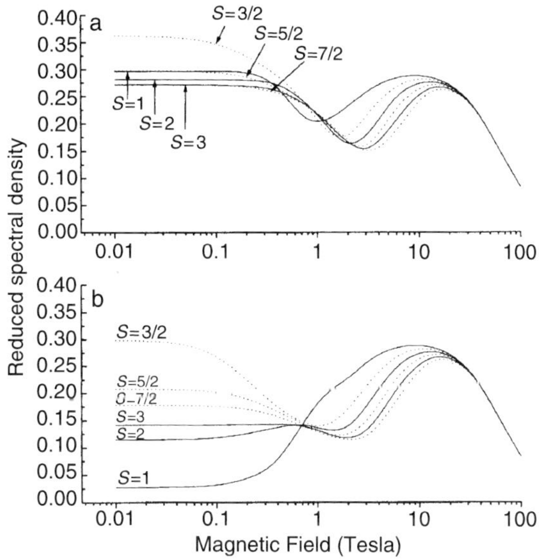

FIG. 7. NMRD profiles calculated for slightly asymmetric, weakly deformable complexes with different electron spin quantum numbers: (a) cylindrically-symmetric ZFS, $E = 0$; (b) maximum rhombicity, $E = D/3$. Reprinted from *J. Magn. Reson.* vol. 146, Nilsson, T.; Kowalewski, J., "Slow motion theory of nuclear spin relaxation in paramagnetic low-symmetry complexes: A generalization to high electron spin", pp. 345–358, Copyright 2000, with permission from Elsevier.

ZFS, in the sense of reducing the low-field PRE. The effect of non-coinciding axes is however less strongly influenced by the S quantum number, as illustrated in Fig. 8 comparing the θ dependence for the cases of $S = 1$ and 3/2 at $E_S = 0$.

The recent versions of the slow motion approach were applied to direct fitting of experimental data for a series of Ni(II) complexes of varying symmetry (*97*). An example of an experimental data set and a fitted curve is shown in Fig. 9. Another application of the slow-motion approach is to provide benchmark calculations against which more approximate theoretical tools can be tested. As an example of work of this kind, we wish to mention the paper by Kowalewski *et al.* (*98*), studying the electron spin relaxation effects in the vicinity and beyond the Redfield limit.

V. Electron Spin Relaxation and the PRE in Some Limiting Cases

The slow-motion theory describes the electron relaxation processes implicitly, through a combined effect of static and transient ZFS, and reorientational and pseudorotational dynamics. This is necessary under very general conditions, but simpler descriptions, appropriate in certain physical limits, can also be useful. In this chapter, we review some work of this type.

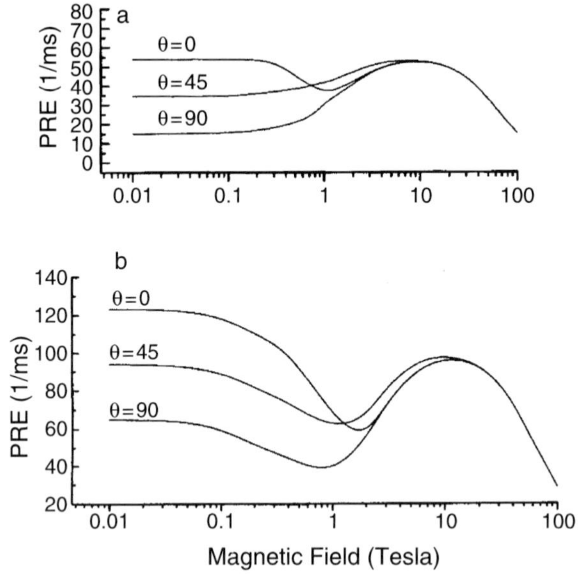

FIG. 8. NMRD profiles calculated for a given set of parameters and different angles θ between the principal axes of the dipole–dipole interaction and the static ZFS. (a) $S = 1$; (reproduced with permission from Nilsson, T.; Svoboda, J.; Westlund, P.O.; Kowalewski, J. *J. Chem. Phys.* **1998**, *109*, 6364–6375. Copyright 1998 American Institute of Physics); (b) $S = 3/2$. (Reprinted from *J. Magn. Reson.* vol. 146, Nilsson, T.; Kowalewski, J., "Slow-motion theory of nuclear spin relaxation in paramagnetic low-symmetry complexes: A generalization to high electron spin", pp. 345–358, Copyright 2000, with permission from Elsevier.)

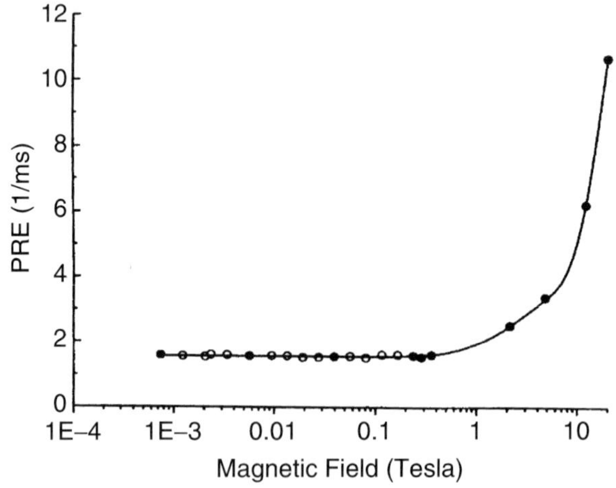

FIG. 9. Fit of the slow-motion theory to the experimental NMRD profile for $Ni(tmc)(H_2O)^{2+}$. Reprinted with permission from Nilsson, T.; Parigi, G.; Kowalewski, J. *J. Phys. Chem.* **2002**, *106*, 4476–4488. Copyright 2002 American Chemical Society.

A. Highly Symmetric Systems and the Redfield Theory for Electron Spin Relaxation

The first limiting case that we wish to discuss is that of highly symmetric $S \geq 1$ systems, i.e., systems where the static ZFS vanishes by symmetry. Such systems are, for example, solvated metal ions with tetrahedral or octahedral configuration. The Bloembergen–Morgan electron spin relaxation theory and the modified Solomon–Bloembergen treatment were formulated for this type of systems. In these theories, one assumes that the electron spin relaxation is characterized by two rate constants, the spin–lattice relaxation rate (T_{1e}^{-1}) and the spin–spin relaxation rate (T_{2e}^{-1}), with the magnetic field dependence given by Eqs. (14) and (15).

The assumption of a single electron spin T_1 and a single T_2 holds usually for $S = 1/2$ and for $S \geq 1$ in certain limits. Let us assume that the instantaneous distortions of the solvation sphere of the ion result in a transient ZFS and that the time-dependence of the transient ZFS can be described by the pseudorotation model, with the magnitude of the transient ZFS equal to Δ_T and the correlation time τ_v. The simple picture of electron relaxation for $S = 1$ is valid if the Redfield condition ($\Delta_T \tau_v \ll 1$) applies. Under the extreme narrowing conditions ($\omega_S \tau_v \ll 1$), the longitudinal and transverse electron spin relaxation rates are equal to each other and to the low-field limit rate τ_{S0}^{-1}, occurring in Eqs. (14) and (15). The low field-limit rate is then given by (*27,86*):

$$\tau_{S0}^{-1} = \frac{\Delta_T^2}{5}[4S(S+1) - 3]\tau_v \tag{52}$$

Within the Redfield limit, but outside of extreme narrowing, the electron spin–lattice and spin–spin relaxation of $S = 1$ system remains single exponential, in analogy with the case of quadrupolar relaxation of nuclear spin with $I = 1$ (*18*). For $S > 1$, the spin relaxation in principle is only single exponential in extreme narrowing, again in analogy with quadrupolar nuclei (*18*). Sticking to this analogy, we can note that the electron spin relaxation becomes nearly-exponential under certain conditions (*99*). Rubinstein, Baram and Luz (*86*) treated the cases of electron spin relaxation for $S = 3/2$ and $S = 5/2$ in the Redfield limit numerically and found that, under near-extreme narrowing conditions, Eqs. (14) and (15), together with Eq. (52), might be applied for "average" relaxation rates. Westlund (*100*) reported a generalization of the SBM theory for the PRE valid for $S > 1$ and outside of extreme narrowing. He used Eqs. (35)–(38) as a starting point. Introducing the decomposition approximation (the electron spin relaxation is uncorrelated with reorientation), he was able to break the average in the correlation function in the integrand of Eq. (38) into a product of rotational part (proportional to $\exp(-\tau/\tau_R)$) and the spin part, $Tr_S\{S_p^{1(L)+}\exp(-i\hat{\hat{L}}_S\tau)S_p^{1(L)}\sigma^T\}$. The spin Liouvillian, $\hat{\hat{L}}_S$, could be expressed in a simplified form:

$$\hat{\hat{L}}_S = \hat{\hat{L}}_{S0} - i\hat{\hat{R}}_{ZFS}^T \tag{53}$$

where $\hat{\hat{L}}_{S0}$ has the same meaning as in Eq. (43) and $\hat{\hat{R}}_{ZFS}$ is a Redfield relaxation superoperator, originating from the transient ZFS.

Using the case of $S = 5/2$ as an illustrative example, he demonstrated that it was possible to derive closed-form analytical expressions for the PRE of the form of the SBM equations times (1 + correction term). For typical parameter values, the effect of the correction term was to increase the prediction of the SBM theory by 5–7%. A similar approach was also applied to the $S = 7/2$ system, such as Gd(III) (*101*), where the correction terms could be larger. For that case, the estimations of the electron spin relaxations rates, obtained in the solution for PRE, were also used for simulations of ESR lineshapes.

B. Slowly-Rotating, Low-Symmetry Systems

High-symmetry systems discussed in the previous section are scarce. In systems with lower symmetry and $S \geq 1$, we must expect a static ZFS, which can have a profound effect on both the electron spin relaxation and the PRE. The treatment of the PRE in systems with static ZFS requires caution. The reorientational motion of the complex modulates the ZFS which can cause the breach of both the Redfield condition for the electron spin relaxation and the assumption that electron spin relaxation and molecular reorientation are statistically independent (the decomposition approximation). One limit where the decomposition approximation is valid is for slowly rotating systems.

Lindner presented an important paper on this type of systems already in mid-1960s (*102*). She considered the whole system as an ensemble of microcrystallites with different orientations of the principal axis of the ZFS tensor with respect to the magnetic field and applied the linear response theory of Kubo and Tomita (*103*). The expression for the nuclear T_1 for the case of $S = 1$ could be written as:

$$T_{1I}^{-1} = \sum_{n \in 0, \pm 1} \int_0^{\infty} \frac{\langle [I_z, H_n^{IL}(t)][H_{-n}^{IL}(0), I_z] \rangle}{\langle I_z^2 \rangle} \exp(in\omega_I t) dt \tag{54}$$

The summation index n has the same meaning as in Eq. (31), i.e., it enumerates the components of the interaction between the nuclear spin I and the remainder of the system (which thus contains both the electron spin and the thermal bath), expressed as spherical tensors. H_n^{IL} are components of the hyperfine Hamiltonian, in angular frequency units, expressed in the interaction representation (*18,19*), with the electron Zeeman and the ZFS in the zeroth order Hamiltonian H_0. The operator $H_n^{IL}(t)$ is evaluated as:

$$H_n^{IL}(t) = \exp(iH_0 t) H_n^{IL}(0) \exp(-iH_0 t) \tag{55}$$

which depends on the orientation of the ZFS with respect to the magnetic field. As a consequence, Eq. (54) describes one particular orientation of the principal axis of the ZFS tensor relative to the magnetic field. In order to get the average nuclear spin–lattice relaxation rate, one has to average over all the orientations in the isotropic space. Lindner's treatment until this stage is exact. In order to get further, however, Lindner introduced an approximation, breaking the average in Eq. (54) and assumed that the electron spin magnetization decays as a simple exponential with the time constant τ_S, independent of reorientation. Lindner also derived corresponding expressions for the spin–spin relaxation rate (*102*).

The original theory of Lindner (*102*) was further extended and refined by Bertini and co-workers (*1,2,104,105*), in order to become applicable to more complex systems. Fukui and co-workers used a similar approach in a study of the effect of ZFS rhombicity on the PRE (*106*). Bertini and co-workers concentrated on the nuclear spin–lattice relaxation problem, which is easier to handle both experimentally and theoretically. The extensions and generalizations provided by this group were summarized in a public-domain computer program (*107*) allowing to treat any spin multiplicities ($S \geq 1$) and including other interactions leading to the splitting of electron spin levels at zero field. Indeed, the electron spin levels may be split due to hyperfine coupling of the unpaired electron(s) to the metal nucleus spin, as well as due to ZFS. At low fields, these effects may be larger than the splitting due to the Zeeman term. Therefore, the eigenstates of the time-independent spin Hamiltonian between which the transitions occur, and the transition frequencies, are calculated by considering the corresponding hyperfine coupling and ZFS Hamiltonian terms together with the Zeeman Hamiltonian. The program permits the calculation of the field dependence of contact and dipolar nuclear relaxation rate enhancements in the slow rotation limit by considering the hyperfine coupling with the metal nucleus and/or the ZFS of any rhombicity in the presence of the Zeeman interaction, for any metal nucleus spin quantum number, any electron spin quantum number and any g tensor anisotropy. Due to the slow-rotation limit approach, the relevant correlation time for nuclear relaxation can be, at least at low fields, only the electron relaxation time or the exchange lifetime. In the original version of the program (*107*), the electron relaxation rate was supposed to be field independent or to have a field dependence as in Eqs. (14) and (52). It should be remembered, however, that Eq. (14) has been derived in the Zeeman limit, and therefore the actual field dependence of the electron relaxation rate can be different in the presence of static ZFS. Many examples of applications are reported in Chapter 3.

Metal ions can be magnetically coupled, when close to each other. In such cases, a further interaction to be considered in the static spin Hamiltonian is the magnetic exchange coupling of a paramagnetic metal ion S_1 to another paramagnetic metal ion S_2. Magnetic coupling gives rise to new energy levels, which can provide new relaxation pathways. Three cases must be discussed, according to the value of the magnetic coupling constant J (in angular frequency units):

(1) $|J| \ll \tau_{S1}^{-1}, \tau_{S2}^{-1}$. If the magnetic coupling is smaller than the electron spin relaxation rates, the latter are not expected to be affected by the magnetic coupling.

(2) $|J| \gg \tau_{S1}^{-1}, \tau_{S2}^{-1}$. In this case, the two metal ions can be considered to have a single set of electron spin relaxation rates. If no additional relaxation mechanisms are established, such common relaxation rates are about equal to the fastest relaxation rates of the uncoupled spins. Actually, calculations indicate the presence of different electron relaxation rates for each level and for each transition. The electron relaxation rates for the pair are the sum of the rates of the two spins, weighted by coefficients depending on the transition (*108*).

(3) $\tau_{S1}^{-1} < |J| < \tau_{S2}^{-1}$. In this case, the fast-relaxing metal ion is considered to be unaffected by the coupling; on the contrary, an additional relaxation mechanism for the slowly-relaxing metal ion is caused by the coupling to the fast relaxing metal ion. It is accounted for by using perturbation theory. The coupling between S_1 and S_2 is, in fact, similar to that between I and S discussed in Section II.B, and the enhancement in the electron spin relaxation rates of the slowly-relaxing ion is provided by an equation equivalent to Eq. (12), with a scalar contribution (with J instead of A_{SC}) and a dipolar contribution (with $C_{DD} = (\mu_0 \gamma_S^2 \hbar / 4\pi \langle r^3 \rangle)$, where $\langle r^3 \rangle$ is the average cube of the interelectronic distance). The correlation times are provided by the electron relaxation times of the fast relaxing ion (*1*).

A similar approach, also based on the Kubo–Tomita theory (*103*), has been proposed in a series of papers by Sharp and co-workers (*109–114*), summarized nicely in a recent review (*14*). Briefly, Sharp also expressed the PRE in terms of a power density function (or spectral density) of the dipolar interaction taken at the nuclear Larmor frequency. The power density was related to the Fourier–Laplace transform of the time correlation functions (*14*):

$$G_r(t) = \langle S_r(t) S_r(0) \rangle = \sum_{\mu} \rho_{\mu\mu}^{eq} \langle \mu | S_r(t) S_r(0) | \mu \rangle \tag{56}$$

where r denotes components of the electron spin vector (rank one tensor), expressed in Cartesian or spherical form ($r = x, y, z$ or $0, \pm 1$) and in the laboratory frame or in the molecule-fixed frame. The index μ enumerates the electron-spin energy levels and $\rho_{\mu\mu}^{eq}$ is the diagonal element of the equilibrium electron spin density matrix, at the high temperature given by $\rho_{\mu\mu}^{eq} = (2S+1)^{-1}$. The time correlation functions give rise to the Lorentzian power density terms centered at the transition frequencies for the electron spin system. In the high-field (Zeeman) limit, Sharp's formalism leads to the Lorentzian terms occurring in the Solomon equation, Eq. (10). The inverse of the correlation time (called dipolar correlation rate), τ_c^{-1}, determines the width of each power band and is a sum of an electronic relaxation rate and the inverse rotational correlation time. More interestingly, the approach of Sharp can also be applied in the limit where the high ZFS dominates over the Zeeman interaction and in the intermediate situations. For the case of the rhombic ZFS acting as the main Hamiltonian, the PRE is given by:

$$T_{1I}^{-1} = R_1 = R_{1\hat{x}} + R_{1\hat{y}} + R_{1\hat{z}} \tag{57}$$

where the fact that the three Cartesian components refer to the molecule-fixed ZFS principal frame is indicated by the superscripting karat. Each of the three contributions to the PRE is given by:

$$R_{1\hat{r}} = \frac{4}{3} C_D r_{IS}^{-6} [1 + P_2(\cos\theta_{\hat{r}})] \sum_{\mu,\nu} \rho_{\mu\mu}^{eq} |\langle\mu|S_r|\nu\rangle|^2 j_{\hat{r}}^{(\mu)}(\omega_{\mu\nu}) \tag{58}$$

C_D is a collection of constants, $P_2(\cos\theta_{\hat{r}})$ is the Legendre polynomial of the angle between the r axis (i.e., one of the principal axes of the ZFS coupling) and the dipole–dipole axis and $j_{\hat{r}}^{(\mu)}(\omega_{\mu\nu})$is the power density at the electron-spin transition frequency between the levels μ and ν. The approach outlined above has been applied to interpret experimental data in model systems, such as complexes of Ni(II) ($S = 1$) (*115*), Co(II), $S = 3/2$ (*116*) and Mn(III) ($S = 2$) (*115*).

The problem of the electron spin relaxation in the early work from Sharp and co-workers (*109–114*) (and in some of its more recent continuation (*115,116*)) was treated only approximately. They basically assume that, for integer spin systems, there is a single decay time constant for the electron spin components, while two such time constants are required for the $S = 3/2$ with two Kramers' doublets (*116*). We shall return to some new ideas presented in the more recent work from Sharp's group below.

Strandberg and Westlund formulated a theory for the PRE in low-symmetry, slowly-rotating systems in the Zeeman-dominated (high field) limit (*117*). The static and transient ZFS was included, only the latter contributing to the electron spin relaxation. The electron spin relaxation was found to be multi-exponential and dependent on the angle between the principal frame of the static ZFS and the magnetic field direction. The electron spin relaxation matrices were derived, assuming an anisotropic version of the pseudorotation model for the transient ZFS, allowing for two different distortional correlation times.

Westlund developed also a theory for PRE in the ZFS-dominated limit for $S = 1$, which included a stringent Redfield-limit approach to the electron spin relaxation in this regime (*118*). Equations (35) and (38) were used as the starting point also in this case. Again, the correlation function in the integrand of Eq. (38) was expressed as a product of a rotational part and the spin part. However, since it is in this case appropriate to work in the principal frame of the static ZFS, the rotational part becomes proportional to $\exp(-\tau/3\tau_R)$ (if τ_R is the correlation time for reorientation of rank two spherical harmonics, then $3\tau_R$ is the correlation time for rank one spherical harmonics) and the spin part is expressed as $Tr_S\{S_p^{(1P)}\exp(-i\hat{\hat{L}}_S\tau)S_p^{(1P)}\sigma^T\}$. The spin Liouvillian is now written as:

$$\hat{\hat{L}}_S = \hat{\hat{L}}_{ZFS}^S - i\hat{\hat{R}}_{ZFS}^T \tag{59}$$

where the static ZFS Liouvillian takes the place of the Zeeman Liouvillian in Eq. (53). Westlund actually also allowed the Zeeman interaction to contribute a relaxation term, which is somewhat problematic in view of the Redfield theory requiring the zero-average perturbation. Westlund derived the electron spin–spin relaxation rates for the three coherences occurring in the three-level $S = 1$ system with rhombic ZFS. The relaxation rates were found to be expressed as linear combinations of spectral densities:

$$J(\omega_{ZFS}) = \frac{\Delta_T^2}{5}\frac{\tau_D}{1+\omega_{ZFS}^2\tau_D^2} \tag{60}$$

taken at zero frequency and at transition frequencies $\omega_{ZFS} = \omega_D + \omega_E$, $\omega_D - \omega_E$, ω_{2E}.

The approach of Westlund (*118*) was generalized to arbitrary S by Nilsson and Kowalewski (*95*). Starting from Eqs. (37) and (38), and assuming that the dynamics of the lattice was contained in the spin part, these authors derived the following expression for in-complex spin–lattice nuclear spin relaxation rate:

$$T_{1I}^{-1} = 2\frac{(C_{DD})^2 S(S+1)}{9}\mathrm{Re}\{s_{-1,-1}^{DD} + 4s_{0,0}^{DD} + s_{1,1}^{DD}\} \tag{61}$$

where the quantities $s_{\sigma,\sigma}^{DD}$ are called electron-spin spectral density functions, evaluated at the nuclear spin Larmor frequency. Assuming the high-temperature description to be valid for the electron spin density operator and limiting the interest to the case of slowly-rotating systems, the electron spectral density functions can be expressed as:

$$s_{\sigma,\sigma}^{DD} = \frac{3}{S(S+1)(2S+1)}\int_0^\infty Tr_S\left\{S_\sigma^{1(P)^+}\left[\exp(-i\hat{\hat{L}}_S\tau)S_\sigma^{1(P)}\right]\right\}\exp(-i\omega_I\tau)d\tau \tag{62}$$

Introducing a superoperator $\hat{\hat{M}}$, given by:

$$\hat{\hat{M}} = -i\hat{\hat{L}}_{Zeeman} - i\hat{\hat{L}}_{ZFS}^S + \hat{\hat{R}}_{ZFS}^T + i\omega_I\hat{\hat{1}} \tag{63}$$

and employing the properties of Laplace transforms, the electron spin spectral densities can be written in a very compact way:

$$s_{\sigma,\sigma}^{DD} = c_\sigma^* \mathbf{M}^{-1} c_\sigma \tag{64}$$

where **M** is the matrix representation of the superoperator $\hat{\hat{M}}$ in a suitable Liouville space basis set and c_σ are the projection vectors from the expansion of the spin tensor operators in the same basis set. The electron-spin spectral density $s_{0,0}^{DD}$ carries information on the dynamics of the z-component of the electron spin (the longitudinal relaxation in the molecule-fixed frame), while $s_{1,1}^{DD}$ and $s_{-1,-1}^{DD}$ describe the transverse relaxation. Note the similarity between Eqs. (61) with (64) on the one hand, and Eq. (46) on the other hand.

Bertini and co-workers (*119*) and Kruk *et al.* (*96*) formulated a theory of electron spin relaxation in slowly-rotating systems valid for arbitrary relation between the static ZFS and the Zeeman interaction. The unperturbed, static Hamiltonian was allowed to contain both these interactions. Such an unperturbed Hamiltonian, H_0, depends on the relative orientation of the molecule-fixed P frame and the laboratory frame. For cylindrically symmetric ZFS, we need only one angle, β, to specify the orientation of the two frames. The eigenstates of $H_0(\beta)$ were used to define the basis set in which the relaxation superoperator $\hat{\hat{R}}_{ZFS}^{T}(\beta)$ is expressed. The superoperator $\hat{\hat{M}}$, the projection vectors and the electron-spin spectral densities (*cf.* Eqs. (62–64)), all become dependent on the angle β. The expression in Eq. (61) needs to be modified in two ways: first, we need to include the cross-terms electron-spin spectral densities, $s_{1,-1}^{DD}$, $s_{1,0}^{DD}$ and $s_{0,-1}^{DD}$. These terms can be proved to be negligible in the limiting cases of low- as well as high magnetic field. Second, also the nuclear spin relaxation enhancement in this extension of Eq. (61) is orientation-dependent and the calculation of the PRE has to contain an averaging step over all possible β values. If the ZFS has a rhombic symmetry, then all the above-mentioned quantities become dependent on one more angle and the situation becomes significantly more complicated. General expressions for arbitrary ZFS tensor have so far been formulated for $S = 1$, but not for higher values of the S quantum number.

The approach of Bertini *et al.* (*119*) and Kruk *et al.* (*96*) was implemented in the software package developed in Florence (*107*) and is called the "modified Florence method". The corresponding program is available in public domain (www.postgenomicnmr.net). The predictions of the ZFS rhombicity-dependence (for $S = 1$) of the "modified Florence" approach are compared with the general theory, with the "original Florence" method (*107*) and with the SBM approach in Fig. 10, taken from the work of Bertini *et al.* (*119*).

The "modified Florence" program is well-suited for fitting the experimental NMRD profiles for slowly-rotating complexes of gadolinium(III), an $S = 7/2$ ion characterized by relatively low ZFS, whose electron spin relaxation can be considered to be in the Redfield limit. An example of fitting an NMRD profile for aqueous protons, using different methods, for a protein adduct of a Gd(III) chelate capable of accommodating one water molecule in the first coordination sphere, is displayed in Fig. 11. Other examples will be provided in Chapter 3.

C. Rapidly-Rotating Systems

When the reorientational motion is rapid and the ZFS averaged over rapid motions (distortions, collisions) is non-zero, the validity of the decomposition

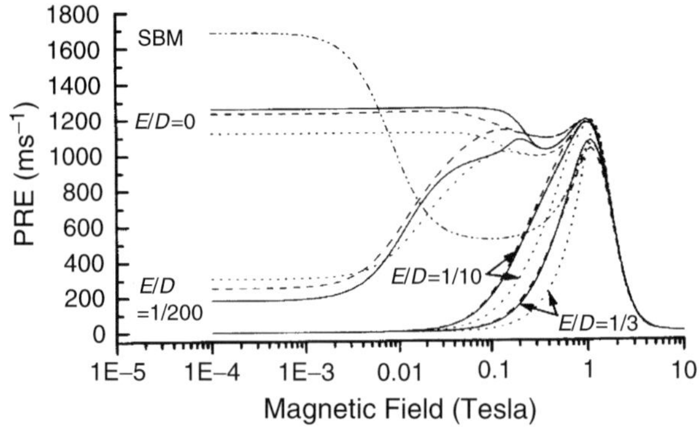

FIG. 10. Calculated NMRD profiles for $S = 1$ for a given parameter set and different theoretical approaches. Solid line: slow motion theory; dotted line: original Florence model; dashed line: modified Florence model; dotted–dashed line: SBM. Reproduced with permission from Bertini, I.; Kowalewski, J.; Luchinat, C.; Nilsson, T.; Parigi, G. *J. Chem. Phys.* **1999**, *111*, 5795–5807. Copyright 1999 American Institute of Physics.

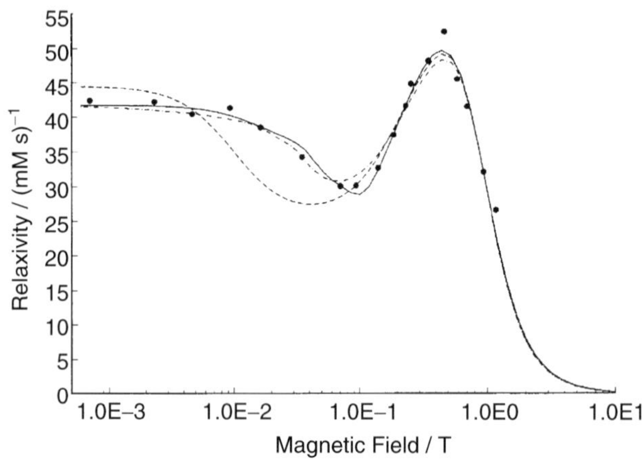

FIG. 11. Experimental NMRD profile for GdDOTA(BOM)$_3$-BSA in aqueous solution and the least squares fits obtained using the modified Florence approach (solid line), original Florence model (dashed–dotted line) and the SBM (dashed line). From Kruk, D.; Nilsson, T.; Kowalewski, J. *Phys. Chem. Chem. Phys.* **2001**, *3*, 4907–4917. Reproduced by permission of the PCCP Owner Societies.

approximation becomes problematic. Besides the general slow-motion theory (Section IV), three ways to circumvent the problem have been proposed so far.

Rast and co-workers have dealt with the electron spin relaxation problem for ions with half-filled shells ($3d^5$, $S = 5/2$, such as Mn(II) or $4f^7$, $S = 7/2$, such as Gd(III)) in the high-field limit (*120*). They allowed for a static ZFS, containing also terms with tensorial rank four and six, along with the rank-two term discussed before. The static ZFS was assumed to be modulated by rotational diffusion. In addition, they included the rank-two transient ZFS, modulated by fast distortions due to collisions with the surrounding solvent. The two random processes were assumed uncorrelated with each other and the Redfield limit conditions were assumed to apply for both mechanisms. The electron spin relaxation matrix elements could then be expressed as sums of terms corresponding to the complex rotation and distortion and the EPR lineshapes could be calculated. In another work from the same group, the theory was applied for interpreting variable-temperature, multiple-field EPR spectral lineshapes for two Gd(III) complexes (*121*). The theory of Rast *et al.* has so far mainly been applied to the PRE problem for the case of intermolecular (outer-sphere) relaxation, which will be reviewed in Section VII. For the inner-sphere relaxation, it was used in the recent study by Borel *et al.*, who simplified the step from the electron spin relation to the PRE in a manner reminiscent of the MSB equations (*122*).

We call the second category of theoretical tools for dealing with rapidly rotating systems the "spin-dynamics methods". The models within this category will be reviewed in Section VI.

The recent work by Kruk and Kowalewski (*123*) belongs to the third group of methods. In this paper, the authors considered the fast rotational motion as a source of modulation of both the static ZFS (which acts as a relaxation mechanism for the electron spin) and as an origin of the stochastic variation of the electron spin-nuclear spin dipole–dipole interaction, leading to nuclear spin relaxation. The situation was investigated in some limiting cases. In the limit of the magnetic field close to zero, the theory of Nilsson and Kowalewski (*95*) was recovered. The inner-sphere relaxation in the high-field limit was investigated carefully and analytical expressions for the PRE were obtained corresponding to the case analogous to that treated on the electron-spin relaxation level by Rast *et al.* (*120*). If we assume the Redfield condition to apply to the rotational modulation of the static ZFS, $\Delta_S \tau_R \ll 1$, then the resulting contributions to the electron spin relaxation rates will be much smaller than the inverse rotational correlation time. It is therefore possible to separate the spin- and space-dependent terms in the spectral density expression similar to Eq. (38) and the decomposition approximation becomes valid. Kruk and Kowalewski considered a couple of experimental examples where NMRD profiles were available for aqueous solutions of a small Gd(III) complex, as well as for the corresponding protein adduct (*123*). The cases of slowly rotating protein adducts were treated using the theory described in Section V.B. For the small, rapidly rotating complex, Gd-EDTA$^-$, the new theory was applied, keeping most of the parameters at the values obtained for the macromolecular complex and adjusting only the rotational correlation time and the exchange lifetime (*123*). The NMRD profiles for both cases and the fitted curves are shown in Fig. 12. In another recent communication from the same laboratory, the analogous case of small and large Mn(II) complexes was considered (*124*). In that study, the authors discussed several regimes corresponding to different rates of rotation. For the case of complexes between Mn(II) chelates and proteins, the authors found that in certain situations it was reasonable to include moderately fast reorientational motion into the time correlation function for the

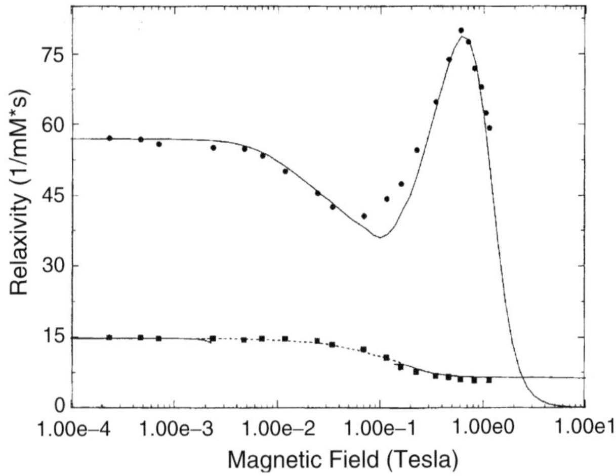

FIG. 12. Experimental and calculated NMRD profiles for $GdEDTA^-$ in aqueous solution in the presence (upper curve) and absence (lower curve) of bovine serum albumin. Reprinted from *J. Magn. Reson.* vol. 162, Kruk, D.; Kowalewski, J., "Nuclear Spin Relaxation in Paramagnetic Systems ($S > 1$) under Fast Rotation Conditions", pp. 229–240, Copyright 2003, with permission from Elsevier.

combined lattice in a way reminiscent of Eq. (11), even while the electron spin relaxation properties were described allowing for multi-exponential decays of electron spin magnetization and averaging over orientations.

In yet another recent investigation, Kruk and Kowalewski considered the case when the static ZFS was smaller than the transient ZFS and the latter term should be considered as the unperturbed Hamiltonian at low magnetic fields (*125*). The validity conditions for the theory derived in that case were rather difficult to realize in experimentally relevant situations. The aqueous solution of Ni(II), a difficult case treated previously by the slow-motion theory (*92,93*), was however found to be possible to describe in a reasonable way using the new approach (*125*).

D. Electron Spin Dynamics in the Equilibrium Ensemble

Sharp and Lohr proposed recently a somewhat different point of view on the relation between the electron spin relaxation and the PRE (*126*). They pointed out that the electron spin relaxation phenomena taking a non-equilibrium ensemble of electron spins (or a perturbed electron spin density operator) back to equilibrium, described in Eqs. (53) and (59) in terms of relaxation superoperators of the Redfield theory, are not really relevant for the PRE. In an NMR experiment, the electron spin density operator remains at, or very close to, thermal equilibrium. The pertinent electron spin relaxation involves instead the thermal decay of time correlation functions such as those given in Eq. (56). The authors show that the decay of the $G_r(\tau)$ (r denotes the electron spin vector components) is composed of a sum of contributions associated with individual eigenstates, α, of the main Hamiltonian combining the electron Zeeman and the static ZFS interaction. The decays corresponding to different components of the spin vector, *r*, and to different levels, α, are not coupled and the central quantities of the theory are the simple exponential decay rate constants:

$$\left(\tau_{S,r}^{(\alpha)}\right)^{-1} = \langle\alpha|S_r^2|\alpha\rangle^{-1} \sum_{\alpha',\beta,\beta'} R_{\alpha\alpha',\beta\beta'} \langle\beta|S_r|\beta'\rangle\langle\alpha'|S_r|\alpha\rangle \tag{65}$$

The Redfield matrix elements are defined in full analogy with the case where the "conventional" electron spin relaxation processes in a non-equilibrium ensemble are considered, but the rates $\left(\tau_{S,r}^{(\alpha)}\right)^{-1}$ are in general different from the non-equilibrium relaxation rates. Equation (65) was derived for one particular orientation; in a slow-rotation situation it must be averaged over molecular orientations. The approach was used to discuss the qualitative aspects of the PRE in $S = 3/2$ Co(II) complexes, characterized by very large static zero-field splittings. It should be noticed that Sharp and co-workers reported earlier an experimental study of one of such Co(II) complexes (*116*), and found it necessary to invoke two different electron spin relaxation for the two Kramers' doublets ($M_S = \pm 1/2$ and $M_S = \pm 3/2$) for the system. In another recent communication, Sharp (*127*) derived a simplified version of the theory, where the level-specific rates of Eq. (65) were averaged over various levels to produce the laboratory-frame quantities τ_{S1} and τ_{S2}, similar to those occurring in the Bloembergen–Morgan theory (*27*).

VI. Spin Dynamics Models

The problem of a strong coupling between the electron spin system and the classical degrees of freedom (rotation in the first place), as expressed in Eq. (29), can also be treated in another way. This class of methods to circumvent the limitations of the Redfield regime is the subject of this section.

Odelius and co-workers reported some time ago an important study involving a combined quantum chemistry and molecular dynamics (MD) simulation of the ZFS fluctuations in aqueous Ni(II) (*128*). The *ab initio* calculations for hexa-aquo Ni(II) complex were used to generate an expression for the ZFS as a function of the distortions of the idealized T_h symmetry of the complex along the normal modes of E_g and T_{2g} symmetries. An MD simulation provided a 200 ps trajectory of motion of a system consisting of a Ni(II) ion and 255 water molecules, which was analyzed in terms of the structure and dynamics of the first solvation shell of the ion. The fluctuations of the structure could be converted in the time variation of the ZFS. The distribution of eigenvalues of ZFS tensor was found to be consistent with the rhombic, rather than axial, symmetry of the tensor, which prompted the development of the analytical theory mentioned above (*89*). The time-correlation function (tcf) for the ZFS was also computed. The ZFS tcf was found to be strongly non-exponential and to decay on sub-picosecond time scale. This observation was the origin of some of the work on the connection between electron spin relaxation and complex vibrations, to be reviewed in Section VIII.

An important point, to be stressed in the present context, was developed in a subsequent paper from the same group (*129*). The starting point of that study was the ZFS trajectory, the time sequence of magnitudes and orientations of the ZFS tensor. The ZFS trajectory could simply be converted into a spin Hamiltonian varying randomly in discrete time steps, to which a Zeeman interaction of arbitrary magnitude could also be added. A method for calculating the tcf for the electron spin, analogous to Eq. (56), was proposed based on solving numerically the time-dependent Schrödinger equation with the Hamiltonian. Given the electron spin time correlation functions, it was then possible to calculate the electron- and nuclear-spin relaxation and their dependence on the magnetic field. The calculated field-dependence of the spin–lattice relaxation time for the electron spin ($S = 1$), T_{1e}, is shown in Fig. 13.

Abernathy and Sharp employed a similar idea, although in a more simplified form (*130*). They also worked in terms of a spin Hamiltonian varying with time in discrete steps and let the Hamiltonian contain the Zeeman and the ZFS interactions. They assumed, however, that the ZFS interaction was constant in the molecule-fixed (P) frame and that variation of the Hamiltonian originated only from fluctuation of the P frame with respect to the laboratory frame. These fluctuations were described in terms of Brownian reorientational motion, characterized by a time interval, τ_r, (related to the rotational correlation time τ_R) and a Gaussian distribution of angular steps.

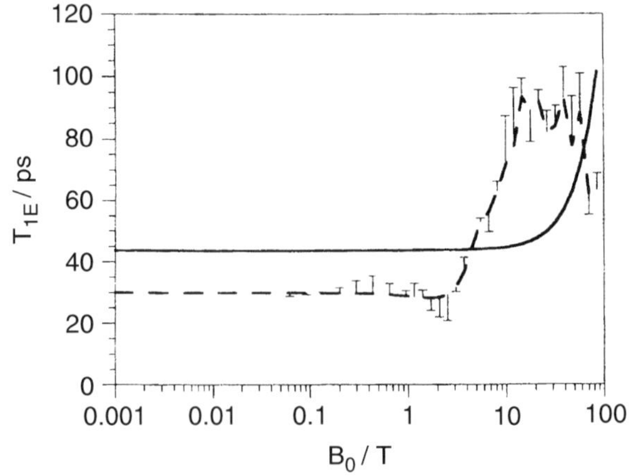

FIG. 13. Predicted magnetic field dependence of the electron spin lattice relaxation time. Solid line: pseudorotation model; dashed line: spin dynamics calculation. Reproduced with permission from Odelius, M.; Ribbing, C.; Kowalewski, J. *J. Chem. Phys.* **1996**, *104*, 3181–3188. Copyright 1996 American Institute of Physics.

The electron-spin time-correlation functions of Eq. (56) were evaluated numerically by constructing an ensemble of trajectories containing the time dependence of the spin operators and spatial functions, in a manner independent of the validity of the Redfield limit for the rotational modulation of the static ZFS. Before inserting thus obtained electron-spin time-correlation functions into an equation closely related to Eq. (38), Abernathy and Sharp also discussed the effect of distortional/vibrational processes on the electron spin relaxation. They suggested that the electron spin relaxation could be described in terms of simple exponential decay rate constant τ_S^{-1}, expressed as a sum of a rotational and a distortional contribution:

$$\tau_S^{-1} = \tau_{S,r}^{-1} + \tau_{S,v}^{-1} \tag{66}$$

where the rotational part was calculated using the spin-dynamics method and the distortional part had the field dependence given by Eq. (14). Abernathy and Sharp concentrated in their first spin-dynamics paper on the case of intermolecular (outer-sphere) PRE (*130*) and we shall return to some of their results in Section VII.

The spin-dynamics method was applied to the intramolecular PRE in the case of aqueous and methyl protons in the Ni(II)(acac)$_2$(H$_2$O)$_2$ complex (acac = 2,4-pentanedione) (*131,132*). The two kinds of protons are characterized by a different angle between the principal axis of the static ZFS and the dipole–dipole axis. The ratio, ρ, of the proton relaxation rates in the axial (the DD principal axis coinciding with the ZFS principal axis) and the equatorial (the DD principal axis perpendicular to the ZFS principal axis) positions takes on the value of unity in the Zeeman limit and up to four in the ZFS limit. A similar spin-dynamics analysis of the NMRD data for a Mn(II) complex has also been reported (*133*).

Rast *et al.* (*134*) proposed a similar approach using dynamically-sampled Monte Carlo procedure (*135*). Both the static ZFS (including higher-order terms) and the transient quadratic ZFS were included, and the method was applied to simulate the electron spin relaxation and the EPR spectra of the Gd(III) complex [Gd(DOTA)H$_2$O]$^-$ (DOTA = 1,4,7,10 tetrakis(carboxymethyl)-1,4,7,10 tetraazacyclododecane) at various fields and temperatures. The validity of the Redfield approximation was investigated and found satisfactory for the EPR data at the magnetic fields used in that technique. At very low magnetic fields, $B_0 \leq 0.01$ T, interesting in the context of NMRD measurements, the Redfield theory results were found to differ notably from the Monte Carlo approach.

VII. Outer-Sphere Relaxation

PRE effects can also be seen for nuclear spins belonging to the solvent or other molecules present in solution, that essentially do not enter the first coordination sphere of the paramagnetic metal ion. This mechanism is referred to as outer-sphere relaxation and is usually less important than the inner-sphere PRE. In the case of outer-sphere PRE, the modulation of the dipole–dipole interaction (outer-sphere PRE due to scalar coupling is usually not considered, since this interaction is transmitted through chemical bonds) arises through the relative translational diffusion of the paramagnetic species, the electron spin dynamics and molecular tumbling. Since no complex formation is assumed, the chemical exchange effects do not need to be considered, but a proper description of the electron spin dynamics is of importance. The outer-sphere case is more difficult to describe and it is relatively less frequently studied.

A. General Aspects

Classical studies of the relaxation processes, caused by translational diffusion, have been presented in the early days by Abragam (*18*), Torrey (*136*) and Pfeifer (*137*). Abragam (*18*) found, by solving the diffusion equation, the following form of the correlation function for the stochastic function $D^2_{0,m}(\Omega)r^{-3}$ under translational diffusion of two spins 1/2:

$$\left\langle \frac{D^{2*}_{0,m}[\Omega(0)]}{r^3(0)} \frac{D^2_{0,m}[\Omega(\tau)]}{r^3(\tau)} \right\rangle = \frac{3}{8}\frac{N}{d^3} \int_0^\infty \left[J_{3/2}(u)\right]^2 \exp\left(-\frac{D_{12}}{d^2}u^2 t\right) \frac{du}{u} \tag{67}$$

where N is the density of spins per unit volume, D_{12} is the relative translational diffusion coefficient of two identical molecules (sum of their self-diffusion coefficients) in which the spins reside. The symbol d is the distance of closest approach of the two spins, equal to twice the molecular radius. Instead of D_{12}, it is in some situations convenient to use the diffusional correlation time, $\tau_D = d^2/D_{12}$. $J_{3/2}(u)$ is a Bessel function. The spectral densities derived from this time correlation function enter the expressions for nuclear spin relaxation rates in the usual way (*18*). Torrey (*136*) studied two cases of isotropic translation diffusion: one in which the jump distance had a probability distribution, and the other in which it was constant. The Poisson distribution of the probability that n jumps take place in time t was incorporated into the description, leading to the distribution of translational correlation times. Pfeifer extended the treatment of Torrey to include the effects of electron spin relaxation (*137*).

An important theoretical development for the outer-sphere relaxation was proposed in the 1970s by Hwang and Freed (*138*). The authors corrected some earlier mistakes in the treatment of the boundary conditions in the diffusion equation and allowed for the role of intermolecular forces, as reflected in the *IS* radial distribution function, $g(r)$. Ayant *et al.* (*139*) proposed, independently, a very similar model incorporating the effects of molecular interactions. The same group has also dealt with the effects of “spin eccentricity” or translation-rotation coupling (*140*).

We defer the discussion of the effects of $g(r)$ until Section VII.C and begin with the special case, referred as a force-free diffusion, with a uniform distribution of electron spins outside the distance of closest approach with respect to the nuclear spin. Under the assumption of the reflecting-wall boundary condition at $r_{IS} = d$, Hwang and Freed found the closed analytical form of the correlation function for translation diffusion (*138*):

$$\left\langle \frac{D^{2*}_{0,1-q}[\Omega_{IS}(0)]}{r^3_{IS}(0)} \frac{D^2_{0,1-p}[\Omega_{IS}(\tau)]}{r^3_{IS}(\tau)} \right\rangle = \frac{4\pi}{5}\frac{1}{d^3} N f(\tau)\delta_{p,q} \tag{68}$$

where

$$\begin{aligned} f(\tau) &= \frac{18}{\pi}\int_0^\infty \frac{u^2}{81 + 9u^2 - 2u^4 + u^6} \exp\left(-\frac{D_{12}}{d^2}u^2\tau\right) du \\ &= \int_0^\infty \xi(u) \exp\left(-\frac{D_{12}}{d^2}u^2\tau\right) du \end{aligned} \tag{69}$$

Freed extended the Hwang-Freed model by including electron relaxation (*141*). The description of the electron spin relaxation in that formulation was very simple, and the approach can be considered as a counterpart of the Solomon–Bloembergen–Morgan description of the inner-sphere relaxation. The single electron relaxation rate, used in the Freed model, corresponds to the extreme narrowing limit of the Bloembergen–Morgan (BM) theory, *cf*. Eq. (52). In the same paper, Freed examined the limiting situation of slow electron spin relaxation and proposed an analytical expression for the spectral density corresponding to the correlation function of Eq. (68):

$$J(\omega) = \frac{1 + 5z/8 + z^2/8}{1 + z + z^2/2 + z^3/6 + 4z^4/81 + z^5/81 + z^6/648} \tag{70}$$

with $z = (2\omega\tau_D)^{1/2}$. This formulation of Ayant's (*139*) and Freed's (*141*) force-free diffusion model was popularized by Polnaszek and Bryant (*142*) and is often used together with the equations that follow for R_{1p} and R_{2p} (*1,2*), which are an outer-sphere counterpart of the Solomon equation, Eq. (10), and its analog for the transverse relaxation.

$$R_{1p} = \frac{32\pi}{405}\left(\frac{\mu_0}{4\pi}\right)^2 1000 N_A[M] \frac{\gamma_I^2\gamma_S^2\hbar^2 S(S+1)}{d(D_M + D_L)}[7J(\omega_S) + 3J(\omega_I)] \tag{71}$$

$$R_{2p} = \frac{16\pi}{405}\left(\frac{\mu_0}{4\pi}\right)^2 1000 N_A[M] \frac{\gamma_I^2\gamma_S^2\hbar^2 S(S+1)}{d(D_M + D_L)}[4J(0) + 13J(\omega_S) + 3J(\omega_I)] \tag{72}$$

The Avogadro number, N_A, times 1000 times the molar concentration of the metal complex, [*M*], replaces here the spin density *N* (in units of m^{-3}) of Eqs. (67) and (68).

Bayburt and Sharp (*143*) formulated a low-field theory (i.e. a theory for the case of ZFS dominating over the electron spin interaction) for the outer-sphere relaxation, treating also the electron spin relaxation in the simplified manner expressed by Eq. (52). That model predicted only a weak dependence of the PRE on the magnitude of the static ZFS and its application to the cases of high static ZFS is problematic.

B. Outer-Sphere PRE and Electron Relaxation: Recent Development

Theoretical models for outer-sphere nuclear spin relaxation in paramagnetic systems, including an improved description of the electron spin relaxation, have been developed intensively for the last couple of years. They can be treated as counterparts of the models of inner-sphere PRE, described in the Section V.B and V.C.

Kruk and co-workers formulated, in analogy with the inner-sphere case, the expression for the nuclear spin relaxation rate of solvent nuclei, $T_{1,OS}^{-1}$, as the real part of the complex spectral density, $K_{1,1}^{OS}$, taken at negative nuclear spin Larmor frequency (*96,144*):

$$T_{1,OS}^{-1} = 2\mathrm{Re}\left\{K_{1,1}^{OS}(-\omega_I)\right\} = 2\mathrm{Re}\int_0^\infty G_{1,1}^{OS}(-\tau)\exp(-i\omega_I\tau)\,d\tau \tag{73}$$

Under the assumption of dominant dipole–dipole interaction between the nuclear and electron spins, the outer-sphere correlation function, $G_{1,1}^{OS}(-\tau)$, is given by the equation:

$$G_{1,1}^{OS}(-\tau) = Tr_L\left\{T_1^{1,OS+}\left[\exp\left(-i\hat{\hat{L}}_L\tau\right)T_1^{1,OS}\right]\rho_L^{eq}\right\} \tag{74}$$

where the lattice tensor operator $T_1^{1,OS}$ in the laboratory frame takes the form similar to that of Eq. (32), which can be written as:

$$T_1^{1,DD} = C^{OS}\sqrt{30}\sum_q \begin{pmatrix} 1 & 2 & 1 \\ 1-q & q & -1 \end{pmatrix} S_q^{1(L)}\frac{D_{0,1-q}^2(\Omega_{IS})}{r_{IS}^3} \tag{75}$$

with $C^{OS} = \frac{\mu_0\hbar\gamma_I\gamma_S}{4\pi}$. The vector $\vec{r}_{IS} = \vec{r}_I - \vec{r}_S$ describes stochastically fluctuating relative positions of the spins *I* and *S*, while Ω_{IS} represents the angles specifying the direction of $\vec{r}_{IS}$ in the laboratory frame. The derivation of the spectral density, $K_{1,1}^{OS}$, is similar to the procedure for the inner-sphere relaxation. The main difference is that, for the outer-sphere relaxation, the relative distance $\vec{r}_{IS}$ is a stochastic variable. For the outer-sphere case, the Liouville superoperator $\hat{\hat{L}}_L$ includes (besides the Zeeman Liouvillian $\hat{\hat{L}}_{S0}$, the commutators with the static and transient parts of the ZFS coupling, the rotational diffusion of the complex ($\hat{\hat{L}}_R$) and distortions in its geometry ($\hat{\hat{L}}_v$)) a classical Markov operator describing relative translational diffusion of the interacting spins *I-S*, ($\hat{\hat{L}}_T$):

$$\hat{\hat{L}}_L = \hat{\hat{L}}_{S0} + \hat{\hat{L}}^S_{ZFS} + \hat{\hat{L}}^T_{ZFS} + \hat{\hat{L}}_R + \hat{\hat{L}}_v + \hat{\hat{L}}_T \tag{76}$$

Formulated in the laboratory frame, the outer-sphere spectral density $K^{OS}_{1,1}(-\omega_I)$ has the form:

$$K^{OS}_{1,1}(-\omega_I) = 30\left(C^{OS}\right)^2 \sum_{q\in(-1,0,1)} \sum_{p\in(-1,0,1)} \begin{pmatrix} 2 & 1 & 1 \\ 1-q & q & 1 \end{pmatrix} \begin{pmatrix} 2 & 1 & 1 \\ 1-p & p & 1 \end{pmatrix}$$
$$\times \int_0^\infty Tr_L\left\{ S_q^{1(L)^+} \frac{D^{2*}_{0,1-q}(\Omega_{IS})}{r^3_{IS}} \left[\exp\left(-i\hat{\hat{L}}_L\tau\right) S_p^{1(L)} \frac{D^2_{0,1-p}(\Omega_{IS})}{r^3_{IS}} \right] \rho_L^{eq} \right\}$$
$$\times \exp(-i\omega_I\tau)d\tau \tag{77}$$

In all the approaches mentioned below, it is assumed that the correlation function can be factorized into a product of correlation functions for the three degrees of freedom: rotational motion, translational diffusion and electron spin dynamics.

An analytical theory of the outer-sphere PRE for slowly rotating systems with an arbitrary electron spin quantum number *S*, appropriate at the limit of low field, has been proposed by Kruk *et al.* (*144*). The theory deals with the case of axial as well as rhombic static ZFS. In analogy to the inner sphere case (*95*), the PRE for the low field limit could be expressed in terms of the electron spin spectral densities $s^{OS}_{m,m}$:

$$T^{-1}_{1,OS} = \frac{16\pi}{9}\left(C^{OS}\right)^2 \frac{N_A 1000[M]}{d^3} S(S+1)\mathrm{Re}\left(s^{OS}_{-1,-1} + s^{OS}_{0,0} + s^{OS}_{1,1}\right) \tag{78}$$

As in Eq. (64), the electron spin spectral densities could be evaluated by expanding the electron spin tensor operators in a Liouville space basis set of the static Hamiltonian. The outer-sphere electron spin spectral densities are more complicated to evaluate than their inner-sphere counterparts, since they involve integration over the variable *u*, in analogy with Eqs. (68) and (69). The main simplifying assumption employed for the electron spin system is that the electron spin relaxation processes can be described by the Redfield theory in the same manner as for the inner-sphere counterpart (*95*). A comparison between the predictions of the analytical approach presented above, and other models of the outer-sphere relaxation, the Hwang and Freed model (HF) (*138*), its modification including electron spin

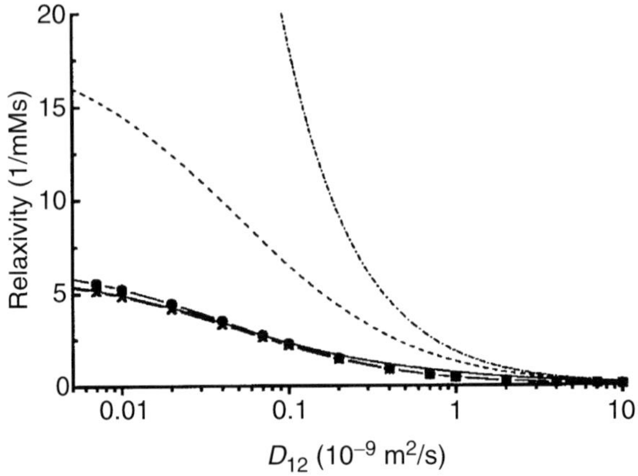

FIG. 14. The outer-sphere relaxivity at zero magnetic field as a function of relative diffusion coefficient for $S = 1$. Reproduced with permission from Kruk, D.; Nilsson, T.; Kowalewski, J. *Mol. Phys.* **2001**, *99* , 1435–1445. Copyright 2001 Taylor and Francis Ltd (http://www.tandf.co.uk/journals/tf/00268976.html).

relaxation (HFS) (*141*) and the model of Bayburt and Sharp (BS) (*143*), is presented in the Fig. 14 for $S = 1$.

A more general theory for outer-sphere paramagnetic relaxation enhancement, valid for an arbitrary relation between the Zeeman coupling and the axial static ZFS, has been developed by Kruk and co-workers (*96*), in the same paper which dealt with the inner-sphere case. The static ZFS was included, along with the Zeeman interaction in the unperturbed Hamiltonian. The general expression for the nuclear spin–lattice relaxation rate of the outer-sphere nuclei was written in terms of electron spin spectral densities, $s^{OS}_{m,k}$, as:

$$T^{-1}_{1,OS}(-\omega_I) = \frac{16\pi}{9}\left(C^{OS}\right)^2 \frac{N_A 1000[M]}{d^3} S(S+1)\cdot$$
$$\mathrm{Re}\left\{s^{OS}_{-1,-1} + s^{OS}_{0,0} + s^{OS}_{1,1} + 2s^{OS}_{1,-1} + 2s^{OS}_{0,-1} + 2s^{OS}_{1,0}\right\} \quad (79)$$

Compared with Eq. (78), one needed in this case also spectral densities, $s^{OS}_{m,k}$, for $m \neq k$. In addition, the evaluation of spectral densities involved averaging over the angle β, specifying the relative orientation of the molecule-fixed P frame and the laboratory frame, prior to the integration over the variable u. The description of translation motion was based on the force-free Hwang and Freed approach, represented by the $\xi(u)$ function defined in Eq. (69). In the low-field limit the description simplifies to the analytical approach mentioned above (*144*). The description of Kruk *et al.* (*96*) has been incorporated into the general framework of the computer program developed by Bertini *et al.* (*107*). In Fig. 15, we show the outer-sphere NMRD profiles of integer spins $S = 1$, 2, 3 and their comparison with the Hwang and Freed approach (*138*).

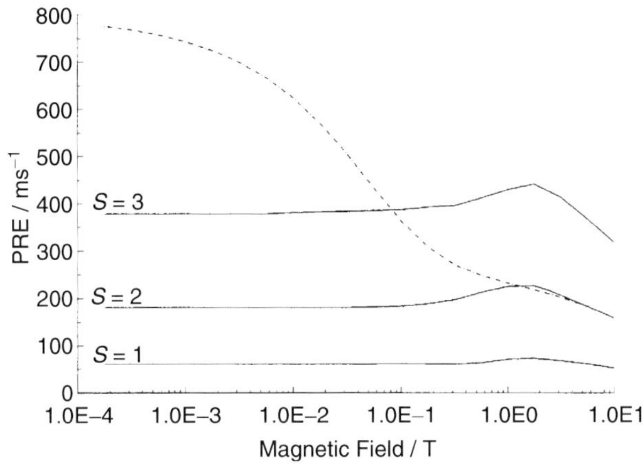

FIG. 15. Calculated outer-sphere NMRD profiles for integer spins ($S = 1, 2, 3$) for a large static ZFS. The dashed line is obtained using the Hwang and Freed model for $S = 2$. From Kruk, D.; Nilsson, T.; Kowalewski, J. *Phys. Chem. Chem. Phys.* **2001**, *3*, 4907–4917. Reproduced by permission of the PCCP Owner Societies.

If the electron spin relaxation effects are important, then the Hwang and Freed theory deviates from the present model. If the translational diffusion is very fast, it masks the relaxation effects, and the theory simplifies to the Hwang and Freed force-free diffusion approach. Experimental NMRD profiles for slowly-rotating complexes of gadolinium(III), $S = 7/2$, (*96*) and manganese(II), $S = 5/2$, (*124*) were interpreted taking into account both the inner-sphere and the outer-sphere contribution. As an example, we show in Fig. 16 the fitting of the NMRD profile for the complex GdDOTA$(BOM)_3$-BSA in aqueous solution. The role of the outer-sphere contribution is modest, but non-negligible.

Models for the outer-sphere PRE, allowing for faster rotational motion, have been developed, in analogy with the inner sphere approaches discussed in the Section V.C. The outer-sphere counterpart of the work by Kruk *et al.* (*123*) was discussed in the same paper. In the limit of very low magnetic field, the expressions for the outer-sphere PRE for slowly rotating systems (*96,144*) were found to remain valid for an arbitrary rotational correlation time τ_R. New, closed-form expressions were developed for outer-sphere relaxation in the high-field limit. The Redfield description of the electron spin relaxation in terms of spectral densities incorporated into that approach, was valid as long as the conditions $\Delta_S^2\tau_R^2 \ll 1$ and $\Delta_T^2\tau_D^2 \ll 1$ were fulfilled. The validity regime of the outer-sphere PRE model by Kruk and Kowalewski (*123*) was the same as for the inner-sphere PRE theory for fast-rotating complexes, contained in the same paper. In the low-field limit, it is determined by the condition: $H_{ZFS}^S \gg H_{ZFS}^T$, while the validity condition in the high-field limit is: $H_{Zeeman} \gg H_{ZFS}^S + H_{ZFS}^T$. One comment concerning the relative roles of the inner- and outer-sphere contributions should be made in the present context. In the discussion of Fig. 16 above, we noticed that the outer-sphere

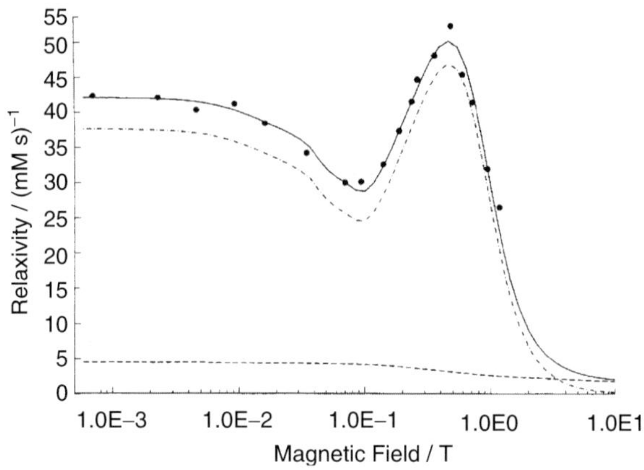

FIG. 16. Experimental and calculated NMRD profiles for GdDOTA(BOM)$_3$-BSA in aqueous solution. Dashed line: outer-sphere contribution; dashed–dotted line inner-sphere contribution; solid line: inner-plus outer-sphere. From Kruk, D.; Nilsson, T.; Kowalewski, J. *Phys. Chem. Chem. Phys.* **2001**, *3*, 4907–4917. Reproduced by permission of the PCCP Owner Societies.

contribution was in that case rather modest. It is worthwhile to notice that, as the molecular tumbling becomes faster, the efficiency of the inner-sphere mechanism decreases, while the outer-sphere contribution does not change much. This may in fact result in the outer-sphere contribution to the observed PRE in the case of fast rotating systems becoming rather important. The outer-sphere contribution to the PRE was included in the analysis of NMRD profiles of small and large Gd(III) (*123*) and Mn(II) complexes (*124*).

Abernathy and Sharp (*130,145*) treated the intermediate regime, when the reorientation of the paramagnetic species is in-between the slow- and fast-rotations limits. They applied the spin-dynamics method, described in Section VI, to the case of outer-sphere relaxation and interpreted NMRD profiles for non-aqueous solvents in the presence of complexes of Ni(II) ($S = 1$) and Mn(III) ($S = 2$).

C. Outer-Sphere Relaxation and Intermolecular Forces

The considerations based on the force-free diffusion model describe the case of non-interacting species, a highly idealized picture of chemical systems in solution. The more complicated chemical reality requires more sophisticated tools. There are, in principle, two ways to deal with the ligand/solvent molecules outside of the first coordination sphere, but interacting with the paramagnetic complex. One of them invokes the concept of second-sphere relaxation. Botta has recently reviewed this approach (*146*). Briefly, one assumes that the molecules carrying nuclear spin of interest reside in well-defined second-sphere complexes. From the theoretical point of view, the second sphere is just like the inner sphere, except for the longer distance to the paramagnetic center.

The second category of methods uses a more general approach, based on fundamental concepts in statistical mechanics of the liquid state. As mentioned above, the Hwang and Freed theory (*138*) and the work of Ayant *et al.* (*139*) allow for the presence of intermolecular forces by including in the formulation the radial distribution function, $g(r)$, of the nuclear spins with respect to the electron spins. The radial distribution function is related to the effective interaction potential, $V(r)$, or the potential of mean force, $W(r)$, between the spin-carrying particles through the relation (*138,139*):

$$g(r) = \exp[-W(r)/k_B T] = \exp[-V(r)] \tag{80}$$

Kruk and Kowalewski combined the theory allowing for the radial distribution with their Redfield-limit description of the electron spin relaxation (*147*). Including the $g(r)$ in the theory led to a more complicated form of the function $f(\tau)$ of Eq. (69), which becomes dependent on the $g(r)$, as well as on the propagator $P(\mathbf{r_0}, 0/\ \mathbf{r}, \tau)$. The rest of the theory remains unchanged with respect to the presentation in sections VII.A–VII.B. The propagator was computed using the Smoluchowski equation:

$$\frac{\partial P(\mathbf{r}, \tau/\mathbf{r_0}, 0)}{\partial \tau} = D_{12}\nabla[\nabla P(\mathbf{r}, \tau/\mathbf{r_0}, 0) + P(\mathbf{r_0},\ 0/\mathbf{r},\ \tau)\nabla V(r)] \tag{81}$$

The Smoluchowski equation was solved numerically, following the approach of Hwang and Freed (*138*). Kruk and Kowalewski called their approach "diffuse second sphere" (DSS), as opposed to the "ordered second sphere" (OSS), where a specific, constant *IS* distance in the second sphere is postulated, in the inner-sphere like manner. They used an experimental NMRD data set for a $Gd(III)(C_{11}\ DOTP)^{5-}$ complex, bound to a protein in order to slow down its rotation, reported by Caravan *et al.* (*148*). There is no space for water molecules in the first coordination sphere of Gd(III), but the complex is highly charged and can be expected to interact strongly with surrounding water. The NMRD profile could be reproduced using the OSS with two or three waters in the second sphere. More interestingly, it was also possible to obtain a reasonable theoretical NMRD profile, using the DSS approach with the $g(r)$ for the aqueous $Gd(III)(DOTP)^{5-}$ complex (i.e., a similar complex, but without the protein and the C_{11}-chain linker), calculated using molecular dynamics (MD) simulations by Borel *et al.* (*149*), *cf.* Fig. 17. Thus calculated DSS profile is compared with the experimental one in Fig. 18.

The group in Grenoble has used the radial distribution function approach in a series of papers on intermolecular relaxation. We wish to mention in particular some of their papers from the 1990s, where the radial distribution functions were obtained through different approximate methods and a relatively simple description of the electron spin relaxation was applied (*150–154*). This work has also been reviewed (*155,156*). In a recent communication from the same group, the improved description of the electron spin relaxation in Gd(III) complexes (*120,121*) was included in the model and applied for

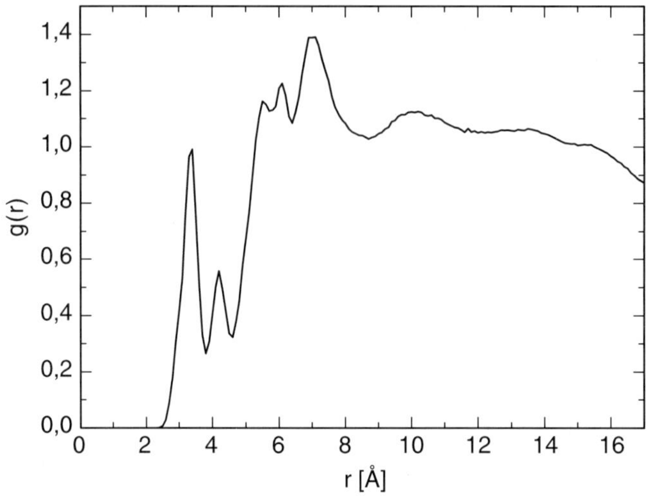

FIG. 17. Gd-aqueous proton radial distribution function for the aqueous solution of the Gd(III)(DOTP)$^{5-}$ complex (after Borel, A.; Helm, L.; Merbach, A.E. *Chemistry – A European Journal* **2001**, *7*, 600–610).

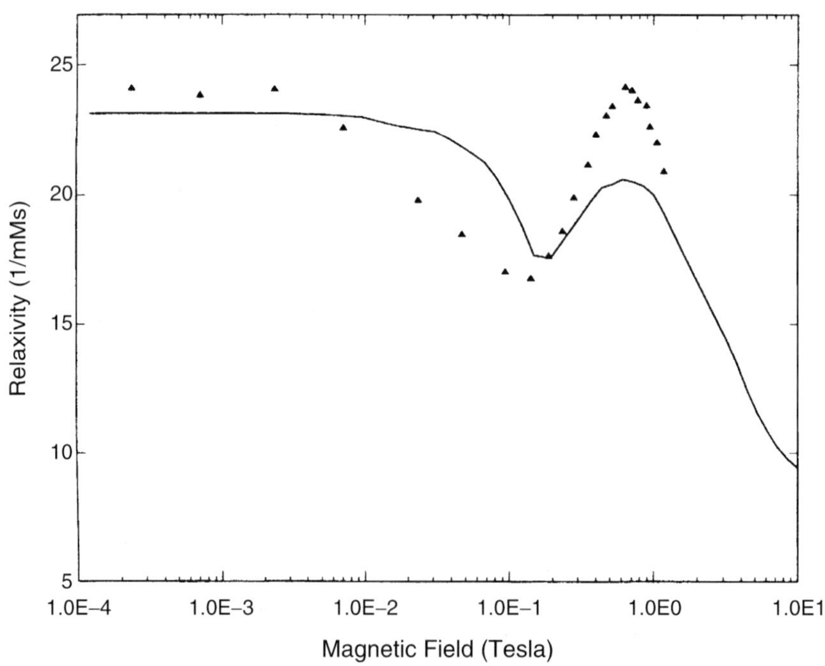

FIG. 18. NMRD profile for Gd(III)(C_{11}-DOTP)$^{5-}$-HSA calculated using the radial distribution function of Fig. 17. Reproduced with permission from Kruk, D.; Kowalewski, J. *J. Chem. Phys.* **2002**, *117*, 1194–1200. Copyright 2002 American Institute of Physics.

interpretation of proton NMRD data in $(CH_3)_4N^+$ in solution of Gd^{3+} in heavy water (*157*). [*Note added in proof*: An important extension of this work appeared recently (Fries, P.H.; Ferrante, G.; Belorizky, E.; Rast, S. *J. Chem. Phys.* **2003**, *119*, 8636–8644).]

D. Chemical Exchange

The chemical exchange processes transfer a spin-carrying molecule between the first (or, possibly, second) coordination sphere of the paramagnetic and the bulk solution. The motivation for including chemical exchange in the "outer-sphere" section is that the exchange can be considered as a long jump along the radial distribution function, from a maximum at a short distance to the flat section at long distances, *cf.* Fig. 17. The chemical exchange can influence the PRE phenomena in two ways. First, the exchange lifetime can act as a correlation time. In an important early paper, Wennerström (*158*) showed stringently that if the exchange occurs with complete randomization of interactions, then it simply implies multiplying the relevant correlation function for the lattice by an exponentially decaying function with a time constant equal to the lifetime of the spin in the complex. This fact was even earlier used by Solomon and Bloembergen (*21,22*) and by Abragam (*18*). It is reflected in the occurrence of τ_M in Eq. (11) of this review. This simple picture of chemical exchange was also considered in early formulations of the slow motion theory (*26,77,78*) and in comparisons between the slow motion theory and the work of Bertini *et al.* (*159*).

The second role of the chemical exchange phenomena can be seen in Eq. (2): the exchange lifetime competes with the in-complex nuclear spin–lattice relaxation time and can become a limiting factor in the attainable PRE. This aspect of the problem is highly relevant in practical consideration in the case of Gd(III) complexes as a potential contrast agent, because the water exchange in these systems is not too fast. This issue is considered to be outside of the scope of this article and we refer to recent literature on the subject (*5,160*) and to other contributions in this volume.

VIII. Molecular Vibrations, Electron Spin Relaxation and the PRE

The main difficulty in theoretical predictions of the PRE effects is caused by the description of electron spin relaxation. The transient ZFS interaction is usually considered as present in every complex with $S \geq 1$ and its modulation is assumed to provide the dominant mechanism for the electron spin relaxation. The commonly used pseudorotation model (*27,85,86*), described earlier, has two advantages: it captures the essential physics of electron spin relaxation (i.e., the fact that relaxation can be caused by motions faster than the overall rotation of the paramagnetic complex, provided that these motions displace the principal axis of the ZFS), and it leads to relatively straightforward mathematical formulation. The assumption that the collision-induced motion of the ligands affects only the orientation of the principal frame of the ZFS interaction is, however, a clear oversimplification. Allowing for the rhombic transient ZFS (*89,97,125*) increases the flexibility of the pseudorotation model, but does not really change this fundamental fact. The amplitude of the ZFS tensor may be stochastically time dependent because of its relation to the structure of the complex. Atomistic simulations of the ZFS fluctuations for $Ni(H_2O)_6^{2+}$ ($S=1$), performed by Odelius *et al.* (*128*), showed that the damped vibrations of E_g and T_{2g} symmetries could generate the transient ZFS interaction. The symmetry-distorting motions were found on a very rapid time-scale relative to typical values of the correlation time describing the ZFS dynamics, predicted by the pseudorotational model.

A model of the ZFS coupling removing the restriction of its constant amplitude and allowing both processes, the stochastic variations of the internal coordinates and the rotational diffusion, to modulate the ZFS interaction was proposed by Westlund and co-workers (*13,85,88,91*). According to this model, the ZFS interaction provided the coupling between the electron spin variables, the stochastic time-dependent distortion coordinates and the reorientational degrees of freedom by the expression:

$$H_{ZFS}(t) = \sum_{n\in(0,\pm1,\pm2)} \sum_{k\in(0,\pm1,\pm2)} (-1)^n S^2_{-n} h_k[q(t)] D^2_{k,n}[\Omega_{ML}(t)] \tag{82}$$

The functions $h_k[q(t)]$ were dependent on the geometry of the complex and were related to the local symmetry-distorting motions by a set of normal coordinates, $q(t) = [q_1(t), ..., q_N(t)]$, through the Taylor expansion:

$$h_k[q(t)] = \sum_i \langle h_k(q_i^0)\rangle + \sum_i \left\langle \frac{\partial h_k}{\partial q_i} \right\rangle \Delta q_i(t) + \sum_{i,j} \left\langle \frac{\partial h_k^2}{\partial q_i \partial q_j} \right\rangle \Delta q_i(t) \Delta q_j(t) + ... \tag{83}$$

where $\Delta q_i(t) = q_i(t) - q_i^0$ gives the deviation of the normal coordinate, $q_i(t)$, from its equilibrium value, q_i^0. The description of the ZFS tensor was restricted to the first order (linear) terms of the above expansion. The dynamics of the normal coordinates was treated as a classical diffusion in a harmonic potential, $V(q_i) = \frac{1}{2} K_i q_i^2$, with the force constant K_i. The conditional probability density, $P(q_i^0, 0|q_i^0, \tau)$, relating the value of q_i at time τ to the value q_i^0 at time zero was assumed to propagate according to the one-dimensional Smoluchowski equation (compare Eq. (81)), applied separately for each vibrational normal mode:

$$\frac{\partial P(q_i^0, 0|q_i^0, \tau)}{\partial \tau} = \Gamma_i P(q_i^0, 0|q_i^0, \tau) \tag{84}$$

where the Smoluchowski operator Γ_i is given by:

$$\Gamma_i = \frac{D_i}{k_B T} \frac{\partial V(q_i)}{\partial q_i} \frac{\partial}{\partial q_i} + \frac{D_i}{k_B T} \frac{\partial^2 V(q_i)}{\partial q_i^2} + D_i \frac{\partial^2}{\partial q_i^2} \tag{85}$$

In this approach, the diffusion constant, D_i, is related to the corresponding characteristic time, τ_i, describing the distortions of the normal coordinate, q_i.

Westlund *et al.* (*85*) used the framework of the general slow-motion theory to incorporate the classical vibrational dynamics of the ZFS tensor, governed by the Smoluchowski equation with a harmonic oscillator potential. They introduced an appropriate Liouville superoperator:

$$\hat{\hat{L}}_D = \hat{\hat{L}}_V - i\hat{\hat{R}}_V \tag{86}$$

where $\hat{\hat{L}}_V$ was derived from the vibrational Hamiltonian and $\hat{\hat{R}}_V$ was a relaxation superoperator describing the vibrational relaxation. Thus, the authors pointed out a hierarchy of events, with the nuclear relaxation being dependent on relaxation in the electron spin subsystem, which in turn depended on the relaxation in the vibrational subsystem. They compared the classical models of the ZFS interaction: the pseudorotation model for the ZFS of constant magnitude and the vibrational model based on the Smoluchowski equation, allowing the magnitude to fluctuate in time. The results of that study were somewhat inconclusive, because of numerical problems. Westlund and Larsson (*91*) tried to apply the one-dimensional classical oscillator model, allowing for the modulation of the magnitude, but not of the orientation, of the ZFS in the molecule-fixed frame, but without real success.

Recently, the vibrational model of the ZFS interaction based on the Smoluchowski equation, has been taken up again by Kruk and Kowalewski (*161*). The system considered was the $Ni(H_2O)_6^{2+}$ complex (of the octahedral symmetry, O_h or T_h) in aqueous solution. The solution for the conditional probability was expressed by expansion in eigenfunctions to the Smoluchowski operator and attained the form of a sum of decaying exponential. The parameters of the model (force constants, exponential decay rates, vibrational amplitudes and the derivatives occurring in Eqs. (83) and (85)) were estimated, based on the results of Odelius *et al.* (*128*) for the normal modes of E_g and T_{2g} symmetry. In this way, the distortions in the complex geometry led not only to a transient ZFS of fluctuating amplitude, they caused also stochastic modulations of the orientation of the ZFS tensor relative to a molecule-fixed frame. Thus, the damped vibrations generated a rhombic transient ZFS of variable amplitude and variable principal direction in the molecule-fixed frame. The molecular tumbling introduced an additional modulation of the orientation of the transient ZFS, which would be negligible in the case of slowly rotating complexes. The classical vibrational approach to the stochastic fluctuations of the transient ZFS was incorporated in two descriptions of the PRE effects. First, the traditional, analytical Solomon–Bloembergen–Morgan (SBM) approach was used. Since the vibrational correlation times were very short, the electron spin relaxation could be described by a Redfield-type theory, in terms of vibrational spectral densities. The spectral densities, $J_n(\omega) = \sum_i J_{n,i}(\omega)$, were obtained by Fourier transformation of the sum of exponentially damped oscillations at normal mode frequencies. At the high-field limit, these spectral densities could be

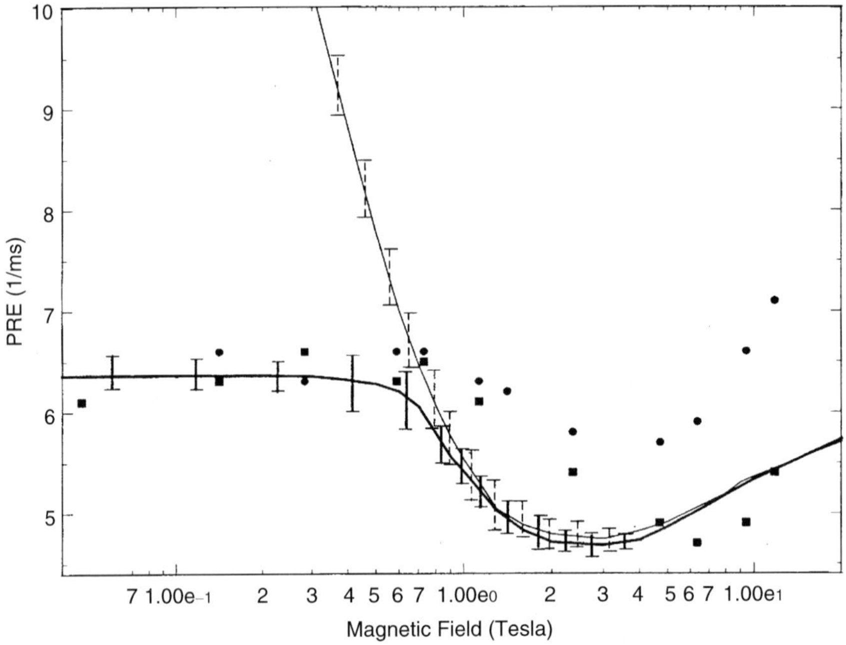

FIG. 19. Experimental NMRD profiles and data calculated using classical vibration model for aqueous Ni(II). Thin line: SBM; thick line: general theory. Reproduced with permission from Kruk, D.; Kowalewski, J. *J. Chem. Phys.* **2002**, *116*, 4079–4086. Copyright 2002 American Institute of Physics.

incorporated into the Bloembergen–Morgan theory for the electron spin relaxation rates. In turn, the electron spin relaxation rates obtained in this way could be used in the modified Solomon–Bloembergen description of the nuclear spin relaxation. Second, a suitably modified version of the general, slow-motion theory was also developed (*161*). The orthonormal basis set, appropriate for the modified slow-motion approach, was formed as a direct product of basis operators for the electron spin system, for the rotational diffusion and for the vibrational motion. At high magnetic field, the MSB approach and the modified general theory agreed well with each other, while the MSB approach did not work correctly when extended to low magnetic field. The results of the general theory showed a remarkable agreement with the experimental PRE data for the $Ni(H_2O)_6^{2+}$ complex in aqueous solution, as can be seen in Fig. 19. The classical vibrational model is, from the physical point of view, more realistic than the commonly used pseudorotational model, but its major problem is that the number of parameters increases drastically.

We wish to stress at this point that, in the approach discussed above (*161*), the vibrational motion has been treated classically, as Markov processes. Typical vibrational energies are two orders of magnitude higher than the typical ZFS energies and, for the static magnetic field less than 10 T, one has also $\omega_V >> \omega_S$, where ω_V is the energy of vibrational transitions in angular frequency units. Thus, from the electron spin point of view, the quantum nature of the bath should be taken into account, and a quantum description of the vibrational motion seems to be a logical and interesting step for obtaining better understanding of the vibrational origin of the electron spin relaxation.

Altshuler and Valiev pointed out for the first time the role of vibrational motions as a possible origin of electron spin relaxation already in the late 1950s (*162*). The authors treated the vibrational bath quantum-mechanically. The approach was based on the second-order perturbation treatment. The Hamiltonian, describing the system consisting of a paramagnetic spin coupled to a bath with vibrational degrees of freedom, was expressed as the sum of a main interaction, H_0, and a perturbation, in accordance with Eq. (5). The unperturbed part of the total Hamiltonian was assumed to contain the coordinates of the electron spin and the spin variables. The interaction between the electron spin system and the bath arose through the stochastically fluctuating normal coordinates of the complex. The transition probabilities between the unperturbed energy levels l and k were calculated as proportional to the appropriate spectral densities $J(\omega_{lk})$. With the assumption of a single-exponential decay of the correlation functions for the perturbing interaction, $H_1(t)$, the transition probabilities were expressed as linear combinations of Lorentzians, taken at transitions frequencies. Different Lorentzians had different weights, but contained the same characteristic time constant of the exponential decay for every vibrational mode. It was further assumed that the mean square amplitudes for all the normal modes were equal. One can argue that the assumed equivalence of all modes is not realistic. Discussing this approach, one should also point out that the perturbation $H_1(t)$ is considered as independent of the spin variables, while the main Hamiltonian, H_0 does not include the normal coordinates.

The quantum alternative for the description of the vibrational degrees of freedom has been commented by Westlund *et al.* (*85*). The comments indicate that, to get a reasonable description of the field-dependent electron spin relaxation caused by the quantum vibrations, one needs to consider the first as well as the second order coupling between the spin and the vibrational modes in the ZFS interaction, and to take into account the lifetime of a vibrational state, T_{1V}, as well as the time constant, T_{2V}, associated with a width of vibrational transitions. A model of nuclear spin relaxation, including the electron spin subsystem coupled to a quantum vibrational bath, has been proposed (*163*). The contributions of the T_{2V} and T_{1V} vibrational relaxation (associated with the linear and the quadratic term in the Taylor expansion of the ZFS tensor, respectively) to the electron spin relaxation was considered. The description of the electron spin dynamics was included in the calculations of the PRE by the SBM approach, as well as in the framework of the general slow-motion theory, with appropriate modifications. The theoretical predictions were compared once again with the experimental PRE values for the $Ni(H_2O)_6^{2+}$ complex in aqueous solution. This work can be treated as a quantum-mechanical counterpart of the classical approach presented in the paper by Kruk and Kowalewski (*161*).

IX. Concluding Remarks

The last decade has witnessed a rapid development of both theoretical ideas and NMR relaxation measurements (high-resolution as well as NMRD) for a variety of paramagnetic systems in solutions. It is our hope that this review will stimulate further work in the field. The theoretical developments high-lighted in this review are centered around the progress of our understanding of various aspects of electron spin relaxation processes in multilevel systems and their connection to the nuclear spin relaxation and its field dependence. Most of the work is based on relatively simple assumptions concerning the dynamics of the coupling between the electron spins and the remaining degrees of freedom (the rotational diffusion and pseudo-rotation models), but the ideas about the role of vibrational motions are also making an interesting progress, summarized in Section VIII of this review.

For the inner-sphere PRE, the basic theoretical understanding has reached a certain maturity and high-level calculations, such as those described in Section IV, are now possible for a variety of situations. Conceptually and computationally simpler theoretical tools, developed recently for slowly-rotating systems, allow reasonably successful and efficient interpretation of experimental data for macromolecular solutions along the avenues summarized in Section V. Low-molecular weight, fast-rotating complexes remain a more difficult challenge, because of the strong coupling between electron spin and rotational degrees of freedom. Here, the development described in Section VI may prove to be fruitful.

The outer-sphere relaxation enhancement is another challenging field where further progress is needed. In addition to all the problems met for the inner-sphere cases, one has here to deal with the translational degrees of freedom and with the effects intermolecular forces have upon them. Several important developments were presented during the recent years, as described in Section VII, but much remains to be done.

A general trend which could be noticed over the last few years and which may be expected to develop further in the near future involves a closer coupling between the use of general tools of computational chemistry (*ab initio* and semi-empirical quantum chemistry, statistical-mechanical simulations) and relaxation theory. When applied to model systems, the computational chemistry methods have the potential of providing new insights on how to develop theoretical models, as well as of yielding estimates of the parameters occurring in the models.

Acknowledgments

The work on theory of relaxation in paramagnetic systems performed at Stockholm University has been supported by the Swedish Natural Science Research Council, Swedish Research Council and the Wenner-Gren Foundations. We wish to acknowledge the courtesy of the publishers who permitted reproductions of the figures published in their journals.

References

1. Bertini, I.; Luchinat, C.; Parigi, G. "*Solution NMR of Paramagnetic Molecules*"; Elsevier: Amsterdam, **2001**, pp. 1–376.
2. Banci, L.; Bertini, I.; Luchinat, C. "*Nuclear and electron relaxation. The magnetic nucleus-unpaired electron coupling in solution*"; VCH: Weinheim, **1991**, pp. 1–208.
3. Bertini, I.; Luchinat, C.; Parigi, G. *Progr. NMR Spectrosc.* **2002**, *40*, 249–273.
4. Krause, W. Ed. "*Contrast Agents I. Magnetic Resonance Imaging*"; Springer: Berlin, 2002, pp. 1–249.
5. Toth, E. Merbach, A. E. Eds. "*The Chemistry of Contrast Agents in Medical Magnetic Resonance Imaging*"; Wiley & Sons. Chichester, **2001**, pp. 1–471.
6. Hausser, K. H.; Brunner, H. "*Encyclopedia of Nuclear Magnetic Resonance*"; Eds. Grant, D. M.; Harris, R. K.; Wiley: Chichester, **1996**, pp. 1864–1869.
7. Atkins, P. W.; Friedman, R. S. "*Molecular Quantum Mechanics*"; 3rd edn; Oxford University Press: Oxford, **1997**, pp. 1–545.
8. Abragam, A.; Bleaney, B. "*Electron Paramagnetic Resonance of Transition Metal Ions*"; Oxford University Press: Oxford, **1970**, pp. 133–216.
9. Swift, T. J.; Connick, R. E. *J. Chem. Phys.* **1962**, *37*, 307–320.
10. Luz, Z.; Meiboom, S. *J. Chem. Phys.* **1964**, *40*, 2686–2692.
11. Kowalewski, J.; Nordenskiöld, L.; Benetis, N.; Westlund, P. O. *Progr. NMR Spectrosc.* **1985**, *17*, 141–185.
12. Kowalewski, J. "*Encyclopedia of Nuclear Magnetic Resonance*"; Eds. Grant, D. M.; Harris, R. K.; Wiley: Chichester, **1996**, pp. 3456–3462.
13. Westlund, P. O. "*Dynamics of Solutions and Fluid Mixtures by NMR*"; Ed. Delpuech, J.-J; Wiley: Chichester, **1995**, pp. 173–229.
14. Sharp, R. R.; Lohr, L.; Miller, J. *Prog. NMR Spectrosc.* **2001**, *38*, 115–158.
15. Solomon, I. *Phys. Rev.* **1955**, *99*, 559–565.
16. Brink, D. M.; Satchler, G. R. "*Angular Momentum*"; Clarendon Press: Oxford, **1979**, pp. 1–170.
17. Redfield, A. G. *Adv. Magn. Reson.* **1965**, *1*, 1–32.
18. Abragam, A. "*The Principles of Nuclear Magnetism*"; Oxford University Press: Oxford, **1961**, pp. 264–353.
19. Slichter, C. P. "*Principles of Magnetic Resonance*"; Springer: Berlin, **1989**, pp. 145–218.
20. Goldman, M. *J. Magn. Reson.* **2001**, *149*, 160–187.
21. Solomon, I.; Bloembergen, N. *J. Chem. Phys.* **1956**, *25*, 261–266.
22. Bloembergen, N. *J. Chem. Phys.* **1957**, *27*, 572–573.
23. Connick, R. E.; Fiat, D. *J. Chem. Phys.* **1966**, *44*, 4103–4107.
24. Reuben, J.; Reed, G. H.; Cohn, M. *J. Chem. Phys.* **1970**, *52*, 1617–1617.
25. Gueron, M. *J. Magn. Reson.* **1975**, *19*, 58–66.
26. Benetis, N.; Kowalewski, J.; Nordenskiöld, L.; Wennerström, H.; Westlund, P. O. *Mol. Phys.* **1983**, *48*, 329–346.
27. Bloembergen, N.; Morgan, L. O. *J. Chem. Phys.* **1961**, *34*, 842–850.
28. Powell, D. H.; Merbach, A. E.; Gonzalez, G.; Brücher, E.; Micskei, K.; Ottaviani, M. F.; Köhler, K.; Von Zelewsky, A.; Grinberg, O. Y.; Lebedev, Y. S. *Helv .Chim. Acta* **1993**, *76*, 2129.
29. Caravan, P.; Toth, E.; Rockenbauer, A.; Merbach, A. E. *J. Am. Chem. Soc.* **1999**, *121*, 10403–10409.
30. Seibig, S.; Toth, E.; Merbach, A. E. *J. Am. Chem. Soc.* **2000**, *122*, 5822–5830.
31. Sternlicht, H. *J. Chem. Phys.* **1965**, *42*, 2250–2251.
32. Bertini, I.; Luchinat, C.; Vasavada, K. V. *J. Magn. Reson.* **1990**, *89*, 243–254.
33. Toth, E.; Helm, L.; Kellar, K. E.; Merbach, A. E. *Chem. Eur. J.* **1999**, *5*, 1202–1211.
34. Dunand, F. A.; Toth, E.; Hollister, R.; Merbach, A. E. *J. Biol. Inorg. Chem.* **2001**, *6*, 247–255.
35. Dunand, F. A.; Borel, A.; Merbach, A. E. *J. Am. Chem. Soc.* **2002**, *124*, 710–716.
36. Waysbort, D.; Navon, G. *J. Chem. Phys.* **1975**, *62*, 1021–1026.
37. Waysbort, D.; Navon, G. *J. Chem. Phys.* **1978**, *68*, 3704–3707.
38. Gottlieb, H. P. W.; Barfield, M.; Doddrell, D. M. *J. Chem. Phys.* **1977**, *67*, 3785–3794.

39. Kowalewski, J.; Laaksonen, A.; Nordenskiöld, L.; Blomberg, M. *J. Chem. Phys.* **1981**, *74*, 2927–2930.
40. Nordenskiöld, L.; Laaksonen, A.; Kowalewski, J. *J. Am. Chem. Soc.* **1982**, *104*, 379–382.
41. Sahoo, N.; Das, T. P. *J. Chem. Phys.* **1989**, *91*, 7740–7748.
42. Sahoo, N.; Das, T. P. *J. Chem. Phys.* **1990**, *93*, 1200–1208.
43. Lata, K. R.; Sahoo, N.; Das, T. P. *J .Chem. Phys.* **1991**, *94*, 3715–3721.
44. Wilkens, S. J.; Xia, B.; Volkman, B. F.; Weinhold, F.; Markley, J. L.; Westler, W. M. *J. Phys. Chem.B* **1998**, *102*, 8300–8305.
45. Mispelter, J.; Momenteau, M.; Lhoste, J.-M. "*Biological Magnetic Resonance: NMR of Paramagnetic Molecules*"; Eds. Berliner, L. J.; Reuben, J.; Plenum Press: New York, **1993**, pp. 299–355.
46. Bertini, I.; Donaire, A.; Luchinat, C.; Rosato, A. *Proteins Struct. Funct. Genet.* **1997**, *29*, 348–358.
47. Ma, L.; Jørgensen, A. M. M.; Sorensen, G. O.; Ulstrup, J.; Led, J. J. *J. Am. Chem. Soc.* **2000**, *122*, 9473–9485.
48. Jeener, J. *Adv. Magn. Reson.* **1982**, *10*, 1–51.
49. Ernst, R. R.; Bodenhausen, G.; Wokaun, A. "*Principles of Nuclear Magnetic Resonance in one and two dimensions*"; Oxford University Press: London, **1987**, pp. 19–44.
50. Werbelow, L. G.; Grant, D. M. *Adv. Magn. Reson.* **1977**, *9*, 189–299.
51. Goldman, M. *J. Magn. Reson.* **1984**, *60*, 437–452.
52. Vega, A. J.; Fiat, D. *Mol. Phys.* **1976**, *31*, 347–355.
53. Bertini, I.; Capozzi, F.; Luchinat, C.; Nicastro, G.; Xia, Z. *J. Phys. Chem.* **1993**, *101*, 198–201.
54. Roch, A.; Muller, R. N.; Gillis, P. *J. Chem. Phys.* **1999**, *110*, 5403–5411.
55. Gillis, P.; Roch, A.; Brooks, R. A. *J. Magn. Reson.* **1999**, *137*, 402–407.
56. Bertini, I.; Kowalewski, J.; Luchinat, C.; Parigi, G. *J. Magn. Reson.* **2001**, *152*, 103–108.
57. Werbelow, L.; Thevand, A. *J. Magn. Reson. Ser. A* **1993**, *101*, 317–319.
58. Bertini, I.; Luchinat, C.; Tarchi, D. *Chem. Phys. Lett.* **1993**, *203*, 445–449.
59. Bertini, I.; Luchinat, C.; Piccioli, M.; Tarchi, D. *Concepts Magn. Reson.* **1994**, *6*, 307–335.
60. Qin, J.; Delaglio, F.; La Mar, G. N.; Bax, A. *J. Magn. Reson. Ser. B* **1993**, *102*, 332–336.
61. Wimperis, S.; Bodenhausen, G. *Mol. Phys.* **1989**, *66*, 897–919.
62. Mäler, L.; Mulder, F. A. A.; Kowalewski, J. *J. Magn. Reson. Ser. A* **1995**, *117*, 220–227.
63. Bertini, I.; Luchinat, C.; Rosato, A. *Chem. Phys. Lett.* **1996**, *250*, 495–504. Erratum: (1997) 265, 677.
64. Ghose, R.; Prestegard, J. H. *J. Magn. Reson.* **1997**, *128*, 138–143.
65. Desvaux, H.; Gochin, M. *Mol. Phys.* **1999**, *96*, 1317–1333.
66. Werbelow, L. "*Encyclopedia of Nuclear Magnetic Resonance*"; Eds. Grant, D. M.; Harris, R. K.; John Wiley & Sons: Chichester, **1996**, pp. 1776–1783.
67. Brüschweiler, R. *J. Chem. Phys.* **1996**, *105*, 6164–6167.
68. Desvaux, H.; 2003. Personal Communication.
69. Bertini, I.; Luchinat, C.; Parigi, G. *Concepts Magn. Reson.* **2002**, *14*, 259–286.
70. Freed, J. H.; Bruno, G. V.; Polnaszek, C. F. *J. Phys. Chem.* **1971**, *75*, 3385–3399.
71. Freed, J. H.; Bruno, G. V.; Polnaszek, C. *J. Chem. Phys.* **1971**, *55*, 5270–5281.
72. Freed, J. H. "*Spin-Labelling: Theory and Applications*"; Ed. Berliner, L. J.; Academic Press: New York, **1976**, pp. 53–132.
73. Lynden-Bell, R. M. *Mol. Phys.* **1971**, *22*, 837–851.
74. Kubo, R. *J. Math. Phys.* **1963**, *4*, 174–183.
75. Hwang, L. P.; Ju, C.-Y. *J. Chem. Phys.* **1985**, *83*, 3775–3789.
76. Wang, P. L.; Lee, J.-H.; Huang, S.-M.; Hwang, L. P. *J. Magn. Reson.* **1987**, *73*, 277–286.
77. Benetis, N.; Kowalewski, J.; Nordenskiöld, L.; Wennerström, H.; Westlund, P. O. *Mol. Phys.* **1983**, *50*, 515–530.
78. Benetis, N.; Kowalewski, J.; Nordenskiöld, L.; Wennerström, H.; Westlund, P. O. *J. Magn. Reson.* **1984**, *58*, 261–281.
79. Westlund, P. O.; Wennerström, H.; Nordenskiöld, L.; Kowalewski, J.; Benetis, N. *J. Magn. Reson.* **1984**, *59*, 91–109.
80. Griffith, J. S. "*The Theory of Transition-Metal Ions*"; Cambridge University Press: Cambridge, **1961**, pp. 320–377.

81. Langhoff, S. R.; Kern, C. W. "*Applications of Electronic Structure Threory*"; Ed. Schaefer III, H. F.; Plenum Press: New York, **1977**, pp. 381–437.
82. Watson, R. E.; Blume, M. *Phys. Rev.* **1965**, *139A*, 1209–1210.
83. Lanczos, C. *J. Res. Nat. Bureau Standards* **1952**, *49*, 33–53.
84. Benetis, N.; Kowalewski, J. *J. Magn. Reson.* **1985**, *65*, 13–33.
85. Westlund, P. O.; Benetis, N.; Wennerström, H. *Mol. Phys.* **1987**, *61*, 177–194.
86. Rubinstein, M.; Baram, A.; Luz, Z. *Mol. Phys.* **1971**, *20*, 67–80.
87. Kowalewski, J.; Larsson, T.; Westlund, P. O. *J. Magn. Reson.* **1987**, *74*, 56–65.
88. Westlund, P. O.; Larsson, T. P.; Teleman, O. *Mol. Phys.* **1993**, *78*, 1365–1384.
89. Svoboda, J.; Nilsson, T.; Kowalewski, J.; Westlund, P. O.; Larsson, P. T. *J. Magn. Reson. Ser. A* **1996**, *121*, 108–113.
90. Friedman, H. L.; Holz, M.; Hertz, H. G. *J. Chem. Phys.* **1979**, *70*, 3369–3383.
91. Westlund, P. O.; Larsson, P. T. *Acta Chem. Scand.* **1991**, *45*, 11–18.
92. Larsson, T.; Westlund, P. O.; Kowalewski, J.; Koenig, S. H. *J. Chem. Phys.* **1994**, *101*, 1116–1128.
93. Nilsson, T.; Svoboda, J.; Westlund, P. O.; Kowalewski, J. *J. Chem. Phys.* **1998**, *109*, 6364–6375.
94. Nilsson, T.; Kowalewski, J. *J. Magn. Reson.* **2000**, *146*, 345–358.
95. Nilsson, T.; Kowalewski, J. *Mol. Phys.* **2000**, *98*, 1617–1638. Erratum: *Mol. Phys.* 2001, 99, 369.
96. Kruk, D.; Nilsson, T.; Kowalewski, J. *Phys. Chem. Chem. Phys.* **2001**, *3*, 4907–4917.
97. Nilsson, T.; Parigi, G.; Kowalewski, J. *J. Phys. Chem.* **2002**, *106*, 4476–4488.
98. Kowalewski, J.; Luchinat, C.; Nilsson, T.; Parigi, G. *J. Phys. Chem. A* **2002**, *106*, 7376–7382.
99. Halle, B.; Wennerström, H. *J. Magn. Reson.* **1981**, *44*, 89–100.
100. Westlund, P. O. *Mol. Phys.* **1995**, *85*, 1165–1178.
101. Strandberg, E.; Westlund, P. O. *J. Magn. Reson. Ser. A* **1996**, *122*, 179–191.
102. Lindner, U. *Ann. Phys.(Leipzig)* **1965**, *16*, 319–335.
103. Kubo, R.; Tomita, K. *J. Phys. Soc. Japan* **1954**, *9*, 888–919.
104. Bertini, I.; Luchinat, C.; Mancini, M.; Spina, G. "*Magneto-structural correlations in exchange-coupled systems*"; Eds. Gatteschi, D.; Kahn, O.; Willett, R. D.; Reidel Publishing Company: Dordrecht, **1985**, pp. 421–461.
105. Bertini, I.; Luchinat, C.; Mancini, M.; Spina, G. *J. Magn. Reson.* **1984**, *59*, 213–222.
106. Fukui, H.; Miura, K.; Matsuda, H. *J. Magn. Reson.* **1990**, *88*, 311–319.
107. Bertini, I.; Galas, O.; Luchinat, C.; Parigi, G. *J. Magn. Reson. Ser. A* **1995**, *113*, 151–158.
108. Bertini, I.; Galas, O.; Luchinat, C.; Parigi, G.; Spina, G. *J. Magn. Reson.* **1998**, *130*, 33–44.
109. Sharp, R. R. *J. Chem. Phys.* **1990**, *93*, 6921–6928.
110. Sharp, R. R. *J. Magn. Reson.* **1992**, *100*, 491–516.
111. Sharp, R. R. *J. Chem. Phys.* **1993**, *98*, 912–921.
112. Sharp, R. R. *J. Chem. Phys.* **1993**, *98*, 2507–2515.
113. Sharp, R. R. *J. Chem. Phys.* **1993**, *98*, 6092–6101.
114. Bovet, J.-M.; Sharp, R. R. *J. Chem. Phys.* **1993**, *99*, 18–26.
115. Abernathy, S. M.; Miller, J. C.; Lohr, L. L.; Sharp, R. R. *J. Chem. Phys.* **1998**, *109*, 4035–4046.
116. Miller, J. C.; Abernathy, S. M.; Lohr, L. L.; Sharp, R. R. *J. Phys. Chem. A* **2000**, *104*, 9481–9488.
117. Strandberg, E.; Westlund, P. O. *J. Magn. Reson.* **1999**, *137*, 333–344.
118. Westlund, P. O. *J. Chem. Phys.* **1998**, *108*, 4945–4953.
119. Bertini, I.; Kowalewski, J.; Luchinat, C.; Nilsson, T.; Parigi, G. *J. Chem. Phys.* **1999**, *111*, 5795–5807.
120. Rast, S.; Fries, P. H.; Belorizky, E. *J .Chem. Phys.* **2000**, *113*, 8724–8735.
121. Rast, S.; Borel, A.; Helm, L.; Belorizky, E.; Fries, P. H.; Merbach, A. E. *J. Am. Chem. Soc.* **2001**, *123*, 2637–2644.
122. Borel, A.; Yerly, F.; Helm, L.; Merbach, A. E. *J. Am. Chem. Soc.* **2002**, *124*, 2042–2048.
123. Kruk, D.; Kowalewski, J. *J. Magn. Reson.* **2003**, *162*, 229–240.
124. Kruk, D.; Kowalewski, J. *J. Biol. Inorg. Chem.* **2003**, *8*, 512–518.
125. Kruk, D.; Kowalewski, J. *Mol. Phys.* **2003**, *101*, 2861–2874.
126. Sharp, R. R.; Lohr, L. L. *J. Chem. Phys.* **2001**, *115*, 5005–5014.

127. Sharp, R. R. *J. Magn. Reson.* **2002**, *154*, 269–279.
128. Odelius, M.; Ribbing, C.; Kowalewski, J. *J. Chem. Phys.* **1995**, *103*, 1800–1811.
129. Odelius, M.; Ribbing, C.; Kowalewski, J. *J. Chem. Phys.* **1996**, *104*, 3181–3188.
130. Abernathy, S. M.; Sharp, R. R. *J. Chem. Phys.* **1997**, *106*, 9032–9043.
131. Sharp, R. R.; Abernathy, S. M.; Lohr, L. L. *J. Chem. Phys.* **1997**, *107*, 7620–7629.
132. Miller, J. C.; Abernathy, S. M.; Sharp, R. R. *J. Phys. Chem.A* **2000**, *104*, 4839–4845.
133. Miller, J. C.; Sharp, R. R. *J. Phys. Chem.A* **2000**, *104*, 4889–4895.
134. Rast, S.; Fries, P. H.; Belorizky, E.; Borel, A.; Helm, L.; Merbach, A. E. *J. Chem. Phys.* **2001**, *115*, 7554–7563.
135. Itzkowitz, M. S. *J. Chem. Phys.* **1967**, *46*, 3048–3056.
136. Torrey, H. C. *Phys. Rev.* **1953**, *92*, 962–969.
137. Pfeifer, H. *Ann. Phys.(Leipzig)* **1961**, *8*, 1–8.
138. Hwang, L. P.; Freed, J. H. *J. Chem. Phys.* **1975**, *63*, 4017–4025.
139. Ayant, Y.; Belorizky, E.; Alizon, J.; Gallice, J. *J. Phys.(Paris)* **1975**, *36*, 991–1004.
140. Ayant, Y.; Belorizky, E.; Fries, P. H.; Rosset, J. *J. Phys. (Paris)* **1977**, *38*, 325–337.
141. Freed, J. H. *J. Chem. Phys.* **1978**, *68*, 4034–4037.
142. Polnaszek, C. F.; Bryant, R. G. *J. Chem. Phys.* **1984**, *81*, 4038–4045.
143. Bayburt, T.; Sharp, R. R. *J. Chem. Phys.* **1990**, *92*, 5892–5899.
144. Kruk, D.; Nilsson, T.; Kowalewski, J. *Mol. Phys.* **2001**, *99*, 1435–1445.
145. Abernathy, S. M.; Sharp, R. R. *J. Phys. Chem.A* **1997**, *101*, 3692–3698.
146. Botta, M. *Eur. J. Inorg. Chem.* 2000, 399–407.
147. Kruk, D.; Kowalewski, J. *J. Chem. Phys.* **2002**, *117*, 1194–1200.
148. Caravan, P.; Greenfield, M. T.; Li, X.; Sherry, A. D. *Inorg. Chem.* **2001**, *40*, 6580–6587.
149. Borel, A.; Helm, L.; Merbach, A. E. *Chemistry – A European Journal* **2001**, *7*, 600–610.
150. Jeannin, M.; Belorizky, E.; Fries, P. H.; Gorecki, W. *J. Phys. II Fr.* **1993**, *3*, 1511–1533.
151. Fries, P. H.; Belorizky, E.; Bourdin, N.; Cinget, F.; Gagnaire, D.; Gorecki, W.; Jeannin, M.; Vottero, P. *Theochem-J. Mol. Struct.* **1995**, *330*, 335–345.
152. Sacco, A.; Belorizky, E.; Jeannin, M.; Gorecki, W.; Fries, P. H. *J. Phys.II Fr.* **1997**, *7*, 1299–1322.
153. Vigouroux, C.; Bardet, M.; Belorizky, E.; Fries, P. H.; Guillermo, A. *Chem. Phys. Lett.* **1998**, *286*, 93–100.
154. Favier, A.; Chapon, D.; Fries, P. H.; Rast, S.; Belorizky, E. *Chem. Phys. Lett.* **2000**, *320*, 49–52.
155. Belorizky, E.; Fries, P. H.; Rast, S. *Compt. Rend. Acad. Sci. Ser IIC* **2001**, *4*, 825–832.
156. Fries, P. H.; Richardi, J.; Rast, S.; Belorizky, E. *Pure Appl. Chem.* **2001**, *73*, 1689–1703.
157. Rast, S.; Belorizky, E.; Fries, P. H.; Travers, J. P. *J. Phys. Chem. B* **2001**, *105*, 1978–1983.
158. Wennerström, H. *Mol. Phys.* **1972**, *24*, 69–80.
159. Bertini, I.; Luchinat, C.; Kowalewski, J. *J. Magn. Reson.* **1985**, *62*, 235–241.
160. Aime, S.; Botta, M.; Fasano, M.; Terreno, E. *Accounts of Chemical Research* **1999**, *32*, 941–949.
161. Kruk, D.; Kowalewski, J. *J. Chem. Phys.* **2002**, *116*, 4079–4086.
162. Al'tshuler, S. A.; Valiev, K. A. *Soviet Physics JETP* **1959**, *35*, 661–668.
163. Kruk, D., Kowalewski, J.; Westlund, P. O. *J. Chem. Phys.* **2004**, *121*, 2215–2227.

^{1}H NMRD PROFILES OF PARAMAGNETIC COMPLEXES AND METALLOPROTEINS

IVANO BERTINI[a,*], CLAUDIO LUCHINAT[b] and GIACOMO PARIGI[b]

[a]CERM and Department of Chemistry, University of Florence, via L. Sacconi, 6, I-50019 Sesto Fiorentino, Italy

[b]CERM and Department of Agricultural Biotechnology, University of Florence, p.le delle Cascine 28, I-50144 Florence, Italy

I. From the NMRD Profile to the Electron Relaxation Mechanism

Relaxometry is a unique tool to obtain information on structural and dynamic features of molecules and on electron relaxation. Direct information on the electron-spin relaxation through EPR measurements may not be easy to obtain at room temperature, especially if the electron spin relaxation is very rapid, because the lineshape is too broad. Even for systems undergoing relatively slow electron spin relaxation at room temperature, it may be difficult to directly measure electron relaxation through EPR because (i) the line is broadening by coupling with other nuclei, and (ii) pulse spectroscopies require very short pulses. Nuclear Magnetic Relaxation Dispersion (NMRD) profiles are obtained by acquiring relaxation rates as a function of the magnetic field. Relaxometry is performed on ^{1}H, ^{2}H, ^{17}O nuclei; in this review we will focus on ^{1}H measurements.

ADVANCES IN INORGANIC CHEMISTRY
VOLUME 57 ISSN 0898-8838 / DOI 10.1016/S0898-8838(05)57003-X

Relaxometric measurements on solutions of paramagnetic solutes are often a convenient way for obtaining the nuclear relaxation rates and the electron relaxation rates, both as functions of the magnetic field. The knowledge of these rates is a precious source of information on the mechanisms responsible for relaxation. In fact, spin relaxation is connected with the (random) atomic and molecular motions which modulate the interactions between the spin magnetic moment and the local fluctuating magnetic fields. The latter are usually generated by the motion of the magnetic moments associated with other nuclei and unpaired electrons, by fluctuations of electric dipoles, etc.

In the case of nuclear spins in a paramagnetic substance, the largest source of nuclear relaxation is often the modulation of the coupling to the electron-spin magnetic dipole (hyperfine coupling, see Eq. (1) of Chapter 2) (*1*). Fluctuations in the hyperfine coupling energy can arise by changes in (a) the metal-nucleus distance r, (b) the $\mathbf{r}$ orientation, and (c) sudden changes in the orientation of the M_z or the M_{xy} components of the electron-spin magnetic moment. Physically, these correspond to (a) conformational changes within the molecule or detachment of a ligand bearing the nucleus of interest, e.g., a coordinated water molecule, (b) reorientation of the complex with respect to the external magnetic field, and (c) longitudinal and transverse electron-spin relaxation. Each of the above processes, characterized by a correlation time called τ_M, τ_R or T_{ie} ($i=1,2$), respectively, may contribute to nuclear relaxation.

As stated in Section II.B of Chapter 2, the actual correlation time for electron-nuclear dipole–dipole relaxation, is dominated by the fastest process among proton exchange, rotation, and electron spin relaxation. It follows that if electron relaxation is the fastest process, the proton correlation time τ_c is given by electron-spin relaxation times T_{ie}, and the field dependence of proton relaxation rates allows us to obtain the electron relaxation times and their field dependence, thus providing information on electron relaxation mechanisms. If motions faster than electron relaxation dominate τ_c, it is only possible to set lower limits for the electron relaxation time, but we learn about some aspects on the dynamics of the system. In the remainder of this section we will deal with systems where electron relaxation determines the correlation time.

A. Dependence of the NMRD Profiles on the Electron Relaxation Parameters

A.1. Systems Without Static ZFS

Let us first examine the ^{1}H NMRD profiles of systems with correlation times $\tau_{ci}=T_{ie}$ assuming the Solomon–Bloembergen–Morgan (SBM) theory (see Section II.B of Chapter 2) (*2–5*) is valid, and in the absence of contact relaxation. We report here the relevant equations for readers' convenience:

$$T_{1M}^{-1} = \frac{2}{15} S(S+1) C_{DD}^2 \left[\frac{\tau_{c2}}{1+(\omega_S - \omega_I)^2 \tau_{c2}^2} + \frac{3\tau_{c1}}{1+\omega_I^2 \tau_{c1}^2} + \frac{6\tau_{c2}}{1+(\omega_S + \omega_I)^2 \tau_{c2}^2} \right] \quad (1)$$

$$T_{2M}^{-1} = \frac{1}{15} S(S+1) C_{DD}^2 \left[4\tau_{c1} + \frac{3\tau_{c1}}{1+\omega_I^2 \tau_{c1}^2} + \frac{\tau_{c2}}{1+(\omega_S - \omega_I)^2 \tau_{c2}^2} + \frac{6\tau_{c2}}{1+\omega_S^2 \tau_{c2}^2} + \frac{6\tau_{c2}}{1+(\omega_S + \omega_I)^2 \tau_{c2}^2} \right] \quad (2)$$

with $C_{DD} = (\mu_0 \gamma_I \gamma_S \hbar / 4\pi r^3)$, where S is the electron-spin quantum number, γ_I and γ_S are the nuclear and electron magnetogyric ratios, r is the interspin distance, and $\tau_{ci}^{-1} = \tau_M^{-1} + \tau_R^{-1} + T_{ie}^{-1}$ $(i = 1, 2)$. The functional form of Eq. (1) immediately suggests that relaxivity can be a useful tool for providing information on r and τ_{ci}. The Lorentzian $\tau/(1 + \omega^2\tau^2)$ terms provide a field dependence of the NMRD profiles called "dispersion". Figure 1 shows the profiles for some values of the parameters and decreasing τ_{ci}. Two dispersions are always apparent in a 7 to 3 ratio, corresponding to the ω_S and ω_I dispersions, respectively, separated by about three orders of magnitude ($\omega_S/\omega_I = 658$ for I being a proton spin). The position of the dispersion is related to the value of the correlation times and moves toward lower fields with increasing τ_{ci} (T_{ie}).

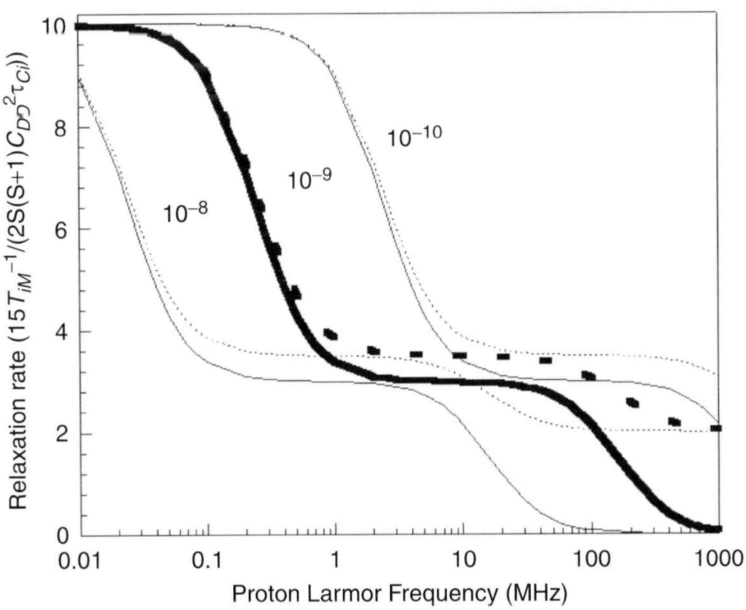

FIG. 1. Longitudinal (solid lines) and transverse (dotted lines) nuclear relaxation rate profiles for different values of the nuclear correlation time ($\tau_c = 10^{-8}, 10^{-9}, 10^{-10}$ s), assumed constant with the field, as predicted by the Solomon equation.

T_{ie} are also expected to be field-dependent. Their field dependence can be described by two parameters: the electron relaxation time at low fields τ_{s0}, and the correlation time for the electron relaxation mechanism τ_v (see Eq. (14) of Chapter 2) (*5*). However, τ_{s0} usually depends on τ_v (see Eq. (52) of Chapter 2). Therefore, it is preferable to select two different parameters for describing the field dependence of electron relaxation. For $S > 1/2$ systems, in case the electron relaxation is due to modulation of a time dependent transient zero-field splitting, Δ_t, (pseudorotational model), the Bloembergen–Morgan equations are obtained (*5,6*):

$$T_{1e}^{-1} = \frac{\Delta_t^2}{25}[4S(S+1)-3]\left[\frac{\tau_v}{1+\tau_v^2\omega_S^2} + \frac{4\tau_v}{1+4\tau_v^2\omega_S^2}\right] \quad (3)$$

$$T_{2e}^{-1} = \frac{\Delta_t^2}{50}[4S(S+1)-3]\left[3\tau_v + \frac{5\tau_v}{1+\tau_v^2\omega_S^2} + \frac{2\tau_v}{1+4\tau_v^2\omega_S^2}\right] \quad (4)$$

and the relevant parameters are Δ_t and τ_v.

Figure 2 shows the nuclear dipolar longitudinal and transverse relaxation rates as a function of the magnetic field for (A) constant Δ_t and increasing τ_v, and (B) increasing Δ_t and constant τ_v. In the former case, T_{ie} decreases with increasing τ_v at low fields and up to the occurrence of the $\omega_S\tau_v$ dispersion. Therefore the low field rate decreases and the proton $\omega_S T_{2e}$ dispersion moves at larger frequencies. A peak appears at high fields because the electron relaxation time increases as soon as $\omega_S\tau_v > 1$, due to the dispersion terms in the equation for electron relaxation ((Eqs. (3) and (4)), and later the proton ω_I dispersion occurs. The peak moves toward lower fields for increasing values of τ_v because both the $\omega_S\tau_v$ and the $\omega_I T_{1e}$ dispersions occur at lower frequency. It can be noted (see Eqs. (3) and (4)) that in the SBM model T_{1e} is equal to T_{2e} and it is constant at low fields, before the beginning of the $\omega_S\tau_v$ dispersion; at high fields the profiles depend only on

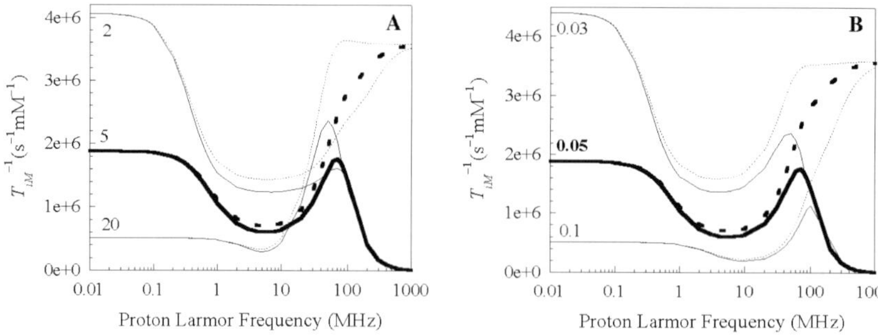

FIG. 2. Field dependence of the dipolar contribution to T_{1M}^{-1} (solid lines) and T_{2M}^{-1} (dotted lines), calculated with the following parameters: $S=5/2$, $\tau_R=3\times10^{-9}$ s, $r=2.8$ Å and (A) constant $\Delta_t=0.05$ cm^{-1} and different $\tau_v=2\times10^{-12}$, 5×10^{-12} and 20×10^{-12} s, and (B) constant $\tau_v=5\times10^{-12}$ s and different $\Delta_t=0.03$, 0.05 and 0.10 cm^{-1}.

T_{1e}, the $\omega_S T_{2e}$ dispersion having already occurred when the $\omega_S \tau_\nu$ dispersion occurs. Actually, the validity of the SBM theory is assured only within the Redfield limit (see Section IV.A of Chapter 2) (7), i.e., in case the energy of the coupling between the spin and the lattice, $E/\hbar$ (in frequency units), whose modulation is responsible for the spin relaxation, is smaller than the inverse of the correlation time, τ_c, for the modulation of the coupling itself, $E/\hbar \ll \tau_c^{-1}$. This determines for $T_{ie} > \tau_v$ (1).

In case τ_ν is kept constant and Δ_t increases (Fig. 2B), low field relaxivity decreases and the $\omega_S T_{2e}$ dispersion moves at larger frequency, as a result of the decreased electron relaxation time.

Longitudinal and transverse nuclear relaxation profiles differ in the high field part. In fact, the equation for the transverse nuclear relaxation rates contains a non-dispersive term, depending only on τ_{c1}. Therefore the transverse relaxation does not go to zero at high fields, as longitudinal relaxation does, but increases because T_{1e} increases (until it increases to the point where it becomes longer than τ_R or τ_M).

In the presence of contact contribution to nuclear relaxation, the NMRD profile results as a sum of the dipolar and contact relaxation rates. The profile of contact relaxation as a function of field is characterized by the presence of only one dispersion (Fig. 3), corresponding to the $\omega_S T_{2e}$ dispersion (Eqs.(5) and (6)), in the hypothesis that $\tau_{ei} = T_{ie}$ (see Section II.B of

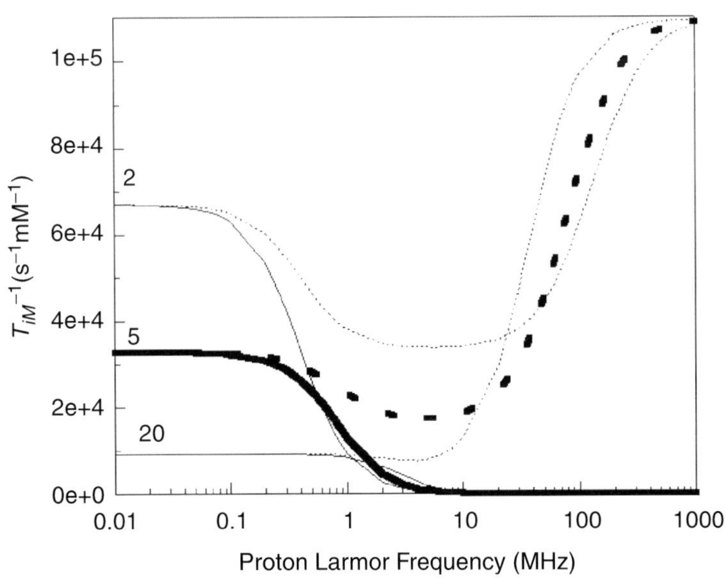

FIG. 3. Field dependence of the contact contribution to T_{1M}^{-1} (solid lines) and T_{2M}^{-1} (dotted lines), calculated with the following parameters: $S = 5/2$, $\tau_M = 2 \times 10^{-9}$ s, $A/h = 0.7$ MHz, $\Delta_t = 0.05$ cm^{-1} and different $\tau_\nu = 2 \times 10^{-12}, 5 \times 10^{-12}$, and 20×10^{-12} s.

Chapter 2):

$$T_{1M}^{-1} = \frac{2}{3}\left(\frac{A_c}{\hbar}\right)^2 S(S+1)\frac{\tau_{e2}}{1+(\omega_S-\omega_I)^2\tau_{e2}^2} \tag{5}$$

$$T_{2M}^{-1} = \frac{1}{3}\left(\frac{A_c}{\hbar}\right)^2 S(S+1)\left(\tau_{e1}+\frac{\tau_{e2}}{1+(\omega_S-\omega_I)^2\tau_{e2}^2}\right) \tag{6}$$

The functional form of the nuclear longitudinal relaxation immediately suggests that the contact contribution can provide the values of the contact coupling constant A_c and of $T_{2e} = \tau_{s0}$, provided that the lifetime, τ_M, is longer than T_{2e}. No information on the field dependence of electron relaxation can be achieved. On the contrary the functional form of transverse nuclear relaxation contains a non-dispersive term, T_{1e}. The latter, as we have seen for the dipolar contribution, increases with increasing the field (Fig. 3), and therefore the nuclear contact transverse relaxation also increases with increasing the field. Its measurement is thus informative on the τ_ν value.

Actually, it was shown that even in the presence of transient ZFS as a result of collisions with solvent molecules in solution, for S > 1 there are many electron longitudinal and transverse relaxation times (*6*). For $S = 3/2$ systems, for instance, there are two electron longitudinal relaxation times and two electron transverse relaxation times, one for the $1/2 \leftrightarrow -1/2$ transition and one for the $1/2 \leftrightarrow 3/2$ and $-1/2 \leftrightarrow -3/2$ transitions. For $S = 5/2$ systems, they are three, one for the $1/2 \leftrightarrow -1/2$ transition, one for the $1/2 \leftrightarrow 3/2$ and $-1/2 \leftrightarrow -3/2$ transitions, and one for the $3/2 \leftrightarrow 5/2$ and $-3/2 \leftrightarrow -5/2$ transitions. Since the information that can be obtained from relaxometry is relative to their averages, which are well approximated by the Bloembergen and Morgan equations (*5*) (see Eqs. (3) and (4)), when dealing with the electron relaxation time affecting the NMRD profiles we should speak in terms of "effective" electron relaxation time. Note that for $S = 1$ systems, only one longitudinal and one transverse relaxation time is defined, corresponding to the $0 \leftrightarrow \pm 1$ transitions.

The Florence NMRD program (*8*) (available at www.postgenomicnmr.net) has been developed to calculate the paramagnetic enhancement to the ^{1}H NMRD profiles due to contact and dipolar nuclear relaxation rate in the slow rotation limit (see Section V.B of Chapter 2). It includes the hyperfine coupling of any rhombicity between electron-spin and metal nuclear-spin, for any metal-nucleus spin quantum number, any electron-spin quantum number and any **g** tensor anisotropy. In case measurements are available at several temperatures, it includes the possibility to consider an Arrhenius relationship for the electron relaxation time, if the latter is field independent,

$$T_{ie} = A\exp(B/T) \tag{7}$$

or for τ_v, if a field dependent electron relaxation time is due to modulation of transient ZFS,

$$\tau_v = A\exp(B/T). \tag{8}$$

The latter expression can also be written as

$$\tau_v = \tau_v^{298}\exp\left(\frac{E_V}{R}\left(\frac{1}{T} - \frac{1}{298.15}\right)\right) \tag{9}$$

where E_v is the activation energy related to τ_v (*9*).

A.2. Systems with Static ZFS

When $S > 1/2$, the effect of spin–orbit coupling is that of producing splitting of the S manifold at zero magnetic field (Fig. 4A). This effect, called static zero-field splitting, removes the degeneracy of the electron spin levels in the absence of magnetic field. In this chapter, when speaking of ZFS we always refer to the effects arising from spin–orbit coupling between the ground and excited states, taken in second order; it critically depends on the symmetry of the ligand environment of the metal ion concerned. When the ZFS is large compared to the Zeeman energy, its presence drastically alters the relaxation equations (see Section II.A.5). It is usually described by the two parameters D and E, representing the axial and rhombic components, respectively.

The SBM theory described in Section II.B of Chapter 2 has been developed without considering the presence of static ZFS terms. The consequences are that (i) the transition probability among the energy levels of the systems are different, and therefore the nuclear relaxation rates are different, when the ZFS energy is larger or equal to the Zeeman energy, and that (ii) the Bloembergen and Morgan equations (*5*) (Eqs. (3) and (4)), which describe the effective electron relaxation times, may not be valid. This determines the fact that the SBM theory for $S > 1/2$ systems is often not applicable. Two cases may arise: the first is that the shape of experimental and theoretical profiles is not in agreement with any value of the parameters, the second is that a good fit is possible, but the values of the parameters, and especially of Δ_t and τ_v, are incorrect (*10*). A closed equation for the effective electron relaxation time in the presence of a large static ZFS is not available, and numerical approaches must be employed. The modified Florence NMRD program (*11,12*) (available at www.postgenomicnmr.net) has been developed to calculate the paramagnetic enhancement of the ^{1}H NMRD profiles in the slow rotation limit and the simultaneous presence of both static and transient ZFS, for any electron-spin quantum number.

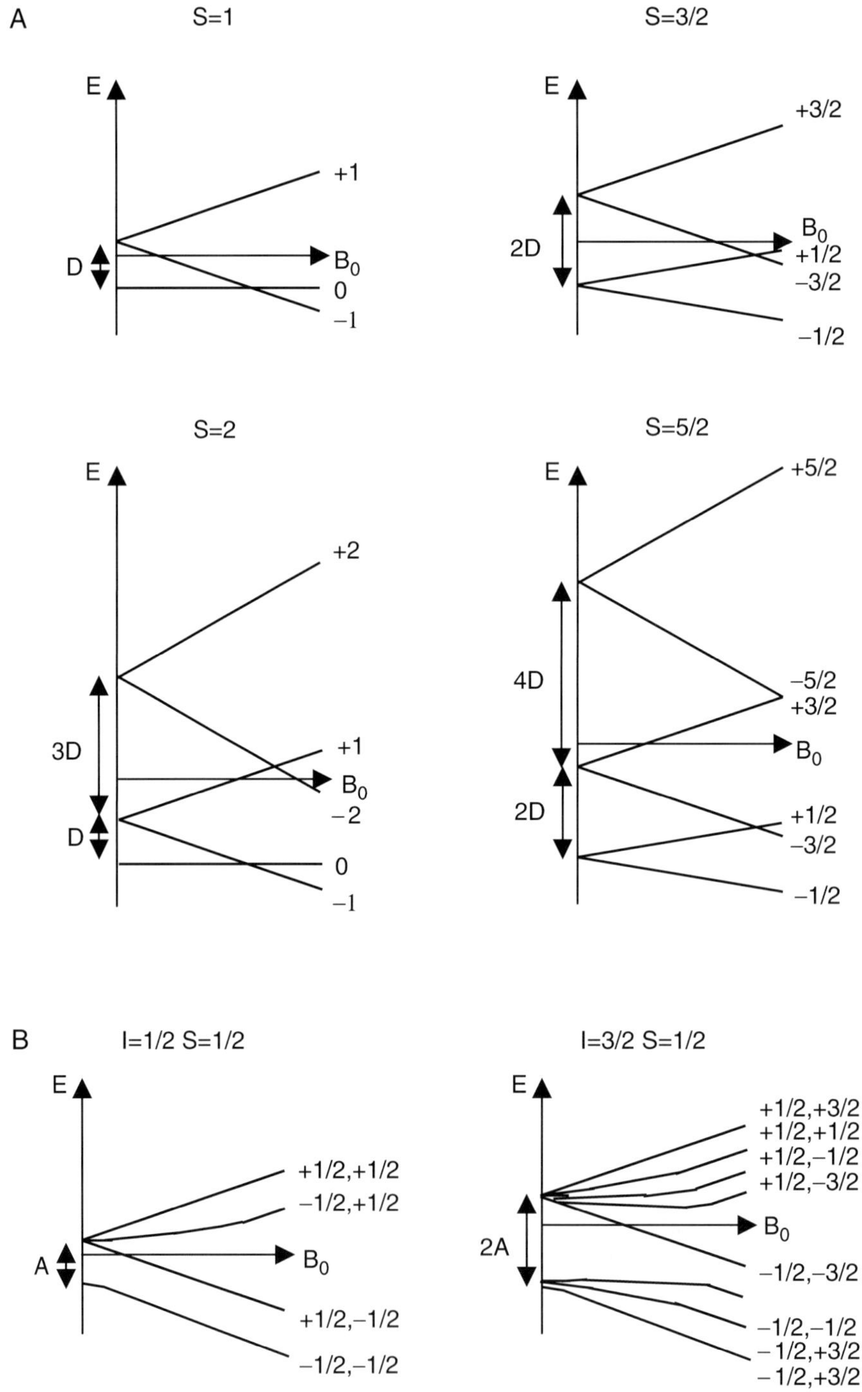

FIG. 4. Effect of (A) axial zero field splitting for the spin systems $S = 1, 3/2, 2,$ and $5/2$ (with $\mathbf{B_0}$ applied along the z direction of the ZFS tensor), and (B) isotropic hyperfine coupling with the metal nucleus for systems with $I = 1/2$, $S = 1/2$ and $I = 3/2$, $S = 1/2$.

B. The Origin of Electron Relaxation in Metal Ions

The most efficient mechanisms producing fluctuating magnetic fields, which induce electron relaxation, are generated by the presence of spin–orbit coupling. Spin–orbit coupling arises from the coupling between the electron-spin magnetic moment and the electronic orbital motion about the charged nucleus. Therefore, it permits exchange of energy between electron spin (Zeeman energy) and atomic motions, such as vibrations or reorientations, i.e., between the spin and the lattice. Spin–orbit coupling can be efficiently modulated by geometrical distortions even in a 3*d* metal ion.

Effective electron relaxation mechanisms for molecules with low-lying energy levels may be of the same type as those operative in the solid state. In solids, electron relaxation is due to the coupling of electrons with vibrational transitions. The lattice is capable of receiving energy from an excited electronic level and dissipating it by creating new phonons. Conversely, phonons can provide the energy for an electronic transition. At room temperature only very few photons have the energy needed for electronic transition between the Zeeman levels. The Raman mechanism (*13,14*) becomes operative when electronic transition energies are too small (between the Zeeman states) or too large (between ground and excited electronic states) with respect to the phonon energy. The Raman mechanism involves an interaction within two phonons whose difference in energy equals the Zeeman energy-splitting of the spin states. The Orbach mechanism (*14*) involves a spin-flip accompanied by a jump from the ground to an excited electronic state. It requires low-lying electronic energy levels, the energy of the most abundant phonons at 300 K ranging between 50 and 1000 cm^{-1}.

The same approach has been developed for systems in solutions (*15,16*), and was found relevant in some cases. The electron relaxation time is described in the cases of Orbach-type mechanism by the equation (*17*)

$$T_{1e}^{-1} = \frac{C}{\delta^2 \tau}[\exp(\hbar\delta/kT) - 1]^{-1} \qquad (10)$$

where $\hbar\delta$ is the excitation energy to the first excited state and τ is a mean correlation time for the intermolecular fluctuations, or for collisions with solvent molecules, usually in the range 10^{-11} to 10^{-12} s. No field dependence in the electron relaxation rate is accounted for by this relaxation mechanism. However, there is no satisfactory theory to describe Orbach relaxation at room temperature, when countless vibrational transitions become active. Equation (10) indicates that the electron relaxation time must decrease with increasing temperature. Such a behavior is in contrast with what is experimentally observed in many systems. A model was thus introduced (called pseudorotational model) where random fluctuations, caused by Brownian motion, in the vibrational normal coordinates of the complex are responsible for the modulation of crystal field parameters. The electron relaxation rate is thus proportional to a correlation time τ_v, which decreases with increasing temperature.

In general, fluctuations in any electron Hamiltonian terms, due to Brownian motions, can induce relaxation. Fluctuations of anisotropic **g**, ZFS, or anisotropic **A** tensors may provide relaxation mechanisms. The **g** tensor is in fact introduced to describe the interaction energy between the magnetic field and the electron spin, in the presence of spin–orbit coupling, which also causes static ZFS in $S > 1/2$ systems. The **A** tensor describes the hyperfine coupling of the unpaired electron(s) with the metal nuclear-spin. Stochastic fluctuations can arise from molecular reorientation (with correlation time τ_R) and/or from molecular distortions, e.g., due to collisions (with correlation time τ_v) (*18*), the latter mechanism being usually dominant. The electron relaxation time is obtained (*15*) as a function of the squared anisotropies of the tensors and of the correlation time, with a field dependence due to the term $\tau_v/(1+\omega_S^2\tau_v^2)$.

High symmetry systems without ZFS, **A** or **g** anisotropies can still relax through the same mechanism as long as $S > 1/2$. In fact, solvent molecules can collide with solute molecules and the system may instantaneously be deformed, causing a transient ZFS. The fluctuating (or transient) component of the ZFS is modulated by the distortional motion and is often the predominant mechanism for electron-spin relaxation. This is the so-called pseudo-rotational model, described in detail in Chapter 2 in all its variants. The correlation time τ_v, characteristic of the distortional motion, is a measure of the fluctuation in the ligand framework due to solvent collisions. In macromolecular systems τ_v can be very different in the presence of different paramagnetic metal ions. In fact, it ranges from 10^{-12} s for some HS iron(III)-proteins, to 20×10^{-12} s for gadolinium(III)-proteins, to 50×10^{-12} s for manganese(II)-proteins. This accounts for the different coordination environments of the different metal ions. It has been shown for the case of hexaaquanickel(II) that a transient ZFS is generated by damped vibrational motions of E_g and T_{2g} symmetry (*19*).

Finally, a further relaxation mechanism is the so-called spin-rotation (*20*). When a molecule rotates, the electrons do not rigidly follow the movement of the nuclear framework, especially in the presence of sudden changes in reorientational motion after collisions. Therefore, molecular rotation provides instantaneous formation of an electric dipole, due to the imbalance of rotating charges, which can interact with both the electronic and nuclear spins. Fluctuations in this interaction may cause electron (as well as proton) relaxation. Since the corresponding correlation time is usually much longer than ω_S^{-1}, it results in a field independent electron relaxation rate. This mechanism is expected to be important only in small complexes, T_{1e} being directly proportional to the reorientational time of the complex, and thus very long relaxation times are accounted for in macromolecules (*20,21*).

The Orbach-type process as well as the collisional process (inducing either ZFS, **g** anisotropy or hyperfine coupling modulation) are mechanisms that can provide electron relaxation independently on reorientation. Electron relaxation is certainly not modulated by reorientational motions when the latter have a mean correlation time much larger than the electron relaxation time itself.

Actually, the least efficient mechanisms are the reorientational mechanisms and spin-rotation, which are expected to account for T_{1e}^{-1} values up to 10^9 s^{-1} and are only detected in the absence of other mechanisms. Modulation of the transient ZFS through solvent bombardment can account for T_{1e}^{-1} values ranging from 10^8 to 10^{11} s^{-1}, depending on the extent of the instantaneous geometrical distortions and spin–orbit coupling. The extent of the splitting of an S manifold depends on the spin–orbit coupling constant, which increases with the atomic number and thus from left to right and from top to bottom in the periodic table. Finally, Raman-type and Orbach-type processes may account for relaxation rates higher than 10^{11} and 10^{12} s^{-1}, respectively. It must be however pointed out that detailed descriptions of room temperature mechanisms for fast electron relaxation are not available.

C. Electron Relaxation for the Different Metal Ions

The electron relaxation time depends on the molecular geometry and on the nature of possible chelating group, besides the nature of the metal ion. In fact it depends on the availability of low-lying excited states, which make spin–orbit coupling more efficient. The T_{1e} values are longer than 10^{-11} s when the excited levels are far from the ground state in energy. The T_{1e} values shorter than 10^{-11} s indicate that there are excited levels at energies comparable with kT.

There are complexes which display a field dependence of the electron relaxation rate, in others the field dependence is not evident. The availability of this information, which comes from ^{1}H NMRD experiments, may thus permit to obtain indications on the most efficient electron relaxation mechanisms in the different systems.

In symmetric complexes where the excited electronic levels are high in energy, for $S > 1/2$, the most efficient electron relaxation mechanism seems to be due to the modulation of transient ZFS with a correlation time independent of τ_R. As already seen, this time is ascribed to the correlation time for the collisions of the solvent molecules, responsible for the deformation of the coordination polyhedron causing transient ZFS. In complexes where a static ZFS is also present, modulation of this ZFS with a correlation time related to τ_R is another possible electron relaxation mechanism.

Metal ions like copper(II), oxovanadium(IV), titanium(III), silver(II) have electron-spin quantum number equal to 1/2, like radicals. Therefore, they have no ZFS. These systems can be divided into three classes, according to the spin–orbit interaction of the paramagnetic center:

(1) Complexes in which the ground state is well-separated from the excited states, so that the spin–orbit coupling is not important. These complexes usually have very long electron relaxation times ($\approx 10^{-7}$ s).
(2) Complexes in which the spin–orbit coupling gives rise to anisotropy in the hyperfine coupling to the metal nucleus and in the **g** tensors.

The electron relaxes through modulation of the **A** and **g** anisotropy. Typical examples are copper(II), oxovanadium(IV) and silver(II) aqua ions. The electronic relaxation times are relatively long (10^{-8}–10^{-10} s at room temperature) and the hyperfine coupling with the metal nuclear spin is usually present. No field dependence of the electron relaxation time is usually evident up to 100 MHz.

(3) Complexes in which there are low-lying excited states, hence allowing the Orbach mechanism to operate efficiently. This causes the electron relaxation time to be faster. Typical examples are titanium(III), with $T_{1e} \approx 10^{-11}$ s, and low spin iron(III), with $T_{1e} \approx 10^{-12}$ s.

In $S > 1/2$ systems, the electronic configuration of the paramagnetic center can dramatically alter the relaxation properties. All systems which have an orbitally non-degenerate ground state (with so-called A symmetry) have the following characteristics: (i) nearly isotropic **g** tensor, with a value similar to that of free electrons, (ii) isotropic **A** tensor, (iii) small transient ZFS. Typical examples are manganese(II), high spin iron(III), chromium(III), nickel(II) in O_h symmetry, gadolinium(III) aqua ions, with $T_{1e} \approx 10^{-9}$–10^{-11} s at low field. The electron relaxation is usually field dependent and the main mechanism for electron relaxation is the modulation of transient ZFS caused by collisions with solvent molecules. Small static ZFS have been estimated for several manganese(II) and gadolinium(III) proteins, and somewhat larger ones for iron(III) compounds. In such low symmetry systems, it is reasonable to expect the magnitude of transient ZFS to be related to that of the static ZFS, as the former can be seen as a perturbation of the latter. As a consequence, systems with increasing static ZFS experience faster electron relaxation rates. Modulation of static ZFS by rotation could be an additional mechanism for relaxation, which may coexist with the collisional mechanism.

Systems with nearly doubly degenerate or triply degenerate (with E or T character in the idealized high symmetry) ground states exhibit an efficient spin–orbit coupling, due to the low-lying energy levels, which allow the Orbach-type mechanism to operate. The electronic relaxation times of these systems are usually short, $T_{1e} \approx 10^{-12}$ s. Typical examples are high-spin cobalt(II), high-spin iron(II), lanthanides(III) except gadolinium(III).

In this section, we describe the behavior of the different metal ions. It may be useful to summarize first some terminologies used to describe the electronic structure. The ground state of the free-ion configuration depends on the maximum values of $L = \sum m_l$ and $S = \sum m_s$. It is indicated by ${}^{2S+1}X$, where X = S, P, D, F depending on the L value equal to 0, 1, 2, 3, respectively. Therefore, a d^1 metal ion has $L = 2$ and $S = 1/2$ and thus a ^{2}D configuration, etc. Other examples are reported in Table I.

The splitting of the free-ion term in octahedral symmetry (O_h symmetry) reduces the degeneracy of the five d orbitals. Three orbitals have energy lower than the other two. This means that if the orbitals are populated by one electron, three degenerate states are possible, according to the three possible positions for the electron in the low-energy levels (T symmetry);

if there are two electrons in three low-energy orbitals, still there are three possible states (T symmetry); if there are three electrons, there is only a non-degenerate ground state (A symmetry); if there are four electrons, and the separation between the lower and the higher sets of orbitals is smaller than the spin-pairing energy (high-spin configuration), one of the four electrons occupies the higher set, and the ground state is doubly degenerate (E symmetry); if there are five electrons, there is only a non-degenerate ground state (A symmetry) if the system is high spin, and three degenerate ground states (T symmetry) if the system is low-spin; and so on. Similar considerations can be made for other idealized symmetries, like tetrahedral (T_d), square planar (D_{4h}), square pyramidal (C_{4v}), trigonal bipyramidal (D_{3h}), etc. The degeneracy of the states in the ligand-field is further removed by real symmetries, which are distorted and not idealized. The splitting of the degenerate states in low-symmetry may result with a separation energy of the order of kT and higher, as it can range up to several thousands of wave numbers (cm^{-1}). When the system is orbitally non-degenerate even in the idealized symmetry (A and B symmetries), the zero-field splitting of the S manifold for $S \geq 1$ ranges from zero to few tens of wave numbers (cm^{-1}). These considerations will guide us in the following sections.

C.1. Copper(II)

Copper(II) has a $3d^9$ electronic configuration, which gives rise to a ^2D free-ion ground state. In ideal octahedral symmetry the ground state is ^2E. However, the departure from the idealized geometry (often of D_{4h} or C_{4v} type, Table I) is so large that the orbitally non-degenerate ground state is separated by several thousands of wave numbers (cm^{-1}) from the first excited state. Therefore, copper(II) behaves as an isolated ground state and exhibits long electron relaxation time, because electron relaxation mechanisms are relatively inefficient. The main relaxation mechanism is probably a Raman-type process and/or modulation of **g** and **A** anisotropy by molecular tumbling for small complexes with $\tau_R \approx 10^{-11}$ s (*22*), or other rotation-independent motions for macromolecules.

Symmetric octahedral copper complexes (ground state ^2E) are known to experience dynamic Jahn–Teller effects. In six-coordinated complexes, instantaneous elongations occur along the three coordination axes, yielding a non-degenerate ground state. Elongation also changes the hyperfine coupling constant between the copper nucleus and the unpaired electrons. It has been proposed that this random process, with a correlation time of about 5×10^{-12} s (mean lifetime of the complex between hops), can thus actually be a further electron relaxation mechanism operative for symmetric complexes, as, for example, the copper aqua ion (*23–25*).

A fascinating alternative model for Cu(II) aqua ion has been recently proposed (*26*), in which the ion is believed to be mostly five-coordinated, and rapidly cycling between square pyramidal and trigonal bipyramidal geometries. In this picture, the correlation time of 5×10^{-12} s would be the time scale for the pyramidal-bypiramidal equilibrium.

Table I

POPULATION OF ELECTRONIC *d* ORBITALS IN DIFFERENT IDEALIZED SYMMETRIES

Electronic configuration	m_l of occupied orbitals	L	S	Free ion configuration	Ligand field O_h (octahedral)		Ligand field T_d (tetrahedral)		Ligand field D_{4h} (square planar) (*)		Ligand field C_{4v} (square pyramidal) (*)		Ligand field D_{3h} (trigonal bipyramidal) (*)	
d^1	2	2	1/2	^{2}D	E_g — — T_{2g} ↑ — —	^{2}T	— — — ↑ —	^{2}E	— — — ↑ —	^{2}E	— — — — ↑	^{2}A	— — — ↑ —	^{2}E
d^2	2,1	3	1	^{3}F	— — ↑ ↑ —	^{3}T	— — — ↑ ↑	^{3}A	— — — ↑ ↑	^{3}A	— — ↑ — ↑	^{3}A	— — — ↑ ↑	^{3}A
d^3	2,1,0	3	3/2	^{4}F	— — ↑ ↑ ↑	^{4}A	↑ — — ↑ ↑	^{4}T	— — ↑ ↑ ↑	^{4}A	— — ↑ ↑ ↑	^{4}A	— ↑ — ↑ ↑	^{4}E
d^4	2,1,0,−1	2	2	^{5}D	↑ — ↑ ↑ ↑	^{5}E	↑ ↑ — ↑ ↑	^{5}T	— ↑ ↑ ↑ ↑	^{5}A	— ↑ ↑ ↑ ↑	^{5}A	— ↑ ↑ ↑ ↑	^{5}A
d^5	2,1,0,−1,−2	0	5/2	^{6}S	↑ ↑ ↑ ↑ ↑ HS	^{6}A	↑ ↑ ↑ ↑ ↑	^{6}A	↑ ↑ ↑ ↑ ↑	^{6}A	↑ ↑ ↑ ↑ ↑	^{6}A	↑ ↑ ↑ ↑ ↑	^{6}A
					— — ↑↓ ↑↓ ↑ LS (S = 1/2)	^{2}T	↑ — — ↑↓ ↑↓	^{2}T						

d^6	2,2,1,0,−1,−2	2	2	^{5}D	↑ ↑ ↑↓ ↑ ↑	^{5}T	↑ ↑ ↑ ↑↓ ↑	^{5}E	↑ ↑ ↑ ↑↓ ↑	^{5}E	↑ ↑ ↑ ↑ ↑↓	^{5}A	↑ ↑ ↑ ↑↓ ↑	^{5}E
d^7	2,2,1,1,0,−1,−2	3	3/2	^{4}F	↑ ↑ ↑↓ ↑↓ ↑ HS ↑ – ↑↓ ↑↓ ↑↓ LS (S = 1/2)	^{4}T ^{2}E	↑ ↑ ↑ ↑↓ ↑↓	^{4}A	↑ ↑ ↑ ↑↓ ↑↓	^{4}A	↑ ↑ ↑↓ ↑ ↑↓	^{4}E	↑ ↑ ↑ ↑↓ ↑↓	^{4}A
d^8	2,2,1,1,0,0,−1,−2	3	1	^{3}F	↑ ↑ ↑↓ ↑↓ ↑↓	^{3}A	↑↓ ↑ ↑ ↑↓ ↑↓	^{3}T	↑ ↑ ↑↓ ↑↓ ↑↓	^{3}A	↑ ↑ ↑↓ ↑↓ ↑↓	^{3}A	↑ ↑↓ ↑ ↑↓ ↑↓	^{3}E
d^9	2,2,1,1,0,0,−1,−1,−2	2	1/2	^{2}D	↑↓ ↑ ↑↓ ↑↓ ↑↓	^{2}E	↑↓ ↑↓ ↑ ↑↓ ↑↓	^{2}T	↑ ↑↓ ↑↓ ↑↓ ↑↓	^{2}A	↑ ↑↓ ↑↓ ↑↓ ↑↓	^{2}A	↑ ↑↓ ↑↓ ↑↓ ↑↓	^{2}A

* (*)high spin only

In order to obtain information on the electron relaxation time in copper aqua ion, measurements should not be performed in water solution, because the correlation time for proton relaxation is in that case the reorientational time, which is much smaller than T_{1e}. ^{1}H NMRD profiles should be actually acquired in ethylene glycol solution and at temperatures lower that room temperature, so that the reorientational time increases one–two orders of magnitude (see Section II.B). In this way T_{1e} of the order of 10^{-9} s can be estimated.

The correlation time obtained for copper(II) containing macromolecules and proteins in water solution, where τ_R is larger than 10^{-8} s, is indeed the electron relaxation time. The shape of the ^{1}H NMRD profiles in such systems can be very different from one another and from what is predicted from the SBM theory predictions, because of the presence of hyperfine coupling between the unpaired electron and the metal nucleus ($I = 3/2$), but all of them exhibit the $\omega_I T_{1e}$ dispersion between 10 and 100 MHz at room temperature. This corresponds to electron relaxation time values in the 10^{-8}–10^{-9} s range (see Table II). As an example, the ^{1}H NMRD profiles of the copper(II) protein superoxide dismutase are shown in Fig. 5 at different temperatures (*27*). Copper(II) in this protein sits in a distorted tetragonal coordination environment (type II copper protein). The profiles show the ω_S and ω_I dispersions, the rate between the relaxivity values before and after the ω_S dispersion being much different from 10:3. Low temperature profiles also show a relaxivity peak at about 5 MHz, which can be accounted for by considering the hyperfine coupling to the metal nuclear spin effects (see Section II.B).

No field dependence in the electron relaxation time was ever found in the investigated region between 0.01 and 100 MHz of proton Larmor frequency, or at 800 MHz when high resolution is achieved (*28*). It was shown that T_{1e} is essentially independent of the reorientational time of the macromolecule and the viscosity of the solution. Therefore, rotation independent mechanisms have to be operative. We also find that T_{1e} decreases with increasing temperature, as also shown in Fig. 5.

When copper is bound to one sulfur atom of a cysteine and two nitrogens of two histidines in an essentially tetrahedrally distorted – trigonal ligand environment (type I copper proteins), the excited levels are low in energy, and the τ_s values are reduced to about 5×10^{-10} s (*29*). Examples are blue copper proteins, like ceruloplasmin and azurin, and copper(II) substituted liver alcohol dehydrogenase (*30–32*).

As an example on the relationship between proton relaxivity, electron relaxation and coordination environment, we report the case of azurin and its mutants. The relaxivity of wild type azurin is very low (Fig. 6) due to a solvent-protected copper site, the closest water being found at a distance of more than 5 Å from the copper ion. The fit, performed with the Florence NMRD program, able to take into account the presence of hyperfine coupling with the metal nucleus ($A_{\parallel} = 62 \times 0^{-4}$ cm^{-1}, see Section II.B) indicates T_{1e} values of 8×10^{-10} s. Although the metal site in azurin is relatively inaccessible, several mutations of the copper ligands open it up to the solvent. The ^{1}H NMRD profiles indicate the presence of water coordination for the

Table II

NMRD PARAMETERS FOR SOME METAL IONS AT 298 K

Metal ion	I	S		τ_{s0} (s)	$\Delta_{t\ \mathrm{aquaion}}$ (cm^{-1})	τ_v (s)	**A**, ZFS tensors in proteins (cm^{-1})	A_c/h_{aquaion} (MHz)	Main electron relaxation mechanism	Ref.
Cu^{2+}	3/2	1/2	d^9	$1–10 \times 10^{-9}$ (type I: 5×10^{-10})			$A_{\|\|} \approx 0.013–0.017$ (type I: $A_{\|\|} \approx 0.005$)	0–0.2 (*)	Raman-type, g, A-anisotropy	(*22,23,86, 111–113*)
VO^{2+}	7/2	1/2	d^1	$5–100 \times 10^{-10}$			$A_{\|\|} \approx 0.015–0.017$	2.1	A-anisotropy	(*58*)
Ti^{3+}		1/2	d^1	$1–10 \times 10^{-11}$				4.5	Orbach	(*61*)
Mn^{2+}	5/2	5/2	d^5	$0.1–5 \times 10^{-9}$	0.015	$2–50 \times 10^{-12}$	$D = 0.01–1$, $A_{\|\|} \approx 0.01$	0.65	ZFS modulation	(*39,114,115*)
Mn^{3+}		2	d^4	$1–10 \times 10^{-11}$			D = 0.01–0.5			
HS Fe^{3+}		5/2	d^5	$0.9–10 \times 10^{-10}$	0.095	$1–10 \times 10^{-12}$	$D = 0.1–3$	0.43	ZFS modulation	(*36,114, 116,117*)
LS Fe^{3+}		1/2	d^5	$0.1–1 \times 10^{-12}$						
HS Fe^{2+}		2	d^6	10^{-12}			D > 3		Orbach	
Cr^{3+}		3/2	d^3	$5–10 \times 10^{-10}$	0.11	$1–10 \times 10^{-12}$	D = 0.01–1	2.1	ZFS modulation	(*112,114*)
Co^{2+}	7/2	3/2	d^7	$1–10 \times 10^{-12}$			$D > 5$	0.4 (*)	Orbach ZFS modulation	(*47,112, 115,118*)
Ni^{2+}		1	d^8	$1-10 \times 10^{-12}$		$1–10 \times 10^{-12}$	$D = 1–5$	0.2 (*)	ZFS modulation Orbach	(*52,112, 119,120*)
Gd^{3+}		7/2	f^7	$1-2 \times 10^{-10}$	0.03–0.03	$10–50 \times 10^{-12}$	$D = 0.001–0.05$		ZFS modulation	(*47,94,118*)
Ln^{3+}				$0.1–1 \times 10^{-12}$					Orbach, Curie	(*14,67,88, 121–123*)

* Data not obtained from NMRD.

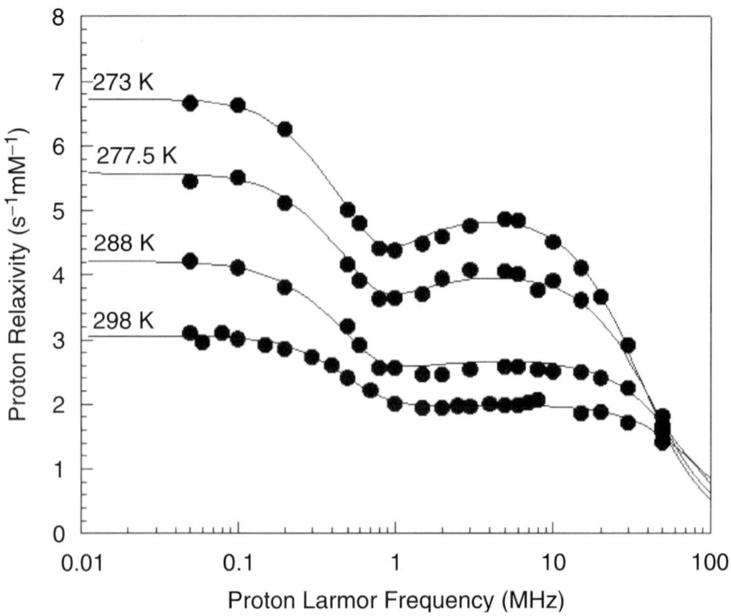

FIG. 5. Water ^1H NMRD profiles for solutions of superoxide dismutase at various temperatures (*124*). The solid lines are best-fit curves obtained with the inclusion of the effect of hyperfine coupling with the metal nucleus (*27*).

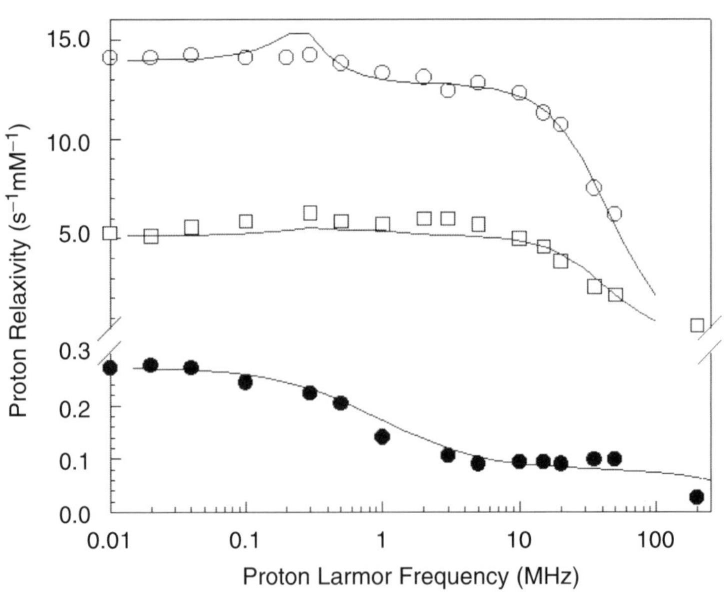

FIG. 6. Paramagnetic enhancements to water ^1H NMRD profiles for solutions of wild type azurin (●), and its His117Gly (○) and His46Gly (□) derivatives at 278 K and pH 7.5 (*32*).

His117Gly (two waters) and His46Gly (one water) mutants (*32*) (Fig. 6). The $\omega_I T_{1e}$ dispersion, between 10 and 100 MHz, is shifted to a lower Larmor frequency than that of the wild type azurin. This means larger T_{1e} values, which are calculated to be 5×10^{-9} and 15×10^{-9} s for the His117Gly and His46Gly, respectively. The fit was performed again by considering the presence of hyperfine coupling with the metal nuclear spin, as needed to account for the disappearance of the ω_S dispersion, and indicated from EPR measurements. The differences in water coordination produce changes in the geometry, which seems to be the main factor contributing to the variations in the electron relaxation of copper. These different coordination environments determine different geometries, ranging from the trigonal (type I) copper site of the wild type azurin to the tetragonal (type II) sites of the mutants. A correlation is thus present between the electron relaxation time and the geometry of the metal site: as the tetragonal character decreases, the relaxation becomes significantly faster.

C.2. *Iron*

(a) *High spin iron(III).* High spin iron(III) has one unpaired electron in each of the five *d* orbitals, and is therefore a ^{6}S state. The ground state is a sextuplet with an orbitally non-degenerate ground state (^{6}A), and thus electron relaxation is not efficient. As there are no excited states with the same multiplicity, spin–orbit coupling can only occur to second order and is relatively weak, introducing a relatively small zero-field splitting (*33*). Modulation of ZFS through solvent bombardment is expected to be the main electron relaxation mechanism. This accounts for the fact that the larger the splitting, the higher the electron relaxation rate (*34*). In quasi-symmetrical complexes, when the axial zero-field splitting parameter, *D*, is small, the electron relaxation time T_{1e} can be as long as 10^{-9} s (*35*). In porphyrin complexes with one axial ligand or with two weak ligands, where *D* can be of the order of 10 cm^{-1}, T_{1e} decreases to about 10^{-11} s, due to the consequent increase in Δ_t.

Figure 7 shows the ^{1}H NMRD profiles of water solutions of $Fe(H_2O)_6^{3+}$ in 1 M perchloric acid at 298 K and in a glycerol-water mixture (*36*). Only one dispersion is observed at about 7 MHz. It corresponds to a correlation time $\tau_c \approx 3 \times 10^{-11}$ s. The small increase of relaxivity above 20 MHz indicates that a field dependent T_{1e} is influential in the determination of τ_c at high field (see also Section II.C). From the fit to the SBM theory, τ_v is estimated to be around 5×10^{-12} s at room temperature, a value commonly found for small complexes in water solution and of the order expected for the mean lifetime between collisions with solvent molecules. The fit also provides a value for $\Delta_t = 0.095$ cm^{-1}, so that τ_{s0} is calculated to be 9×10^{-11} s at room temperature. By increasing the viscosity through glycerol–water mixture, it is shown that the relative influence of τ_R on τ_c with respect to T_{1e} becomes lower and lower with the increase in relaxivity in the high-field region being more and more evident. The fit of the profile acquired in the glycerol solution, performed by assuming that *r*, A_c, and Δ_t are not affected by the presence

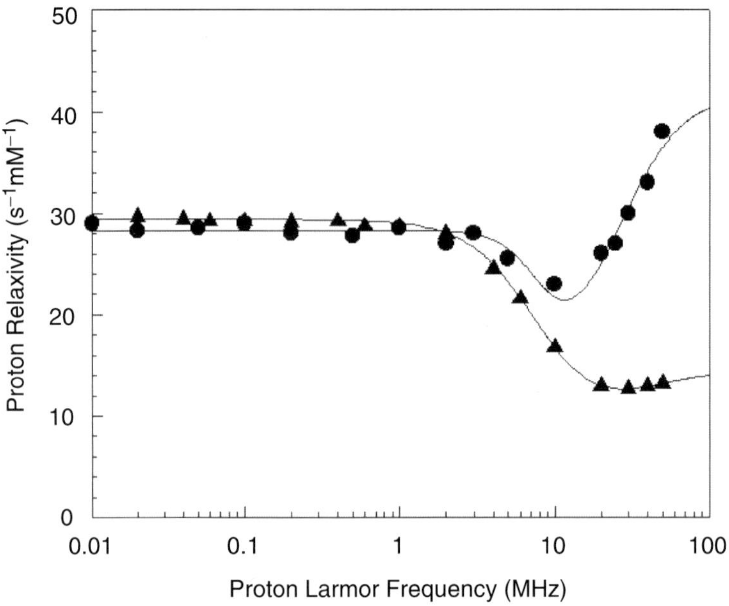

FIG. 7. Paramagnetic enhancements to solvent ^{1}H NMRD profiles for $Fe(H_2O)_6^{3+}$ solutions at 298 K with (▲) pure water and (•) 60% glycerol. The lines represent the best fit curves using the Solomon–Bloembergen–Morgan equations [Eqs. (1)–(6)] (*36*).

of glycerol, provides a value for τ_v of 1.4×10^{-11} s, i.e., larger than in aqueous solution. This increase is a common feature observed in many other systems, like nickel(II) aqua ion, gadolinium(III) aqua ion, manganese(II) aqua ion and in manganese(II)-proteins. The efficiency of the solvent bombardment in such systems is in fact reduced in the presence of glycerol. Some iron-containing proteins seem to represent an exception to this behavior, despite their long τ_R. In fact, profiles have been acquired, for instance of methemoglobin (see Section II.C) (*37*) or catalase (*38*), where no field dependence of electron relaxation is evident up to 50 MHz. A possible explanation could be a switch in the dominating electron relaxation mechanism between iron(III) aqua ion and iron(III) proteins. The large ZFS present in proteins may in fact favor other field independent electron relaxation mechanisms.

The water proton NMRD profiles of diferric iron transferrin (*35*) (Fig. 8) indicate a field dependent electron relaxation time with τ_v of about 1.2×10^{-11} s. The fit must be done by taking into account a static ZFS, which results in a D parameter of about 0.2 cm^{-1} and $E/D = 1/3$. The presence of ZFS is indicated by the additional inflection in the profile at about 10 MHz. In the fit, two sets of electron relaxation times (relative to ZFS axes and to external magnetic field axes) were taken into account to describe the low and the high field experimental data, the SBM theory being inadequate to describe the field dependence of the electron relaxation over the whole range of frequencies. Effective electron relaxation times of the order of 10^{-10} s have

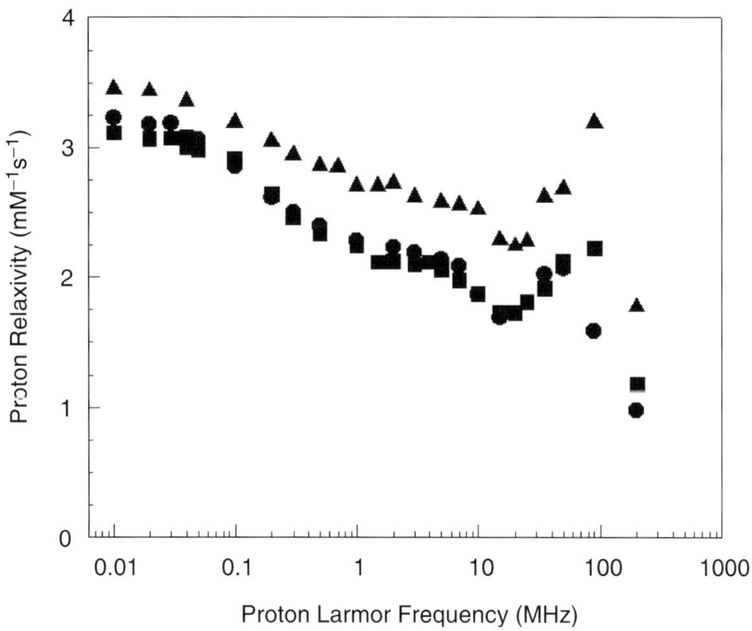

FIG. 8. Paramagnetic enhancements to water ^{1}H NMRD profiles for solutions of diferric transferrin at (●) 278 K, (■) 293 K, (▲) 308 K (*35*).

been obtained at low fields, and of about 10^{-9} s at intermediate fields, before the $\omega_s\tau_v$ dispersion occurs.

(b) *Low spin iron(III).* Low-spin iron(III) complexes, which have a ^{2}T configuration in octahedral symmetry, are characterized by low-lying excited states; therefore, Orbach processes are likely to be very efficient and short relaxation times are expected. Low-spin iron(III) occurs with strong ligands and often with hexacoordination. The ^{1}H NMRD profile of the complex $Fe(CN)_6^{3-}$ indicates $T_{1e} \leq 10^{-13}$ s (*1*). In heme complexes the T_{1e} is about 10^{-12} s when the apical ligands are His/His, His/Cys or His/CN; it is somewhat longer in the cases of His/OH^- and Cys/H_2O.

(c) *High spin iron(II).* Excited states in high-spin iron(II) are close in energy for the same reasons as in the case of low-spin iron(III). Therefore, the electron relaxation times are rather short ($\tau_s \approx 10^{-12}$ s). Consistently, the NMRD profile of $Fe(H_2O)_6^{2+}$, obtained from a Mohr salt solution $(NH_4)_2Fe(SO_4)_2 \cdot 6H_2O$), reported in Fig. 9, does not exhibit any dispersion below 50 MHz of proton Larmor frequency.

C.3. *Manganese*

(a) *Manganese(II).* Manganese(II) has a $3d^5$ electronic configuration, like high-spin iron(III). The ground state is a sextet with an orbitally nondegenerate ground state (^{6}A), and thus electron relaxation is not efficient. Degeneracy is only removed by ZFS due to second order spin–orbit coupling

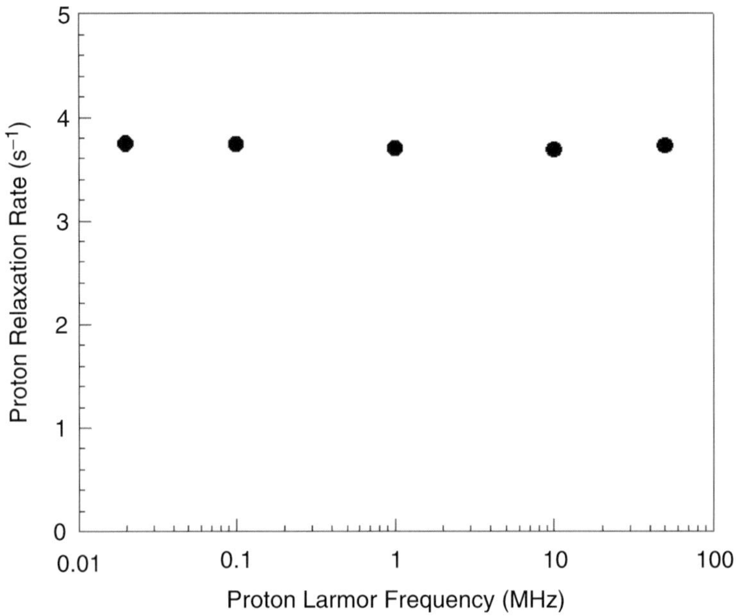

FIG. 9. Water ^1H NRMD profiles for 10 mM solutions of $(NH_4)_2Fe(SO_4)_2 \cdot 6H_2O$.

with excited states that arise from other free-ion terms of lower multiplicity. The ZFS in manganese(II) systems is in the range 0–1 cm^{-1}, and it is thus generally smaller than that in high-spin iron(III). This is because, compared to iron(III), the spin–orbit coupling constant is smaller due to the smaller charge of the manganese(II) ion, and because excited states are closer in manganese(II) than in iron(III). In any case, modulation of the transient ZFS is the most efficient relaxation mechanism. Typical electron relaxation times at low fields are around 10^{-9}–10^{-10} s.

The ^1H NMRD profile of $Mn(H_2O)_6^{2+}$ in water solution shows two dispersions (Fig. 10) in the 0.01–100 MHz range of proton Larmor frequency: one, at about 0.05 MHz, due to the contact relaxation, and a second, at about 7 MHz, due to the dipolar relaxation (*39*). The correlation time for contact relaxation is the electron relaxation time, whereas the correlation time for dipolar relaxation is the reorientational time ($\tau_R = 3.2 \times 10^{-11}$, in accordance with the value expected for hexaaquametal(II) complexes). This accounts for the different positions of the two dispersions in the profile. From a best fit of longitudinal and transverse proton relaxation profiles, the electron relaxation time is described by the parameters $\Delta_t = 0.02$–0.03 cm^{-1} and $\tau_v = 2$–3×10^{-12} s (providing $\tau_{s0} = 3.5 \times 10^{-9}$ s). The measurement of the transverse proton relaxation rate at high fields, in fact, permits to obtain the field dependence of the electron relaxation time from the contact contribution to relaxation. The constant of the contact interaction is calculated to be equal to 0.65 MHz.

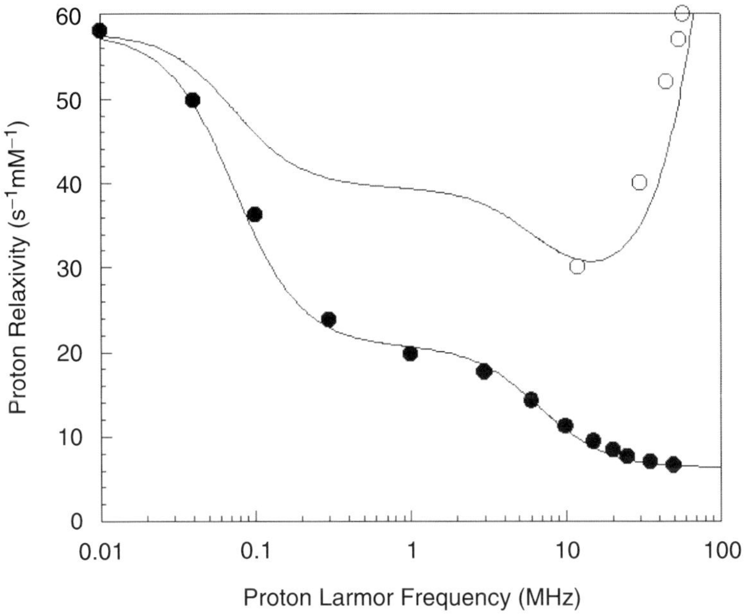

FIG. 10. Water ¹H NMRD longitudinal (•) and transverse (○) profiles for $Mn(H_2O)_6^{2+}$ solutions at 298 K (*39*).

As an example of manganese(II)-containing proteins, the ¹H NMRD profiles obtained with manganese(II) bound to concanavalin A (*40*) are reported in Fig. 11. The profiles are practically flat up to 1 MHz, when a peak appears. Both contact and dipolar relaxation are now dispersed at the same frequency, because the correlation time for both processes is the electron relaxation time, being shorter than the reorientational correlation time at low field. The peak observed at high fields is due to the increase in the electron relaxation time until τ_R becomes dominant, and the final dispersion is due to the ω_I dispersion. It is actually possible to fit the profiles without considering the presence of static ZFS, but it results in wrong values (of the order of a factor 2–3) of τ_M, τ_R, τ_v, Δ_t, and, above all, of the hydration number. The fit performed with including ZFS with $D = 0.04$ cm^{-1} provides an indication of 6–7 protons at 2.8 Å from the manganese(II) ion, with τ_v about 5×10^{-11} s and τ_{s0} about 3×10^{-10} s at room temperature; the fit performed according to the SBM theory indicates the presence of 13–15 protons at 2.8 Å, with $\tau_\nu = 6.9 \times 10^{-11}$ s and $\tau_{s0} = 1.1 \times 10^{-10}$ s (*41*). In fact, the main effect of ZFS is that of decreasing the proton relaxivity at low fields without altering the relaxivity at high fields (see Section II.A.5), thus affecting the best-fit values of all parameters.

(b) *Manganese(III)*. Manganese(III) is a d^4 ion, and generally it is high spin, with $S = 2$. Low symmetry and Jahn–Teller effects provide an orbitally non-degenerate ground state with the ZFS of the resulting quintet being in general ≤ 1 cm^{-1}, except for porphyrin derivatives, where it is larger (*42*). The electron relaxation time in hexaaqua complexes is $\leq 10^{-11}$ s (*43*), and about

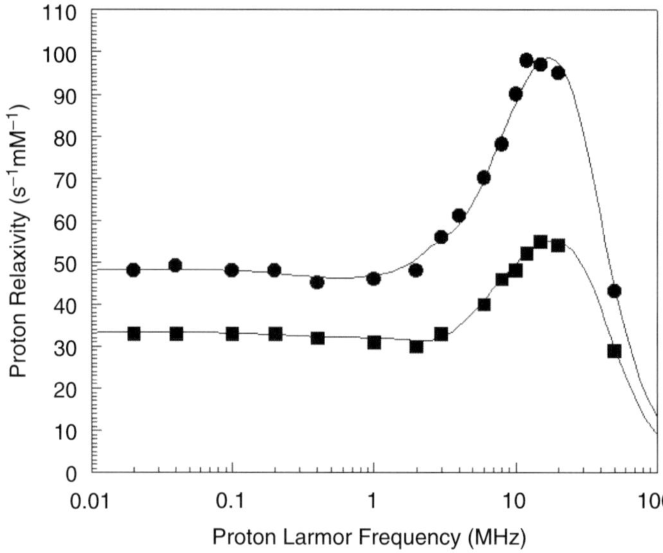

FIG. 11. Water ^{1}H NMRD profiles for solutions of Mn^{2+} Con A at 298 (•) and 278 (■) K (*40*). The solid curves are calculated with $D = 0.04\ cm^{-1}$.

10^{-10} s in manganese(III)-porphyrins (*44*). In the latter complexes the electron relaxation time is relatively longer because the two other excited levels (which contain one electron only) are farther from the three levels at lowest energy (each of them containing one electron). The NMRD profiles of manganese(III) porphyrins (Fig. 12) are characterized by a peak at high

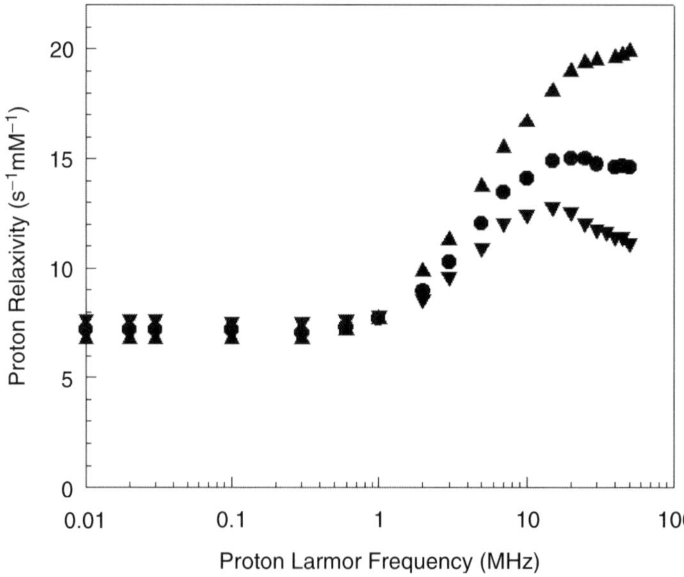

FIG. 12. Water ^{1}H NMRD profiles for solutions of the tetraphenylsulfanyl porphyrin ($TPPS_4$) manganese(III) complex at 278 (▲), 293 (•), and 308 (▼) K (*125*).

fields. The latter can be fit according to the SBM theory with a field dependent electron relaxation time; however, Sharp *et al.* (*45,46*) pointed out that such a peak can be ascribed to the effect of rhombic ZFS for integer spins. The rising dispersion thus corresponds to Zeeman energy of the order of E, the rhombic component of ZFS (see also Section I.C.5 and II.A.5), and no field dependence for electron relaxation would be detected.

C.4. Cobalt(II)

High-spin cobalt(II) is a d^7 ion with $S = 3/2$ ground state. The splitting of the ^{4}F free-ion term in ground state indicates that there are relatively close excited states in every geometry: in octahedral symmetry it gives rise to a triply degenerate ground state (^{4}T). In six-coordinate complexes, where the six Kramers doublets are spread over a 10^2–10^3 cm^{-1} range, the spin–orbit coupling mechanism is very efficient, and electron relaxation is fast, T_{1e} being around 10^{-12} s. Orbach-type processes, in the presence of low-lying excited states, are in fact expected to be very efficient. In five-coordinate complexes of C_{4v} symmetry the ground state is doubly degenerate and the situation is similar, T_{1e} being slightly longer because the low-lying excited states are relatively farther. In tetrahedral (T_d) and trigonal bipyramidal (D_{3h}) symmetries the ground state is orbitally non-degenerate (^{4}A), the separation in energy between the ground and the excited states is relatively large, and the ZFS is small. The electron relaxation time is longer, i.e., generally one order of magnitude higher in the latter case, the Orbach-type mechanisms being probably less efficient. Modulation of ZFS may also represent a possible relaxation mechanism. The magnitude of the splitting depends on the closeness of the low-lying excited states: the farther they are, the smaller is the ZFS. In all cases, no field dependence of the electron relaxation time is observed up to 100 MHz.

The water proton NMRD of the pseudooctahedral $Co(H_2O)_6^{2+}$ (reported in Fig. 13) shows almost field-independent water proton relaxation rate values in the 0.01–60 MHz region (*47*). Therefore, the $\omega_S\tau_c = 1$ and of course the $\omega_I\tau_c = 1$ dispersions must occur at fields higher than 60 MHz. This provides an upper limit value for T_{1e} equal to 4×10^{-12} s. Such a low T_{1e} value is consistent with the low water proton relaxation rate values. By using the SBM theory, T_{1e} at 298 K can be estimated to be about 10^{-12} s. It can be larger, if the presence of a probable static ZFS is taken into account (*47*). When measurements are performed in highly viscous ethyleneglycol the observed rates are similar to those obtained in water. This suggests that T_{1e} is also similar and, therefore, it is rotation-independent (*47*).

As an example of tetra-coordinate cobalt(II) systems, the NMRD profile of cobalt(II)-substituted carbonic anhydrase (MW 30,000) at high pH is reported (Fig. 14). The metal ion is coordinated to three histidines and to a hydroxide ion (*48*). The NMRD profile shows a $\omega_S\tau_c$ dispersion centered around 10 MHz, which qualitatively sets the correlation time around 10^{-11} s. As the reorientational correlation time of the molecule is much longer, this value is a measure of the effective electronic relaxation time. A quantitative

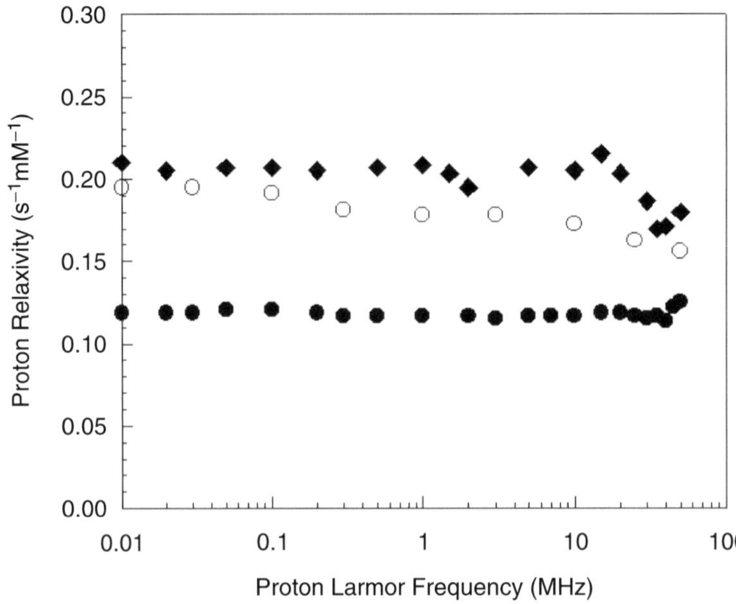

FIG. 13. Paramagnetic enhancements to solvent ^{1}H NMRD profiles for water solution of $Co(ClO_4)_2 \cdot 6H_2O$ at 298 K (○) and for ethyleneglycol solutions at 264 K (◆) and 298 K (●) (*47*).

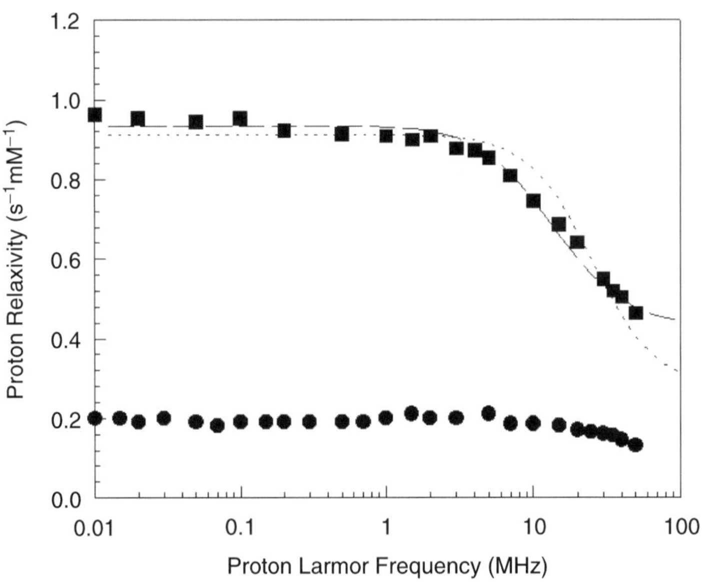

FIG. 14. Paramagnetic enhancements to water ^{1}H NMRD profiles for solutions of cobalt(II) human carbonic anhydrase I at pH 9.9 and 298 K (■) (*48,49*) and for solutions of the nitrate adduct of cobalt(II) bovine carbonic anhydrase II at pH 6.0 and 298 K (●) (*126*). The dashed line shows the best fit profile of the former data calculated with including the effect of ZFS, whereas the dotted line shows the best fit profile calculated without including the effect of ZFS.

analysis of the profile actually requires consideration of the presence of zero-field splitting ($D \geq 10$ cm^{-1}), which is known to be sizeable in cobalt(II) complexes. It accounts for the smoother shape of the dispersion curve with respect to the Lorentzian shape predicted in the SBM theory (*49*), and provides a more realistic value for the metal-proton distance.

If another ligand is added in the fifth coordination site and water is maintained in the coordination sphere (*50*), as it happens in the nitrate derivative, the water ^{1}H NMRD profile dramatically decreases because the effective electron relaxation time decreases by at least one order of magnitude (*47,51*).

Cobalt(II) is generally low spin ($S = 1/2$) in planar (D_{4h}) compounds and in some square pyramidal (C_{4v}) compounds. Its electron relaxation time is much longer (10^{-9}–10^{-10} s) because of the high energy of the first excited state and the absence of ZFS.

C.5. *Nickel(II)*

Nickel(II) is a d^8 ion, with $S = 1$. As already discussed for cobalt(II) systems, the ligand-field configuration depends on the coordination number and geometry of the system. Octahedral nickel(II) has an orbitally non-degenerate ground state (^{3}A) and tetrahedral nickel(II) has an orbitally triply degenerate ground state (^{3}T). The low-lying excited states present in the latter systems indicate that Orbach-type processes are likely to be very efficient and short relaxation times are expected. Pseudo-octahedral nickel(II) has excited states high in energy (about 10,000 cm^{-1} for the first excited state). This means that in pseudo-octahedral nickel(II) Orbach-type relaxation mechanisms are expected to be inefficient, and the modulation of the ZFS by reorientation or collisions may become dominant. Actually, pseudo-octahedral nickel(II) relaxes significantly slower than pseudo-tetrahedral nickel(II).

The ^{1}H NMRD profile of the hexaaqua nickel(II) complex is independent of the magnetic field up to about 100 MHz, as shown in Fig. 15. An increase in the proton relaxivity is then observed for higher fields, ascribed to a field dependence of the electron relaxation time, caused by fluctuations of the zero-field splitting. A value of T_{1e} around 3×10^{-12} s can be estimated from the SBM theory, or, more appropriately, around 10^{-11} s if ZFS, expected to be around 3 cm^{-1} (*52*), is taken into account in the fit. Data have been analyzed using the slow-motion theory (see Section IV of Chapter 2), which does not provide any value for the electron relaxation time, and indicates $\tau_v = 1.7$–1.9×10^{-12} s, $\tau_R = 8$–9×10^{-12} s and $\Delta_t = 3.2$–3.9 cm^{-1} at 324 K by applying the coupled or extended pseudo-rotational model (*53*).

Water ^{1}H NMRD profiles acquired for other complexes and proteins always exhibit the same features of hexaaqua nickel(II). As an example, we report here the profile of the hexa-coordinate nickel(II)-substituted bovine carbonic anhydrase II (*54,55*) (Fig. 15). As in the aqua complex, (i) the low-field profile is flat, (ii) no dispersion appears, the ω_s dispersion being quenched in $S = 1$ complexes with large static ZFS (*56*) (see Section I.A.5) and the

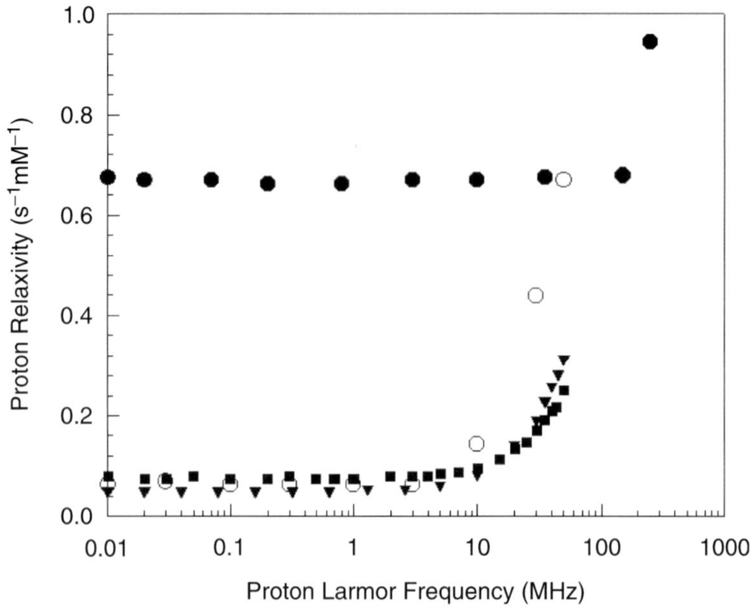

FIG. 15. Paramagnetic enhancements to solvent ^{1}H NMRD profiles for $Ni(H_2O)_6^{2+}$ solutions in pure water at 298 K (●), and in ethyleneglycol at 264 (▼) and 298 (■) K (*47*), and for water solutions of nickel(II) bovine carbonic anhydrase II at pH 6.0 and 298 K (○) (*54*).

ω_I dispersion occurring outside the accessible frequency range, (iii) a sharp rise occurs at high field. Again, the latter is due to the field dependence of the electron relaxation, and to the removal of the relaxivity quenching present at low fields and due to the presence of rhombic ZFS in integer spin systems. Such removal is caused by the fact that Zeeman energy becomes larger than the rhombic ZFS term (*45,46*). In case the peak is fit by applying the pseudo-rotational model causing field-dependent electron relaxation, it can be noted that in proteins the correlation time, τ_v, for the modulation of the ZFS is longer than that in the aqua ion, as indicated by the fact that the increase in proton relaxivity begins at lower fields.

C.6. Oxovanadium(IV)

Oxovanadium(IV) is a d^1 ion, and, as for all $S = 1/2$ ions, ZFS modulation cannot be present and the electron relaxation times are expected to be long unless in the presence of low-lying excited states. The oxovanadium(IV) aqua ion is forced to adopt a C_{4v} rather than O_h symmetry, and thus the ground state is orbitally non-degenerate (^{2}A). Under these conditions, the first excited state is expected to be high in energy, and it has been indeed estimated to be more than 10,000 cm^{-1} higher than the ground state (*57*).

The ^{1}H NMRD profiles of $VO(H_2O)_5^{2+}$ in pure water and in the presence of glycerol are shown in Fig. 16 (*58*). The profile acquired in pure water is

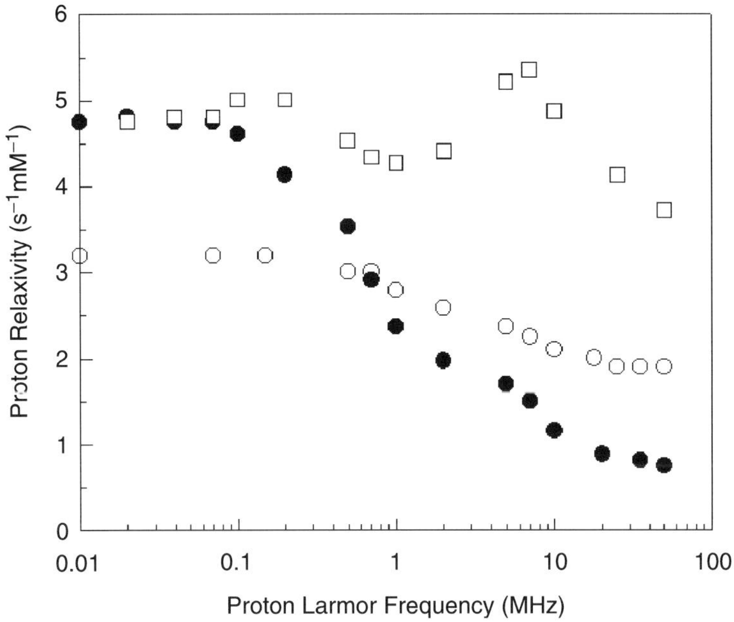

FIG. 16. Paramagnetic enhancements to solvent ^{1}H NMRD profiles for water solutions of $VO(H_2O)_5^{2+}$ at 308 K (●), and for solutions with 60% glycerol (○), and with 87% glycerol (□) (*58*).

similar to that of the manganese(II) aqua ion: the first dispersion is ascribed to the contact relaxation, the second to the dipolar relaxation. It is thus possible to estimate T_{1e} around 5×10^{-10} s from the position of the first dispersion. Electron relaxation mechanisms have been proposed to be dynamic Jahn–Teller effect, modulation of hyperfine coupling with the metal nucleus, and spin-rotation. By increasing viscosity, τ_R increases, and thus the second dispersion moves towards the first dispersion (*44*). In a 87% glycerol solution, it is evident from the peak in the high-field region that τ_R becomes longer than the electron relaxation time, the latter becoming the correlation time for nuclear relaxation. The observed peak can in fact be accounted for with the field-dependent electron relaxation time. The profiles could be reproduced by taking into account the presence of hyperfine coupling with the metal nucleus. The profiles also show that the effective electron relaxation time must slow down with increasing viscosity, because τ_ν increases. This results in a sizeable decrease in relaxivity at low field with a moderate increase of viscosity, due to a decrease in the contact contribution, in turn determined by T_{1e}. The fact that the peak relaxivity in the higher viscosity profile does not increase much with respect to the low-field relaxivity, although τ_R increases sizeably, is an indication of τ_M-limited situation.

Long values of T_{1e}, consistently with expectations, were obtained for some VO-protein compounds. Analysis of the ^{1}H NMRD profile of bis-oxovanadium(IV) transferrin (*59*), for instance, provides a value for τ_c of

2×10^{-8} s from the position of the observed dispersion. Such a correlation time is expected to be dominated by the electron relaxation time (see Section II.E.2). No field dependence for T_{1e} has been observed up to 50 MHz. In the same coordination environment, electron relaxation in VO-proteins is about one order of magnitude lower than in the Cu-proteins, due to the stronger spin–orbit interaction of the latter ion.

C.7. Titanium(III)

Titanium(III) is also a d^1 metal ion, and therefore is expected to have relatively long electron relaxation times. However, electron relaxation is somewhat faster than for copper(II) and oxovanadium(IV), because the different ligand fields of the metal ions split the energy levels in a different way. The first excited states (E_g state) for titanium hexaaqua ion have been estimated to be around 2000 cm^{-1} higher than the ground state (T_{2g} state) (*60*). The ^1H NMRD profiles of water solutions of $Ti(H_2O)_6^{3+}$ at different temperatures are shown in Fig. 17. One dispersion only is displayed and the ratio between the relaxivity values before and after the dispersion is much larger than 10/3: this can be accounted for by considering the presence of contact and dipolar contributions to relaxivity with the same correlation time. This indicates that the effective electron relaxation time is around 3×10^{-11} s at room temperature (*61*), as is τ_R for hexaaqua 3*d* metal ions. No field dependence in the electron relaxation time has been detected up to 600 MHz. Different from other metal ions (like oxovanadium(IV), manganese(II),

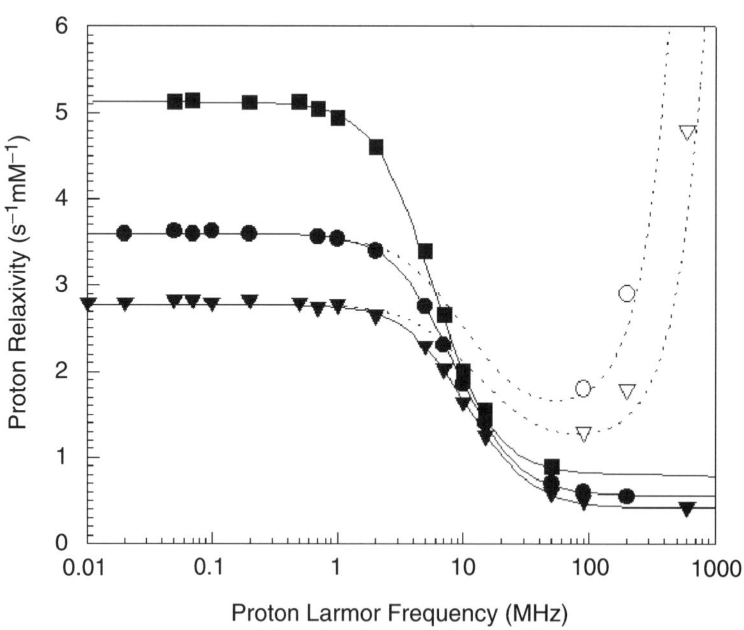

FIG. 17. Water ^1H NMRD longitudinal profiles for $Ti(H_2O)_6^{3+}$ solutions at 278 K (■), 293 K (●) and 308 K (▼), and transverse profiles at 293 K (○) and 308 K (▽) (*61*).

chromium(III), iron(III), gadolinium(III)), the effective electron relaxation time decreases with increasing temperature (as it happens for copper(II)). Such behavior is consistent with what is expected for an electron relaxing through an Orbach-type mechanism (see (Eq. 10)), so that the electron relaxation time is linearly proportional to the mean correlation time for intermolecular fluctuations (*61*). Measurements performed on a solution containing 65% glycerol-d_8 indicate an electron relaxation time six times longer than in pure water. This is consistent with an Orbach-type mechanism, if the correlation time for electron relaxation is viscosity-dependent.

C.8. Chromium(III)

Chromium(III) has a ^4A ground state in pseudo-octahedral symmetry. The absence of low-lying excited states excludes fast electron relaxation, which is in fact of the order of 10^{-9}–10^{-10} s. The main electron relaxation mechanism is ascribed to the modulation of transient ZFS. Figure 18 shows the ^{1}H NMRD profiles of hexaaqua chromium(III) at different temperatures (*62*). The position of the first dispersion, in the 333 K profile, indicates a correlation time of 5×10^{-10} s. Since it is too long to be the reorientational time and too fast to be the water proton lifetime, it must correspond to the electron relaxation time, and such a dispersion must be due to contact relaxation. The high field dispersion is the ω_S dispersion due to dipolar relaxation, modulated by the reorientational correlation time $\tau_R = 3 \times 10^{-11}$ s. According to the Stokes–Einstein law, τ_R increases with decreasing temperature, and

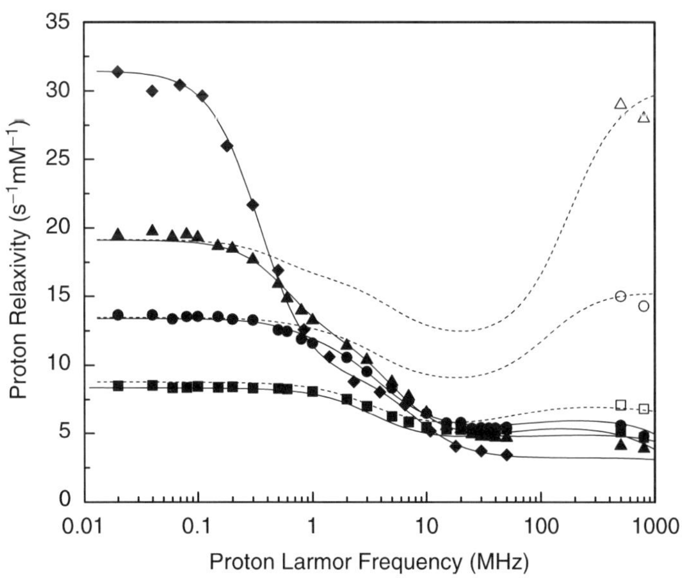

FIG. 18. Water ^{1}H NMRD profiles for $Cr(H_2O)_6^{3+}$ solutions at pH 0 and 278 (■), 298 (●), 313 (▲), and 333 (◆) K. Solid symbols indicate T_1^{-1} measurements and open symbols T_2^{-1} measurements. The solid lines represent the best fit profiles of T_1^{-1} dashed lines indicate the best fit profiles of T_2^{-1}.

correspondingly the position of this second dispersion moves toward lower fields. The measurements also show that (i) the position of the high field dispersion moves toward higher fields with decreasing temperature, thus indicating that the effective electron relaxation time at low fields decreases with decreasing temperature, and (ii) the transverse relaxation rates increase sizably at high fields, thus indicating that the effective electron relaxation time is field-dependent. Both these features are in qualitative accordance with the SBM theory. The fit provided a value for the constant of contact interaction, A_c/h, around 2 MHz. In the low temperature profiles the contact dispersion disappears: this feature is related to the occurrence of slow exchanging regime for water protons, and will be discussed in details in Section II.E.4.

Longitudinal relaxation measurements do not allow to establish whether the electron relaxation is constant or field-dependent. Transverse relaxation measurements, on the contrary, prove that the effective relaxation time is indeed field-dependent (see Sections I.A.1 and II.A.4). The occurrence of a field-dependent electron relaxation time is confirmed by the longitudinal relaxation measurements in glycerol solution (*62*), as the typical high field relaxivity peak appears, with Δ_t of 0.11 cm^{-1} and τ_ν of 7×10^{-12} s at 298 K.

C.9. Gadolinium(III)

Gadolinium(III) is an f^7 ion with an ^{8}S ground term. Each of the f orbitals is occupied by one electron. In analogy to manganese(II), which has a ^{6}S ground term and one electron per d orbital, and displays the longest electronic relaxation times among the 3d metal ions, gadolinium(III) has the longest electronic relaxation times of all the lanthanides ($T_{1e} \geq 2 \times 10^{-10}$ s, at room temperature) (*63,64*). As for manganese(II), modulation of the transient ZFS, due to second order spin–orbit coupling with excited states of lower multiplicity arising from different free-ion terms, is again the dominant electron relaxation mechanism.

Figure 19 shows the NMRD profile of gadolinium(III) in water and in ethyleneglycol solutions (*47*). The dispersion observed in water solution corresponds to a τ_c equal to 5×10^{-11} s, which can be ascribed to τ_R, but a good fit of the profile at high fields can be obtained only by introducing a contribution of a field-dependent T_{1e}. The reorientational correlation time and the effective electron relaxation time are thus of the same order of magnitude above 10 MHz. The fitting provides $\Delta_t = 0.035$–0.058 cm^{-1}, $\tau_\nu = 0.7$–1.6×10^{-11} s (and thus $\tau_{s0} = 1.2 \times 10^{-10}$ s), $\tau_R = 4.5 \times 10^{-11}$ s and 8 coordinated water molecules with metal proton distances of 3.1 Å. Simultaneous fits of ^{17}O, ^{1}H NMRD and EPR measurements have also been performed (*65*), by considering electron spin relaxation due to the combination of static and transient ZFS. In viscous solution the correlation time for proton relaxation coincides with the electron relaxation time, and the high field peak due to its field dependence appears. The correlation time for electron relaxation, τ_v, in ethyleneglycol solution is longer than in water. This is in accordance with the fact that collisions of solvent molecules with the ion are slowed down in

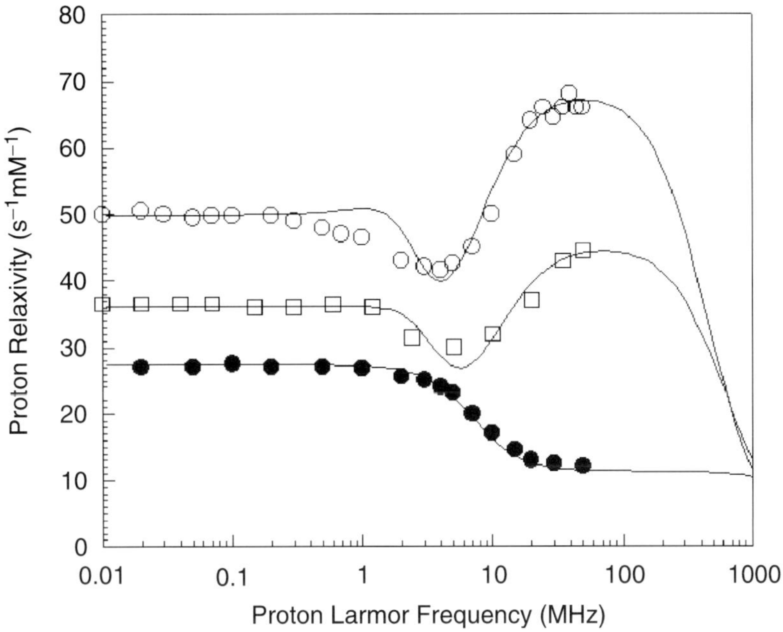

FIG. 19. Solvent ^{1}H NMRD profiles for water solutions of $GdCl_3$ at 298 K (●) and for ethyleneglycol solutions of $GdCl_3$ at 298 K (○) and 312 K (□) (*47*).

viscous solvents. Actually, in order to have accurate values the fit should be performed by taking into account the presence of both static and transient ZFS in fast rotating systems. Data were also fit using a spectral density function of the transient ZFS interaction where modulation is allowed with two correlation times (representing the time-scales for collision-induced distortions of the coordination symmetry, and for the non-zero partially averaged ZFS) and an order parameter (*66*).

In macromolecular systems, the field dependent electron relaxation time is the correlation time for dipolar relaxation. This is responsible for the common features of all gadolinium(III) containing macromolecule profiles, which are (i) the presence of the ω_S dispersion at about 1 MHz, (ii) the following peak centered at about 20–50 MHz. In fact τ_v seems to be always around 15–25 × 10^{-12} s at room temperature without any dependence on reorientational time, i.e., on the molecular weight. As an example we report the ^{1}H NMRD profile of solution containing gadolinium(III) complexed with Zn(II)-Ca(II)-concanavalin A (Fig. 20). Since gadolinium(III) often exhibits the presence of a small static ZFS, the SBM theory is inadequate to fit the experimental profiles, and programs taking into account the effect of both static and transient ZFS have to be employed (see Fig. 11 of Chapter 2). Second generation contrast agents are often designed to bind to macromolecules in vivo, to exploit the favorable relaxation features of gadolinium(III) protein complexes (see Chapter 4).

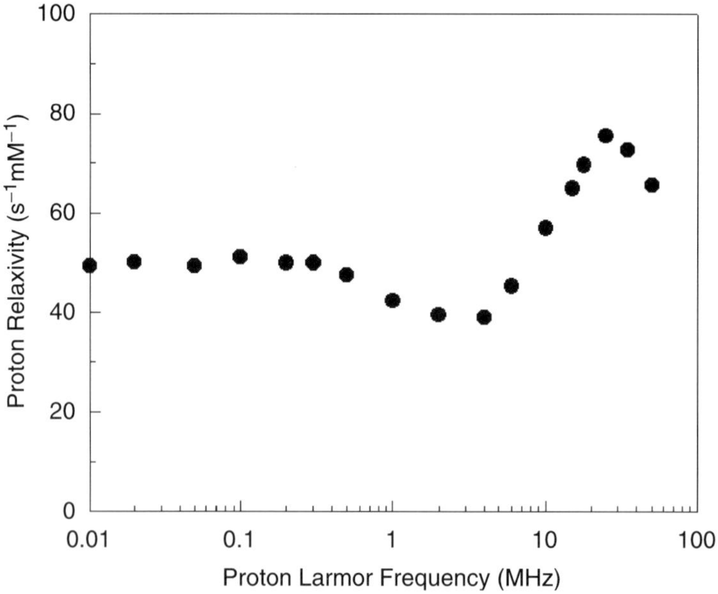

FIG. 20. Water ^{1}H NMRD profile for solutions of gadolinium(III) complexed with Zn^{2+}-Ca^{2+}-concanavalin A at pH 6.4 and 298 K (*127*).

C.10. Other Lanthanides

All lanthanide ions, with the exception of gadolinium(III) and europium(II), are likely to be relaxed by Orbach-type processes at room temperature. In fact, the f^n configurations ($n \neq 7$) of lanthanides(III) give rise to several free-ion terms that upon strong spin–orbit coupling, provide several closely spaced energy levels. Table III reports the multiplicity of the ground levels, which varies from 6 to 17, and is further split by crystal field effects.

The ^{1}H NMRD profiles for lanthanides aqua ions are reported in Fig. 21 (*67*). It is apparent that in all cases proton relaxivity is small and field independent up to 50 MHz, thus indicating that τ_c is indeed provided by the electron relaxation time, being much smaller than τ_R. The high field increase in relaxivity is due to the Curie relaxation, which depends on the square of the external magnetic field (see Section III.A of Chapter 2). Actually, Curie relaxation can only be expected relevant when $T_{1e} \ll \tau_R$, as it occurs for lanthanides, and it was observed also in small complexes (*68,69*). Lanthanides are among the few examples where Curie relaxation is observable on nuclear T_1 and not only on T_2, due to their unusually large J value and extremely short T_{1e}. The very small field dependence of proton relaxivity showed by Pr^{3+}, Yb^{3+} and Sm^{3+} aqua ions is due to the relatively small effective magnetic moments of such ions. The total paramagnetic relaxation rate is thus obtained as the sum of the dipolar and Curie contributions. The fit was made with Eq. (22) of Chapter 2 by using the proton-lanthanide distance according to the X-ray data, the experimentally available magnetic moments (*70*), and a coordination number of water molecules equal to 9 for

Table III

SOME ELECTRONIC PROPERTIES OF LANTHANIDE IONS

Ion	Configuration	$^{2S+1}L_J$ of ground state (multiplicity in parentheses)	g_J[a]
Ce^{3+}	$4f^1$	$^2F_{5/2}$ (6)	6/7
Pr^{3+}	$4f^2$	$^3H_4$ (9)	4/5
Nd^{3+}	$4f^3$	$^4I_{9/2}$ (10)	8/11
Pm^{3+}	$4f^4$	$^5I_4$ (9)	3/5
Sm^{3+}	$4f^5$	$^6H_{5/2}$ (6)	2/7
Eu^{3+}(Sm^{2+})	$4f^6$	$^7F_0$ (1)	-
Gd^{3+}(Eu^{2+})	$4f^7$	$^8S_{7/2}$ (8)	2
Tb^{3+}	$4f^8$	$^7F_6$ (13)	3/2
Dy^{3+}	$4f^9$	$^6H_{15/2}$ (16)	4/3
Ho^{3+}	$4f^{10}$	$^5I_8$ (17)	5/4
Er^{3+}	$4f^{11}$	$^4I_{15/2}$ (16)	6/5
Tm^{3+}	$4f^{12}$	$^3H_6$ (13)	7/6
Yb^{3+}	$4f^{13}$	$^2F_{7/2}$ (8)	8/7

[a]Calculated according to the equation: $g_J = 1 + \frac{J(J+1) - L(L+1) + S(S+1)}{2J(J+1)}$. The equation does not hold for f^6 ions.

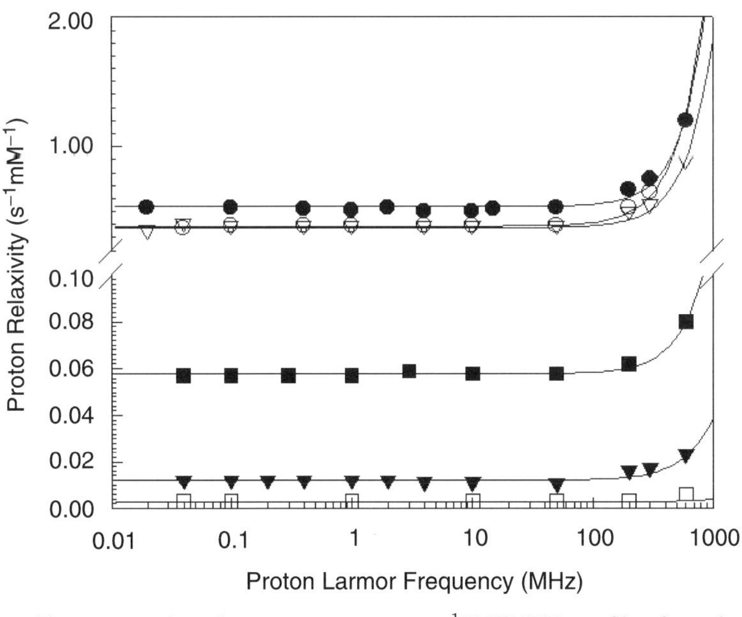

FIG. 21. Paramagnetic enhancements to water ^{1}H NMRD profiles for solutions of some lanthanide aqua ions at 298 K: (□) Sm, (▼) Pr, (■) Yb, (▽) Er, (○) Ho, (•) Dy (*67*).

Table IV

DYNAMIC AND STRUCTURAL PARAMETERS FOR SOME LANTHANIDE AQUA IONS AT 298 K (*67*)

	Pr	Sm	Dy	Ho	Er	Yb
Coord. No.	9	9	8	8	8	8
r (Å)	3.208	3.166	3.111	3.101	3.083	3.065
τ_R (10^{-12} s)	63	67	63	65	61	68
T_{1e} (10^{-12} s)	0.25	0.16	0.99	0.81	0.78	0.48

the lanthanide ions in the first half and 8 in the second half of the series (*71*). The best-fit values obtained for the reorientational time and the electron relaxation time with qualitative inclusion of the effect of a strong static ZFS are reported in Table IV (*71*).

(a) *Europium(II).* The ^{1}H NMRD profiles of europium(II) aqua ions are similar to those of gadolinium(III) aqua ions, although relaxivity at room temperature is about half. Electron relaxation times for europium(II) aqua ions are longer than those for gadolinium(III). In fact, whereas the transient ZFS is found similar (0.056 cm^{-1}), τ_υ is surprisingly found to be only 1×10^{-12} s at 298 K, thus resulting in an effective electron relaxation time at low fields equal to 7.5×10^{-10} s. The decrease in relaxivity is ascribed to the shorter reorientational time (16×10^{-12} s) (*72*). The electron relaxation time of europium(II)-DTPA is described by the parameters $\Delta_t = 0.07$ cm^{-1} and $\tau_\upsilon = 14 \times 10^{-12}$ s, providing a relatively fast electron relaxation, with $\tau_{s0} = 3.4 \times 10^{-1}$ s (*9*).

II. From the NMRD Profile to the Structural and Dynamic Parameters

The enhancement in water proton relaxation rate due to the presence of a paramagnetic complex in solution at a given field is a linear function of the concentration of the paramagnetic metal ions. For a fixed concentration, the enhancement depends on a number of parameters including (a) the magnetic field, (b) the nature of the metal ion, (c) the availability of coordination sites for solvent molecules, (d) the temperature, (e) the size of the paramagnetic molecule, (f) the presence of solutes that increase the viscosity of the solution. We know the identity of the metal ion present in the paramagnetic complex and its concentration, besides the temperature and viscosity of the solution. The acquisition of the water proton relaxation rate enhancements as a function of the magnetic field can thus easily allow the study of the presence of water molecules coordinated to the metal ion.

This is not all. In fact, the structural and dynamic information at a molecular level that are contained in the NRMD profiles of a paramagnetic system comprise: geometrical parameters (distances and angles), which determine the position of protons relative to the paramagnetic site; the number of protons in the first coordination sphere, and possibly in the second coordination sphere, as well as the distance of closest approach of the diffusing water protons; dynamic parameters such as τ_R and τ_M, besides the electron relation time and its field dependence, as already discussed; the presence and the extent of second-order effects like zero-field splitting and hyperfine coupling to the metal nucleus.

The analysis of the paramagnetic enhancement of the relaxation rates must be done after subtraction of the diamagnetic contribution from the relaxation rates of the paramagnetic sample, obtained by performing measurements on the diamagnetic analog. It is customary to refer to 1 mM concentration of paramagnetic substances to define the relaxivity of the sample.

A. Dependence of the NMRD Profiles on the Different Parameters of the Molecule

A.1. Metal Hydration

^{1}H NMRD profiles depend on the presence and distance of protons close to the paramagnetic site. The Solomon equation (Eq. 1) indicates that the paramagnetic enhancement to the relaxation rate for such protons is proportional to r^{-6}: this is expected to ensure a good precision in determining the distance between those protons and the metal ion. However, ^{1}H NMRD measurements are performed by measuring the relaxation rates of solvent water protons. Therefore, the acquired profiles are given by the sum of the contributions from all protons, in the hypothesis that all protons are fast-exchanging, i.e., their exchange rate is faster than their relaxation rate. Therefore, relaxometry measurements can only provide a value for $\sum_i n_i/r_i^6$, where n is the number of equivalent protons and r is their distance. The sum is extended to all protons around the paramagnetic center. This means that a value for the proton(s)-metal ion distance can be obtained only on the assumption that the number of protons around the metal ion is known and that all of them are at the same distance. *Vice versa*, if the distance of coordinated water protons is fixed in a small range by coordination chemistry considerations, information on the hydration number is obtained.

Typically, the solvent proton relaxivity is determined by protons of water molecules (or hydroxide ions) interacting with the paramagnetic metal ion, giving rise to the so-called first coordination sphere. Weaker interactions may also be present between the first coordination sphere and farther water molecules. In some cases, this interaction can be strong enough to give rise to a second coordination sphere (*73,74*). If the water molecules involved have exchange rates longer than the diffusional time, they contribute to the observed rate. Note that the dependence on r^{-6} helps, in the sense that the protons closer to the metal ion (if present) actually determine the measured relaxation rates, because farther protons provide much smaller (and often neglected) contributions.

Diffusive water protons should also be taken into account for considering all contributions to the observed rates (see Section II.A.7).

The picture is much more complicated in the presence of slow-exchanging protons with different lifetimes (see Section II.A.3). In the hypothesis that the closest protons are exchanging with a very low rate, for instance, and second sphere water protons are fast-exchanging, the latter will provide the largest contribution to relaxation (for an example see, Section II.C).

In the presence of static ZFS or anisotropic hyperfine coupling with the metal nuclear spin, the Solomon equation does not hold anymore, and much more complicated expressions need to be used (see Section V.B of Chapter 2). In these cases, the paramagnetic contribution to the relaxation rate of coordinated protons depends on the position of the protons with respect to the ZFS or hyperfine coupling tensor (see Section II.A.5 and II.A.6). Therefore, all contributions should be calculated independently by assuming a particular position for each proton. Otherwise the obtained best-fit parameters will be approximate. Furthermore, if fits of systems affected by hyperfine coupling to the metal nuclear spin or by static zero-field splitting are performed without considering these effects, some parameters can be affected by substantial errors. The decreased low field relaxation rates due to hyperfine coupling or ZFS, for instance, could be accounted for by a smaller r value or a larger hydration (examples are reported in Sections I.C.3 and I.C.4).

Finally, non-rigid reorientation of the macromolecule causes a decrease in the relaxation rates with respect to what is calculated on the assumption of rigid rotation (see Section II.B.2). When this effect is not taken into account, this decrease is usually reflected in a longer proton–metal ion distance.

From all these considerations, it results that it is usually very easy to obtain information on the presence or absence of protons in the first coordination sphere of a paramagnetic metal ion from the analysis of solvent proton relaxivity, but it may be hard to obtain their number, or to have information on second coordination sphere protons.

A.2. The Reorientational Correlation Time, τ_R

The reorientational correlation time can be predicted for spherical rigid particles, according to the Stokes–Einstein equation (*75–77*):

$$\tau_R = \frac{4\pi\eta a^3}{3kT} = \frac{\eta MW}{dN_A kT} \tag{11}$$

where η is the viscosity ($\mathrm{kg s^{-1} m^{-1}}$) of the solvent, a is the radius of the molecule, assumed spherical, MW is the molecular weight, ($\mathrm{kg\ mol^{-1}}$), d is the density ($\mathrm{kg\ m^{-3}}$) of the molecule (usually taken equal to 10^3), N_A is Avogadro's constant, k is the Boltzmann constant and T is the temperature. At room temperature and in water solutions, τ_R is about 3×10^{-11} s in hexaaqua metal complexes, about 10^{-8} s in small proteins ($MW \approx 25$–30 kg mol^{-1}), 10^{-6} s for larger macromolecules ($MW \approx 2.5$–3×10^3 kg mol^{-1}). The value of τ_R may also be estimated from the ^{1}H NMRD profile of a diamagnetic analog of the metal complex of interest. In fact, in diamagnetic systems the nuclear relaxation times are usually determined by dipolar coupling with other nuclei in the same molecule, the correlation time for the interaction being the reorientational correlation time of the molecule itself (in the absence of chemical exchange) (*78*).

In order to reduce the possibility of having different sets of parameter values providing the same relaxativity profiles, sometimes it is convenient to perform measurements at several temperatures, and to fit the available relaxation rates at all temperatures simultaneously. In this way we can constrain some parameters (like r, the ZFS parameters D and E (see Section II.A.5), the hyperfine coupling parameters $A_\parallel$ and $A_\perp$ (see Section II.A.6), the constant of contact interaction A_c (see Section II.A.4), and, possibly, Δ_t) to have the same values for all temperatures, whereas other parameters (like the reorientational correlation time, the lifetime, the correlation time for electron relaxation) are allowed to vary with temperature, following a given law. This largely reduces the number of unknowns in the fitting procedure. As far as the reorientational correlation time is concerned, an Arrhenius relationship with temperature is expected due to the temperature dependence of viscosity

$$\tau_R = A\exp(B/T) \tag{12}$$

so that only the values of the two parameters A and B need to be determined from the fit.

In the previous discussion, the electron-nucleus spin system was assumed to be rigidly held within a molecule isotropically rotating in solution. If the molecule cannot be treated as a rigid sphere, its motion is in general anisotropic, and three or five different reorientational correlation times have to be considered (*79*). Furthermore, it was calculated that free rotation of water protons about the metal ion-oxygen bond decreases the proton relaxation time in aqua ions of about 20% (*79*). A general treatment for considering the presence of internal motions faster than the reorientational correlation time of the whole molecule is the Lipari–Szabo model–free treatment (*80*). Relaxation is calculated as the sum of two terms (*81*), of the type

$$J(\omega) = S^2 \frac{\tau_R}{1 - \omega^2\tau_R^2} + (1 - S^2)\frac{\tau_f}{1 - \omega^2\tau_f^2} \tag{13}$$

where S is the generalized order parameter, $\tau_f^{-1} = \tau_R^{-1} + \tau_l^{-1}$, and τ_l is the correlation time for the fast metal-proton vector reorientation. In this approach, S^2 and τ_l are two new adjustable parameters. The generalized order parameter is a measure of the degree of spatial restriction of motion. Completely restricted motion means $S^2 = 1$, completely isotropic internal motion means $S^2 = 0$. However, the order parameter can vanish even if the motion is not isotropic. For instance, in the case of free rotation about a fixed axis,

$$S = \frac{3\cos^2\alpha - 1}{2} \tag{14}$$

where α is the semiangle of the cone. Therefore, S is equal to 0 when α is equal to the magic angle (54.7°). In the case of rotation within a cone of semiangle α,

$$S = \frac{\cos^2\alpha + \cos\alpha}{2} \tag{15}$$

The model–free approach applies every time there is evidence of local motions or when the value obtained for τ_R is smaller than the value expected from the Stokes–Einstein equation.

Alternatively, in order to take into account the effects of rotational diffusion of a water molecule around the metal-oxygen axis, a rotational correlation time for the metal-H vector was considered as an additional parameter besides the longer overall reorientational time (*82*).

A.3. The Lifetime, τ_M

The lifetime τ_M generally acts as a limiting factor for the propagation of the effect from the paramagnetic site to the solvent nuclei. In the absence of exchange, there is no paramagnetic contribution to the relaxation rate of solvent nuclei. In the presence of fast exchange, on the contrary, the paramagnetic contribution to the observed solvent relaxation rate is proportional to the relaxation rates of the coordinated nuclei, the mole fraction of the latter being the weightage factor.

Many systems are in an intermediate regime, i.e., $\tau_M \approx T_{1M}$. In those cases, the measured paramagnetic enhancement of the solvent relaxation rate is given by (Eq. (2) of Chapter 2):

$$T_{1P}^{-1} = \frac{P_M}{\tau_M + T_{1M}} \tag{16}$$

where P_M is the mole fraction of ligand nuclei in bound positions. The result on the shape of the profile is that all dispersions appear less evident, the ratio between the relaxation rates before and after the dispersions being closer to 1 (see Fig. 22), thus masking the effects related to paramagnetism and making it more difficult to obtain information from the profiles. In any case the measured relaxation rates are smaller.

Temperature dependence of the relaxation rate is a precious tool to exploit if dealing with the intermediate exchanging regime. In fact, τ_M is expected to have a temperature dependence described by an Eyring relationship:

$$\tau_M = \frac{A}{T}\exp(B/T). \tag{17}$$

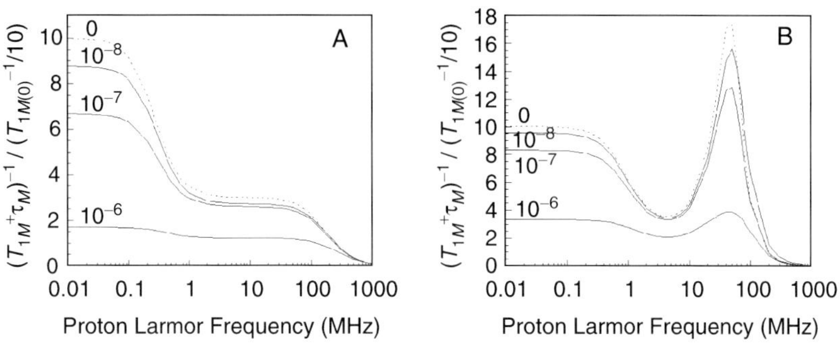

FIG. 22. Effect of the lifetime on the NMRD profiles. Conditions: S = 5/2, (A) twelve protons at 2.85 Å, $\tau_s = 10^{-9}$ s, or (B) two protons at 2.7 Å, $\Delta_t = 0.04$ cm^{-1}, $\tau_v = 10^{-11}$ s, and different values of τ_M (as labeled in the figure, in s).

where A and B are adjustable parameters, or, equivalently,

$$\tau_M^{-1} = \frac{kT}{h}\exp\left(\frac{-\Delta G^{\neq}}{RT}\right). \tag{18}$$

where $\Delta G^{\neq}$ is the free energy of activation for the dissociation process. Therefore, τ_M decreases significantly when the temperature increases. T_{1M} may either decrease or increase with increasing temperature, depending on the magnetic field and on the correlation time which modulates the relaxation process and its temperature dependence: if τ_c is the reorientational time, T_{1M} increases at low fields; if τ_c is a field independent effective electron relaxation time, T_{1M} increases at low fields; if τ_c is the field dependent effective electron relaxation time, and Δ_t is temperature independent, T_{1M} decreases at low fields and increases in correspondence of the high field relaxivity peak. Therefore, a strong increase in the solvent relaxation rates (proportional to $(\tau_M + T_{1M})^{-1}$) when temperature increases is a clear indication of τ_M being the limiting factor.

A.4. The Contact Coupling Constant, A_c

Table II reports the contact coupling constant for different aqua ion systems at room temperature. The contact coupling constant is a measure of the unpaired spin density delocalized at the coordinated protons. The values were calculated from the analysis of the contact contribution to the paramagnetic enhancements of relaxation rates in all cases where the correlation time for dipolar relaxation is dominated by τ_R and $T_{1e} > \tau_R$. In fact, in such cases the dispersion due to contact relaxation occurs earlier in frequency than the dispersion due to dipolar relaxation. In metalloproteins the contact contribution is usually negligible, even for metal ions characterized by a large contact contribution in aqua ion systems. This is due to the fact that the dipolar contribution is much larger because the correlation time increases by orders of magnitude, and τ_R becomes longer than T_{1e}. Under these conditions, the electron relaxation time is the correlation time for both contact and dipolar interactions. Systems with effective electron relaxation time smaller than the reorientational time usually have a negligible contact contribution, as can be easily verified by taking into account the values of all constants in the equations for dipolar and contact contributions (see Eqs. (1) and (5)), and considering the fact that A_c/h is usually smaller than 1 MHz.

The presence of contact contribution can be more easily evaluated from analysis of the transverse relaxation rate profiles, rather than the longitudinal relaxation rate profiles, in case the effective electron relaxation time is field-dependent. In fact, the non-dispersive term present in the equation for contact transverse relaxation rate (see (Eq. 6)) is responsible for a large increase in the rate with increasing the field, which begins just after the ω_S dispersion (an example is shown in Section II.E.4). Also for a field independent effective electron relaxation time, inspection of transverse relaxation rate with respect to longitudinal relaxation rate just after the ω_S dispersion (and before the ω_I dispersion) indicates the presence of a contact contribution. In fact, the non-dispersive term is responsible for a large difference between transverse and longitudinal relaxation rates (their ratio being larger than 7/6).

For nuclei other than protons the dipolar term is smaller, due to the smaller value of γ_N, and the contact term may be larger, in case of directly coordinated nuclei. Therefore, contact relaxation may more often be the dominant contribution to nuclear relaxation.

A.5. The Zero-Field Splitting

In general the NMRD profiles are affected also by other parameters characterizing the electron spin system such as the *g*-factor, the hyperfine coupling with the metal nuclear spin (for $I \geq 1/2$ systems) and the ZFS (for $S > 1/2$ systems).

The static ZFS, which is present in low-symmetry complexes, affects mainly the energy level fine structure. It is described by axial and rhombic components, D and E. Its effects on nuclear relaxation depend on two angles, θ and ϕ, defining the position of the nucleus with respect to the ZFS principal tensor axes. Figure 23 shows the dispersion profiles for different values of S, D, E and θ. Many such examples are reported in Chapter 2.

A major effect of the introduction of static ZFS is that of reducing the amplitude of the $\omega_S\tau_c$ dispersion, because the energy of most electronic spin transitions increases and can be much larger than the Zeeman energy, and larger than $\hbar\tau_c^{-1}$, so that for those transitions $\omega_S\tau_c \gg 1$ at all magnetic fields with dominant ZFS energy. Actually, we need to distinguish between integer and half-integer S systems. In fact, when S is a half-integer, there is always at least the $1/2 \leftrightarrow -1/2$ transition occurring at or close to the Zeeman energy, since the energy of these levels is not affected by the presence of ZFS. Therefore, the $\omega_S\tau_c$ dispersion (even if smaller) is present in the profile. When S is an integer, all transitions occur with energy altered by the ZFS (see Fig. 4A), and thus the $\omega_S\tau_c$ dispersion is totally quenched.

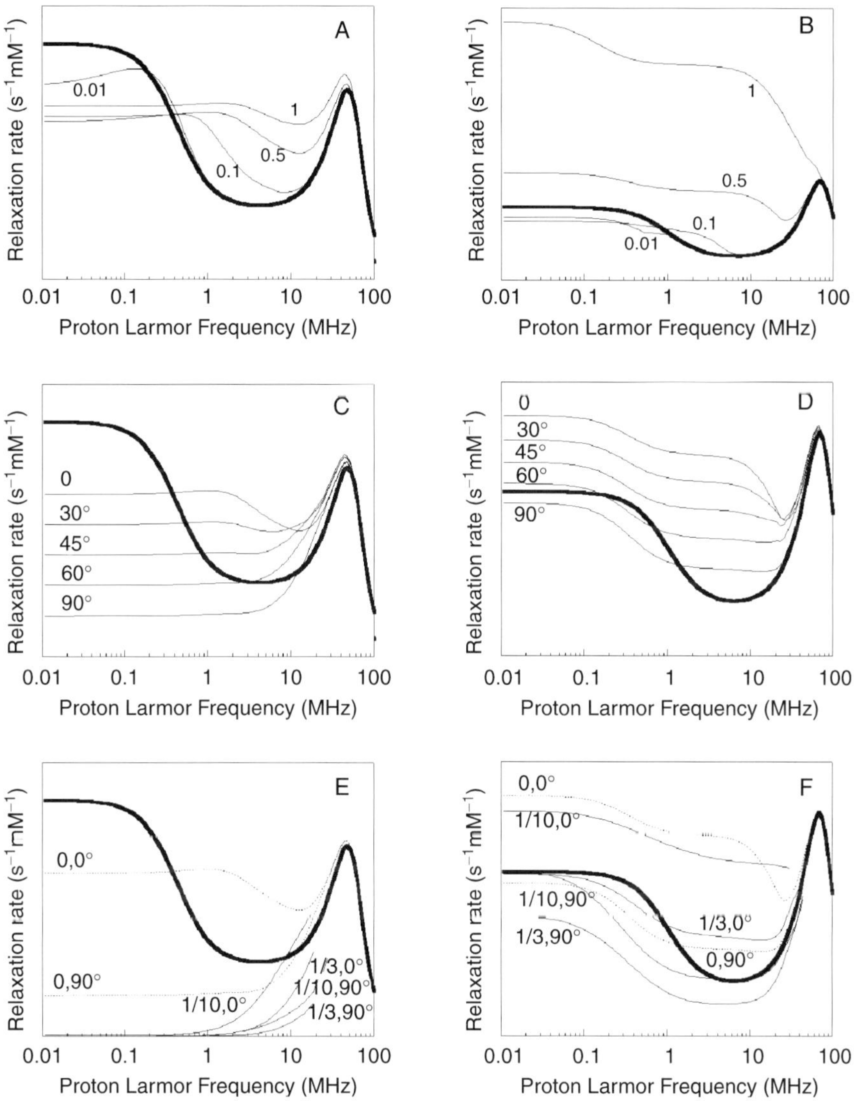

FIG. 23. Effect of static ZFS on the NMRD profiles for integer (S = 1) electron spin quantum numbers (panels A, C, and E) and half-integer ($S = 3/2$) electron spin quantum numbers (panels B, D, and F). Conditions: $\Delta_t = 0.1\ \mathrm{cm}^{-1}$, $\tau_v = 5 \times 10^{-12}$ s; (A) and (B): different values of the axial ZFS ($D = 0.01, 0.1, 0.5, 1\ \mathrm{cm}^{-1}$) and $\theta = 0$; (C) and (D): different angles ($\theta = 0, 30, 45, 60, 90°$) and $D = 0.5\ \mathrm{cm}^{-1}$; (E) and (F) different rhombicity of the ZFS ($E + D = 0.5\ \mathrm{cm}^{-1}$; $E/D = 0$, 1/10, 1/3) and different angles ($\theta = 0$, 90°). The profile in bold shows the relaxation rate values calculated without including ZFS (Solomon profile).

The presence of ZFS may also cause the occurrence of a further dispersion in the plot of relaxation rate as a function of the magnetic field, corresponding to the transition from the dominant ZFS limit to the dominant Zeeman limit (see Section I.C.2.a).

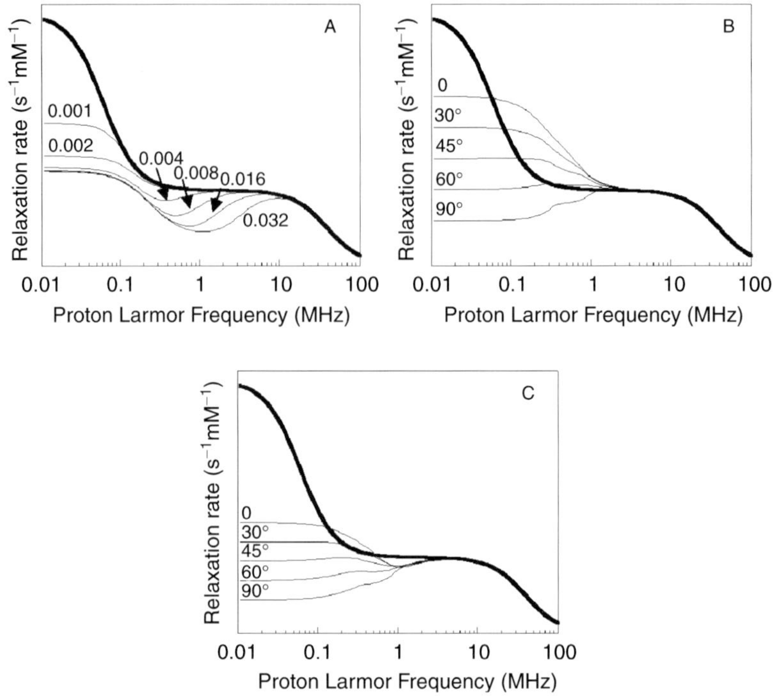

FIG. 24. Effect of the hyperfine coupling with the metal nuclear spin on the NMRD profiles, calculated for $S = 1/2$ and $I = 3/2$. Conditions: $\tau_c = 4 \times 10^{-9}$ s and (A) different values of the isotropic hyperfine coupling constant ($A_{||} = A_{\perp} = 0.001, 0.002, 0.004, 0.008, 0.016, 0.032$ cm^{-1}); (B) $A_{||} = 0.016$ cm^{-1}, $A_{\perp} = 0$ and different angles ($\theta = 0, 30, 45, 60, 90°$); (C) $A_{||} = 0.016$ cm^{-1}, $A_{\perp} = 0.005$ cm^{-1} and different angles ($\theta = 0, 30, 45, 60, 90°$). The profile in bold shows the relaxation rate values calculated without including hyperfine coupling with the metal nucleus effects (Solomon profile).

Further complications arise due to the effect of the rhombicity of ZFS, described by the parameter E. E can vary in the range $-1/3 \leq E/D \leq 1/3$. We only note here that rhombicity causes a drop to zero of the relaxation rate for integer S systems, whereas its effect is minor for half-integer S systems. For further details we refer to the review by Sharp *et al.* (*45*).

A.6. The Hyperfine Coupling with the Metal Nucleus

The hyperfine interaction between the electron spin and the metal nuclear spin affects the electron energy levels (see Fig. 4B), and thus the transition probabilities. The shape of the ^{1}H NMRD profiles can be very different according to the different values of $A_{||}$, $A_{\perp}$ and the position of the proton with respect to the molecular frame defined by the hyperfine $\boldsymbol{A}$ tensor, similarly to what is seen in the case of ZFS. Figure 24 shows the dispersion profiles for different values of S, $A_{||}$, $A_{\perp}$, and θ. Again, the general effect is that of reducing the relaxation rate at all fields corresponding to a Zeeman energy smaller than the hyperfine coupling energy. We note that, at variance with ZFS, the hyperfine coupling with the metal nuclear spin affects the relaxivity profiles also in the case of I and S equal to 1/2 and for isotropic systems ($A_{||} = A_{\perp}$) (*1,27*).

A.7. Diffusing Water Protons

Molecular hydration in solution is described not only by the inner-sphere water molecules (first and second coordination spheres, see Section II.A.1) but also by solvent water molecules freely diffusing up to a distance of closest approach to the metal ion, d. The latter molecules are responsible for the so-called outer-sphere relaxation (*83,84*), which must be added to the paramagnetic enhancement of the solvent relaxation rates due to inner-sphere protons to obtain the total relaxation rate enhancement, $T^{-1}_{1obs} - T^{-1}_{1dia}$:

$$T^{-1}_{1obs} - T^{-1}_{1dia} = P_M \sum_i (T_{1Mi} + \tau_{Mi})^{-1} + T^{-1}_{1os}. \tag{19}$$

The classical equation for T^{-1}_{1os} is provided in Section VII.A of Chapter 2. It depends only on the spin quantum number S, on the molar concentration of paramagnetic metal ions, on the distance d, and on a diffusion coefficient D, which is the sum of the diffusion coefficients of both the solvent molecule (D_L) and the paramagnetic complex (D_M), usually much smaller. The outer-sphere relaxivity calculated with this equation at room temperature and in pure water solution, by assuming d equal to 3 Å, is shown in Fig. 25. It appears that the dispersions do not have the usual Lorentzian form.

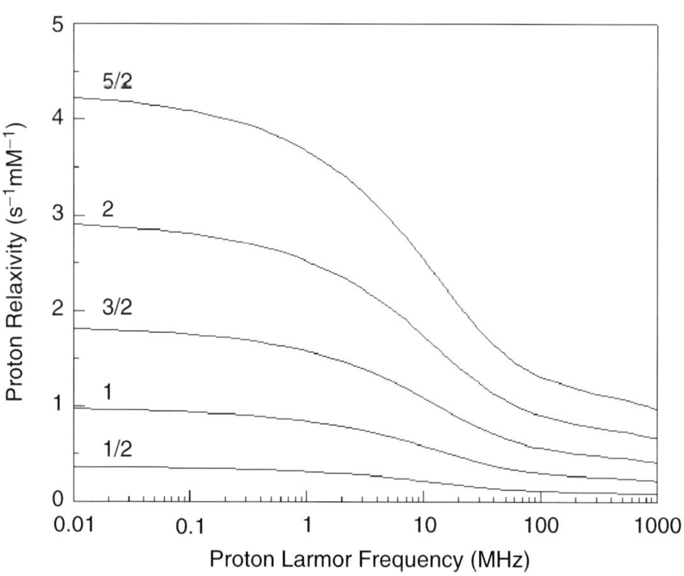

FIG. 25. Outer-sphere contribution to the water proton relaxation rate, calculated for a distance of closest approach, d, of 3 Å and different values of S (1/2, 1, 3/2, 2, 5/2).

This model is not completely accurate as it does not incorporate the dependence on the electron relaxation time (see Section VII.B of Chapter 2). However, it provides the correct order of magnitude of the outer-sphere relaxation in case the effective electron relaxation time is not shorter than the diffusional correlation time $\tau_D = d^2/D$, and thus is a good estimate if our investigation is focused on a much larger contribution from inner-sphere relaxation. On the contrary, if outer-sphere relaxation is of the same order of magnitude of inner-sphere relaxation, the correct equations taking into account the electron-spin relaxation effects and ZFS terms need to be considered (see Section VII of Chapter 2). A good approximation is provided by considering the following definition of z to be inserted into the spectral density $J(\omega)$ derived in the Freed's force-free diffusion model (*85*).

$$z = \sqrt{2\left(\omega\tau_D + \frac{\tau_D}{T_{1e}}\right)}. \tag{20}$$

B. Copper Complexes and Proteins

Figure 26 shows the water proton NMRD profile of the copper(II) aqua ion at 298 K (*86*). Only one dispersion appears, and the relaxivity values before

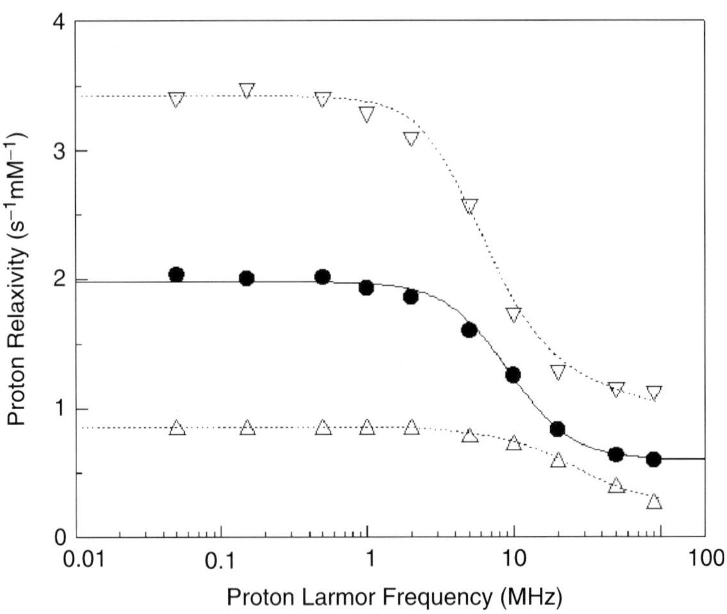

FIG. 26. Paramagnetic enhancements to water ^{1}H NMRD profiles for $Cu(H_2O)_6^{2+}$ solutions at 298 K (•), 278 K (▽), and 338 K (Δ). The lines represent the best-fit curves obtained using the Solomon equation [Eq. (1)], with a Cu–H distance of 0.27 nm and $\tau_c = 2.6 \times 10^{-11}$ s (298 K), 4.0×10^{-11} s (278 K) and 0.95×10^{-11} s (338 K).

and after the dispersion are in the 10 to 3 ratio. This means that the profile can be satisfactorily fit with the Solomon equation (Eq. (1)), the observed dispersion being the $\omega_S\tau_c$ dispersion. The fit performed on the two parameters r and τ_c indicates that the distance between the 12 (10) water protons and the paramagnetic center, r, is equal to 2.8 (2.7) Å, and that the correlation time τ_c is equal to 3×10^{-11} s. This value is of the order of what is expected from the Stokes–Einstein equation (Eq. (11)) for an aqua ion. Since the electron relaxation time for copper is many orders of magnitude longer and also the lifetime is expected to be longer (*23*), we conclude that the correlation time is the reorientational time. From the analysis of this profile we can also exclude any possible contact relaxation contribution, because otherwise the corresponding $\omega_S T_{2e}$ dispersion would have been present in the NMRD profile. Actually, the unpaired electron occupies an E_g orbital, and no delocalization is thus expected on the water molecule. In fact, one fully occupied water molecular orbital of σ type directly overlaps with the T_{2g} metal orbital set (*87*). Profiles acquired at different temperature warrant that the system is in a fast exchange regime, i.e., $\tau_M \ll T_{1M}$.

When the temperature is lowered and/or the viscosity of the solution is increased by using glycerol-water mixtures as solvent, the reorientational correlation time increases. Since the reorientational time is the correlation time for nuclear relaxation, the effects on the ^{1}H NMRD profile (Fig. 27) are: (i) higher relaxivity values at low frequencies; (ii) a shift toward lower

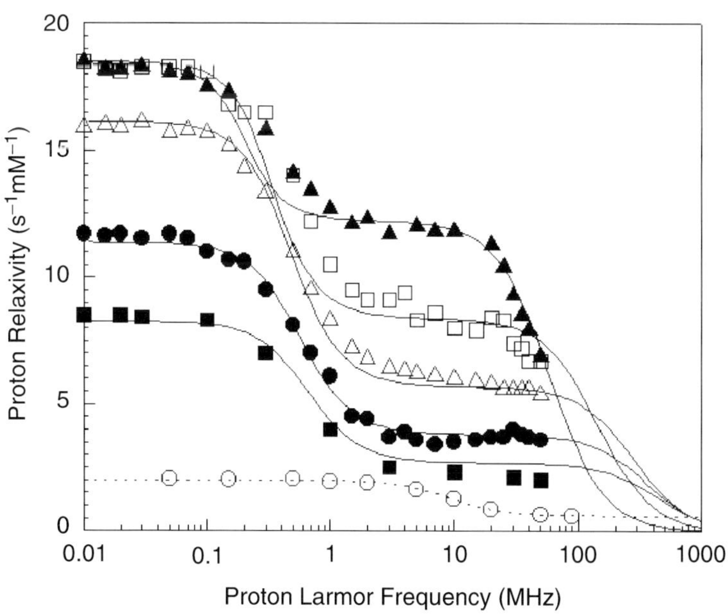

FIG. 27. Paramagnetic enhancements to solvent ^{1}H NMRD profiles for ethyleneglycol solutions of $Cu(ClO_4)_2 \cdot 6H_2O$ at 264 (▲), 278 (□), 288 (△), 298 (•), and 312 (■) K as compared to the profile for water solutions at 298 K (○) (*47*). The solid lines are best fit curves obtained using an isotropic A value.

frequencies of the $\omega_s\tau_c$ dispersion; (iii) the appearance of a second dispersion, the $\omega_I\tau_c$ dispersion, at high fields. However, with increasing the reorientational time, the relaxivity values before and after the $\omega_s\tau_c$ dispersion are in a ratio increasingly longer than the 10 to 3 ratio predicted by the Solomon equation. This is related to the fact that τ_R is increased so much that the correlation time is now determined by the effective electron relaxation time. Under such conditions, in fact, electron energy levels and their transition probability must be calculated by considering, besides the Zeeman interaction energy, also the hyperfine coupling between metal ion (copper(II) has $I = 3/2$) and the unpaired electron ($S = 1/2$). As shown in Section II.A.6 the effect on the shape of the relaxivity profile can be quite dramatic, in the range of frequencies where the hyperfine coupling energy is larger than the Zeeman energy, i.e., $Ach \geq g_e\mu_B B_0$, where A is the electron-copper nucleus hyperfine coupling constant in cm^{-1}, and thus for proton Larmor frequency smaller than $Ac\gamma_I/\gamma_S$. The profiles were fit with an isotropic hyperfine constant of 0.0026 cm^{-1}.

^1H NMRD profiles of copper(II) proteins should thus be always fit by considering the effects of hyperfine coupling between the unpaired electron and the metal nucleus. The values of $A_{\|}$ and $A_{\perp}$ can be obtained by EPR measurements or directly from the relaxation profile. Typical $A_{\|}$ values for type II copper containing proteins at room temperature are between 120 and 190×10^{-4} cm^{-1}, as measured in copper bovine carbonic anhydrase II and its anion adducts, Cu_2Zn_2 superoxide dismutase and adducts, Cu_2 transferrin, Cu_2 alkaline phosphatase, Cu phthalate dioxygenase (*88*), whereas $A_{\perp}$ is usually very small, up to 40×10^{-4} cm^{-1}. The fit performed for the already shown ^1H NMRD profiles of the Cu_2Zn_2 superoxide dismutase solution at 273, 278, 288, and 298 K (Fig. 5), for instance, indicates the presence of an axial water molecule, coordinated to the copper ion with a Cu – H distance of 3.4 Å, and provides values for the electron relaxation time ranging from 4.6×10^{-9} s at 273 K to 1.8×10^{-9} s at 298 K, $A_{\|} = 137 \times 10^{-4}$ cm^{-1} and $A_{\perp}$ negligible (*27*), in agreement with EPR measurements (*89*).

The ^1H NMRD profiles of the protein copper(II)-CopC at 278, 288, and 303 K are shown in Fig. 28 (*28*). Only the $\omega_S\tau_c$ dispersion is detected in the investigated range of frequency. The fit performed with the Florence NMRD program provided a value for the correlation time τ_c of 0.109×10^{-12} $e^{3012/T}$ s (where T is the temperature), corresponding to 5×10^{-9} s at 278 K and 2×10^{-9} s at 303 K, $A_{\|}$ equal to 177×10^{-4} cm^{-1}, $A_{\perp} \approx 0$, $\theta = 42°$ and two water protons at 3.5 Å. Both electron relaxation time and reorientational time are probably contributing to the value of τ_c, as τ_R is estimated around 5×10^{-9} s at 303 K. It is worth noting that it was necessary to acquire the profiles at different temperatures and to fit them simultaneously to obtain a unique fit. In fact, each profile can be fit independently with more than one set of parameters.

In copper(II) systems, a small (10–15%) g-anisotropy is always present. Calculations show that such modest g-anisotropy has negligible effect on the dispersion curves, thus affecting the resulting best fit parameters only

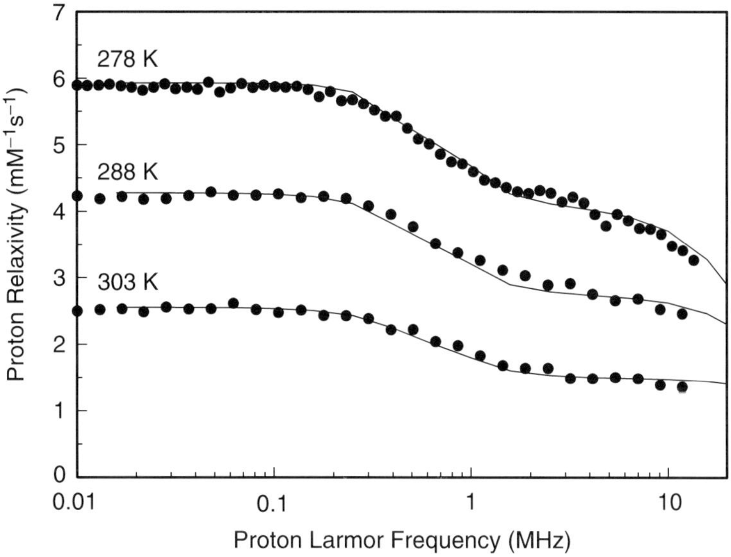

FIG. 28. Water ^{1}H NMRD profiles for solutions of Cu(II)-CopC obtained at three different temperatures (*28*).

slightly (*27*). The Florence NMRD program can, anyway, take into account *g*-anisotropy effects.

C. IRON COMPLEXES AND PROTEINS

The ^{1}H NMRD profiles of water solutions of $Fe(H_2O)_6^{3+}$ in 1 M perchloric acid at 278, 288, 298, 308 K are shown in Fig. 29 (*36*). The profiles are characterized by the following features: (i) one dispersion only is displayed; (ii) proton relaxivity slightly increases at low fields and decreases at high fields with increasing temperature; (iii) proton relaxivity increases with field in the region between 10 and 50 MHz; such an increase is more evident for low temperature. All these features can be rationalized by concluding that the correlation time is influenced by a field dependent electron relaxation time, according to the SBM theory. The position of the dispersion indicates a correlation time around 3×10^{-11} s. Since this value is similar to what is expected for the reorientational time of the complex, it means that both reorientational and electron relaxation time are contributing to the value of τ_C for dipolar relaxation. On the other hand, under these conditions also the correlation time for contact relaxation has about the same value, and thus the dispersion relative to contact relaxation would be superimposed on the dispersion relative to dipolar relaxation. Contact relaxation can be estimated from the measurement of the transverse proton relaxation rates. Actually, the increase in the transverse relaxation rate with field indicates clearly that a contact contribution is present (see Section II.A.4). Finally, we must consider the possibility that protons are not in the fast exchange regime, as it can be easily understood from the fact that at low fields, proton

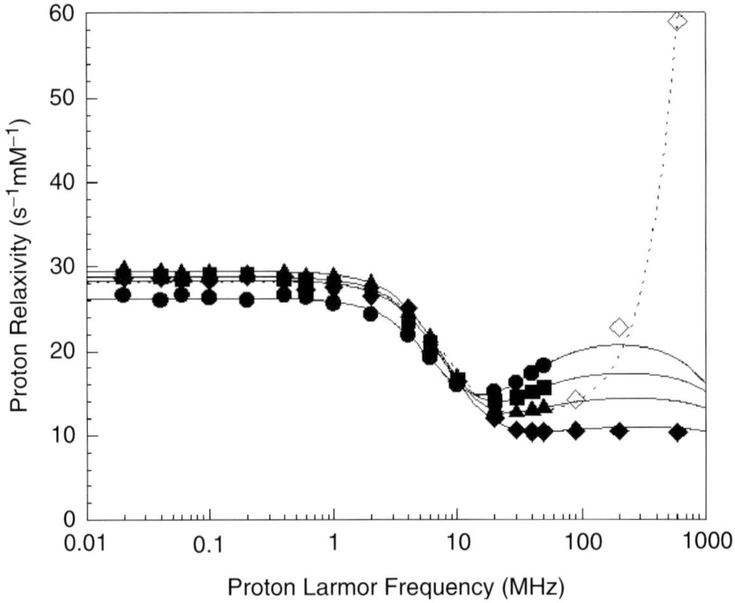

FIG. 29. Water ^{1}H NMRD longitudinal profiles for $Fe(H_2O)_6^{3+}$ solutions at 278 (●), 288 (■), 298 (▲), and 308 (◆) K. High field transverse relaxivity data at 308 K (◇) are also shown. The lines represent the best fit curves using the Solomon–Bloembergen–Morgan equations (Eqs. (1–6)) (*36*).

relaxivity increases with increasing temperature, and the ratio between the values before and after the dispersion is smaller than 10:3. The simultaneous fit of all the profiles does not converge to an unique solution. This ambiguity was resolved by using the values for τ_M obtained from independent experiments (*90*): τ_M was fixed at 2.0×10^{-7} s at 308 K and 1.7×10^{-6} s at 278 K. The latter value is indeed about the same value of T_{1M} at low field at the same temperature. The obtained best fit values to the SBM equations are the following: $r = 2.62$ Å, $A_c/h = 0.43$ MHz, $\Delta_t = 0.095$ cm^{-1}, $\tau_R = 1.1 \times 10^{-10}$ s at 278 K and 3.9×10^{-11} s at 308 K, $\tau_v = 6.2 \times 10^{-12}$ s at 278 K and 3.5×10^{-12} s at 308 K. The metal–proton distance, r, obtained in the fit is indeed rather short. This can be due partially to the fact that any outer-sphere contribution to relaxation was neglected, and to the fact that second coordination sphere protons have been also neglected. Water molecules in the second coordination sphere are in fact expected to be strongly hydrogen bonded to the inner shell of the coordinated water molecules. This can also account for the fact that the reorientational time is slightly larger than what is expected from the Stokes–Einstein equation (see also Section II.E.4). By increasing pH above 3, proton relaxivity decreases markedly as a result of the formation and precipitation of a variety of hydroxides.

In macromolecules, slow exchange effects often quench the relaxivity (Fig. 30) (*37*) even in the presence of water molecules directly coordinated to iron(III) (*91*). For instance, in methemoglobin the relaxation rates are attributable to one water molecule coordinated to the paramagnetic center

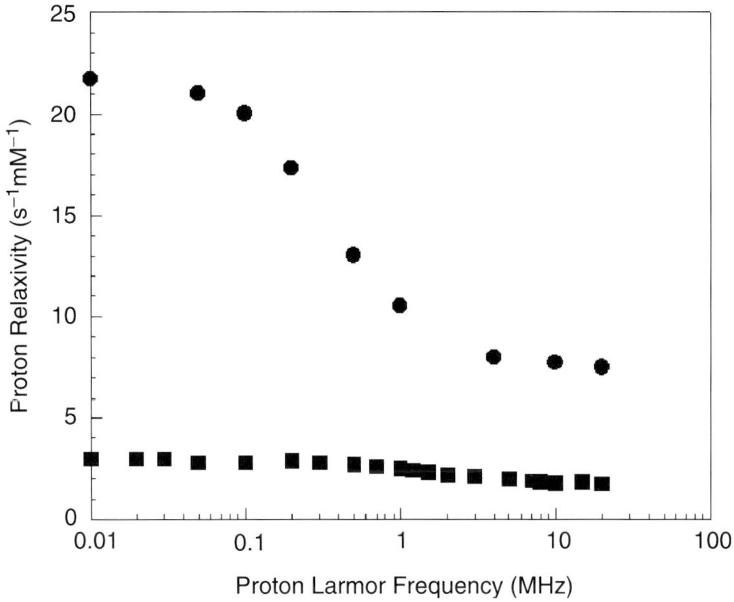

FIG. 30. Water ^{1}H NMRD profiles for solutions of methemoglobin (■) and fluoro-methemoglobin (●) at 279 K (*37*).

in slow exchange regime, besides the outer-sphere contribution. The fluoride derivative of methemoglobin is a clear example of a relaxation profile in the presence of fast exchanging water molecules (Fig. 30) (*37,91*). In fact, in fluoro-methemoglobin the water molecule is replaced by fluoride, which is H-bonded to a water molecule. In this case, the second sphere water in fast exchange provides a higher relaxivity than that of the directly coordinated but slowly exchanging water in methemoglobin.

The proton relaxivity profiles of oxidized rubredoxin (Fig. 31) have been analyzed in terms of outer-sphere and inner-sphere dipolar relaxation, and indicate that exchangeable water protons must be at distances larger than 4 Å to account for such low rates (*92*). The position of the dispersion provides a value for the electron relaxation time of the order of 70×10^{-12} s at room temperature. The fit was performed, as necessary in slowly rotating systems with $S > 1/2$, by taking into account the presence of static ZFS (> 1.5 cm^{-1}). Analogous relaxation measurements performed on the C6S rubredoxin variant, where a hydroxide ion is coordinated to the metal ion, show dramatically higher relaxation rate values, ascribed to increased hydration and electron relaxation time of the order of 150×10^{-12} s at room temperature. The analysis excludes, in any case, the presence of first-sphere exchangeable protons, thus ruling out fast exchange of the coordinated hydroxide proton (*92*).

D. Manganese Complexes and Proteins

The ^{1}H NMRD profiles of manganese(II) aqua ion and of manganese(II)-concanavalin A have already been analyzed in Section I.C.3. The NMRD

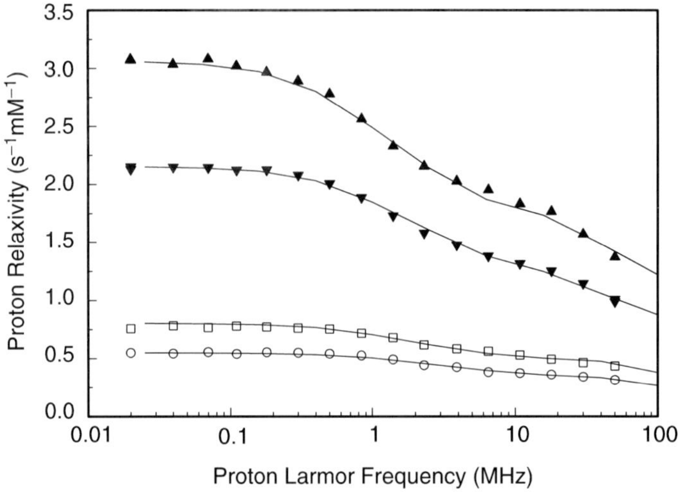

FIG. 31. Paramagnetic enhancements to water ^{1}H NMRD profiles for solutions of wild type (open symbols) and C6S (filled symbols) rubredoxin at 298 K (○ and ▼ respectively) and 283 K (□ and ▲ respectively) (*92*).

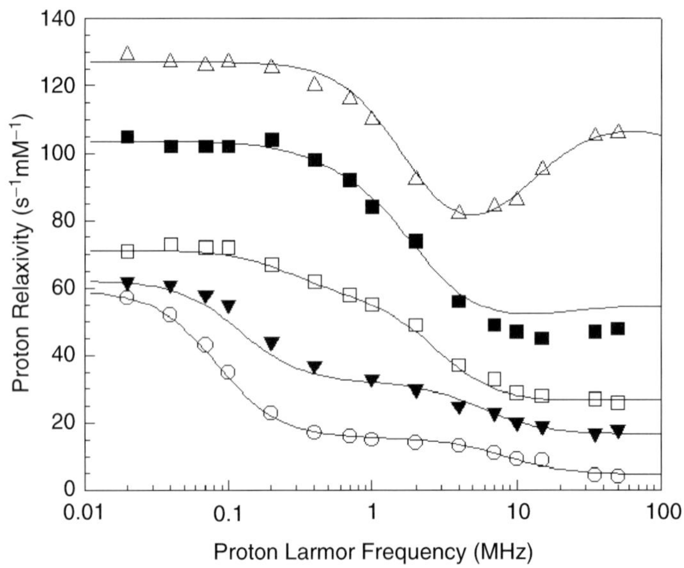

FIG. 32. Solvent ^{1}H NMRD profiles for $Mn(H_2O)_6^{2+}$ solutions at 308 K in pure water (○) and with increasing amounts of d^8-glycerol: 35% (▼), 55% (□), 65% (■), 75% (△) (*39*).

profiles of manganese(II) at different concentrations of d^8 glycerol (*39*) are reported in Fig. 32. The main features that can be observed on the profiles with increasing viscosity are: (i) the relaxivity increases, (ii) the contact dispersion progressively disappears, (iii) the dipolar dispersion moves toward lower fields, (iv) a peak appears at high fields. These features can be explained by the fact that the reorientational time increases, thus increasing the correlation time for dipolar relaxation, and, as a result, the dipolar contribution to proton relaxivity. At the same time the correlation time for electron relaxation, τ_v, increases, thus resulting in a decrease in the effective electron relaxation time at low fields, and thus of the contact contribution. At some point the correlation time for dipolar relaxation becomes the effective electron relaxation time, so that its field dependence produces the peak arising at high fields. The fit was obtained with constant $\Delta_t = 0.03$ cm^{-1}, $A_c/h = 0.64$ MHz and $r = 2.78$ Å, and it was found that both τ_r and τ_v increase linearly with viscosity, pointing out that modulation of the transient ZFS by collision with solvent is the dominant source of relaxation up to very high viscosity. The range for τ_v spreads from 6×10^{-12} s for the pure water solution to 5×10^{-11} s for a solution containing 65% w/w of glycerol, at 288 K (*39*).

At variance with the aqua ion, in most manganese(II) proteins and complexes the contact contribution to relaxation is found negligible. This is clearly the case for MnEDTA (Fig. 33), the relaxivity of which indicates the presence of the dipolar contribution only, and one water molecule bound to the complex (*93*). Actually the profile is very similar to that of GdDTPA (see Chapter 4), and is provided by the sum of inner-sphere and outer-sphere contributions of the same order. The relaxation rate of MnDTPA is accounted for by outer-sphere relaxation only (see Section II.A.7), no water molecules being coordinated to the complex (*94*).

At variance with the manganese(II) concanavalin A, the relaxivity of which increases with increasing temperature at low field, indicating the

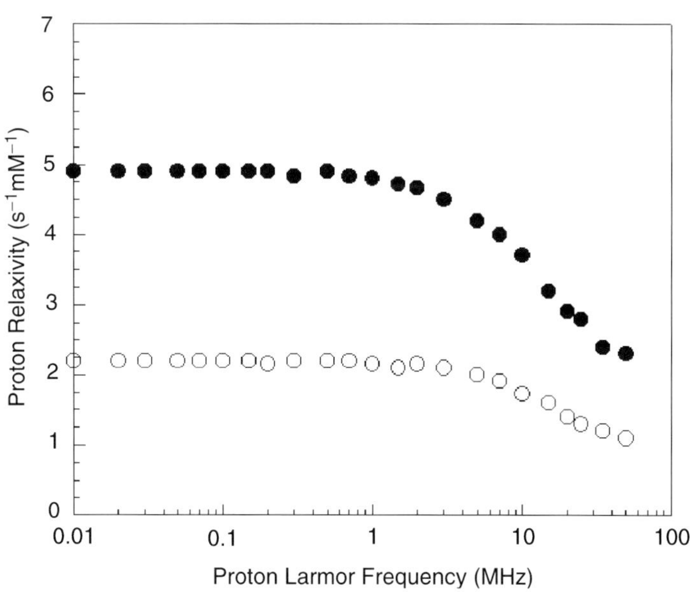

FIG. 33. Water ^{1}H NMRD profiles for MnEDTA (•) and MnDTPA (○) solutions at 308 K.

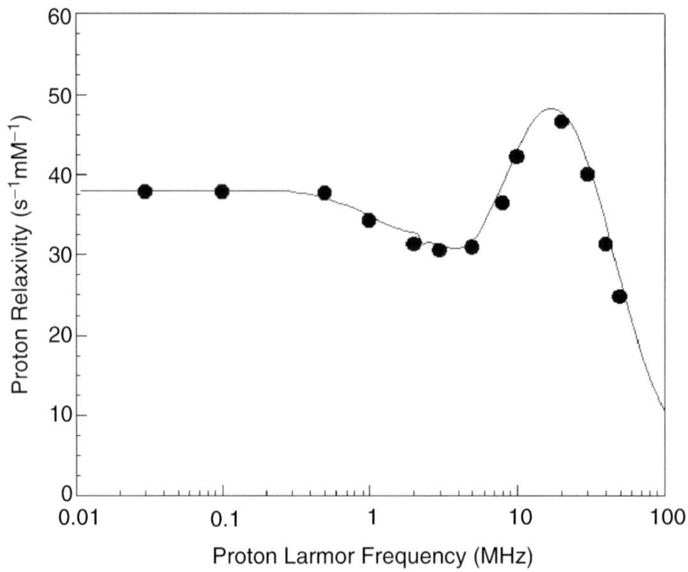

FIG. 34. Water ^{1}H NMRD profiles for $Mn(EDTA)(BOM)_2$-HSA solutions at 298 K *(96,97)*.

presence of coordinated protons in slow exchange, in manganese(II) lentil lectin (a protein with high degree of structural similarity with concavalin A) relaxivity is due to fast-exchanging protons, located far from the metal ion, as shown by the low rates *(95)*, lower than 4 s^{-1} mM^{-1}. In Fig. 34 we report the ^{1}H NMRD profile of $[Mn(EDTA)(BOM)_2]^{2-}$ bound to human serum albumin (HSA) *(96)*. Again, the profile is characterized by the following features: (i) the presence of a dispersion at about 0.2–1 MHz, due to the dipolar $\omega_S T_{2e}$ dispersion in the presence of static ZFS, (ii) an increase in relaxivity with increasing the field from 1 to 10 MHz, due to the field dependence of T_{1e}, (iii) a subsequent decrease due to the $\omega_I T_{1e}$ dispersion. The best-fit analysis performed with the modified NMRD program indicates that one fast-exchanging water molecule is coordinated to the metal ion, the electron relaxation is determined by the parameters $\Delta_t = 0.012$ cm^{-1} and $\tau_v = 4 \times 10^{-11}$ s, and static ZFS is present with $D = 0.05$ cm^{-1} *(96,97)*. This set of parameters can also account for the NMRD profile of $Mn(EDTA)(BOM)_2$ in water solution, if data are properly analyzed by taking into account fast rotation conditions *(97)*. It is to be noted that the fit of $Mn(EDTA)(BOM)_2$-HSA data performed with the SBM theory would have provided somewhat wrong values *(95)*.

E. OTHER TRANSITION METAL IONS COMPLEXES AND PROTEINS

E.1. Nickel(II)

^{1}H NMRD profiles of nickel(II) complexes in solution have been acquired to obtain information on relevant dynamic and structural parameters of the investigated systems, and thus on the magnitude of transient ZFS and on τ_v, on the water proton distance from the paramagnetic center, and on the magnitude and rhombicity of the large static ZFS, typically present in nickel(II) systems. Nickel(II) has integer spin quantum number, $S=1$. This accounts for the fact that very low relaxivity values, down to zero, at low field, can be due to the large rhombicity of the static ZFS. No dispersion is observed at low and medium fields, due to the fast electron relaxation time, and to the absence of the ω_S dispersion expected for all integer spin systems with static ZFS (see Section II.A.5). The profiles are thus characterized by being flat and with low values of relaxivity up to almost 50 MHz, when a sharp increase in relaxivity is detected (see Section I.C.5). These systems are often outside the Redfield limit for electron relaxation (see Section IV.A of Chapter 2) and thus, an appropriate slow-motion theory should be applied. As an example, the ^{1}H NMRD profile of $Ni(dpm)_2(aniline\text{-}d_5)_2^{2+}$ was fit according to the pseudorotational model, and the following parameters were obtained: axial ZFS, $D = 5$ cm^{-1}, $\Delta_t = 1.7$ cm^{-1}, $\tau_v = 11 \times 10^{-12}$ s, $r = 3.5$ Å (*98*). Figure 9 of Chapter 2 reports the ^{1}H NMRD profile for $Ni(tmc)(H_2O)^{2+}$ at 298 K; fitting parameters are $D = 6$ cm^{-1}, $E = 0.93$ cm^{-1}, $\Delta_t = 6$ cm^{-1}, $\tau_v = 0.6 \times 10^{-12}$ s, $r = 3.1$ Å. As expected, low relaxivity values are due to ZFS rhombicity. The obtained Δ_t and τ_v values indicate very large flexibility or deformability in this complex (*98*).

E.2. Oxovanadium(IV)

The NMRD profiles of $VO(H_2O)_5^{2+}$ at different temperatures are shown in Fig. 35 (*58*). As already seen in Section I.C.6, the first dispersion is ascribed to the contact relaxation, and is in accordance with an electron relaxation time of about 5×10^{-10} s, and the second to the dipolar relaxation, in accordance with a reorientational correlation time of about 5×10^{-11} s. A significant contribution for contact relaxation is actually expected because the unpaired electron occupies a T_{2g} orbital, which has the correct symmetry for directly overlapping the fully occupied water molecular orbitals of σ type (*87*). The analysis was performed considering that the four water molecules in the equatorial plane are strongly coordinated, whereas the fifth axial water is weakly coordinated and exchanges much faster than the former. The fit indicates a distance of 2.6 Å from the paramagnetic center for the protons in the equatorial plane, and of 2.9 Å for those of the axial water, and a constant of contact interaction for the equatorial water molecules equal to 2.1 MHz. With increasing temperature, the measurements indicate that the electron relaxation time increases, whereas the reorientational time decreases.

The NMRD profile of bis-oxovanadium(IV) transferrin displays two dispersions (Fig. 36) (*59*). The one at about 10 MHz is attributed to the ω_I dispersion, providing in this way a value for τ_c equal to 2×10^{-8} s, which is ascribed to the electron relaxation time with possible contributions from τ_M, since the reorientational time of the protein is of the order of $2\text{–}3 \times 10^{-7}$ s. The fit requires considering the presence of hyperfine coupling with

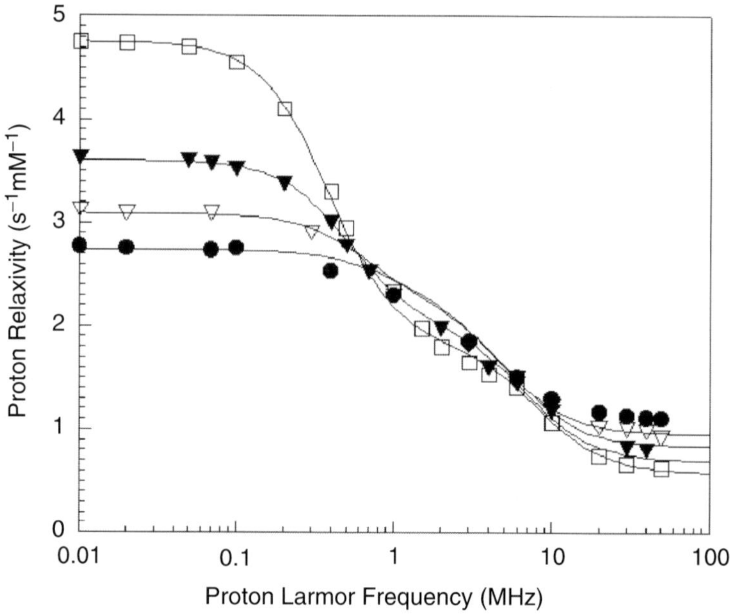

FIG. 35. Paramagnetic enhancements to water ^{1}H NMRD profiles for $VO(H_2O)_5^{2+}$ solutions at (●) 278 K, (▽) 288 K, (▼) 298 K, and (□) 308 K. The solid lines represent the best-fit curves using the Solomon–Bloembergen–Morgan equations (Eqs. (1 and 2, 5 and 6)) (*58*).

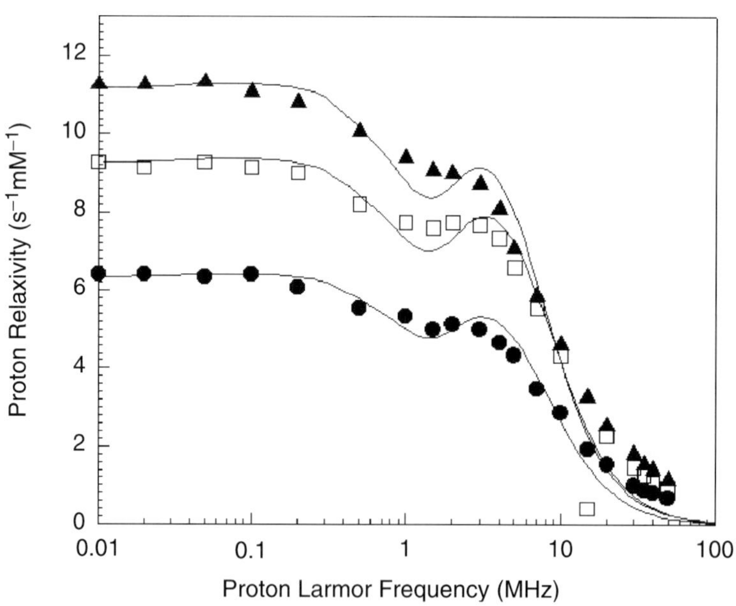

FIG. 36. Paramagnetic enhancements to water ^{1}H NMRD profiles for solutions of VO^{2+}-tranferrin at pH 8 and (▲) 281 K, (□) 298 K, and (●) 311 K (*59*).

the metal nucleus ($A_{||} = 170 \times 10^{-4}$ cm^{-1} and $A_{\perp} = 60 \times 10^{-4}$ cm^{-1}), which largely affects the shape of the low field region, and indicates the presence of two water protons at 3.3 Å.

E.3. *Titanium(III)*

The ^{1}H NMRD profiles of water solution of $Ti(H_2O)_6^{3+}$ have been shown in Section I.C.7 and have been already discussed. We only add here that the best fit procedures provide a constant of contact interaction of 4.5 MHz (*61*), and a distance of the twelve water protons from the metal ion of 2.62 Å. If a 10% outer-sphere contribution is subtracted from the data, the distance increases to 2.67 Å, which is a reasonably good value. The increase at high fields in the R_2 values cannot in this case be ascribed to the non-dispersive term present in the contact relaxation equation, as in other cases, because longitudinal measurements do not indicate field dependence in the electron relaxation time. Therefore they were related to chemical exchange contributions (see Eq. (3) of Chapter 2) and indicate values for τ_M equal to 4.2×10^{-7} s and 1.2×10^{-7} s at 293 and 308 K, respectively.

E.4. *Chromium(III)*

The ^{1}H NMRD profile of chromium(III) aqua ion (Fig. 18) is characterized by slow exchanging water protons, as clearly shown by the fact that the solvent proton relaxivity at low fields increases with increasing the temperature. The occurrence of slow exchange hinders any increase in relaxivity below 300 K, thus explaining the fact that the contact dispersion disappears in the low temperature profiles, whereas it is well shown in the high temperature profiles, as already discussed in Section I.C.8.

The smaller contribution to solvent proton relaxation due to the slow exchanging regime also allows detection of second and outer sphere contributions (*62*). In fact outer-sphere and/or second sphere protons contribute less than 5% of proton relaxivity for the highest temperature profile, and to about 30% for the lowest temperature profile. The fact that they affect differently the profiles acquired at different temperature influences the best-fit values of all parameters with respect to the values obtained without including outer and second sphere contributions, and not only the value of the first sphere proton-metal ion distance (as it usually happens for the other metal aqua ions). A simultaneous fit of longitudinal and transverse relaxation rates provides the values of the distance of the 12 water protons from the metal ion (2.71 Å), of the transient ZFS (0.11 cm^{-1}), of the correlation time for electron relaxation (about 2×10^{-12} s at room temperature), of the reorientational time (about 70×10^{-12} s at room temperature), of the lifetime (about 7×10^{-6} s at room temperature), of the constant of contact interaction (2.1 MHz). A second coordination sphere was considered with 26 fast exchanging water protons at 4.5 Å from the metal ion (*99*), and the distance of closest approach was fixed in the range between 5.5 and 6.5 Å.

Interestingly, the reorientational time is about 2–3 times larger than expected for a hexaaqua ion. Indeed, the second sphere water molecules increase the apparent weight of the hydrated ion by about three times the molecular weight of the hexaaquachromium(III) ion considered without second sphere water molecules. In the approximation of the Stokes–Einstein model [Eq. (11)], this would imply that the lifetime of these water molecules is still longer than the reorientational time.

The presence of second-sphere water molecules could be considered also for other metal aqua ions, like iron(III) and oxovanadium(IV) aqua ions, where the reorientational time is found to be longer than expected. However, in the other cases τ_R increases much less than for the chromium(III) aqua ion, thus suggesting that second-sphere water molecules are more labile, their lifetime being of the order of the reorientational time.

F. Gadolinium Complexes and Proteins

As an example of behavior of a typical Gd-complex and Gd-macromolecule we discuss here the NMRD profiles of a derivative of Gd-DTPA with a built-in sulfonamide (SA) and the profile of its adduct with carbonic anhydrase (see Fig. 37) (*100*). Other systems are described in Chapter 4. The profile of Gd-DTPA-SA contains one dispersion only, centered at about 10 MHz, and can be easily fit as the sum of the relaxation contributions from two inner-sphere water protons and from diffusing water molecules. Both the reorientational time and the field dependent electron relaxation time contribute to the proton correlation time. The fit performed with the SBM theory, without

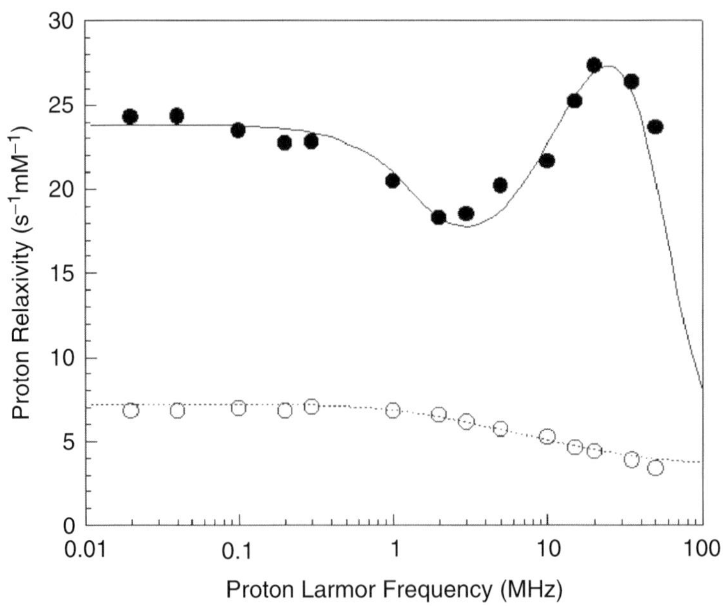

Fig. 37. Water ^1H NMRD profiles for solutions of Gd-DTPA-SA at 298 K in the absence (○) and in the presence (•) of 1.36 mM of Carbonic Anhydrase (*100*).

including the effect of a static ZFS, provides the values for Δ_t equal to 0.042 cm^{-1}, for τ_v equal to 17×10^{-12} s, and for τ_M equal to 0.50×10^{-6} s, by considering one regularly coordinated water molecule (r = 3.1 Å). These values are very similar to those generally provided for complexes used or under development as contrast agents for MRI. It has been discussed in Section V.C of Chapter 2 that some parameters change relevantly if the fit is performed by considering the effect of static ZFS and the proper theory for fast-rotating systems ($\Delta_t = 0.018$ cm^{-1}, $\tau_v = 17 \times 10^{-12}$ s, $D = 0.02$ cm^{-1}, r = 2.7 Å) (*101*). In any case Δ_t seems to be related to the molecular symmetry of the complex. Lower molecular symmetry may correspond to an increased transient ZFS, and thus to lower effective electron relaxation times at low fields (*102*). In fact, the spin–orbit coupling is increased in a less symmetrical structure.

The ^{1}H NMRD profile of the protein adduct shows a largely increased relaxivity, with the dispersion moved at about 1 MHz and a relaxivity peak in the high field region. This shape is clearly related to the fact that the field dependent electron relaxation time is now the correlation time for proton relaxation even at low fields. The difference in relaxivities before and after the dispersion is in this case very small, and therefore the profile cannot be well fit with the SBM theory, and the presence of a small static ZFS must be taken into account (*103*). The best fit parameters obtained with the Florence NMRD program are: $D = 0.01$ cm^{-1}, $\Delta_t = 0.017$ cm^{-1}, $\tau_v = 18 \times 10^{-12}$ s, and $\tau_M = 0.56 \times 10^{-6}$ s. Such values are clearly in agreement with those obtained with fast-motion theory (*101*).

Investigators should also check, when fitting the relaxivity profile of gadolinium(III) complexes, that electron relaxation is within the Redfield limit ($T_{1e} > \tau_v$), as otherwise good fits can be still obtained but with parameters that may be wrong by orders of magnitude (*10*).

III. Magnetic Coupled Systems

A. Nuclear and Electron Relaxation

An electron spin can relax by coupling with a neighboring electron spin. Therefore, when a paramagnetic metal ion interacts with a second paramagnetic metal ion, the electron relaxation rates of the two metal ions may be dramatically affected. If S_1 and S_2 are the two spins coupled by a scalar interaction, new spin levels will be established due to the interaction, with total S' varying in unitary steps from $|S_1 - S_2|$ to $S_1 + S_2$. The energies of these spin levels are given by (*1*)

$$E_i = \frac{J}{2} S'(S' + 1), \tag{21}$$

where J is the magnetic coupling constant. The temperature-dependent population P_i of any spin level of energy E_i is given by the Boltzmann law:

$$P_i = \frac{\exp(-E_i/kT)}{\sum_i \exp(-E_i/kT)}. \tag{22}$$

The J values may range from fractions of wavenumbers up to 10^3 cm^{-1}. Therefore, in many cases a significant difference in the population of the various S' levels is observed.

In the case of a nucleus sensing only one metal ion, with spin S_1, the relaxation rate enhancement of a that nucleus due to the dipole–dipole interaction is given by

$$T_{1M}^{-1} = \frac{2}{15}\left(\frac{\mu_0}{4\pi}\right)^2 \frac{\gamma_I^2 g_e^2 \mu_B^2}{r^6} \times \frac{\sum_i\left[C_i^2 S_i'(S_i'+1)(2S_i'+1)\exp(-E_i/kT)\left(\frac{7\tau_{c2}}{1+\omega_S^2\tau_{c2}^2}+\frac{3\tau_{c1}}{1+\omega_I^2\tau_{c1}^2}\right)\right]}{\sum_i\left[(2S_i'+1)\exp(-E_i/kT)\right]} \tag{23}$$

where τ_{ci}^{-1} is the sum of rotational, exchange and electronic relaxation (T_{1e}^{-1} = longitudinal, T_{2e}^{-1} = transverse) rates, and

$$C_i = \left[S_i'(S_i'+1) + S_1(S_1+1) - S_2(S_2+1)\right]/\left[2S_i'(S_i'+1)\right] \tag{24}$$

Analogously, on the assumption that the establishment of magnetic interactions within a cluster does not change the spin density distribution on the ligands nor on the metal ion, the equation for contact relaxation is:

$$T_{1M}^{-1} = \frac{2}{3}\frac{A_M^2}{\hbar^2} \frac{\sum_i\left[C_i^2 S_i'(S_i'+1)(2S_i'+1)\exp(-E_i/kT)\left(\frac{\tau_{c2}}{1+\omega_S^2\tau_{c2}^2}\right)\right]}{\sum_i\left[(2S_i'+1)\exp(-E_i/kT)\right]} \tag{25}$$

where τ_{ci}^{-1} is the sum of exchange and electronic relaxation rates, and A_M is the contact coupling constant of the nucleus with the single ion when it is not magnetically coupled. If the nucleus senses the other metal ion also, then the total relaxation rate enhancement will be given by the sum of the contributions due to each interaction.

The values of the electron relaxation rates of the coupled metal ion strongly depend both on the relative electron relaxation rates of the isolated ions and on the value of the magnetic coupling constant J. When the absolute value of J (expressed as $|J|/\hbar$) is smaller than both electronic relaxation rates, no effect on the electronic relaxation of the pair is expected. When $|J|/\hbar > T_{1e(1)}^{-1}$ (electron relaxation rate of the first ion) but smaller than $T_{1e(2)}^{-1}$ (electron relaxation rate of the second ion), from first order perturbation theory it is obtained, for the dipolar and the contact contributions, respectively,

$$\Delta T_{1e(1)}^{-1} = \frac{2}{15}\left(\frac{\mu_0}{4\pi}\right)^2\left(\frac{1}{\hbar}\right)^2 \frac{g_e^4\mu_B^4 S_2(S_2+1)}{\langle r^3\rangle^2} \times \left(\frac{T_{2e(2)}}{1+(\omega_{S1}-\omega_{S2})^2 T_{2e(2)}^2} + \frac{3T_{1e(2)}}{1+\omega_{S1}^2 T_{1e(2)}^2} + \frac{6T_{2e(2)}}{1+(\omega_{S1}+\omega_{S2})^2 T_{2e(2)}^2}\right) \quad (26)$$

and

$$\Delta T_{1e(1)}^{-1} = \frac{2}{3} S_2(S_2+1)\left(\frac{J}{\hbar}\right)^2 \frac{T_{2e(2)}}{1+(\omega_{S1}-\omega_{S2})^2 T_{2e(2)}^2} \quad (27)$$

where $\Delta T_{1e(1)}^{-1}$ is the enhancement in the electronic relaxation rate of the slow relaxing ion, ω_{S1} and ω_{S2} are the Larmor frequencies of the M_1 and M_2 ions, $T_{ie(2)}$ the electronic relaxation rate of the fast relaxing ion, which acts as the correlation time for the interaction, and $\langle r^3\rangle$ is the average cube of the inter-electronic distance. The contact interaction is often dominant, except for very small J values and electron–electron distances.

With decreasing $\langle r^3\rangle$ or increasing J, $T_{1e(1)}$ may approach $T_{1e(2)}$. Actually, both electron-electron interactions may often be so strong that the Redfield limit is reached, i.e., the resulting relaxation rate for the slower relaxing metal ion is as large as the relaxation rate for the faster relaxing metal ion, which is the correlation time for relaxation of the former ion. In the limiting case of magnetic coupling constant much larger than the electron relaxation rates of both interacting metal ions Eqs. (26) and (27) do not hold anymore. This limiting case has been treated (*104*) by calculating the linewidths and the lifetimes in coupled systems as a function of the relaxation rates of the isolated spins, and effective electron relaxation times have been obtained (Table V) in the high ($|J| \ll kT$) and low temperature ($|J| \gg kT$) limits. Note that in the latter case the relaxation rate enhancement also depends on the sign of J, i.e., on the coupling being antiferromagnetic ($J > 0$) or ferromagnetic ($J < 0$).

One should be aware that this overall treatment is expected to be strictly valid only in a few ideal cases. In fact, in practical cases spin–orbit coupling causes ZFS, and the S' levels are split at zero magnetic field. Furthermore, new relaxation pathways may be operative among the new energy levels provided by magnetic coupling. New transitions involving spin levels can occur through coupling with the lattice, i.e., solvent collisions, solvent fluctuations, and molecular tumbling. Therefore, the above quantitative considerations may be regarded as guidelines for interpretation of experimental evidences and predictions of relaxation trends in magnetic coupled systems.

Table V

Effective electron relaxation rates for J-coupled systems (A: 1/2-1/2, B: 1/2-1, C: 1/2-3/2, and D: 1-1 case) in the HT (high temperature), AF (antiferromagnetic, $|J| \gg kT$ and $J > 0$) and F (ferromagnetic, $|J| \gg kT$ and $J < 0$) limits. Calculations are performed for the nucleus, N, interacting with either one of the two spins (*104*)

	$N\text{-}S_A$			$N\text{-}S_B$		
	HT	AF	F	HT	AF	F
A.						
$T_{1e(1)}^{-1}$	$1/2R_A$ $+1/2R_B$	diamagnetic state	$1/2R_A$ $+1/2R_B$	$1/2R_A$ $+1/2R_B$	diamagnetic state	$1/2R_A$ $+1/2R_B$
$T_{1e(2)}^{-1}$	$3/4R_A$ $+3/4R_B$		$3/4R_A$ $+3/4R_B$	$3/4R_A$ $+3/4R_B$		$3/4R_A$ $+3/4R_B$
B.						
$T_{1e(1)}^{-1}$	$107/198R_A$ $+112/99R_B$	$13/18R_A$ $+14/9R_B$	$47/90R_A$ $+49/45R_B$	$73/126R_A$ $+77/63R_B$	$13/18R_A$ $+14/9R_B$	$47/90R_A$ $+49/45R_B$
$T_{1e(2)}^{-1}$	$71/99R_A$ $+187/99R_B$	$7/9R_A$ $+7/3R_B$	$32/45R_A$ $+83/45R_B$	$46/63R_A$ $+125/63R_B$	$7/9R_A$ $+7/3R_B$	$32/45R_A$ $+83/45R_B$
C.						
$T_{1e(1)}^{-1}$	$9/16R_A$ $+179/96R_B$	$11/16R_A$ $+2R_B$	$43/80R_A$ $+147/80R_B$	$331/560R_A$ $+2123/1120R_B$	$11/16R_A$ $+2R_B$	$43/80R_A$ $+147/80R_B$
$T_{1e(2)}^{-1}$	$43/48R_A$ $+215/64R_B$	$3/4R_A$ $+117/32R_B$	$37/40R_A$ $+33/10R_B$	$69/80R_A$ $+7677/2240R_B$	$3/4R_A$ $+117/32R_B$	$37/40R_A$ $+33/10R_B$
D.						
$T_{1e(1)}^{-1}$	$5/4R_A$ $+5/4R_B$	diamagnetic state	$23/20R_A$ $+23/20R_B$	$5/4R_A$ $+5/4R_B$	diamagnetic state	$23/20R_A$ $+23/20R_B$
$T_{1e(2)}^{-1}$	$7/4R_A$ $+7/4R_B$		$69/40R_A$ $+69/40R_B$	$7/4R_A$ $+7/4R_B$		$69/40R_A$ $+69/40R_B$

B. Examples

We have seen that copper(II) is a slowly relaxing metal ion. Magnetic coupling of copper to a fast relaxing metal ion increases the electron relaxation rate of copper, as clearly shown by the ^{1}H NMRD profiles of tetragonal copper(II) complexes reacting with ferricyanide (*105*) (Fig. 38). The electron relaxation time, estimated from the relaxation rate of the water protons coordinated to the copper ion, is 3×10^{-10} s, a factor of 10 shorter than in the absence of ferricyanide.

In Cu_2Zn_2 superoxidedismutase (SOD), a dimeric enzyme with a dimeric Cu,Zn site in each monomer, a water molecule is coordinated to each copper ion. The ^{1}H NMRD of water solutions of Cu_2Zn_2SOD indicates a copper-oxygen distance of 2.4 Å and an electron relaxation time of 2×10^{-9} s.

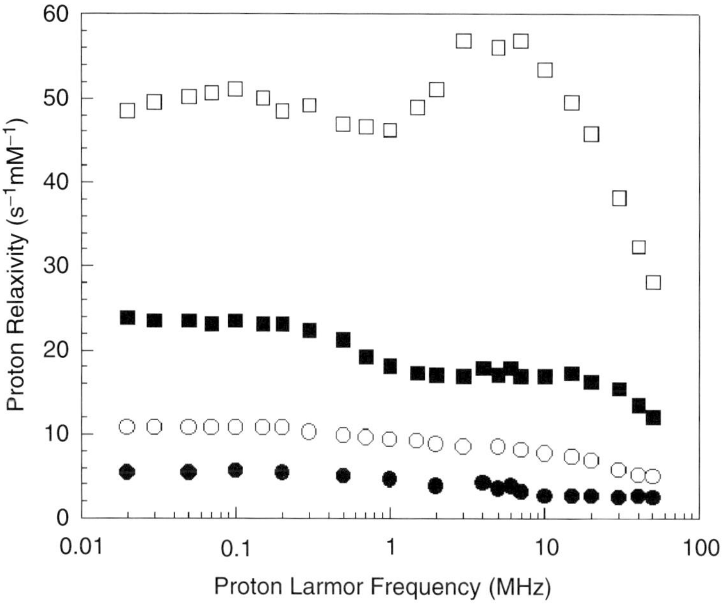

FIG. 38. Water ^{1}H NMRD profiles for solutions of the adduct of copper(II) bis(ethylenediamine) with diamagnetic $Co(CN)_6^{3-}$ (■) and with paramagnetic $Fe(CN)_6^{3-}$ (•) at 278 K (open symbols) and 298 K (filled symbols) (*105*).

The zinc ion can be substituted by paramagnetic ions, making SOD a good test case for the theories of relaxation in dimers. When zinc is substituted with another copper(II) ion, the ^{1}H NMRD profile decreases by about a factor of two (Fig. 39) (*106*). In the latter case J is 52 cm^{-1} and all the S' levels are almost equally populated at room temperature. Actually, the copper-oxygen distance for the native copper is expected to be the same, and the second copper ion does not interact with any water molecule. The factor two is thus accounted for by a halved paramagnetic contribution to the nuclear relaxation rates, according to the theoretical predictions (*1*) for coupled systems containing equal metal ions and $J \ll kT$ (see Eq. (23)). In fact, in the case of two $S = 1/2$ ions, the $S' = 1$ level is threefold degenerate, $C_1 = 1/2$ and $P_1 = 3/4$. When zinc is substituted by nickel(II) ($S = 1$) or cobalt(II) ($S = 3/2$) ions, the nuclear relaxation rates decrease sizably, because the electron relaxation time is very close to that typical of the latter ions (*104*). For CuCoSOD ($J = 17$ cm^{-1}), the average $T_{1e(1)}^{-1}$ is predicted to be 9/16 R_{Cu} + 179/96 R_{Co} (see Table V), where $R_{Cu} = 5 \times 10^8$ s^{-1} and $R_{Co} = 10^{11}$ s^{-1} are the electron relaxation rates of the uncoupled ions Cu(II) and Co(II), respectively. It results that $T_{1e(1)}^{-1}$ is about twice R_{Co}. The experimental $T_{1e(1)}^{-1}$ value ranges between 1 and 2×10^{11} s^{-1}, in good agreement with the predictions (*104*).

When $|J|$ is of the order of hundreds of wavenumbers, i.e., it is of the order of kT, then the excited S' levels are not fully populated. In homodimers antiferromagnetically coupled, the ground state is diamagnetic ($S' = 0$). If only one metal ion is sensed by the resonating nucleus, nuclear relaxation

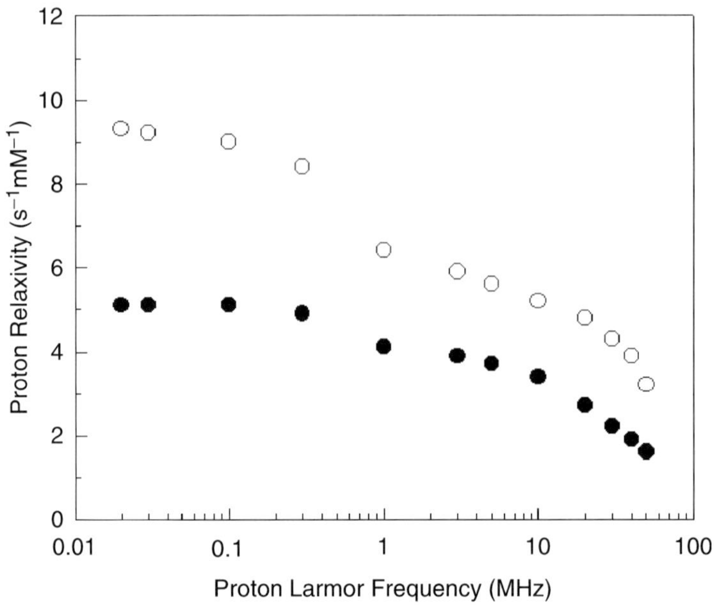

FIG. 39. Water ^{1}H NMRD profiles for solutions of Cu_2Zn_2-SOD (○) and Cu_2Cu_2-SOD (•) (*106*).

rate is expected to be smaller than in uncoupled systems of a factor larger than two. For $|J| \gg kT$, the nuclear relaxation enhancement will decrease substantially, and eventually approach zero (diamagnetic case in Table V).

Oxidized Fe_2S_2 ferredoxins, containing two equivalent iron atoms, with $J = 400\ cm^{-1}$, show sharper NMR lines with respect to the monomeric iron model provided by oxidized rubredoxin (*107–109*), due to the decreased Boltzmann population of the paramagnetic excited states. For reduced ferredoxins ($S_1 = 5/2$, $S_2 = 2$), with $J = 200\ cm^{-1}$, the ground state is paramagnetic ($S' = 1/2$) (*110*). A smaller decrease in linewidth is expected. However, the fast electron relaxation rates of the iron(II) ion cause both ions to relax faster, and the linewidths in the dimer are sharp.

IV. Conclusions

In the last ten years several novel theories for the analysis of ^{1}H NMRD profiles have been proposed for different paramagnetic systems. These have resulted in novel equations and computer programs for numerical calculations. In this review we tried to summarize the classical theory as well as some recent theoretical treatments, and to show their applicability to different metal ion systems by selecting some experimental profiles which exhibit the most characteristic features. In the last few years, dedicated instruments, called field-cycling relaxometers, have appeared on the market (see Chapter 9). These instruments now allow to measure nuclear relaxation rates on strong NMR signals over fields that can range smoothly between a few kHz up to about 50 MHz. On the other hand the availability of spectrometers working at very high magnetic fields (up to 900 MHz) allows researchers to obtain relaxation profiles covering ≥ 5 orders of magnitude in proton Larmor frequency, thus allowing a more informative analysis of the data, and a better understanding of high field effects.

The NMRD technique is thus a precious tool to obtain information on, for instance, water-metal ion ligands and on the electron relaxation time of the paramagnetic species. Understanding electron relaxation is important to be able to predict the NMR behavior of paramagnetic compounds at high field and to master high resolution NMR experiments on paramagnetic systems. Further perspectives are represented by applications of relaxometry for in vitro and in vivo measurements, in order to efficiently monitor the relaxing properties of the compounds in their "working" environment, the most relevant field being that of the characterization of novel and specialized contrast agents for MRI (see Chapter 4).

Acknowledgments

This work was supported by the European Union (contract HPRI-CT-2001-50028).

References

1. Bertini, I.; Luchinat, C.; Parigi, G. "*Solution NMR of Paramagnetic Molecules*"; Elsevier: Amsterdam, **2001**, pp. 1–376.
2. Solomon, I. *Phys. Rev.* **1955**, *99*, 559.
3. Solomon, I.; Bloembergen, N. *J. Chem. Phys.* **1956**, *25*, 261.
4. Bloembergen, N. *J. Chem. Phys.* **1957**, *27*, 572.
5. Bloembergen, N.; Morgan, L. O. *J. Chem. Phys.* **1961**, *34*, 842.
6. Rubinstein, M.; Baram, A.; Luz, Z. *Mol. Phys.* **1971**, *20*, 67.
7. Redfield, A. G. *Res. Dev.* **1957**, *1*, 19.
8. Bertini, I.; Galas, O.; Luchinat, C.; Parigi, G. *J. Magn. Reson. Ser. A* **1995**, *113*, 151.
9. Seibig, S.; Toth, E.; Merbach, A. E. *J. Am. Chem. Soc.* **2000**, *122*, 5822.
10. Kowalewski, J.; Luchinat, C.; Nilsson, T.; Parigi, G. *J. Phys. Chem. A* **2002**, *106*, 7376.
11. Bertini, I.; Kowalewski, J.; Luchinat, C.; Nilsson, T.; Parigi, G. *J. Chem. Phys.* **1999**, *111*, 5795.
12. Kruk, D.; Nilsson, T.; Kowalewski, J. *Phys. Chem. Chem. Phys.* **2001**, *3*, 4907.
13. Van Vleck, J. H. *Phys. Rev.* **1940**, *57*, 426.
14. Orbach, R. *Proc. R. Soc. London,Ser.* **1961**, *A264*, 458.
15. Kivelson, D. *J. Chem. Phys.* **1960**, *33*, 1094.
16. Al'tshuler, S. A.; Valiev, K. A. *Soviet Physics JETP* **1959**, *35*, 661.
17. Kivelson, D. *J. Chem. Phys.* **1966**, *45*, 1324.
18. Bruno, G. V.; Harrington, J. K.; Eastman, M. P. *J. Phys. Chem.* **1977**, *81*, 1111.
19. Kruk, D.; Kowalewski, J. *J. Chem. Phys.* **2002**, *117*, 1194.
20. Atkins, P. W.; Kivelson, D. *J. Chem. Phys.* **1966**, *44*, 169.
21. Nyberg, G. *Mol. Phys.* **1967**, *12*, 69.
22. Bertini, I.; Luchinat, C.; Brown III, R. D.; Koenig, S. H. *J. Am. Chem. Soc.* **1989**, *111*, 3532.
23. Powell, D. H.; Helm, L.; Merbach, A. E. *J. Chem. Phys.* **1991**, *95(12)*, 9258.
24. Freed, J. H.; Kooser, R. G. *J. Chem. Phys.* **1968**, *49*, 4715.
25. Kooser, R. G.; Volland, W. V.; Freed, J. H. *J. Chem. Phys.* **1969**, *50*, 5243.

26. Pasquarello, A.; Petri, I.; Salmon, P. S.; Parisel, O.; Car, R.; Toth, E.; Powell, D. H.; Fischer, H. E.; Helm, L.; Merbach, A. E. *Science* **2001**, *291*, 856.
27. Bertini, I.; Briganti, F.; Luchinat, C.; Mancini, M.; Spina, G. *J. Magn. Reson.* **1985**, *63*, 41.
28. Arnesano, F.; Banci, L.; Bertini, I.; Felli, I. C.; Luchinat, C.; Thompsett, A. R. *J. Am. Chem. Soc.* **2003**, *125*, 7200.
29. Holm, R. H.; Kennepohl, P.; Solomon, E. I. *Chem. Rev.* **1996**, *96*, 2239.
30. Koenig, S. H.; Brown, R. D. *Ann. N. Y. Acad. Sci.* **1973**, *222*, 752.
31. Andersson, I.; Maret, W.; Zeppezauer, M.; Brown III, R. D.; Koenig, S. H. *Biochemistry* **1981**, *20*, 3424.
32. Kroes, S. J.; Salgado, J.; Parigi, G.; Luchinat, C.; Canters, G. W. *JBIC* **1996**, *1*, 551.
33. Abragam, A.; Bleaney, B. "*Electron Paramagnetic Resonance of Transition Metal Ions*"; Clarendon Press: Oxford, **1970**, p. 152.
34. La Mar, G. N.; Walker, F. A. *J. Am. Chem. Soc.* **1973**, *95*, 1782.
35. Bertini, I.; Galas, O.; Luchinat, C.; Messori, L.; Parigi, G. *J. Phys. Chem.* **1995**, *99*, 14217.
36. Bertini, I.; Capozzi, F.; Luchinat, C.; Xia, Z. *J. Phys. Chem.* **1993**, *97*, 1134.
37. Koenig, S. H.; Brown III, R. D.; Lindstrom, T. R. *Biophys. J.* **1981**, *34*, 397.
38. Hershberg, R. D.; Chance, B. *Biochemistry* **1975**, *14*, 3885.
39. Bertini, I.; Briganti, F.; Luchinat, C.; Xia, Z. *J. Magn. Reson.* **1993**, *101*, 198.
40. Banci, L.; Bertini, I.; Briganti, F.; Luchinat, C. *J. Magn. Reson.* **1986**, *66*, 58.
41. Meirovitch, E.; Luz, Z.; Kalb, A. J. *J. Am. Chem. Soc.* **1974**, *96*, 7538.
42. Brackett, G. C.; Richards, P. L.; Caughey, W. S. *J. Chem. Phys.* **1971**, *54*, 4383.
43. Luz, Z.; Fiat, D. *J. Chem. Phys.* **1967**, *46*, 469.
44. La Mar, G. N.; Walker, F. A. *J. Am. Chem. Soc.* **1973**, *95*, 6950.
45. Sharp, R. R.; Lohr, L.; Miller, J. *Prog. NMR Spectrosc.* **2001**, *38*, 115.
46. Abernathy, S. M.; Miller, J. C.; Lohr, L. L.; Sharp, R. R. *J. Chem. Phys.* **1998**, *109*, 4035.
47. Banci, L.; Bertini, I.; Luchinat, C. *Inorg. Chim. Acta* **1985**, *100*, 173.
48. Koenig, S. H.; Brown III, R. D.; Bertini, I.; Luchinat, C. *Biophys. J.* **1983**, *41*, 179.
49. Bertini, I.; Luchinat, C.; Mancini, M.; Spina, G. *J. Magn. Reson.* **1984**, *59*, 213.
50. Bertini, I.; Canti, G.; Luchinat, C.; Scozzafava, A. *J. Am. Chem. Soc.* **1978**, *100*, 4873.
51. Bertini, I.; Canti, G.; Luchinat, C. *Inorg. Chim. Acta* **1981**, *56*, 99.
52. Kowalewski, J.; Larsson, T.; Westlund, P. O. *J. Magn. Reson.* **1987**, *74*, 56.
53. Svoboda, J.; Nilsson, T.; Kowalewski, J.; Westlund, P. O.; Larsson, P. T. *J. Magn. Reson. Ser. A* **1996**, *121*, 108.
54. Bertini, I.; Luchinat, C.; Mancini, M.; Spina, G. "*Magneto-structural correlations in exchange-coupled systems*"; Eds. Gatteschi, D.; Kahn, O.; Willett, R.D.; Reidel Publishing Company: Dordrecht, **1985**; pp. 421–461.
55. Bertini, I.; Borghi, E.; Luchinat, C.; Monnanni, R. *Inorg. Chim. Acta* **1982**, *67*, 99.
56. Lindner, U. *Ann. Phys. (Leipzig)* **1965**, *16*, 319.
57. Bencini, A.; Gatteschi, D. "*Transition Metal Chemistry*"; Eds. Melson, G.A.; Figgis, B.N.; Marcel Dekker, INC: New York and Basel, **1982**; pp. 1–178.
58. Bertini, I.; Luchinat, C.; Xia, Z. *J. Magn. Reson.* **1992**, *99*, 235.
59. Bertini, I.; Briganti, F.; Koenig, S. H.; Luchinat, C. *Biochemistry* **1985**, *24*, 6287.
60. Wilson, R. C.; Myers, R. J. *J. Chem. Phys.* **1976**, *64*, 2208.
61. Bertini, I.; Luchinat, C.; Xia, Z. *Inorg. Chem.* **1992**, *31*, 3152.
62. Bertini, I.; Fragai, M.; Luchinat, C.; Parigi, G. *Inorg. Chem.* **2001**, *40*, 4030.
63. Stephens, E. M.; Grisham, C. M. *Biochemistry* **1979**, *18*, 4876.
64. Epstein, M.; Reuben, J.; Levitzki, A. *Biochemistry* **1977**, *16*, 2449.
65. Borel, A.; Yerly, F.; Helm, L.; Merbach, A. E. *J. Am. Chem. Soc.* **2002**, *124*, 2042.
66. Strandberg, E.; Westlund, P. O. *J. Magn. Reson. Ser. A* **1996**, *122*, 179.
67. Bertini, I.; Capozzi, F.; Luchinat, C.; Nicastro, G.; Xia, Z. *J. Phys. Chem.* **1993**, *101*, 198.
68. Burns, P. D.; La Mar, G. N. *J. Magn. Reson.* **1982**, *46*, 61.
69. Aime, S.; Barbero, L.; Botta, M. *Magn. Res. Imaging* **1991**, *9*, 843.
70. Shannon, R. D.; Prewitt, C. T. *Acta Cryst.* **1969**, *B25*, 925.
71. Helm, L.; Merbach, A. E. *Eur. J. Solid State Inorg. Chem.* **1991**, *28*, 245.
72. Caravan, P.; Toth, E.; Rockenbauer, A.; Merbach, A. E. *J. Am. Chem. Soc.* **1999**, *121*, 10403.
73. Chen, J. W.; Belford, R. L.; Clarkson, R. B. *J. Phys. Chem.* **1998**, *102*, 2117.

74. Botta, M. *Eur. J. Inorg. Chem.* **2000**, 399.
75. Stokes, G. *Trans. Cambridge Philos. Soc.* **1956**, *9*, 5.
76. Einstein, A. "*Investigations on the Theory of the Brownian Movement*"; Dover: New York, **1956**, pp. 19–34.
77. Debye, P. "*Polar Molecules*"; Dover: New York, **1929**, p. 45.
78. Bertini, I.; Fragai, M.; Luchinat, C.; Parigi, G. *Magn. Reson. Chem.* **2000**, *38*, 543.
79. Woessner, D. E. *J. Chem. Phys.* **1962**, *3*, 647.
80. Lipari, G.; Szabo, A. *J. Am. Chem. Soc.* **1982**, *104*, 4546.
81. Alhaique, F.; Bertini, I.; Fragai, M.; Carafa, M.; Luchinat, C.; Parigi, G. *Inorg. Chim. Acta* **2002**, *331*, 151.
82. Dunand, F. A.; Borel, A.; Merbach, A. E. *J. Am. Chem. Soc.* **2002**, *124*, 710.
83. Hwang, L. P.; Freed, J. H. *J. Chem. Phys.* **1975**, *63*, 4017.
84. Polnaszek, C. F.; Bryant, R. G. *J. Chem. Phys.* **1978**, *68*, 4034.
85. Freed, J. H. *J. Chem. Phys.* **1978**, *68*, 4034.
86. Hausser, R.; Noack, F. *Z. Phys.* **1964**, *182*, 93.
87. Luz, Z.; Shulman, R. G. *J. Chem. Phys.* **1965**, *43*, 3750.
88. Bertini, I.; Luchinat, C. *NMR of Paramagnetic Molecules in Biological Systems*; Benjamin/Cummings: Menlo Park, CA, **1986**, pp. 1–319.
89. Liebermann, R. A.; Sands, R. H.; Fee, J. A. *J. Biol. Chem.* **1982**, *257*, 336.
90. Koenig, S. H.; Epstein, M. *J. Chem. Phys.* **1975**, *63*, 2279.
91. Zewert, T. E.; Gray, H. B.; Bertini, I. *J. Am. Chem. Soc.* **1994**, *116*, 1169.
92. Bertini, I.; Luchinat, C.; Nerinovski, K.; Parigi, G.; Cross, M.; Xiao, Z.; Wedd, A. G. *Biophys. J.* **2003**, *84*, 545.
93. Koenig, S. H.; Baglin, C.; Brown III, R. D.; Brewer, C. F. *Magn. Reson. Med.* **1984**, *1*, 496.
94. Koenig, S. H.; Brown III, R. D. *Progr. NMR Spectrosc.* **1990**, *22*, 487.
95. Bhattacharyya, L.; Brewer, C. F.; Brown III, R. D. ; Koenig, S. H. *Biochemistry* **1985**, *24*, 4985.
96. Aime, S.; Anelli, P. L.; Botta, M.; Brocchetta, M.; Canton, S.; Fedeli, F.; Gianolio, E.; Terreno, E. *J. Biol. Inorg. Chem.* **2002**, *7*, 58.
97. Kruk, D.; Kowalewski, J. *J. Biol. Inorg. Chem.* **2003**, *8*, 512.
98. Nilsson, T.; Parigi, G.; Kowalewski, J. *J. Phys. Chem.* **2002**, *106*, 4476.
99. Bleuzen, A.; Foglia, F.; Furet, E.; Helm, L.; Merbach, A.; Weber, J. *J. Am. Chem. Soc.* **1996**, *118*, 12777.
100. Anelli, P. L.; Bertini, I.; Fragai, M.; Lattuada, L.; Luchinat, C.; Parigi, G. *Eur. J. Inorg. Chem.* **2000**, 625.
101. Kruk, D.; Kowalewski, J. *J. Magn. Reson.* **2003**, *162*, 229.
102. Li, C.; Parigi, G.; Fragai, M.; Luchinat, C.; Meade, T. J. *Inorg. Chem.* **2002**, *41*, 4018.
103. Bligh, S. W. A.; Chowdhury, A. H. M. S.; Kennedy, D.; Luchinat, C.; Parigi, G. *Magn. Res. Med.* **1999**, *41*, 767.
104. Bertini, I.; Galas, O.; Luchinat, C.; Parigi, G.; Spina, G. *J. Magn. Reson.* **1998**, *130*, 33.
105. Bertini, I.; Lanini, G.; Luchinat, C.; Mancini, M.; Spina, G. *J. Magn. Reson.* **1985**, *63*, 56.
106. Bertini, I.; Banci, L.; Brown III, R. D.; Koenig, S. H.; Luchinat, C. *Inorg. Chem.* **1988**, *27*, 951.
107. Xia, B.; Westler, W. M.; Cheng, H.; Meyer, J.; Moulis, J.-M.; Markley, J. L. *J. Am. Chem. Soc.* **1995**, *117*, 5347.
108. Bertini, I.; Lanini, G.; Luchinat, C. *Inorg. Chem.* **1984**, *23*, 2729.
109. Benini, S.; Ciurli, S.; Luchinat, C. *Inorg. Chem.* **1995**, *34*, 417.
110. Banci, L.; Bertini, I.; Luchinat, C. *Struct. Bonding* **1990**, *72*, 113.
111. Fugiwara, S.; Hayashu, H. *J. Chem. Phys.* **1965**, *43*, 23.
112. Luz, Z.; Shulman, R. G. *J. Chem. Phys.* **1965**, *43*, 3750.
113. Lewis, W. B.; Alei, M.; Morgan, L. O. *J. Chem. Phys.* **1966**, *44*, 2409.
114. Bloembergen, N.; Morgan, L. O. *J. Chem. Phys.* **1961**, *34*, 842.
115. Ducommun, Y.; Newman, K. E.; Merbach, A. *Inorg. Chem.* **1980**, *19*, 3696.
116. Boere, R. T.; Kidel, R. *Annu. Rep. NMR Spectrosc.* **1982**, *13*, 319.
117. Kennedy, S. D.; Bryant, R. G. *Magn. Reson. Med.* **1985**, *2*, 14.
118. Swift, T. J.; Connick, R. E. *J. Chem. Phys.* **1962**, *37*, 307.
119. Friedman, H. L.; Holz, M.; Hertz, H. G. *J. Chem. Phys.* **1979**, *70*, 3369.

120. Ducommun, Y.; Earl, W. L.; Merbach, A. E. *Inorg. Chem.* **1979**, *18*, 2754.
121. Banci, L.; Bertini, I.; Luchinat, C. "*Nuclear and Electron Relaxation. The Magnetic Nucleus-Unpaired Electron Coupling in Solution*"; VCH: Weinheim, **1991**, pp. 1–208.
122. Alsaadi, B.M.; Rossotti, F.J.C.; Williams, R.J.P. *J. Chem. Soc., Dalton Trans.* **1980**, 2147.
123. Burs, P. D.; La Mar, G. N. *J. Magn. Reson.* **1982**, *46*, 61.
124. Gaber, B. P.; Brown III, R. D.; Koenig, S. H.; Fee, J. A. *Biochim. Biophys. Acta* **1972**, *271*, 1.
125. Koenig, S. H.; Brown III, R. D.; Spiller, M. *Magn. Reson. Med.* **1987**, *4*, 252.
126. Bertini, I.; Luchinat, C. *Adv. Inorg. Biochem.* **1985**, *6*, 71.
127. Koenig, S. H.; Brown III, R. D. *Invest. Radiol. (Suppl. 2)* **1985**, *20*, 297.

Gd(III)-BASED CONTRAST AGENTS FOR MRI

SILVIO AIME[a,*], MAURO BOTTA[b] and ENZO TERRENO[a]

[a]Department of Chemistry IFM and Centre for Molecular Imaging , University of Torino, via P. Giuria, 7, I-10125 Torino, Italy

[b]Department of Environmental and Life Sciences, University of Eastern Piedmont "Amedeo Avogadro", Spalto Marengo, 33, I-15100, Alessandria, Italy

I. General Remarks

The superb spatial resolution and the outstanding capacity of differentiating soft tissues have determined the widespread success of Magnetic Resonance Imaging (MRI) in clinical diagnosis (*1,2*). The main determinants of the contrast in a MR image are the proton relaxation times T_1 and T_2. When there is a poor contrast between healthy and pathological regions due to a very small variation in relaxation times, the use of a contrast agent can be highly beneficial. Contrast agents are chemicals that are able to alter markedly the relaxation times of water protons in the tissues where they

ADVANCES IN INORGANIC CHEMISTRY
VOLUME 57 ISSN 0898-8838 / DOI 10.1016/S0898-8838(05)57004-1

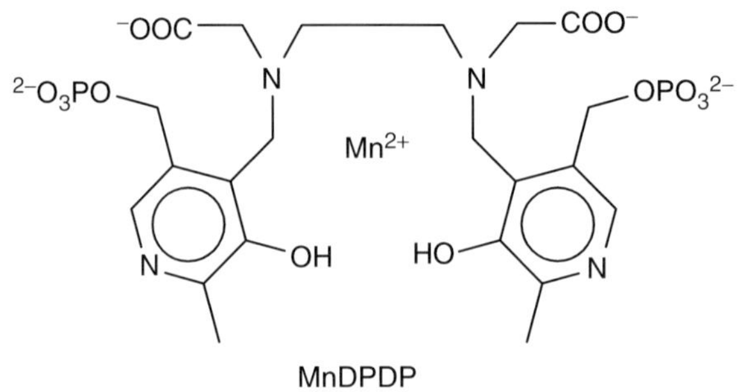

CHART 1.

are distributed. Their use has led to remarkable improvements in medical diagnosis in terms of higher specificity, better tissue characterization, reduction of image artifacts and functional information. According to whether the dominant effect occurs mainly on T_1 or T_2, MRI contrast agents can be classified as positive or negative agents, respectively. The most representative class of T_1-positive agents is represented by paramagnetic Gd(III) chelates, whereas iron oxide particles represent the class of T_2-negative agents (*3*).

Paramagnetic chelates of Mn(II) (five unpaired electrons) have also been considered. The main drawback appears to be related to the stability of these complexes. Since Mn(II) ion is an essential metal, evolution has selected biological structures which are able to sequester Mn(II) ions with high efficiency. Thus, it has been difficult to design Mn(II) chelates that maintain their integrity when administered to living organisms. Actually, MnDPDP (Teslascan®, Amersham, Chart 1) has entered the clinical practice and is recommended as a hepatrotopic agent (*4*). It is the only agent that performs by releasing metal ions to endogenous macromolecules. The huge proton relaxation enhancement brought about by the resulting Mn(II) protein adducts is responsible for the MRI visualization of hepatocytes also at low administered doses of $[MnDPDP]^{4-}$.

Iron oxide particles yield very strong T_2 effects as the result of a dramatic long-range disturbance in magnetic field homogeneity. These agents consist of a crystalline core of superparamagnetic iron(III) oxide (SPIO, maghemite, γ-Fe_2O_3) surrounded by the coating materials like dextran or carboxydextran. The diameter of the iron oxide core is just 3–5 nm whereas the overall particle may be of 50–200 nm size. Two products are available for clinical use: Endorem® (Guerbet) and Resovist® (Schering) (*5,6*). These agents provide excellent (negative) contrast when administered at doses as low as 8–15 μmol/kg body weight. Once administered intravenously, as particles, these agents accumulate in the cells of the reticuloendothelial system. The pharmacodynamic properties of the iron oxide particles are affected by both the size and the overall electric charge. The smaller particles remain in the blood circuit for a time long enough to be considered as blood pool agents for angiographic assays.

Currently about one-third of the MRI scans recorded in clinical settings make use of contrast agents (CA), mainly Gd(III) complexes (*7*). The effectiveness of a Gd(III) complex to act as MRI contrast agent is first assessed by measuring its relaxivity, i.e., the relaxation enhancement of water protons observed for a millimolar solution of the chelated paramagnetic metal. In the past 15 years, a number of articles addressing the relationship between structure/dynamics and relaxivity of Gd(III) complexes have been published. This has led to a substantial advancement of our understanding of the structural, dynamic, and electronic parameters determining the relaxivity of paramagnetic chelates.

In addition to acting as a catalyst for the relaxation processes of tissutal water protons, a potential MRI-CA has to fulfill several requirements related to tolerance, safety, toxicity, stability, osmolality, viscosity, biodistribution, elimination, and metabolism. The currently used Gd(III) chelates are based on polyaminocarboxylate ligands (Chart 2), being either linear or macrocyclic molecules. All these ligands form very stable complexes (Table I) so that the risk of dissociation is so low that the danger of acute toxic effects occurring after injection of gadolinium chelates is practically non-existent with all the products currently in use. Actually, the design of safe Gd(III) chelates deals with both the thermodynamic stability and kinetic stability. Another issue in the stability evaluation is to measure stability in the presence of those metals (Cu(II), Ca(II), and most importantly Zn(II)) whose presence in biological fluids could lead to transmetallation.

The study of the Central Nervous System (CNS) is the primary clinical indication for the use of extracellular Gd(III) agents. The majority of these pathologies are brain tumors, and three quarters of them are represented by metastases occurring in patients undergoing treatment for systemic cancer (Fig. 1). Other brain diseases, such as multiple sclerosis and cerebral injuries can be also investigated by contrast-enhanced MRI.

There are several other indications for the use of CAs outside CNS. For instance, in the diagnosis of breast cancer, MRI with contrast agents is becoming an alternative diagnostic procedure to mammography. Particularly interesting is the dynamic use of contrast agents. The breast is imaged repeatedly over the first few minutes following contrast agent administration, and a graph reporting the increase in signal for a selected region of interest is plotted as a function of time. The kinetics of the distribution of the contrast agents in the extra-vascular space is related to the vascular permeability. Neo-formed vessels functional to the tumor growth display a permeability much higher than the normal capillaries and the corresponding areas are therefore characterized by high signal intensity (*9*). Of course, Dynamic Enhanced MRI is applied to demonstrate the aggressive nature of tumors in several other areas.

For extracellular Gd(III) chelates the dose of 0.1 mmol/kg body weight is the standard dose routinely used in clinical practice. However, in certain instances, the physician may wish to improve visualization by administering another dose some time later, up to a total of 0.3 mmol/kg body weight.

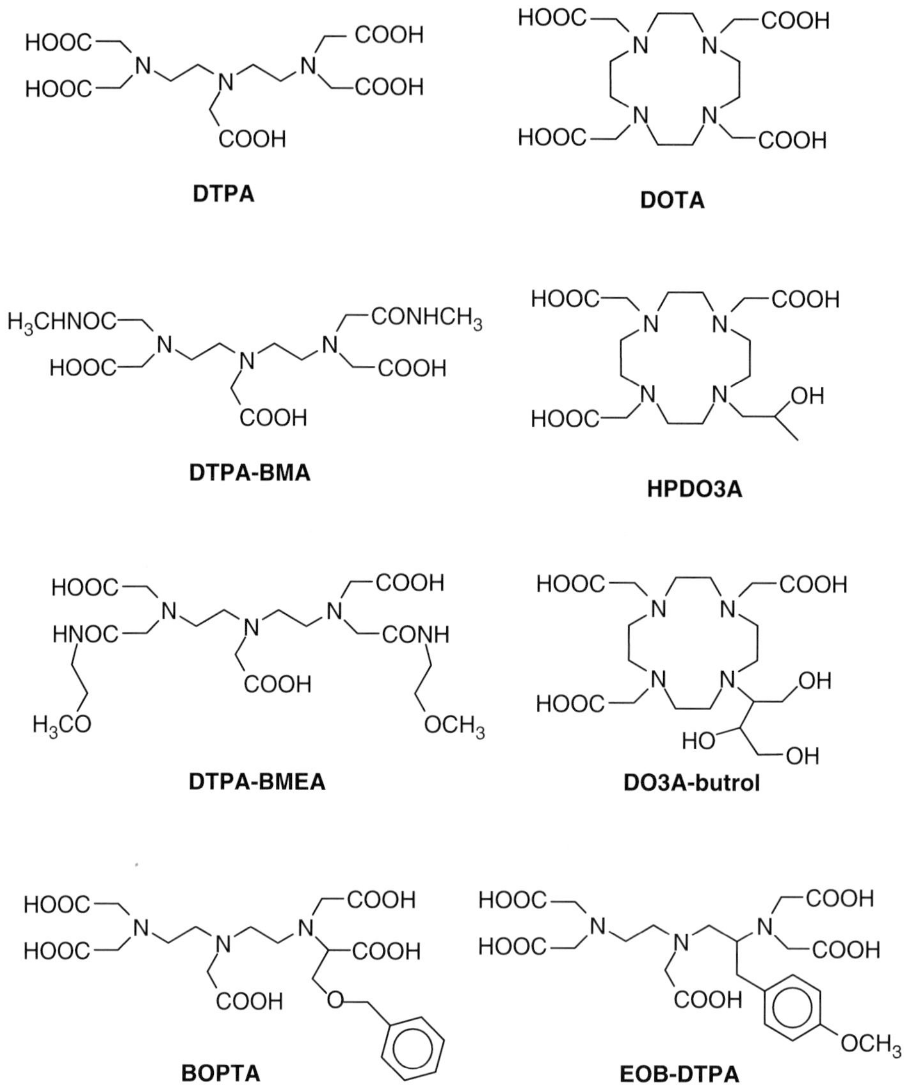

CHART 2.

II. Contributions to the Relaxivity

The contrast enhancing efficiency of a contrast agent is commonly expressed in terms of relaxivity, i.e., the increase in the solvent proton longitudinal and transverse relaxation rates normalized to a one millimolar concentration of the chelated paramagnetic metal (*10–12*):

$$R_i = [CA]r_i + R_{iw} \quad i = 1, 2 \tag{1}$$

Table I

CLINICALLY ACCEPTED Gd(III)-BASED CONTRAST AGENTS

Complex	Brand name	Company	Log K_{GDL} (*8*)
$[GdDTPA(H_2O)]^{2-}$	Magnevist®	Schering	22.5
$[GdDOTA(H_2O)]^{-}$	Dotarem®	Guerbet	24.7
$[GdHPDO3A(H_2O)]$	ProHance®	Bracco	23.8
$[GdDO3A\text{-}butrol(H_2O)]$	Gadovist®	Schering	20.8
$[GdDTPA\text{-}BMA(H_2O)]$	Omniscan®	Nycomed-Amersham	16.8
$[GdDTPA\text{-}BMEA(H_2O)]$	OptiMARK®	Mallinckrodt	16.8
$[GdBOPTA(H_2O)]^{2-}$	MultiHance®	Bracco	22.6
$[GdEOB\text{-}DTPA(H_2O)]^{2-}$	Eovist®	Schering	23.5

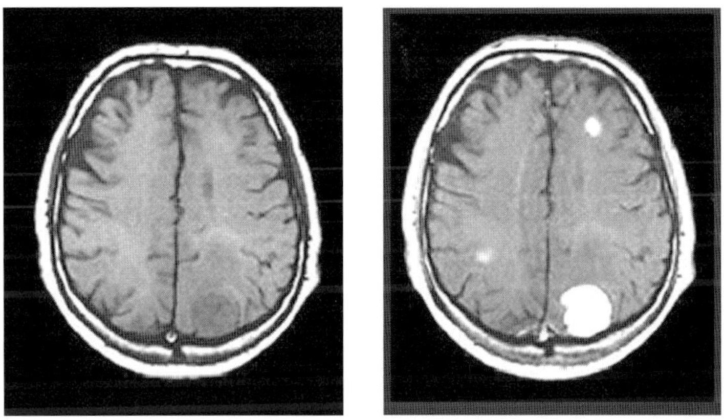

FIG. 1. Metastases in the brain of a patient with brain tumor. The metastases are detected upon intravenous administration of $[GdDTPA(H_2O)]^{2-}$ (right image). Blood vessels (capillaries) are typically quite leaky to small molecules like $[GdDTPA(H_2O)]^{2-}$ allowing them to enter the extracellular space. In the brain, however, the cells forming the walls of capillaries have very tight junctions and prevent small molecules leaving the intravascular space, thereby forming a blood-brain barrier (BBB). Tumors and other pathologies of CNS cause impairment of the BBB, thus allowing the contrast agent to leak from capillaries into extracellular fluid.

In Eq. (1) R_{iw} is the longitudinal ($i = 1$) or transverse ($i = 2$) relaxation rate of the bulk water protons, corresponding to that measured for an analogous diamagnetic solution. In practice, R_{iw} coincides with the value determined for pure water under identical conditions of pH, temperature, and observation frequency. Clearly, the above relation strictly holds only for dilute solutions, in the absence of solute–solute interactions and of variations in viscosity.

The relaxivity enhancement of water protons in the aqueous solutions of paramagnetic complexes arises from time fluctuation of the dipolar coupling between the electron magnetic moment of the metal ion and the nuclear magnetic moment of the solvent nuclei (*13,14*). The dipolar interaction involves both the water molecules that belong to the inner coordination sphere of the metal ion (*inner sphere* contribution) and those water molecules that are not directly bound to the paramagnetic centre but are either diffusing next to the complex (*outer sphere* contribution) or localized in a well defined position with respect to the metal ion *via* hydrogen–bonding interactions with polar groups of the ligand (*second sphere* contribution). Therefore the total relaxivity is expressed as the sum of the three contributions (*10–12*):

$$r_i = r_{ip}^{IS} + r_{ip}^{OS} + r_{ip}^{SS} \qquad (i = 1, 2) \tag{2}$$

The three different mechanisms which traditionally describe the ability of a CA to catalyze the water proton nuclear magnetic relaxation present themselves in different degrees of relative importance in determining the overall relaxivity. This latter attribute primarily depends on the structural properties of the CA, the temperature and the pH of the aqueous solution, and on the observation frequency. The CAs that are currently used in clinical practice are low molecular weight hydrophilic monoaqua Gd(III) complexes (Chart 2) that, at 25°C and 20 MHz, possess a relaxivity of about 4–5 mM^{-1} s^{-1} (*7*). In this case, the *inner* and *outer sphere* contributions are comparable, whereas the relaxation mechanism due to the water molecules in the second hydration sphere of the complexes is neglected or simply incorporated into the *outer sphere* contribution. The increase in molecular dimensions, the presence of suitable hydrogen-bond acceptor groups on the surface of the complex, the variation in the overall electric charge, and the modification of the hydrophilic/hydrophobic domains of the complex are all accompanied by large variations in the relative contributions of the three relaxation pathways and will be discussed in some detail in this chapter.

A. Inner-Sphere Contribution to the Relaxivity

All the Gd(III)-based contrast agents commercially available and those under development present one or more water molecules in their inner coordination sphere. The water molecule(s) is labile and the exchange from the coordination site in the complex and the bulk solvent represents the main source of the solvent relaxation enhancement. Therefore, the *inner sphere* longitudinal relaxivity is given by the following expression:

$$r_{1p}^{IS} = \frac{[CA]q}{55.6} \frac{1}{T_{1M} + \tau_M} \tag{3}$$

where the hydration number q indicates the number of metal-bound water molecules, τ_M is their mean residence lifetime, and T_{1M} is the longitudinal nuclear magnetic relaxation time of the bound water protons. The value of T_{1M} is given by the Solomon–Bloembergen–Morgan (SBM) equations which describe the time fluctuation of the water proton-Gd(III) dipolar coupling as discussed in great detail in the previous two chapters of this book. Briefly, the modulation of the interaction arises from three main factors: rotation of the complex, described in terms of the correlation time τ_R, the residence lifetime of the bound water (τ_M) and electron paramagnetic relaxation (T_{1e} and T_{2e}). The following equations are written in a form suitable and commonly used for the analysis of the relaxivity of Gd(III) complexes:

$$\frac{1}{T_{1M}} = \frac{2}{15}\frac{\gamma_H^2 g^2 S(S+1)\beta^2}{r_H^6}\left[\frac{3\tau_{C1}}{1+\omega_H^2\tau_{C1}^2} + \frac{7\tau_{C2}}{1+\omega_S^2\tau_{C2}^2}\right] \quad (4)$$

$$\frac{1}{\tau_{Ci}} = \frac{1}{\tau_R} + \frac{1}{\tau_M} + \frac{1}{T_{ie}} \qquad (i = 1, 2) \quad (5)$$

$$\left(\frac{1}{T_{1e}}\right)^{\text{ZFS}} = \frac{1}{25}\tau_v^2\{4S(S+1)-3\}\left(\frac{1}{1+\omega_S^2\tau_v^2} + \frac{4}{1+4\omega_S^2\tau_v^2}\right) \quad (6)$$

$$\left(\frac{1}{T_{2e}}\right)^{\text{ZFS}} = \tau_v^2\left[\frac{5.26}{1+0.372\omega_S^2\tau_v^2} + \frac{7.18}{1+1.24\omega_S^2\tau_v^2}\right] \quad (7)$$

In Eqs. (4)–(7) S is the electron spin quantum number, γ_H the proton nuclear magnetogyric ratio, g and β the electronic g factor and Bohr magneton, respectively. r_H is the distance between the metal ion and the protons of the coordinated water molecules, ω_H and ω_S the proton and electron Larmor frequencies, respectively, and τ_R is the reorientational correlation time. The longitudinal and transverse electron spin relaxation times, T_{1e} and T_{2e}, are frequency dependent according to Eqs. (6) and (7), and characterized by the correlation time of the modulation of the zero-field splitting (τ_v) and the mean-square zero-field-splitting energy (Δ^2). The limits and the approximations inherent to the equations above are discussed in detail in the previous two chapters.

The commercial CAs and the several functionalized derivatives based on the DTPA and DOTA basic structures are monohydrated ($q = 1$) complexes with a molecular weight of *ca.* 600–800 Da. that corresponds to rotational correlation times τ_R of about 60–80 ps. For this class of polyaminocarboxylate complexes the residence lifetime τ_M is typically found to be in the range 50–500 ns and $T_{1e} \approx 1$ ns at 0.5 T and thus the inner sphere relaxivity, r_{ip}^{IS}, assumes a value of *ca.* 2.5–3.5 mM^{-1} s^{-1}, at 25°C(*12*). Therefore, as it was recognized early, it is clear that at 0.5 T the overall correlation time τ_{Ci} is largely dominated by the rotational correlation time, whereas the contribution of both the residence lifetime and the electronic relaxation play an almost negligible role. It follows that, with τ_R being roughly proportional to the "effective" molecular radius (Debye-Stokes equation; see Chapter 3), r_{ip}^{IS} is primarily determined by the molecular size of the complexes and shows a fairly good linear dependence on their molecular weight (Fig. 2, Charts 2 and 3). This behavior also suggests and implies that the *outer sphere* contribution is nearly identical for the various complexes considered in Fig. 2

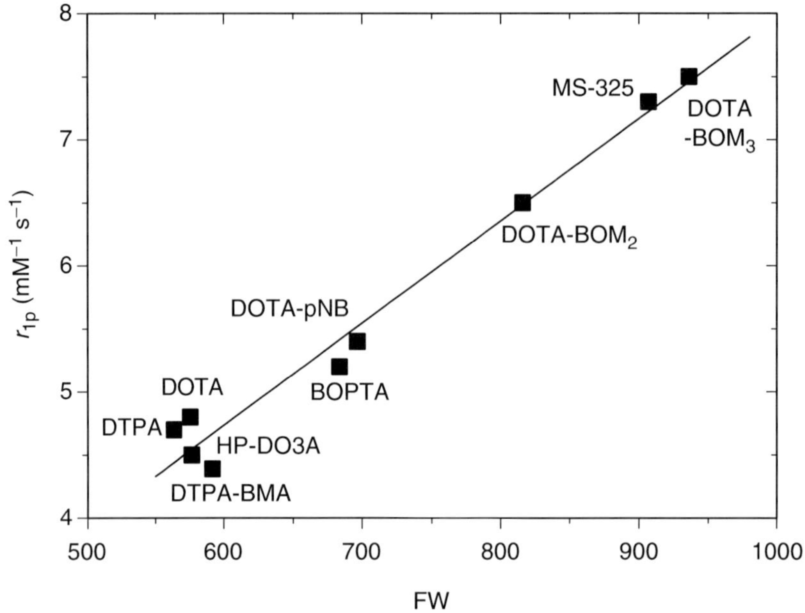

FIG. 2. Plot of the relaxivity (20 MHz an 25°C) for monoaquo polyaminocarboxylate Gd(III) complexes (Charts 2 and 3) versus molecular weight. The correlation coefficient R is 0.988.

and that the second sphere contribution, which has a more pronounced dependence on the structural features of the complexes, is negligible.

Another important structural parameter that influences the *inner sphere* relaxivity is the hydration number q. This represents a scaling factor in Eq. (3) and then a higher number of coordinated water molecules ($q > 1$) provides a clear advantage in terms of relaxivity. The fact that all the commercial Gd(III)-based CAs have $q = 1$ reflects the great importance given to safety and the need to avoid the release of the metal ion under physiological conditions. The use of hepta- or hexadentate ligands would in principle result in Gd(III) complexes with 2 and 3 coordinated water molecules, respectively, but the decrease in the denticity of the ligand is likely to be accompanied by a large decrease in their thermodynamic stability and an increase in their toxicity. However, some stable Gd(III) chelates containing two *inner sphere* water molecules have been identified and are under intense scrutiny. For instance, a new class of complexes have been recently developed by Raymond based on a TREN capping scaffold which connects three hydroxypyridinonate (HOPO) binding units (Chart 4) (*15–17*). The ligands are hexadentate, with six oxygen donor atoms and the Gd(III) complexes feature two coordinated water molecules which are in fast exchange. The complexes present a high thermodynamic stability and are suitable for use in vivo. Their relaxivity is about two times larger than the traditional monoaquo polyaminocarboxylate complexes as a result of their larger hydration number. Also, in this case a

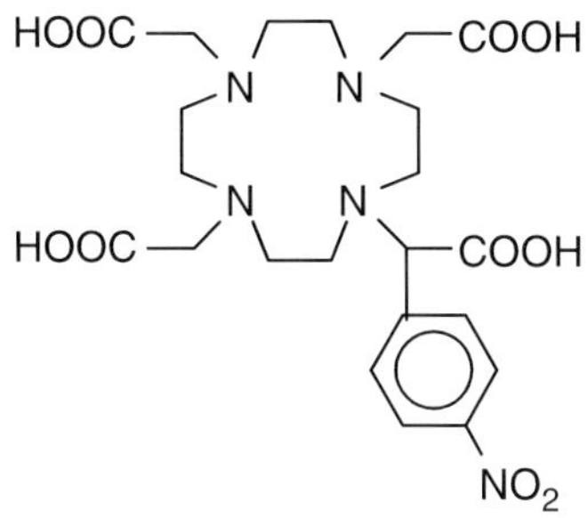

DOTA-pNB

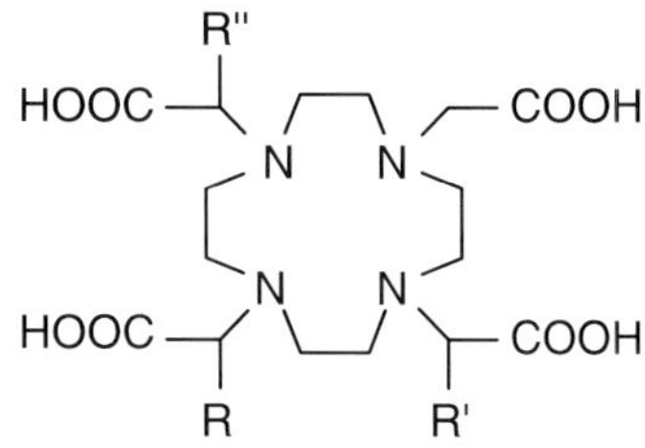

DOTA-$(BOM)_2$
(R = H, R',R'' = $-CH_2OCH_2C_6H_5$)

DOTA-$(BOM)_3$
(R,R',R'' = $-CH_2OCH_2C_6H_5$)

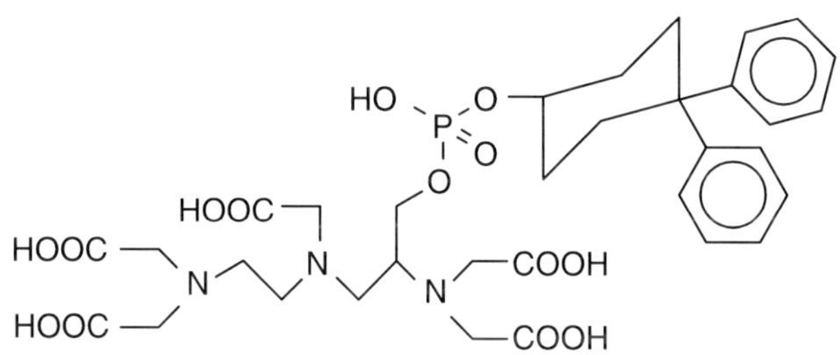

MS-325

CHART 3.

linear relationship has been found between the relaxivity and the molecular weight of the complexes (*17*).

The *inner sphere* relaxivity depends on other important parameters also: the mean residence lifetime of the coordinated water(s), the electronic relaxation times and the Larmor frequency. The correlation time for the water proton exchange, τ_M, coincides with the correlation time for the exchange of the entire water molecule near neutral pH and its value covers a large range of values: from few ns, as for the aqua ion and the HOPO derivatives, up to several μs for the cationic tetraamide derivatives of DOTA (*12,18*). It is a parameter of great relevance because it plays a dual role: it may contribute to the overall correlation time τ_C for the dipolar interaction [Eq. (5)] when the rotation of the complexes is slowed down following an increase in the molecular dimension of the complexes (larger complexes, micelles, dendrimers or macromolecular adducts) and it controls the efficiency of the transmission of the dipolar interaction to the bulk [Eq. (3)]. For complexes endowed with long values of τ_R and T_{ie}, T_{1M} may decrease to such an extent to become comparable to τ_M and in this case (intermediate or slow exchange regime) the water exchange rate

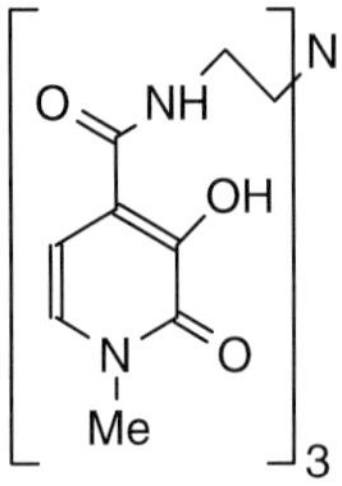

Tren-Me-3,2-HOPO

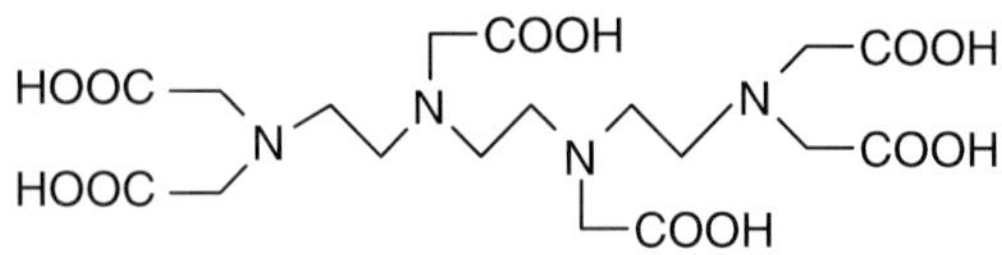

TTHA

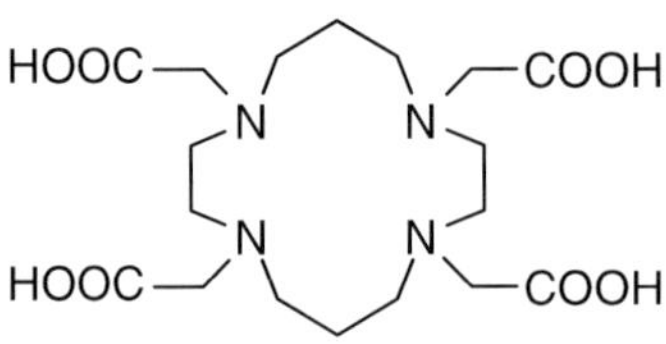

TETA

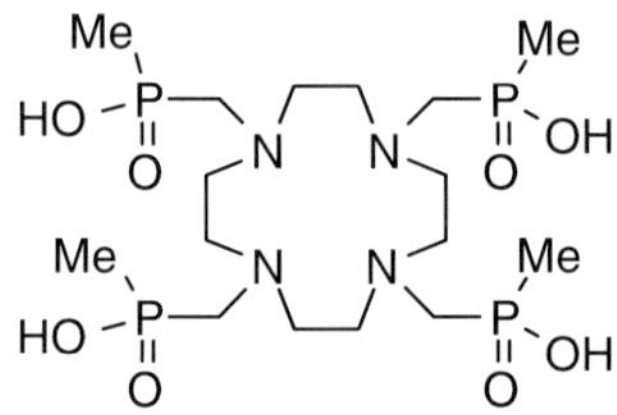

DOTPMe

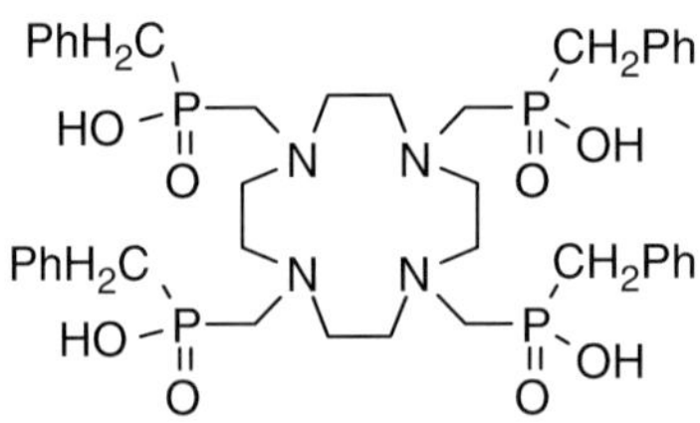

DOTPBz

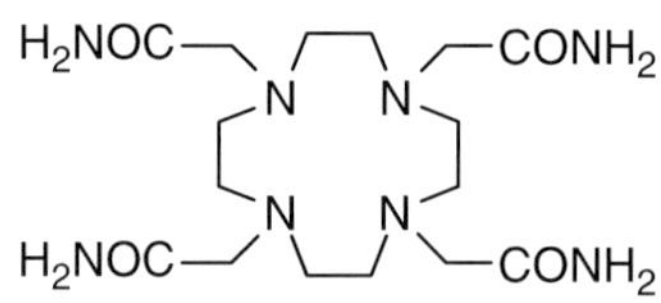

DOTAM

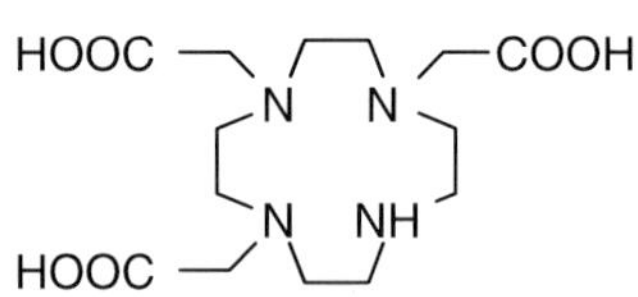

DO3A

CHART 4.

($k_{ex} = 1/\tau_M$) represents a limiting factor for r_1. The first example of this effect was reported for the neutral complex [GdDTPA-BMA(H_2O)] (*19,20*). The replacement of two carboxylate groups of the DTPA ligand with two carboxoamide groups decreases the overall charge on the Gd(III) complex from −2 to zero, and consequently τ_M increases from 303 ns to 2.2 μs. It follows that τ_M is

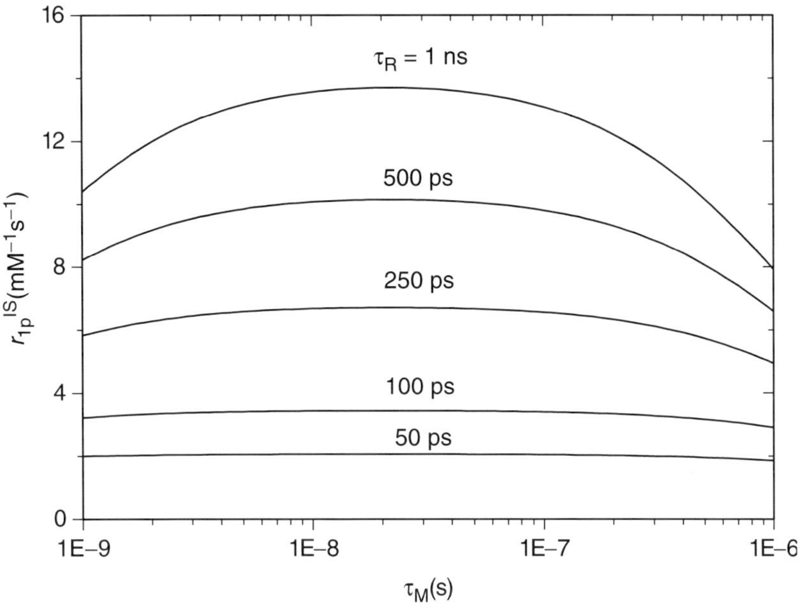

FIG. 3. Simulation curves of the *inner sphere* relaxivity as function of τ_M for a monoaquo Gd(III) complex at 25°C and 20 MHz for increasing values of the rotational correlation time. The other relaxation parameters were set as follows: $\Delta^2 = 5.0 \times 10^{19}\ s^{-1}$, $\tau_V = 15$ ps, $r_H = 3.0$ Å, $q = 1$.

no longer negligible as compared to T_{1M} and the relaxivity, at 20 MHz and 25°C, decreases by *ca.* 7% with respect to $[GdDTPA(H_2O)]^{2-}$. This behavior is remarkably enhanced in the case of the tetraaamide derivatives of DOTA where τ_M becomes so long (*ca.* 20 μs) to dominate the sum ($T_{1M} + \tau_M$) and the inner sphere relaxivity is unusually small, and actually it represents the smaller contribution to the observed relaxivity (*21*).

The dependence of r_{1p}^{IS} (20 MHz and 25°C) on τ_M is graphically represented in Fig. 3 for monoaqua Gd(III) complexes with different values of the rotational correlation times. The limiting effect of the residence lifetime is small for low molecular weight complexes (τ_R = 50–100 ps) and detectable only when $\tau_M \geq 1$ μs. On the other hand, for slowly tumbling complexes the unfavorable effect of long or short τ_M is clearly visible and the need to optimize this parameter for attaining high relaxivities is evident. An optimal value of τ_M at the imaging fields can be set in the range 10–50 ns.

Another important parameter that influences the *inner sphere* relaxivity of the Gd(III)-based contrast agents is the electronic relaxation time. Both the longitudinal and transverse electron spin relaxation times contribute to the overall correlation times τ_{Ci} for the dipolar interaction and are usually interpreted in terms of a transient zero-field splitting (ZFS) interaction (*22*). The pertinent equations [Eqs. (6) and (7)] that describe the magnetic field dependence of $1/T_{1e}$ and $1/T_{2e}$ have been proposed by Bloembergen and Morgan and their validity and limitations are thoroughly discussed in Chapters 2 and 3. Most of the data regarding the electron spin relaxation times for complexes relevant to MRI CAs have been evaluated through the analysis of the magnetic field dependence of the relaxivity according to Eqs. (3)–(7). Although the values derived might be affected by rather large errors due to the limitations of the relaxation theory, the values of Δ^2 and τ_v are used as empirical parameters to find some useful correlations with the structural properties of the complexes. It was observed early that the octadentate ligands DTPA and DOTA form monoaqua complexes with Gd(III) whose solution structures differ in terms of symmetry and stereochemical rigidity (both are lower for $[GdDTPA(H_2O)]^{2-}$). Furthermore, both complexes have a similar molecular dimension (i.e., similar τ_R value) and a very similar value of r and τ_M. However, the electron spin relaxation times show a marked difference, being about six times longer for $[GdDOTA(H_2O)]^-$. This results in a higher relaxivity (*ca.* 65%) for $[GdDOTA(H_2O)]^-$ with respect to $[GdDTPA(H_2O)]^{2-}$ at low magnetic fields (< 1 MHz). This has been explained by a larger value of Δ^2 being associated with Gd^{3+} complexes of low symmetry and high flexibility. The axially symmetric and stereochemically rigid macrocyclic complex Gd(DOTA) has a Δ^2 value of 1.6×10^{19} s^{-1}, whereas the highly fluxional $[GdDTPA(H_2O)]^{2-}$ presents a Δ^2 value of 4.6×10^{19} s^{-1} (*11,23*). This only represents an empirical rule as the chemical nature of the coordinating groups may also play a significant role (e.g., DOTA vs. DOTP (Chart 5)) but, in general, this parameter is shorter for complexes of macrocyclic ligands than in the case of the complexes with the corresponding acyclic ligands (*24*). At the imaging fields, where T_{2e} does influence the relaxivity because the "7 term" [Eq. (4)] has already dispersed, the longitudinal electron spin relaxation is long as compared to the rotational correlation time of the commercial CAs and therefore it has little influence on the *inner sphere* relaxivity. However, for larger complexes the contribution of T_{1e} to τ_{C1} becomes important and a marked dependence of r_{1p}^{IS} on the electron relaxation parameters is observed. In Fig. 4 simulated curves of r_{1p}^{IS} as a function of Δ^2 for a Gd(III) complex with increasing values of the rotational correlation time are reported. For τ_R values longer than 100 ps, r_{1p}^{IS} shows a marked decrease with increasing Δ^2: clearly for the attainment of macromolecular contrast agents of high relaxivity, both T_{1e} and τ_M have to be optimized.

Finally, from Eqs. (4)–(7) it is clear that the inner sphere contribution to the relaxivity strongly depends on the observation frequency. As already discussed in the previous two chapters of this volume, the study of the field dependence of the relaxivity (NMRD technique) represents a powerful tool for extracting the values of the relaxation parameters and its application to the investigation of Gd(III) complexes of relevance as MRI CA will be discussed in Section III.

B. Outer-Sphere Contribution to the Relaxivity

This mechanism may represent a sizeable contribution to the relaxivity of monoaquo Gd(III) complexes at the imaging fields which

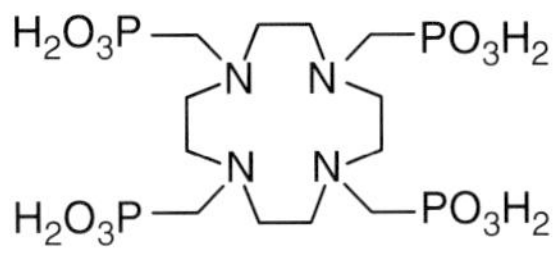

DOTP

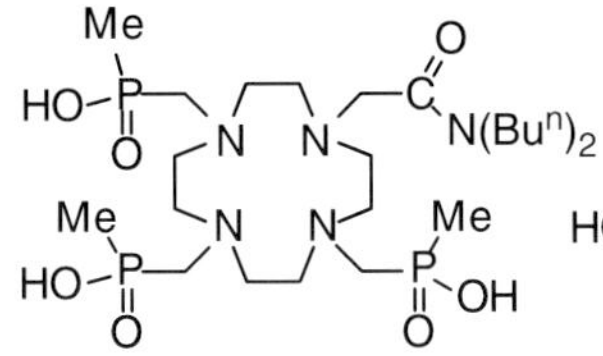

DOTP-MBu$_2$

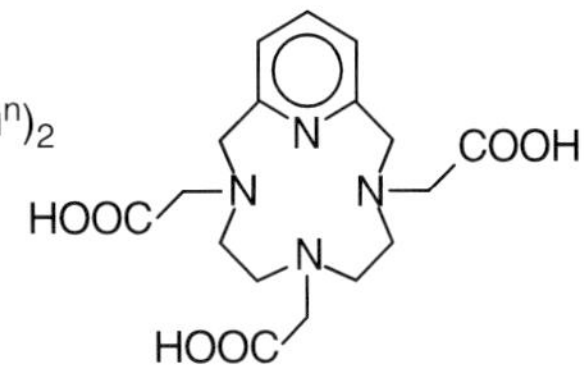

PCTA

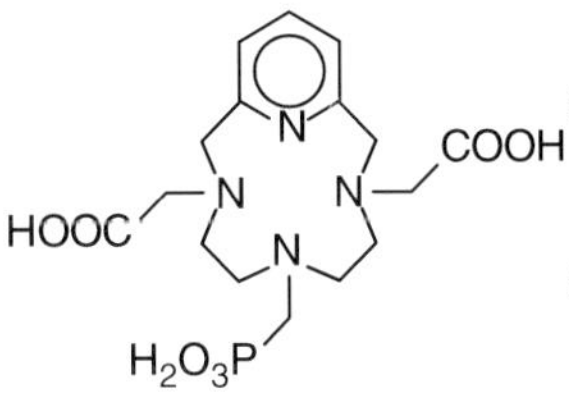

PCTP

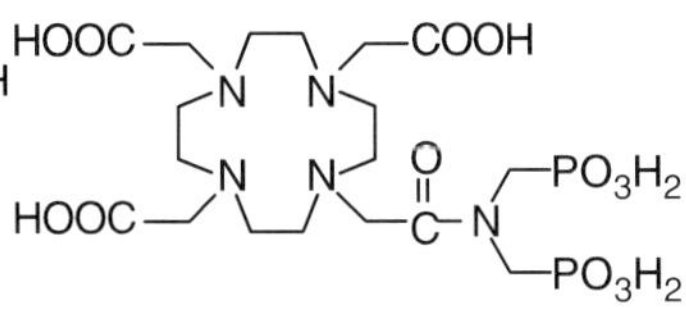

DO3ANP$_2$

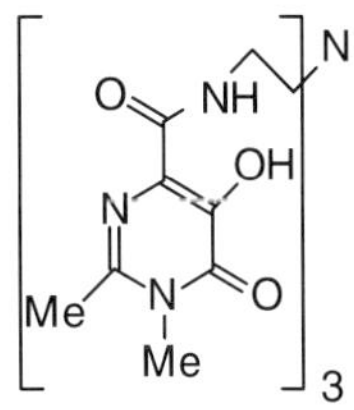

TrenHOPY

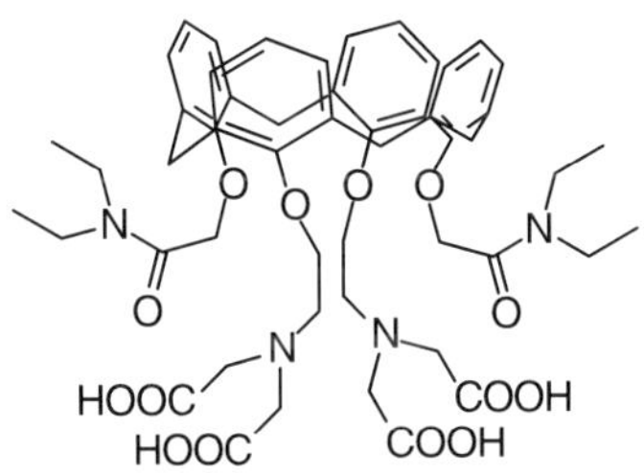

Calix[4]arene-TA

CHART 5

arises from the modulation of the dipolar interaction of the paramagnetic center with the water molecules diffusing next to the surface of the complex. An exhaustive discussion of the available theory is provided in Chapter 2. Here, we only summarize the relevant aspects for MRI contrast agents. The component of the relaxivity due to the r_{ip}^{OS} term is commonly estimated by the expression derived by Freed [Eqs. (8) and (9)] which takes into account the translational diffusive motion of the water molecules, considered as hard spheres, and incorporate the effects of the electron spin relaxation *(25,26)*.

$$r_{ip}^{OS} = C^{OS}\left(\frac{[CA]}{aD}\right)[7J(\omega_S) + 3J(\omega_H)] \tag{8}$$

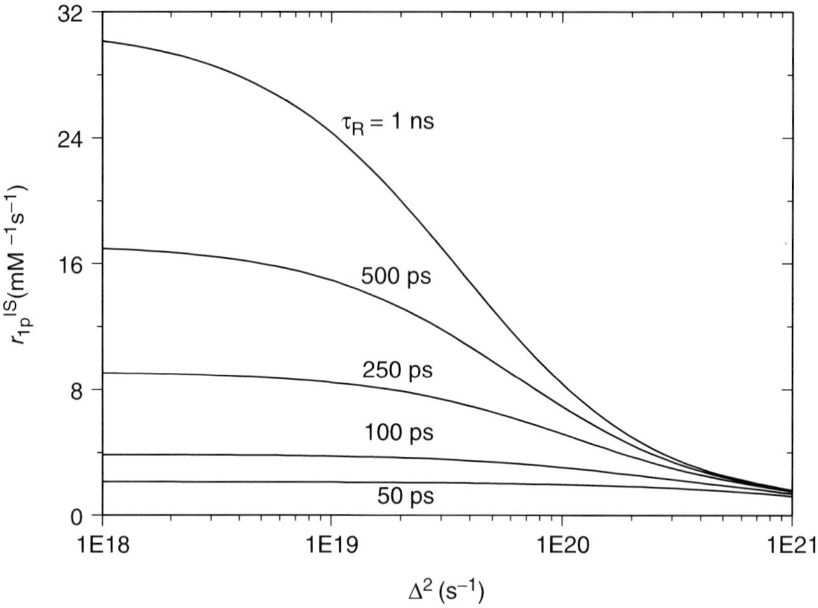

FIG. 4. Simulation curves of the *inner sphere* relaxivity as function of Δ^2 for a monoaquo Gd(III) complex at 25°C and 20 MHz for increasing values of the rotational correlation time. The other relaxation parameters were set as follows: $\tau_M = 100$ ns, $\tau_V = 15$ ps, $r_H = 3.0$ Å, $q = 1$.

$$J(\omega) = \mathrm{Re}\left[\frac{1 + 1/4\left(i\omega\tau_d + \frac{\tau_d}{T_{je}}\right)^{1/2}}{1 + \left(i\omega\tau_d + \frac{\tau_d}{T_{je}}\right)^{1/2} + 4/9\left(i\omega\tau_d + \frac{\tau_d}{T_{je}}\right) + 1/9\left(i\omega\tau_d + \frac{\tau_d}{T_{je}}\right)^{3/2}}\right]$$
$$\text{with } j = 1, 2; \quad \tau_d = a^2/D \tag{9}$$

In the above equations, C^{OS} is a constant (5.8×10^{-10} m^6 mol^{-1} s^{-2}) and the non-Lorentzian spectral density functions $J(\omega_i)$ contain the field dependence on $T_{1,2e}$. At the magnetic fields of interest, r_{ip}^{OS} depends primarily on the distance of water closest approach a, related to the molecular dimension and charge distribution of the complex, and on the relative diffusion coefficent of solute and solvent D. It follows that r_{1p}^{OS} assumes a comparable value (above 10 MHz) for all the small gadolinium chelates of clinical interest. In fact, very similar values of the relaxivity (*ca.* 2.1–2.5 mM^{-1} s^{-1}) have been measured for small Gd(III) chelates with $q = 0$ (Chart 4), such as $[GdTTHA]^{3-}$ (*27,28*), $[GdTETA]^{-}$ (*29*), $[GdDOTPBz]^{-}$ (*30*) and $[GdDOTPMe]^{-}$ (*30*). The small differences in relaxivity have been accounted for primarily in terms of the different dimensions of the complexes which is reflected in different values of the parameter a. On the other hand, at low magnetic fields the relaxivities differ according to the different values of the electron spin relaxation times of the complexes.

It is clearly difficult to separate the *inner* and *outer sphere* contributions to the relaxivity without a magnetic field dependent study and, at high field (> 10 MHz), only a rough estimation of r_{1p}^{OS} is possible based on a judicious choice of the parameter a and a determination of the diffusion coefficient of the complex using Pulsed Gradient Spin Echo (*31*) or electrochemical techniques (*32*). However, some additional information can be extracted by measuring the temperature-dependence of the relaxivity at a fixed frequency, since the two contributions may have a different behavior. The *outer sphere* relaxivity increases by lowering T because of the increase in the relative diffusion coefficient and in $T_{1,2e}$. The temperature dependence of r_{1p}^{IS} depends on the relative values of T_{1M} and τ_M [Eq. (3)] and two limiting situations may occur:

(i) $\tau_M < T_{1M}$ (fast exchange regime); in this case the behavior is dictated by the temperature dependence of T_{1M}, which decreases on decreasing T because τ_C lengthens and r_{1p}^{IS} increases. This situation is illustrated in Fig. 5 (top) for $[GdDOTA(H_2O)]^-$. By assuming an Arrhenius-type behavior of the parameters τ_R, τ_M, τ_V, and D whose values at 25°C are known (along with the other, temperature independent, structural parameters) from analysis of the NMRD profile and ^{17}O NMR data (see Chapter 7), it is possible to fit the temperature dependence of the relaxivity (20 MHz) in terms of the parameters $\Delta H_i (i = M, V, R, D)$. As shown in the figure, the relaxivity increases by decreasing T as a result of the increase of both contributions. Only at temperatures close to zero does r_{1p}^{IS} tend to level off because τ_M becomes so long that it is no longer negligible as compared to T_{1M}.

(ii) $\tau_M \geq T_{1M}$ (slow/intermediate exchange regime); this situation is illustrated in Fig. 5 (bottom) for the cationic $[GdDOTAM(H_2O)]^{3+}$ complex (Chart 4) which presents a τ_M value of 20 μs at 25°C (*36*). The water exchange rate is so slow that $\tau_M > T_{1M}$ for $T < 70$°C and the *inner sphere* contribution decreases with decreasing temperature and tends to zero. In fact $r_{1p}^{IS} < r_{1p}^{OS}$ for $T < 60$°C and the overall relaxivity is sensibly lower than for the corresponding carboxylate $[GdDOTA(H_2O)]^-$ complex.

As already observed in Chapter 2 (VII.C), the model of Freed is based on a force-free diffusion model that does not take into account the interaction of the water molecules with the complex. Clearly, this hypothesis only represents a crude approximation in the case of the polyaminocarboxylate complexes where multiple H–bonding interactions between the water molecules and the polar groups of the ligand are likely to occur. In fact, the commercial CAs are complexes characterized by the presence of well-defined hydrophilic (carboxylate and carboxoamide groups) and hydrophobic regions (ethylenediamino groups), thus the distribution of the water molecules outside the first coordination sphere of the metal ion is presumably highly anisotropic. Borel *et al.* have performed quantum mechanical (QM) calculations in order to obtain the electrostatic potential for macrocyclic and linear Gd(III) complexes and nicely depicted the hydrophilic and hydrophobic regions around the surface of the ligand (*29*). In the axially symmetric DOTA-type systems

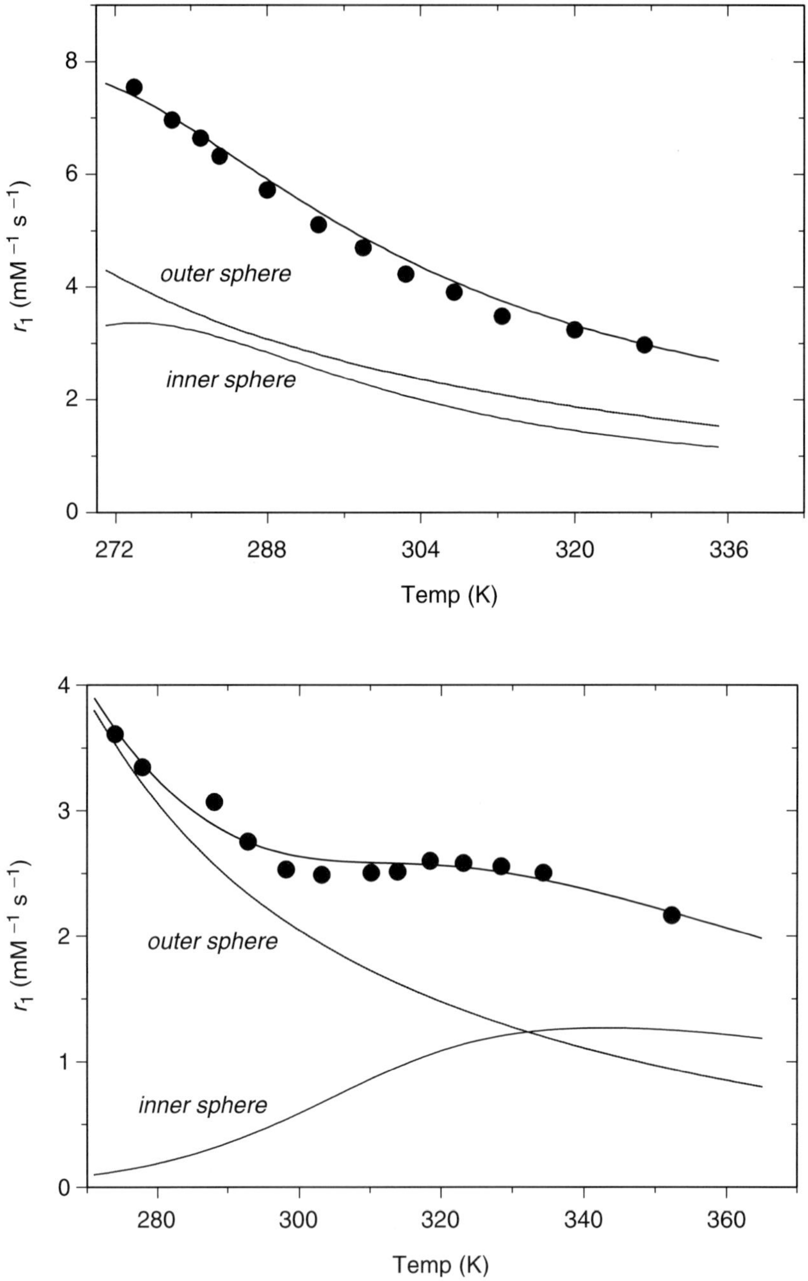

FIG. 5. Temperature dependence of the proton relaxivity (20 MHz) for Na[GdDOTA(H_2O)] (top) and [GdDOTAM(H_2O)]tfl$_3$ (bottom) and individual contributions of the *inner* and *outer* sphere mechanisms.

(DOTA, DO3A (Chart 4), DOTP) the complex may be divided in two hemispheres: one corresponds to the hydrophilic region, that includes the carboxylate (carboxamide/phosphonate) groups and the bound water molecule, the other consists of the hydrophobic domain, identified with the tetraazacyclodecane ring. Then, by using molecular dynamics (MD) calculations the authors have computed the distribution of the *outer sphere* water molecules around the complexes and found a region, at a distance from 3 to 5 Å, where a relatively small number of solvent molecules are H-bonded to the carboxylate groups with a mean residence lifetime of about 20–30 ps. Finally, from the radial distribution functions it was possible to calculate the *outer sphere* relaxivity and an excellent agreement with the experimental data was found.

C. Second-Sphere Contribution to the Relaxivity

It is clear that the Freed's model is not fully adequate for a realistic and physically sound description of the interaction of the outer sphere water molecules with the paramagnetic complexes. The main limitation is the lack of consideration of specific interactions of the solvent molecules with hydrophilic groups of the ligands, i.e., the formation of a second coordination shell. From a practical point of view, the use of a purely translational diffusion model for the evaluation of the *outer sphere* relaxivity often gives reasonably good results as shown by the analysis of the magnetic field dependence of the relaxivity of $q = 0$ complexes like $[GdTTHA]^{3-}$, $[GdTETA]^{-}$, and $[GdDOTPBz]^{-}$ (Chart 4). Presumably, this happens when the residence time of the water molecules in the second hydration shell is close to the value of the self-diffusion of the solvent molecules. In these cases, neglecting r_{ip}^{SS} mostly results in an overestimation of the contribution of r_{ip}^{OS} with perhaps unrealistic values of a, often shorter than the molecular radius of the complex. On the other hand, there are cases where the Freed's model completely fails to account for the *outer sphere* relaxivity. A notable example is represented by $[GdDOTP]^{5-}$, a complex structurally similar to GdDOTA but without any coordinated water molecules, as assessed by ^{17}O NMR data. The relaxivity of $[GdDOTP]^{5-}$ is very similar to that of the monoaquo complexes $[GdDOTA(H_2O)]^{-}$ and $[GdDTPA(H_2O)]^{2-}$, and therefore cannot be explained in terms of the Freed's equation. The presence of four, highly negatively charged phosphonate groups promotes the formation of strong hydrogen bonds with the solvent molecules with a relatively long residence time. Borel has calculated from MD simulations the presence of 4.3 *second sphere* water molecules with a τ_M of 56 ps, which is sensibly longer than for the case of $[GdDOTA(H_2O)]^{-}$ and $[GdDTPA(H_2O)]^{2-}$ (*29*). This example clearly shows how the proper choice of the coordinating groups may affect the number and residence lifetime of the water molecules in the proximity of the paramagnetic center and enhance this contribution to the relaxivity.

Although a clear distinction between r_{ip}^{OS} and r_{ip}^{SS} is not possible, several examples have been recently reported where a contribution of r_{ip}^{SS} was indeed invoked in order to explain the relaxivity data (*33*). The neutral complex $[GdDOTP\text{-}MBu_2]$ (Chart 5) has a hydration number $q = 0$ but a relaxivity

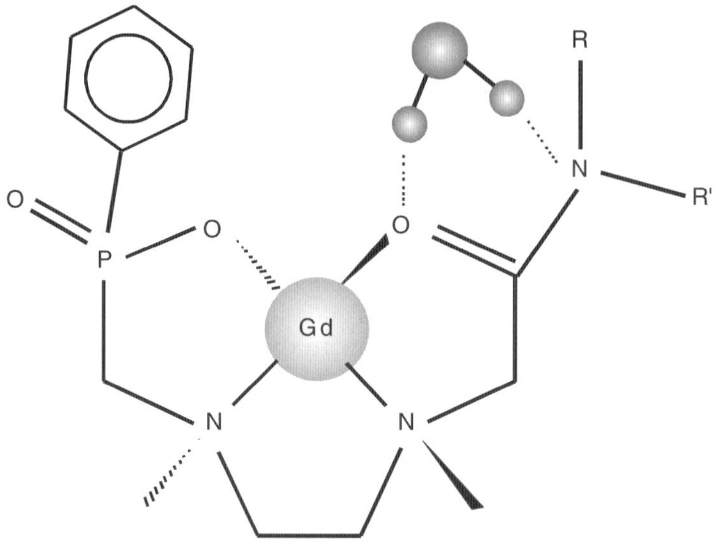

FIG. 6. Schematic representation of the *second sphere* water molecule affecting the relaxivity of the complex [GdDOTMP-MBu$_2$].

sensibly higher than for [GdDOTPBz]$^-$. In this case the use of the translational diffusion model would give either an *a* value unreasonably too short or parameters describing the diffusional dynamics of the solution physically not meaningful. The relaxivity has been explained on the basis of the presence of a tightly bound water molecule in the second coordination sphere of the metal ion at a distance of *ca.* 3.6 Å from the Gd(III) ion (*34*). The interaction was suggested to involve a hydrogen bond between the oxygen and the nitrogen atoms of the carboxoamide group, and a water molecule which is to be localized close enough to the paramagnetic center so as to affect the solvent relaxation (Fig. 6).

Another clear example of a large contribution of the *second sphere* mechanism to relaxivity comes from the comparison of two related macrocyclic complexes: [GdPCTA (H$_2$O)$_2$] and [GdPCTP(H$_2$O)]$^{3-}$ (Chart 5), differing by the presence of three carboxylate (PCTA) or phosphonate (PCTP) groups and by the value of the hydration number *q*, 2 and 1, respectively. The relaxivity of the phosphonate derivative is 7.5 mM^{-1} s^{-1}, at 20 MHz and 25°C whereas a value of 6.9 mM^{-1} s^{-1} is measured for the corresponding carboxylate complex (*35*). In spite of the lower hydration number, [GdPCTP(H$_2$O)]$^{3-}$ has a higher relaxivity which is explained by the presence of tightly bound water molecules to the phosphonate moieties on the surface of the ligand.

A sizeable contribution of the term r_{ip}^{SS} has also been evidenced in the study of the pH-dependence of the relaxivity in the case of the Gd(III) complexes with tetraamide DOTA derivatives. As previously discussed, these cationic monoaqua complexes are characterized by a slow rate of exchange of the coordinated water molecule, which strongly limits the *inner sphere* relaxivity around neutral pH. By increasing the pH, a base-catalyzed prototropic exchange occurs that lowers the effective τ_M value and increases the contribution of r_{ip}^{IS}. However, a further increase in the relaxivity is observed at pH > 10 and this has been accounted for in terms of the stepwise deprotonation of the amide nitrogen atoms that favors the localization in the proximity of the metal center of a few water molecules, bridging through hydrogen bond interactions, the amide group and the bound water (*21,36*). Recently, Sherry has reported a similar example concerning tetraamide DOTA complexes bearing hydroxypyridyl substituents. At high pH the deprotonation of the amides favors the formation of intramolecular acid–base pair interaction with the phenolic protons accompanied by the formation of a highly organized second hydration sphere that increases the relaxivity (*37*).

Water molecules hydrogen-bonded to polar groups of the ligand not only provide an additional relaxation mechanism for the bulk water protons but may affect the relaxivity by changing the value of some key parameters of the *inner sphere* contribution. For example, the complex $[GdDO3ANP_2(H_2O)]^{4-}$ (Chart 5) bears two pendant phosphonic groups and the analysis of the temperature dependence of the water ^{17}O transverse relaxation rate (see Chapter 7) has suggested the presence of a tightly bound water molecule interacting with both the coordinated water and one of the phosphonate groups. The result is that the process of exchange of the *inner sphere* water is rendered more difficult and the τ_M value (1.6 μs) is longer than for related DOTA monoamide complexes (*38*).

It must be observed that the *second sphere* relaxation pathway in principle depends on the same parameters as the *inner sphere* term and then it can be analysed in terms of Eqs. (3)–(7). In this case the parameters are denoted with a prime (τ_C', r_H', etc.). It follows that this contribution is enhanced by slowing down the rotational mobility of the complex, increasing the number of hydrogen-bonded waters and their mean lifetime. In fact we found that the macromolecular adducts with human serum albumin (HSA) of $q = 0$ complexes, such as the tetrabenzylphosphinate DOTA derivative (*30*) and $[GdDOTP]^{5-}$ (*33*), show a marked relaxivity enhancement which is explained by the presence of exchangeable protons on the protein close to the interaction site of the complex and from a network of hydrogen-bonded water molecules in the second coordination sphere of the metal ion. The increase in relaxivity is notable and arises from the long τ_R value of the macromolecular complex and, presumably, the high number of long-lived water molecules in the second hydration shell. Thus the optimization of r_{ip}^{SS} is possible and to this aim the following points should be considered:

(1) The mean lifetime of the interaction of the *second sphere* water molecules is much shorter than that of the coordinated water molecule(s) and it is likely to make a sizeable contribution to τ_C [Eqs. (4) and (5)]. Longer values of τ_M, are associated with a stronger interaction that can be obtained by the design of a ligand with suitable hydrogen-bond acceptor groups with an optimal geometric disposition in the complex so as to minimize their distance from the metal center.

(2) The r_H' distance of the *second sphere* water molecules does not have a fixed value, but it is a flexible parameter which may assume values in the range 3.5 – 4.5 Å. Because of the $1/(r_H')^6$ dependence of the relaxivity, an optimization of this parameter, which depends on structural properties of the complex, would result in enhanced relaxivities.

(3) The number of water molecules in the second hydration layer can be increased by a proper ligand design without modifying its denticity and thus preserving the thermodynamic stability of the Gd(III) complex.

III. Relevant Parameters in Fitting the NMRD Profiles of Contrast Agents

As it has been exhaustively discussed in the previous two chapters, the measurement of the longitudinal relaxation rate as a function of the magnetic field strength, over a wide range of values, represents a powerful means for extracting the large structural and dynamic information content inherent to the dipolar interaction between the solvent and the paramagnetic complex. The relaxivity depends on the magnetic field either directly, through Eqs. (4) and (8), or indirectly through the field dependence of the electronic relaxation times that contribute to the overall correlation time. So, in principle it is possible to separate and evaluate the different contributions to the relaxivity and the underlying processes (rotation, exchange, electron spin relaxation). Unfortunately, in Eqs. (3) and (9) the number of parameters is too large whereas the NMRD curves are rather featureless. It follows that it is not generally possible to obtain a unique set of values from a best-fit of the NMRD profiles or, alternatively, several assumptions have to be made and some of the relaxation parameters need to be either fixed, to physically reasonable and plausible values or independently evaluated by different techniques. In the following sections, we will discuss how the most relevant relaxation parameters affect the NMRD profiles, their dependence on the chemical structure of the complexes and possible alternative techniques for their evaluation. Several excellent reviews have been recently published which cover this subject in detail and therefore we will only critically summarize the important aspects of the field and incorporate some recent results (*12,39,40*).

A. The Hydration Number q

The number of *inner sphere* water molecules has an obvious effect on the relaxivity. Doubling the value of q will imply a doubling of r_{ip}^{IS} and then a substantial increase in the relaxation efficacy of the Gd(III) complex, as shown in Fig. 7. Knowing q is then crucial for a proper analysis of the NMRD profile, but this parameter cannot be extracted from a best fit procedure as, at best, only the term q/r_H^6 can be obtained. From a qualitative point of view, for relatively small Gd(III) complexes and in the absence of

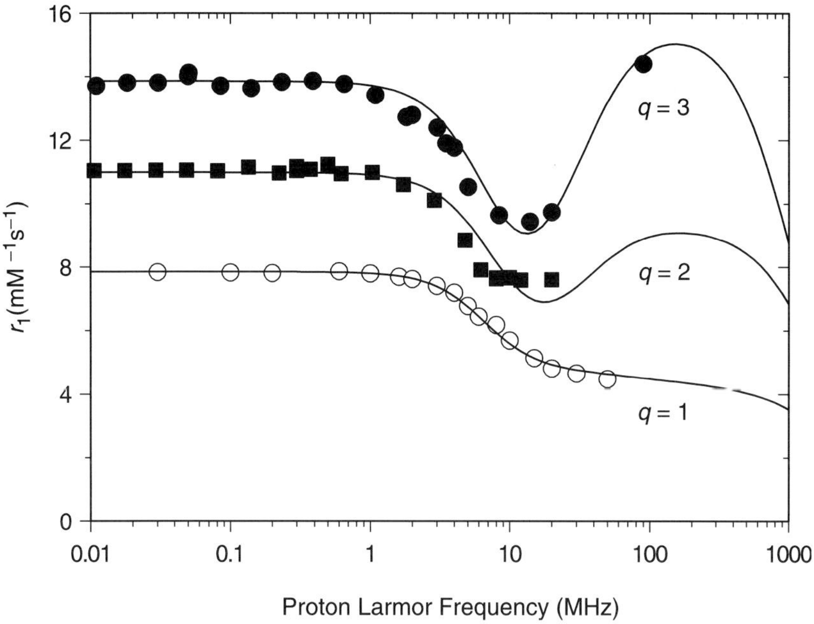

FIG. 7. ^{1}H NMRD profiles (25°C) of three Gd(III) complexes differing in the number q of *inner sphere* water molecules: $[GdDTPA(H_2O)]^{2-}$ (open circles), $[Gdtren\text{-}HOPY(H_2O)_2]$ (squares), $[GdCalix[4]arene(H_2O)_2]^-$ (filled circles).

intermolecular interactions and a large *second sphere* contribution, one could take advantage of the linear relation implicit in Eq. (1) and, at high field where τ_R dominates the correlation time τ_C, obtain an estimate of q from a plot of the type reported in Fig. 2. To this end, the relaxivity values measured at higher temperatures should minimize the possible limiting effect of a slow rate of water exchange. However, the preferable approach is to obtain the hydration number from an independent technique. Several possibilities exist even though none of them refers in particular to the Gd(III) ion.

(a) *luminescence lifetime*. The substitution of D_2O for H_2O induces an increase in the intensity and lifetime of the luminescence of Eu(III) and Tb(III) complexes in solution. This is the result of the quenching of the luminescence occurring *via* energy transfer involving O–H vibrational processes and it is proportional to the number of O–H oscillators in the inner coordination sphere of the metal ion (*41*). Horrocks and Sudnik proposed an empirical equation which allows the estimation of the hydration number from the experimental rate constants measured in D_2O and H_2O (*42*):

$$q = A\left(k_{H_2O} - k_{D_2O}\right) \tag{10}$$

where $A = 4.2$ ms^{-1} for Tb(III) and 1.05 ms^{-1} for Eu(III). More recently, Parker has proposed new equations that consider the contribution of other exchangeable X–H oscillators (N–H and C–H) and of the unbound water molecules (*43*):

$$q_{Eu} = 1.2\left[\left(k_{H_2O} - k_{D_2O}\right) - 0.25\right] \tag{11}$$

$$q_{Tb} = 5\left[\left(k_{H_2O} - k_{D_2O}\right) - 0.06\right] \tag{12}$$

Since Gd(III) is in between Eu(III) and Tb(III) in the periodic table this procedure represents the most accurate, though indirect method for the evaluation of q. One has only to be aware of the possible change in the coordination number across the Ln series and therefore a determination of both q_{Eu} and q_{Tb} is preferable.

(b) *Dy-induced ^{17}O shift.* For the Dy(III) cation the hyperfine shift is largely dominated by the contact term and therefore, without introducing significant errors, there is no need to perform the long procedure for dissecting this contribution from the dipolar one. Furthermore, the contact ^{17}O shifts of the bound water molecules do not depend on the structure of the complex and thus they are directly proportional to the hydration number. Peters *et al.* suggested this method to evaluate q based on the measurement of the ^{17}O shift of the water signal as a function of the concentration of the paramagnetic complex (*44*). A straight line is obtained whose slope is proportional to the q value of the Dy(III) complex. The actual value is then obtained by comparison with the slope of the line obtained with a reference sample, normally the aqua ion $[Dy(H_2O)_8]^{3+}$. Of course, what is obtained by this procedure is the hydration number for the Dy(III) complex which coincides with the value for Gd(III) only if coordination equilibria across the series do not affect q.

Finally, it is worth mentioning that the presence of hydration equilibria could result in effective fractional q values. This can be checked by UV-Vis measurements on the Eu(III) complex as a function of temperature (*12*).

B. The Gd–H Distance r_H

Because of the dependence of the inner sphere relaxivity on $1/r_H^6$, the Gd-water proton distance is extremely important in determining the efficacy of a CA. In Fig. 8 the experimental profile of $[GdDTPA(H_2O)]^{2-}$ is shown and also the best-fit curve obtained with a r_H value of 3.1 Å, compared with two calculated profiles corresponding to r_H values of 2.95 (upper curve) and 3.25 Å. A variation of this parameter of only ± 0.15 Å changes the relaxivity about 16% at low fields. Estimates from X-ray data of the Gd–O distance are affected by some errors as the tilt angle of the water molecule in solution is not defined with precision.

Reported r_H values for Gd(III) chelates in aqueous solution, obtained from the fitting of the NMRD profiles, are typically in the range 2.90–3.15 Å. However, the values derived from this procedure are likely to be rather

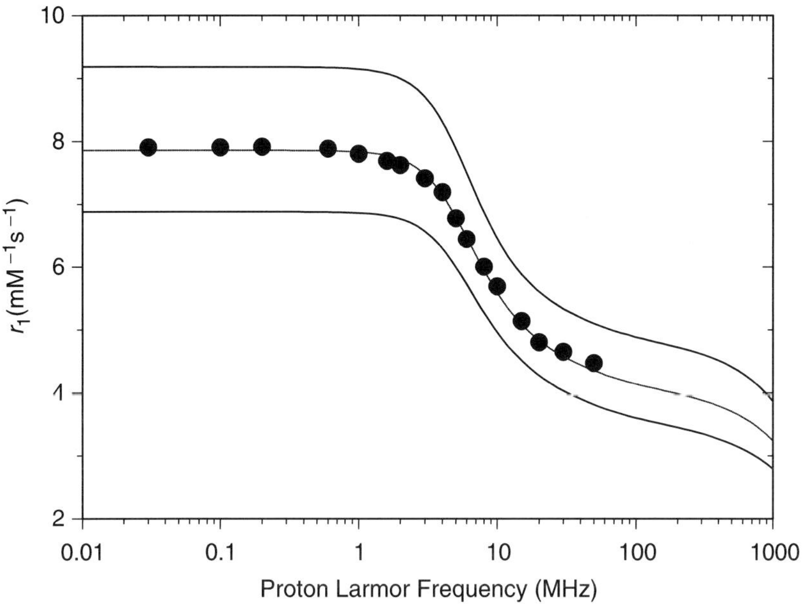

FIG. 8. ^{1}H NMRD profile (25°C) of $[GdDTPA(H_2O)]^{2-}$ (r = 3.1 Å) and simulated curves for a distance r of the bound water molecule of 2.95 (upper curve) and 3.25 (lower curve) Å.

inaccurate because r strongly depends on how precisely the *outer* and *second sphere* contributions have been estimated, and also shows a marked relationship with the overall correlation time τ_C. In practice, during the fitting procedure the parameter r_H is allowed to vary in a narrow range of values, but the differences (±0.30 Å) found among different complexes do not actually reflect a dependence on the distance from the structural features of the Gd(III) chelates but rather an innacuracy in the evaluation of the *inner sphere* contribution. Interestingly, Caravan *et al.* has recently determined this parameter directly for the aqua ion and for a series of four commercial Gd(III)-based contrast agents from pulsed ENDOR spectra and found that the distance was the same (3.1 ± 0.1 Å) for all five complexes (*45*). Thus, at present a relaxivity enhancement through an optimization of this parameter by proper ligand design does not appear an easy task.

C. THE ROTATIONAL CORRELATION TIME τ_R

The molecular reorientational correlation time tends to dominate the overall correlation time of low molecular weight Gd(III) chelates, particularly in the high field region, and therefore represents a key parameter in governing their relaxivity. The effect of the increase in τ_R on the shape and amplitude of the NMRD profiles was understood in detail early on and, as a consequence, the attempts at optimizing the relaxivity were primarily focused on slowing down the rotation by increasing the size of the

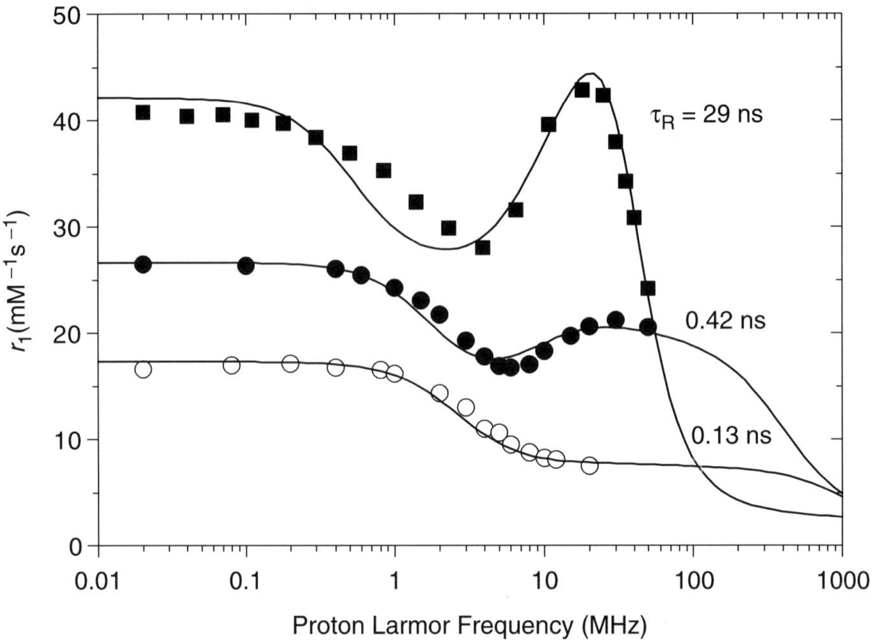

FIG. 9. ^{1}H NMRD profiles (25°C) of $[GdDOTA(BOM)_3(H_2O)]^-$ (open circles), its inclusion complex with β-cyclodextrin (filled circles) and of $[GdDOTA(BOM)_3(H_2O)]$-HSA adduct (squares). The different shapes and amplitudes of the profiles are primarily due to the different rotational correlation times of the paramagnetic complexes.

complexes, attaching the complexes to proteins or exploiting the non-covalent interactions with macromolecular substrates. Figure 9 shows the experimental NMRD profiles of $[GdDOTA(H_2O)]^-$, its trisbenzyloxymethylenic derivative $[GdDOTA\text{-}BOM_3(H_2O)]^-$ (Chart 2), the inclusion complex with β-cyclodextrin of $[GdDOTA\text{-}BOM_3(H_2O)]^-$ and the non-covalent adduct of this complex with human serum albumin (HSA). The relaxivity is significantly enhanced following the increase in τ_R with the characteristic "peak" in the high field region, increasing in intensity, sharpening and shifting towards the lower field, all caused by the increase in τ_R. This effect originates from the field dependence of T_{1e} which becomes visible when τ_R is increased. This parameter is typically obtained from the fit of the experimental NMRD profile and its value is more accurate if good, independent estimation of q and r are available. However, other techniques are available in order to independently estimate the rotational correlation time:

(i) The Debye–Stokes equation (Eq. 13) can provide the value of τ_R for spherical complexes if the microviscosity η and the molecular radius r_{eff} are known with good precision. However, the microviscosity may largely differ from the experimentally accessible macroscoscopic viscosity and the radius is difficult to estimate. The following equation can be used to compare the rotational correlation times of complexes of different sizes, since the value of η is likely to be the same, the ratio of the radii corresponds to the ratio of molecular weights (*12*);

$$\tau_R = 4\pi\eta r_{eff}^3 / 3k_B T \tag{13}$$

(ii) The ^{17}O longitudinal relaxation of the Gd(III)-bound water molecules is dominated by the quadrupolar mechanism which depends on the quadrupolar coupling constant χ^2, the asymmetry parameter η and τ_R:

$$\frac{1}{T_{1q}} = \frac{3\pi^2}{10}\left(\frac{2I+3}{I^2(2I-1)}\right)\chi^2(1+\eta^2/3) \times \left[0.2\left(\frac{\tau_R}{1+\omega_I^2\tau_R^2}\right) + 0.8\left(\frac{\tau_R}{1+4\omega_I^2\tau_R^2}\right)\right] \tag{14}$$

The main difficulty of this method is the unknown value of the term $\chi^2(1+\eta^2/3)$ and therefore the value for pure water is often used. On the other hand, it has the advantage that the measurement is directly performed on a solution of the Gd(III) complex and that the rotational correlation time of the Gd-coordinated water oxygen vector is actually determined (*12*);

(iii) ^{13}C longitudinal relaxation time and NOE measurements on a diamagnetic (Y, La or Lu) analog can also be used to obtain the rotational correlation time of the complexes. With this method the τ_R corresponds to that of a proton of the ligand and thus the assumption is made that this coincides with that of the Gd-coordinated water proton vector. The values obtained with this procedure and by fitting the NMRD curves are quite similar and then this approximation appears to be plausible (*46*). Alternatively, it is possible to label a methylenic group of the ligand with ^{2}H (typically acetate) and then calculate the rotational correlation time from ^{2}H T_1 measurements on a diamagnetic analog of the Gd(III) chelate. For deuterium the longitudinal relaxation rate is dominated by the quadrupolar mechanism and then $1/T_1$ is directly proportional to τ_R (*47*).

It must be noted that in Eqs. (4) and (5) the rotational correlation time refers to the tumbling of the protons of the coordinated water molecule, this could be different from the τ_R value characterizing the whole complex. Recently, Dunand *et al.* have analyzed the internal motion of the bound water molecule of $[LnDOTAM(H_2O)]^{3+}$ complexes by exploiting the Curie contribution to the ^{1}H relaxation rate in $[TbDOTAM(H_2O)]^{3+}$ and the ^{2}H and ^{17}O relaxation rates in $[EuDOTAM(H_2O)]^{3+}$ (*48*). The result is that the rotational correlation time of the Ln–H_{water} vector is shorter than that for the whole complex by a factor of only about 1.5. They concluded that the internal mobility of the bound water molecule should not represent a major limitation in the attainment of high relaxivity in macromolecular complexes. However, the peculiar strength of the Ln–O_{water} bond in DOTAM complexes may not represent the general behavior for intramolecular mobility of the coordinated water in the entire class of monohydrated Ln(III) systems.

D. The Exchange Correlation Time τ_M

The mean residence lifetime of coordinated water protons represents one of the most important parameters that control the relaxivity of Gd(III) complexes (*18*). For several years it had been assumed that τ_M for the low molecular weight polyaminocarboxylate Gd(III) complexes was of the order of a few ns, as found for the octaaqua ion. In 1993, Merbach reported the first direct measurement of the water exchange rate for $[GdDTPA(H_2O)]^{2-}$ and $[GdDOTA(H_2O)]^{-}$ and found that for these CAs the rate of water exchange was nearly three orders of magnitude lower than that for $[Gd(H_2O)_8]^{3+}$ (*49*). This result has obviously stimulated the measurement of this parameter for many other Gd(III) complexes with the results that now, much more is known of the dependence of this parameter on the structural properties of the complexes: charge, coordination number, steric interaction at the bound water coordination site, isomeric composition of the solutions, etc. (*12*). For low molecular weight anionic Gd(III) chelates, the relatively long exchange correlation times do not have an influence on the NMRD profiles at $T > 25°C$. In fact for τ_M values of about 100–500 ns, the fast exchange condition ($\tau_M < T_{1M}$) holds and the contribution to the overall correlation time τ_C is negligible. A limiting effect has been observed from the analysis of the NMRD profile of the neutral complex $[GdDTPA\text{-}BMA(H_2O)]$ (Chart 2) which is characterized by a τ_M value of the order of 2 μs, long enough to become comparable to T_{1M} (intermediate exchange condition) (*19,20*). The effect is even larger for the cationic complex $[GdDOTAM(H_2O)]^{3+}$, which has a bound water molecule residence lifetime of 19 μs (*21*). In Fig. 10 the NMRD profiles of the related complexes $[GdDOTA(H_2O)]^{-}$ and $[GdDOTAM(H_2O)]^{3+}$ are shown, these are characterized by similar relaxation parameters and thus the differences in their relaxivity are only due to their different rates of coordinated water molecule exchange. The situation may differ considerably when dealing with macromolecular complexes since the increased molecular size is accompanied by a marked decrease in the rotational mobility (long τ_R), which results in much shorter values of T_{1M} and thus the systems are likely to be in the slow/intermediate exchange condition. In fact, a limiting effect on the water residence lifetime is often observed, which has not so far permitted the attainment of the high relaxivity values predicted by theory.

The rate of water exchange is directly obtained by measuring the ^{17}O transverse relaxation rate and chemical shift of the water signal in a fairly concentrated (10–50 mM) aqueous solution of the Gd(III) complex. This subject will be discussed in detail in Chapter 7 of this volume. We just want to mention here that what is relevant for the analysis of NMRD curves is the mean residence lifetime of the water protons. In ^{17}O measurements we obtain the residence lifetime of the water molecule itself, but normally the prototropic exchange rate, particularly around neutral pH, is limited by the

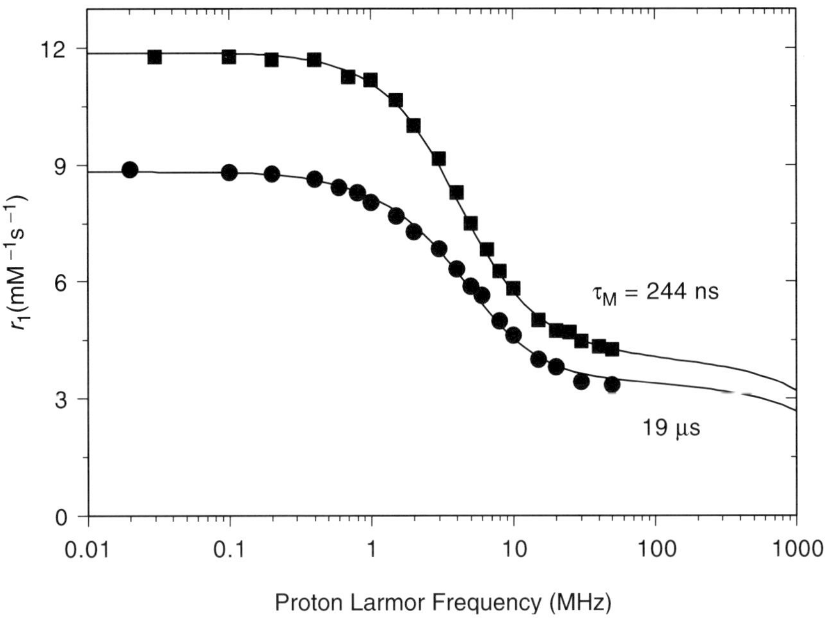

FIG. 10. ^{1}H NMRD profile (25°C) of $[GdDOTA(H_2O)]^-$ (squares) and $[GdDOTAM(H_2O)]^{3+}$ (circles) showing the effect on relaxivity due to a slow rate of water exchange.

exchange rate of the entire water molecule. Nevertheless, examples have been reported where τ_M is so long that the proton exchange dominates. This has been observed for macrocyclic cationic complexes in acidic or basic solutions where acid- and base-catalyzed prototropic exchange processes are operative (*36*).

E. The Electron Spin Relaxation Times $T_{1,2e}$

The longitudinal and transverse electronic relaxation times T_{1e} and T_{2e} are of great importance in the analysis of proton (and ^{17}O) relaxivity. However they are difficult to determine directly by independent techniques. About ten years ago, Powell *et al.* proposed the use of empirical equations that were later on applied in a simultaneous fitting of EPR, ^1H, and ^{17}O NMR data (*22*). Recently, Rast *et al.* developed a new theoretical model which interprets the electron spin relaxation in terms of a static (modulated by rotation of the complex) and the usual dynamic crystal zero-field splitting (*50*). This refined model was shown to successfully reproduce the EPR data of a multifrequency and variable temperature study on the Gd(III) aqua ion and $[GdDOTA(H_2O)]^-$ (*51*). However, the high field EPR spectrometers are not readily accessible. So, commonly the electron spin relaxation parameters for Gd(III)-based contrast agents (Δ^2 and τ_V) are obtained by fitting the NMRD profiles, preferably at different temperatures, or by a simultaneous fitting of NMRD and ^{17}O data. The problem is the validity of the

Solomon–Morgan theory (see Chapters 2, 3, and 8) and thus the calculated values of the parameters might have a limited physical meaning.

IV. Methods for Improving Relaxivity

The attainment of high relaxivity is the primary target for overcoming the sensitivity problems of the MRI technique. As pointed out in the previous sections, the relaxivity of a contrast agent is the result of a complex interplay between several structural and dynamic parameters whose contribution to the observed relaxivity can significantly differ according to the magnetic field strength, temperature, pH, and nature of the paramagnetic center.

This section intends to provide the reader with some examples of how the "high relaxivity challenge" has been tackled for Gd(III) complexes so far. For the sake of clarity, this section has been divided in three sub-sections in relation to the specific relaxation parameter considered for relaxivity enhancement, namely the hydration state of the metal centre, the tumbling rate of the CA, and the exchange rate of the mobile protons coordinated to the paramagnetic center.

A. Increase of the Hydration State

Both the inner- and the second-sphere contributions to the overall relaxivity are directly dependent on the molar fraction of water protons interacting with the paramagnetic center (see Eq. (3)). Therefore, a relaxivity enhancement might be simply obtained by increasing the number of water protons in the coordination shells (inner- and second-) of the Gd(III) ion.

As far as the inner-coordination sphere is concerned, all the Gd(III)-based contrast agents currently used in clinical protocols possess one water molecule coordinated to the metal center ($q = 1$ complexes, see Section II) owing to the presence of eight donor atoms in the structure of their ligands. Hence, the use of hepta- or hexa-coordinating ligands would allow the design of contrast agents with two or three metal-bound water molecules, thus considerably improving the inner-sphere contribution to their relaxivity. Unfortunately, this approach is somewhat limited by two main drawbacks which are related to the increased solvent accessibility at the metal site: (i) the lowering of the thermodynamic and kinetic stability of the metal complex, and (ii) the possibility that anionic species may replace water in the inner coordination shell of the Gd(III) ion (formation of ternary complexes).

The search for Gd(III) complexes with an increased number of inner-sphere water molecules ($q > 1$), though sufficiently stable to be safely used in vivo has led to consider the model compounds shown in Chart 6. The relaxivity values of these complexes are significantly higher with respect to the Gd(III) chelates of similar size and $q = 1$. By taking as reference the r_1 values of $[GdDOTA(H_2O)]^-$ and $[GdDTPA(H_2O)]^{2-}$ (4.7 $s^{-1}mM^{-1}$), the

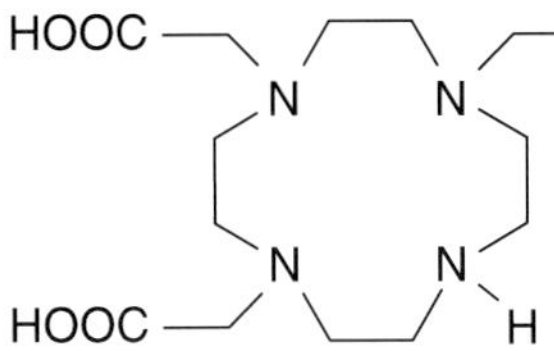

DO3A

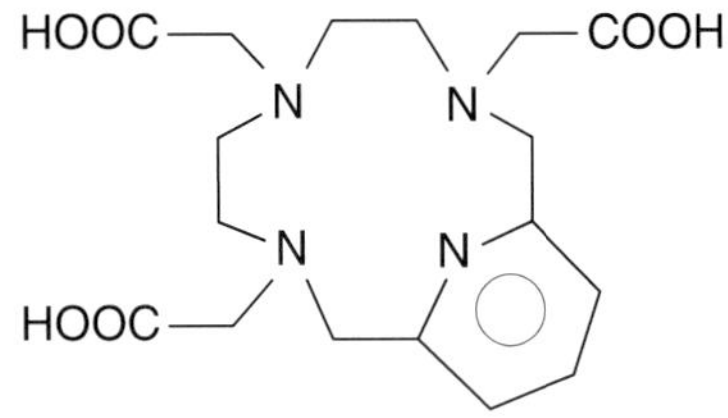

PCTA

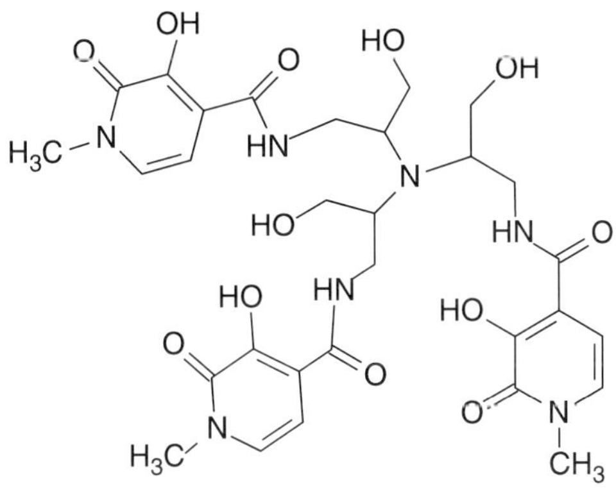

L-Ser_3-TREN-Me_3-3,2-HOPO

CHART 6.

relaxivity enhancement of $[GdDO3A(H_2O)_2]$ (*35*), $[GdPCTA(H_2O)_2]$ (*35*) and [Gd(L-Ser_3-TREN-Me_3-3,2-HOPO$(H_2O)_2$] (*16*) is 28, 47, and 91%, respectively. The higher relaxivity observed for $[GdPCTA(H_2O)_2]$ with respect to $[GdDO3A(H_2O)_2]$ has been attributed to a shorter Gd-H_2O distance (3.1 Å vs. 3.15 Å) for the former complex, whereas the considerable enhancement observed for the hydroxy-pyridinonate derivative is the result of the larger molecular size of this species which determines an increase of *ca.* 90% of its reorientational correlation time τ_R (130 ps vs. 66–70 ps at 25°C for $[GdDO3A(H_2O)_2]$ and $[GdPCTA(H_2O)_2]$, respectively).

As far as the tendency of forming ternary adducts is concerned, the three typologies of ligands behave differently. In fact, PCTA- and HOPO-derivatives do not form ternary species, whereas GdDO3A-like complexes show a marked tendency to interact with a wide array of anionic substrates (see Section V.E) with the consequent decrease in their relaxivity owing to the reduction of q. Nevertheless, this detrimental effect on the relaxivity can be overcome by modifying the surface of the ligand, leaving unchanged the donor set of DO3A. In fact, the affinity towards anionic substrates can be substantially reduced: (i) by increasing the steric hindrance at the metal site, (ii) by introducing negatively charged substituents in the close proximity of the

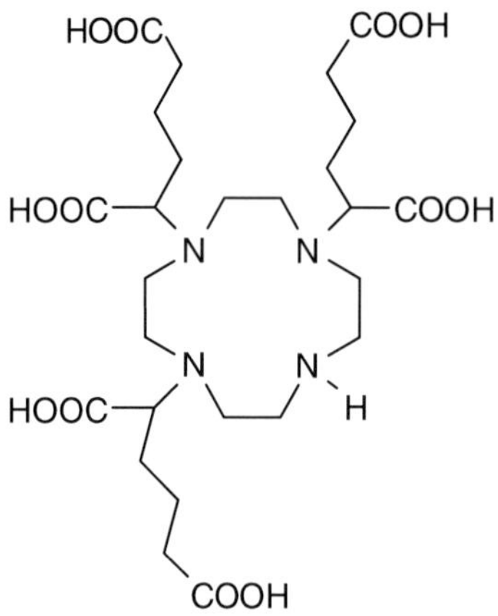

aDO3A

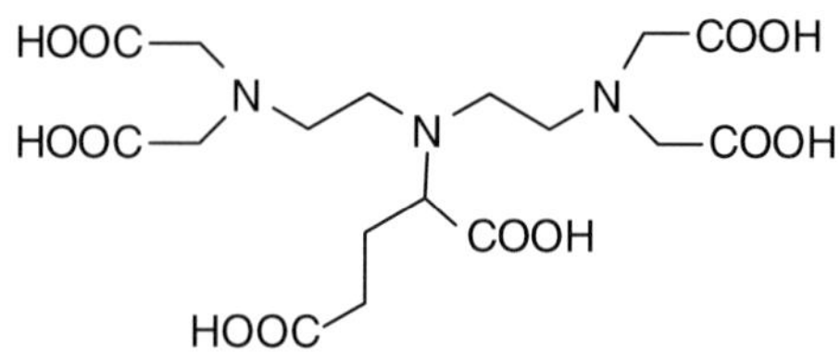

DTPA-Glu

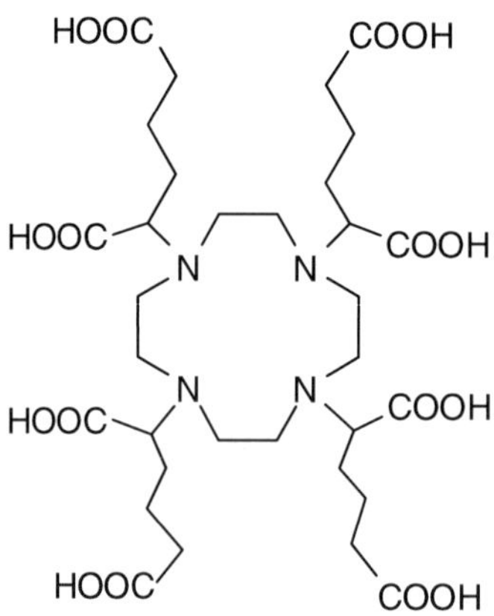

aDOTA

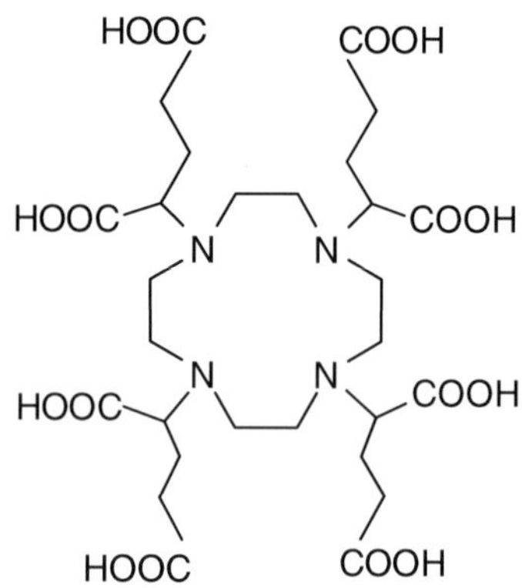

gDOTA

CHART 7.

open coordination shell of the metal complex (*52*), or (iii) by introducing suitable electron-donor substituents in order to reduce the positive charge of the metal center (*53*). As an example, it is worth mentioning the case of a GdDO3A-derivative ($[GdaDO3A(H_2O)_2]^{3-}$, Chart 7) characterized by the presence of three pendant butanoate residues on the coordinating arms of the hepta-dentate ligand (*52*). Interestingly, the relaxivity of this chelate ($12.5\ s^{-1}\ mM^{-1}$, 20 MHz, and 25°C) was 105 % higher than that of the parent $[GdDO3A(H_2O)_2]$. Though this complex has a longer τ_R value owing to its increased molecular size, it is likely that the significant relaxivity enhancement is the result of a contribution arising from the exchange of water protons hydrogen-bonded to the carboxylate groups of the pendant arms (second-sphere effect). It has been proposed that the structuring effect of the free carboxylates on the nature of the second hydration sphere of Gd(III) ion is also the main factor responsible for the extremely short residence lifetime (30 ns at 25°C) observed for the metal-bound water molecules in this complex.

An enhanced relaxivity in the presence of pendant free carboxylate groups has been observed also in other $q = 1$ complexes. For instance, the r_1 value (at 20 MHz and 25°C) of a chelate bearing a free propionate group on the central coordinating acetate arm of the DTPA skeleton ($[GdDTPA\text{-}Glu(H_2O)]^{3-}$, Chart 7) is 30% higher ($r_1 = 6.2\ s^{-1}mM^{-1}$) than the parent $[GdDTPA(H_2O)]^{2-}$ compared to an increase in the molecular weight of only 18%. Similar results were obtained for a couple of tetra-derivatives of $[GdDOTA(H_2O)]^{-}$, $[GdgDOTA(H_2O)]^{5-}$, and $[GdaDOTA(H_2O)]^{5-}$ (Chart 7) containing four free propionate and butanoate groups, respectively. The relaxivity of such chelates is higher than $[GdDOTA(H_2O)]^{-}$ by 55% (7.3 s^{-1} mM^{-1}) and 62% (7.6 s^{-1} mM^{-1}), respectively (*52*). These values are significantly higher than the values expected on the basis of their molecular weight (see Fig. 2), particularly for $[GdgDOTA(H_2O)]^{5-}$ where the free carboxylates are closer to the Gd(III) ion.

The additional contribution to relaxivity arising from exchangeable protons hydrogen-bonded to polar groups present on the surface of a Gd(III) complex may provide useful routes for improving the efficacy of a contrast agent. To this regard, the typical example is represented by the $[GdDOTP]^{5-}$ complex (Chart 5) (*54*). The relaxivity of this complex ($r_1 = 4.7\ s^{-1}\ mM^{-1}$), which does not possess any water molecule bound to the metal center (i.e., $q = 0$), is equal to the value measured for the tetracarboxylic analog $[GdDOTA(H_2O)]^{-}$ where $q = 1$. This suggests that the presence of four phosphonate groups promotes the formation of strong hydrogen bonds with water molecules, whose overall contribution to relaxivity roughly corresponds to that of a single water in the inner coordination sphere of the metal.

The relaxivity enhancement observed upon replacement of a coordinating carboxylate with a phosphonate group may be limited by a decreased hydration of the metal center often encountered with phosphonate containing systems (*54*). A nice compromise between these opposite effects has been found in the $[GdPCP2A(H_2O)_2]^{-}$ complex (Chart 8) (*55*). The replacement of only one carboxylate of $[GdPCTA(H_2O)_2]$ with a phosphonate leads to a relaxivity enhancement of *ca.* 20% (8.3 s^{-1} mM^{-1} vs. 6.9 s^{-1} mM^{-1}) because the number of inner-sphere water molecule remains unchanged ($q = 2$). Furthermore, the introduction of the phosphonate group results in the increase of two orders of magnitude of its thermodynamic stability ($\log K_{GdL} = 23.4$), which is a relevant advantage in view of the possible in vivo applications of this chelate.

The relaxivity enhancement associated with exchangeable protons in the second-coordination sphere of a Gd(III) complex is quite important in the supramolecular adducts of Gd(III)-based systems with proteins. For example, it has been reported that, besides water protons in the second-coordination

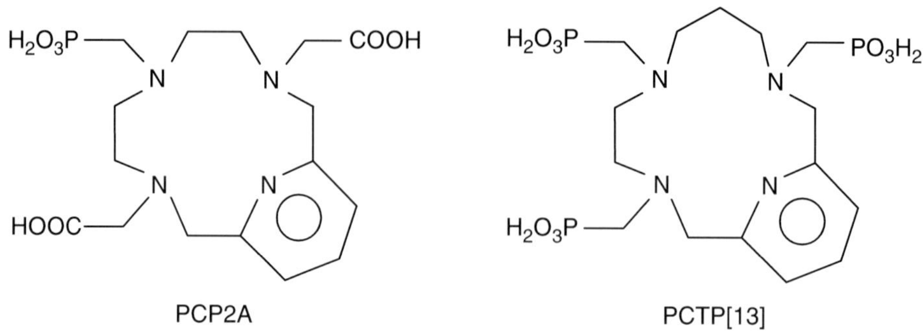

CHART 8.

shell of the metal complex, the strong relaxation enhancement observed for phosphonate-based Gd(III) complexes, like $[GdPCTP[13](H_2O)]^{3-}$ ($q = 1$ complex, Chart 8) , upon non-covalent binding with Human Serum Albumin (r_1 of the adduct of 45 s^{-1} mM^{-1}) is due in part to water protons of the hydration shell of the protein and/or to exchangeable protons of HSA that are in the close proximity of the metal complex (Fig. 11) (*54*). Similar conclusions were also drawn by analyzing the relaxometric behavior of a series of inclusion compounds formed by Gd(III) chelates and β-cyclodextrin oligomers (Poly-β-CD) (*56*). Figure 12 reports the NMRD profile for the macromolecular adduct formed by $[GdDOTA(BOM)_3(H_2O)]^-$ ($q = 1$ complex, Chart 3) and a multimeric (polymerization degree of about 12) β-cyclodextrin derivative. The relaxivity of this adduct ($r_1 = 49.0$ s^{-1} mM^{-1} at 20 MHz) receives a significant contribution (in particular at low magnetic field strengths) that has been ascribed to water protons tightly hydrogen-bonded to the hydroxyl groups on the rims of the β-CD cavities.

Finally, the role of hydration appears highly relevant in determining the outstanding relaxivities shown by Gd-fullerenols and $[GdHPDO3A(H_2O)]$ entrapped into the Apoferritin cavity. In the former case, the high relaxivity showed by water soluble Gd-fullerenols ($q = 0$, r_1 of 62.0 s^{-1} mM^{-1} at 20 MHz and 37°C for $Gd@C_{82}(OH)_{40}$) (*57*) with respect to the non-hydroxylated form (*58*) has been accounted for in terms of an electronic transfer from Gd to the fullerene cage followed by the dipolar coupling to a large number of water molecules hydrogen-bonded to the external -OH groups of the cage (*59*).

$[GdHPDO3A(H_2O)]$ loaded Apoferritin represents another original route to enhance the observed relaxivity. The commercially available Gd(III) complex ($[GdHPDO3A(H_2O)]$, $q = 1$, Chart 2) is entrapped into the protein cavity (up to 8–10 complex units), where water molecules have free access (Fig. 13) (*60*). Interestingly, the relaxivity of this system is about 20 times higher than the values observed for $[GdHPDO3A(H_2O)]$ ($r_1 = 80.0$ s^{-1} mM^{-1} vs. 4.2 s^{-1} mM^{-1} at 20 MHz and 25°C). It has been proposed that such an enhancement, which cannot be accounted for only in terms of the classical inner-/outer-sphere models, arises from the formation of a sort of "relaxing sink" inside the cavity made of: (i) water molecules freely moving inside the cavity or hydrogen-bonded to its inner surface, and (ii) exchangeable protons of the

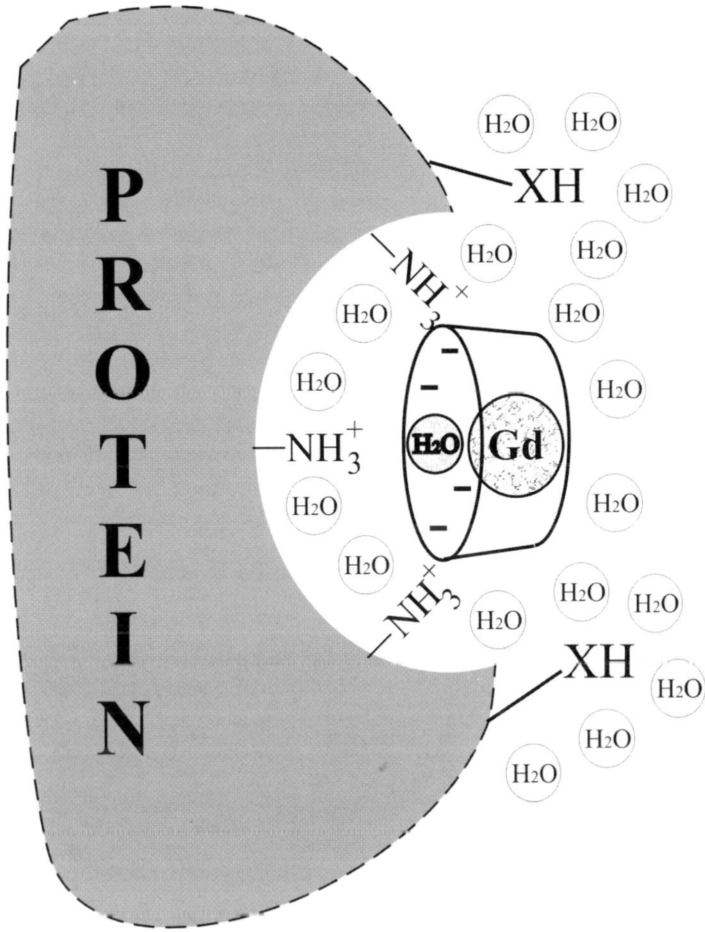

FIG. 11. Schematic representation showing the network of mobile protons close to a negatively charged Gd(III) complex bound to a protein.

protein. This relaxation model is very similar to the one proposed above for describing the relaxivity of a metal complex interacting with a protein surface, but here the effect is likely to be amplified by the spherical surface of the cavity.

B. LENGTHENING OF τ_R

As mentioned in Section III.C, it was realized early on that for Gd(III) agents, a significant gain in relaxivity could be obtained by slowing down the tumbling motion (increasing τ_R) of the metal complex (*10*).

This task can be achieved by increasing the molecular size of the paramagnetic complex. In fact, by considering the paramagnetic probe as an isotropically tumbling sphere, it is expected that τ_R and molecular size (commonly assumed to increase linearly with the molecular weight) are

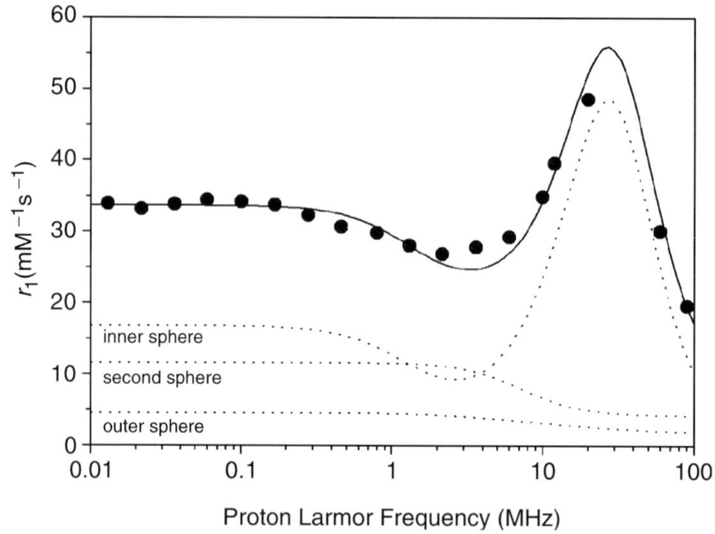

FIG. 12. NMRD profile of GdDOTA(BOM)$_3$/Poly-β-CD adduct (pH 7, 25°C).

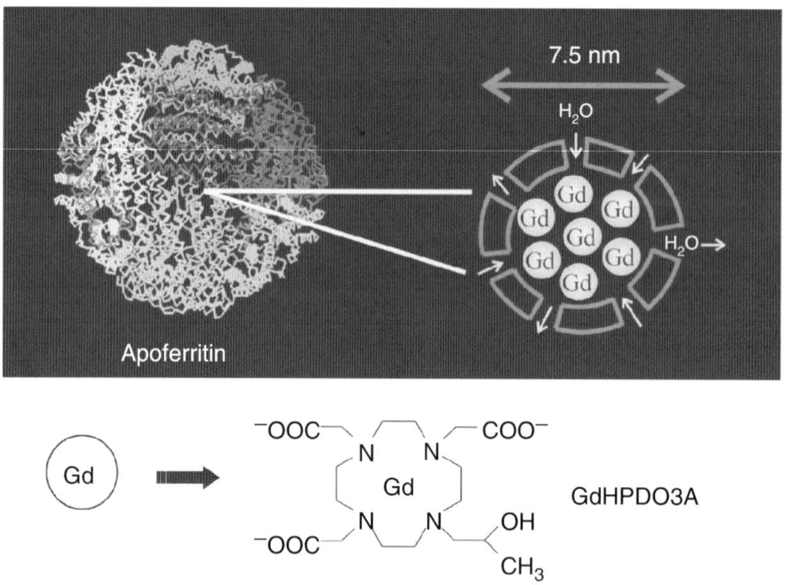

FIG. 13. Schematic representation of [GdHPDO3A(H_2O)] entrapped into the Apoferritin cavity.

directly related when the $\omega_S\tau_{C2} \gg 1$ condition is met (see Eq. (4)). Actually, Fig. 2 nicely demonstrates that for a series of small-sized Gd(III) complexes (with identical q value), relaxivity and molecular weight are linearly correlated, but for larger molecular size (FW > few kDa) the occurrence of anisotropic reorientational motions as well as the dependence of τ_{Ci} on τ_R (and

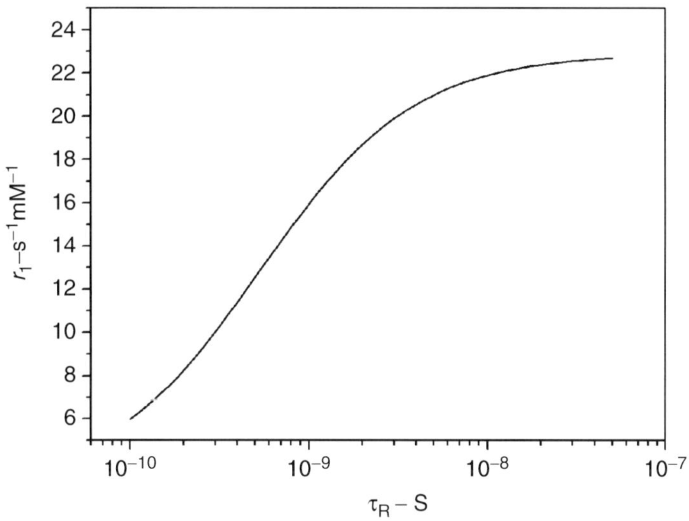

FIG. 14. Theoretical dependence of relaxivity on τ_R calculated on the basis of the classical inner-/outer-sphere model (20 MHz, $q = 1$, $\Delta^2 = 5 \times 10^{19}\,s^{-2}$, $\tau_V = 20$ ps, $\tau_M = 200$ ns, $r_H = 3$ Å, $a = 3.8$ Å, $D = 2.2 \times 10^{-5}\,cm^2\,s^{-1}$).

eventually also the limiting effect of τ_M, see Eqs. (3)–(5) and Fig. 14) makes relaxivity no longer linearly correlated with the rotational motion of the paramagnetic system. Nevertheless, Fig. 14 suggests that the "high-relaxivity challenge" requires slow tumbling Gd(III) complexes endowed with a rotational correlation time longer than nanoseconds. For this reason most of the efforts aimed at developing high relaxivity agents have been focused on the design of large-sized paramagnetic systems in which the Gd(III) complex has been bound covalently or non-covalently to a macromolecular substrate. These two approaches mainly differ in the pharmacokinetic behavior of the resulting construct rather than in its relaxometric properties even if, in principle, the non-covalent route, facilitating the occurrence of multiple bonding between the interacting partner, might lead to longer τ_R values.

Many bio-compatible and low-toxic slowly tumbling substrates have been so far considered as substrate for lengthening τ_R, including proteins (*61*), polyaminoacids (*62–67*), peptide nucleic acids (PNAs) (*68*), cyclic and linear oligo- and poly-saccarides (*56,69–74*), micelles (*75–80*), liposomes (*81–85*), dendrimers (*86–93*), and linear synthetic polymers (*94–96*). In some cases, slow tumbling systems can be also obtained by exploiting auto-aggregation or self-assembling of suitably functionalized Gd(III) chelates (*97–99*).

In principle, the choice of the macromolecular substrate is driven by the clinical application of the CA rather than by the ability of the substrate to enhance the relaxivity of the Gd(III) unit. A typical example showing how the medical need has led to the design of macromolecular agents is represented by the development of blood pool agents (BPAs). The clinical purpose of such diagnostic media, which are designed for Magnetic Resonance Angiography (MRA) protocols, is the visualization of blood vessels for

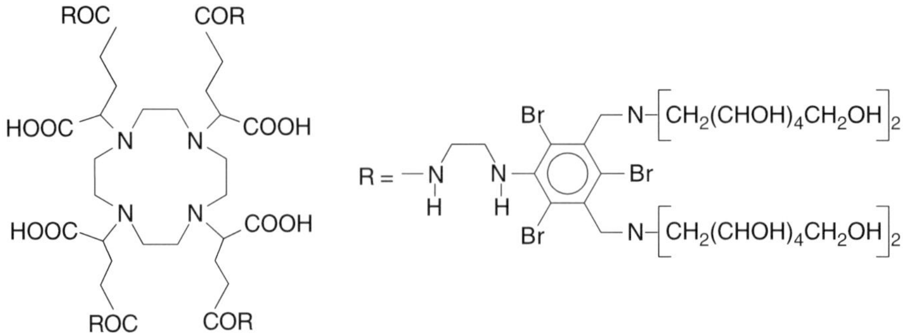

P760

CHART 9.

detecting abnormalities in the vascular system. For this reason, a BPA needs to be characterized by longer plasmatic half-life than the typical extracellular agents, and its distribution in the interstitial space of healthy tissues has to be minimized. Since the diffusion across the vascular endothelium depends on the size of a given compound, the simplest approach for developing BPAs has been the synthesis of large-sized paramagnetic compounds. According to their pharmacokinetic properties, BPAs can be classified into three main classes: (i) Low Diffusion Agents (LDA), in which the diffusion is not completely prohibited, but only slowed down, (ii) Rapid Clearance BP agents (RCBPAs), in which the diffusion is absent (or severely limited), but the elimination still occurs through kidneys, and (iii) Slow Clearance BP Agents (SCBPAs), in which again the diffusion is absent, but the elimination occurs preferentially through the hepatobiliary route (*100*). Roughly, each class is characterized by a given size of the paramagnetic construct. In fact, LDAs have medium molecular size: the reference compound in this class is the hydrophilic compound P760 (Chart 9) (*101–101*) whose relaxivity (*ca.* 25 s^{-1} mM^{-1} at 20 MHz and 37°C) is significantly improved, with respect to the parent compound $[GdDOTA(H_2O)]^-$, following the increase in the rotational time (τ_R of *ca.* 2 ns at 310 K). RCBPAs are larger molecules: the most investigated system is represented by Gadomer 17, a dendrimer-like structure (FW of 17.5 kDa) based on 1,3,5-benzene tricarboxylic core containing 24 units of a monoamide derivative of $[GdDOTA(H_2O)]^-$ (*91*). In spite of the relatively large molecular size of this system, its r_1 value is only 17.3 s^{-1} mM^{-1} (20 MHz, 39°C), likely a consequence of a non-optimal τ_M value for the water molecule coordinated to the Gd(III) ion (*102*). Finally, most of the macromolecular Gd(III)-based agents so far developed belong to the class of SCBPAs. They include Gd(III) chelates covalently or non-covalently bound to: (i) proteins (mostly human serum albumin) or polyaminoacids, (ii) polysaccharides (e.g., dextran), and (iii) linear or branched synthetic polymers (PEG copolymers and dendrimers). Furthermore, the Gd(III) unit can be properly functionally designed in order to be included in the formation of large molecular aggregates like liposomes or micelles. The relaxivity values of SCBPAs can markedly vary very much ranging from about 10 s^{-1} $(mMGd)^{-1}$ (for polysaccharides) (*12*) to around 50 s^{-1} mM^{-1} for some reversible adducts with HSA (*61*). The high variability in the r_1 values of the macromolecular systems depends on a number of reasons, including long τ_M values, faster reorientation of the Gd(III) moieties with respect to the slow tumbling motion of the associated macromolecule or even the superimposition of an internal motion of the coordinated water protons (*48*).

Anyway, it is worth while to remind that, unlike the hydration parameter that has a direct effect on r_1, the elongation of τ_R enhances the relaxivity only up to a given τ_R value (see Fig. 14). Beyond that upper limit, r_1 becomes much less affected by the rotational motion of the complex, being mainly determined (at least for the inner-sphere contribution) by q, τ_M, $T_{1,2e}$, and in turn by the magnetic field strength (see Figs. 9 and 10). Hence, besides the indications associated with its clinical use, the design of slow tumbling Gd(III)-based systems endowed with high relaxivity relies on: (i) the identification of Gd(III) complexes whose τ_M and $T_{1,2e}$ values have been optimized in order to fully exploit the lengthening of τ_R, and (ii) the choice of the proper macromolecular system and binding scheme in order to attain the desired τ_R value without affecting the favorable properties of the Gd(III) unit, and possibly to provide additional contributions (e.g., second-sphere) to the overall relaxivity.

The latter point is rather critical. In fact, whereas long τ_R values can be easily attained through a proper choice of the macromolecular system and the binding modality, it may not be an easy task to keep the relaxometric properties of the paramagnetic moiety unaltered, upon formation of the macromolecular adduct. In fact, this may result in a reduced hydration of the Gd(III) ion as donor groups such as aspartate or glutamate on the surface of a protein can replace coordinated water molecules (*103–104*). When the hydration state is maintained, the occurrence of a marked reduction of the exchange rate of inner-sphere water is very common (see next Section).

Concerning the presence of additional contributions to the relaxivity, it has already been discussed in the previous sub-section that some macromolecules (e.g., proteins) may increase the number of mobile protons in the close proximity of the paramagnetic center. What is important to recall here is that the contribution to the relaxivity arising from such protons is also likely to be amplified by the overall slow tumbling of the system (see as example the GdHPDO3A/Apoferritin construct described in the previous sub-section). It is worth outlining that, as for the attainment of long τ_R values, the occurrence of additional contributions from second-sphere protons is enhanced in the case of non-covalent adducts as this type of binding interaction often results in an extended array of contacts between the metal complex and the surface of the macromolecular substrate.

C. Optimization of τ_M

It appears clearly from the discussion in the previous sections that the design of high relaxivity Gd(III)-based agents needs a fine tuning of the residence lifetime, τ_M, of the water protons coordinated to the paramagnetic center. The simulated curves reported in Fig. 3 indicate that τ_M starts to affect the inner-sphere contribution to r_1, at 20 MHz, only when the rotational mobility of the Gd(III) complex is limited ($\tau_R > 500$ ps). At this field, it has been reported that the optimal τ_M values for fully exploiting the relaxivity enhancement arising from the elongation of τ_R are in the range of few tens of nanoseconds (*11*). For this reason much efforts have been focused, in the last decade, to the understanding of the relevant factors determining τ_M in Gd(III) complexes.

When the lanthanide ion is nona-coordinate (CN = 9), it has been demonstrated that the exchange process occurs mainly through a dissociative pathway in which the transition state is eight-coordinate (*12,105*). Therefore, water residence lifetime is related to the free-energy difference, $\Delta G^{\#}$, between the transition and the ground states. $\Delta G^{\#}$ is usually dominated by the enthalpic term ($\Delta H^{\#}$) that is strongly affected by the strength of the Ln–O bond. The essentially electrostatic nature of this bond justifies the fact that τ_M of isostructural Gd(III) chelates markedly depends on the overall electric charge of the complex, being relatively short for anionic complexes and exceedingly long for cationic ones (*105*). Moreover, $\Delta H^{\#}$ may also be affected by the coordination geometry of the metal complex, by the steric crowding at the water coordination site or by the structural property of the second hydration sphere of the metal ion. The role played by the coordination geometry of the complex is particularly evident in the case of macrocyclic DOTA-like complexes. Such chelates may adopt two distinct coordination geometries: square antiprismatic (SAP) or twisted square antiprismatic (TSAP), and it has been observed that the residence lifetime of the TSAP arrangement is considerably shorter than the SAP geometry (*106–107*). Since the two coordination polyhedrons may interconvert, the τ_M value of a given macrocyclic Gd(III) complex may be correlated to the SAP/TSAP ratio. Hence, the optimization of τ_M for this class of compounds requires the design of complexes that adopt preferentially the TSAP geometry.

Another possible route for optimizing τ_M is to exploit the fact that the water binding site in monocapped square antiprismatic structure, as well as in the more common tricapped trigonal prismatic geometry (i.e., the coordination geometry of $[GdDTPA(H_2O)]^{2-}$ and derivatives), is a sterically demanding position. For example, it is worth to mention the shortening of τ_M across the lanthanide series for isostructural Ln(III) complexes (*108–109*). In fact, this behavior has been rationalized on the basis of the stabilization of the eight-coordinate transition state and/or destabilization of the nine-coordinate ground state following the reduction of the ionic radius, which, in turn, increases the steric constraint at the water binding site. The water accessibility at the metal site, and consequently τ_M, can also be modulated by changing the number (*18*) or the hydrophobic nature (*110*) of the substituents on the pendant arm of DOTA- or DTPA-like ligands. Similar considerations have been made for explaining the short τ_M value observed for the Gd(III) complex of EGTA (Chart 10), where the increased steric hindrance at the

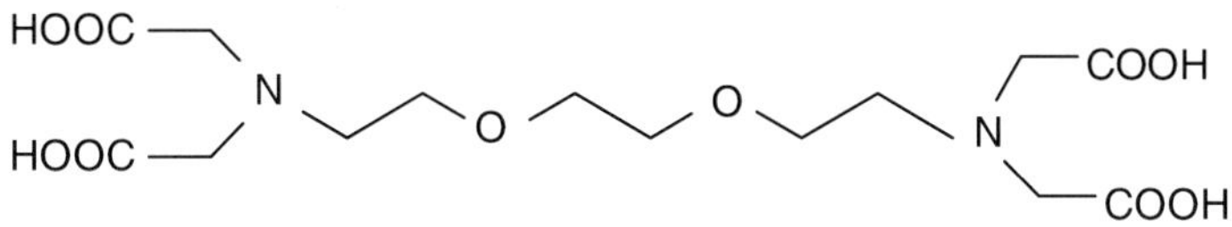

EGTA

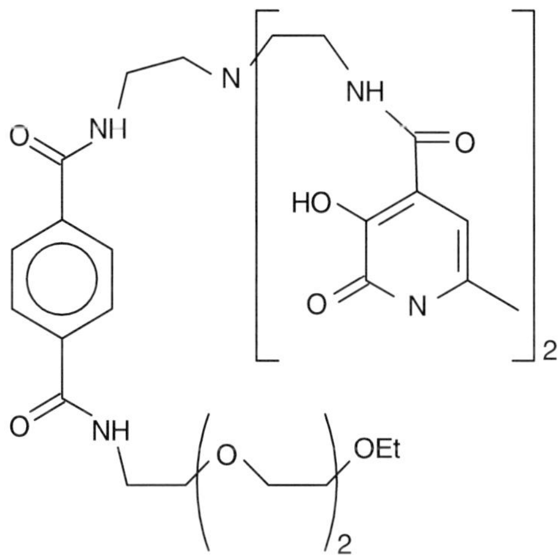

1

CHART 10.

metal site is determined by the coordination of the oxoethylenic bridge of the ligand (*111*).

The role of the structure of the second hydration sphere of Gd(III) complexes in determining τ_M has been emphasized only recently. The case of the $[GdDO3ANP_2(H_2O)]^{4-}$ complex (Chart 5) has already been discussed in Section II, where the presence of a solvent molecule hydrogen-bonded to both the metal coordinated water protons and one of the two pendant phosphonate groups slows down the exchange process (τ_M of 1.6 μs at 25°C *vs.* the typical τ_M values < 1 μs for DOTA-monoamide derivatives) (*38*). Significant changes in τ_M have recently been reported for the tripositive $[GdDOTAM(H_2O)]^{3+}$ complex (Chart 4) in the presence of different counter-anions acting as counterions (*112*). For instance, within the halides series, the residence lifetime values at 25°C follow the order Cl^- (13 μs) > Br^- (5 μs) > I^- (2.3 μs). This finding has been interpreted in terms of the structuring/destructuring effect of the anion towards the second hydration sphere of the metal complex. Anions able to promote ordering of the second hydration sphere (e.g., Cl^-) can give rise to longer τ_M values compared to less charge dense anions (e.g., I^-) that lack the ability to break up the hydrogen bond network between water molecules around the metal center.

For octa-coordinate Gd(III) complexes, the exchange process can occur through an associative (or associative interchange) mechanism. Actually, this mechanism has been often only presumed, but recently it has been experimentally proven (by a variable pressure ^{17}O-NMR study) on a hetero-tripodal complex based on the hydroxo-pyridinonate (HOPO) coordination cage (Chart 10, ligand **1**) (*113*). The nona-coordinate transition state and the octa-coordinate ground state are quite close in energy, thus leading to a very short and close to the optimal τ_M value (around 20 ns at 25°C). It is likely that similar considerations can hold for other fast-exchanging octa-coordinate Gd(III) chelates, like $[GdPCP2A(H_2O)]^-$ (Chart 8). It is worth noting that for this class of compounds an increased steric hindrance at the water site can stabilize the CN = 8 ground state, thus leading to a lengthening of τ_M.

Once a chelating cage endowed with an optimal τ_M value has been selected, the design of a high-relaxivity system requires the reduction of its rotational mobility (see previous sub-section). As this task is usually achieved through the formation of a macromolecular system, it is of fundamental importance to check whether the macromolecular substrate or the binding modality influences the exchange process at the metal site. Addressing this issue, it has been found that in several of the investigated cases, e.g., for some dendrimer-based Gd(III) complexes (*12*), for copolymers containing $[GdDTPA(H_2O)]^{2-}$ bisamide moieties (*94–95*), for a Gd(III) complex forming micelles ($[GdDOTA\text{-}C_{12}(H_2O)]^-$) (*76*), and for the non-covalent adduct between MP-2269 (Chart 11) and HSA (*114*), the exchange process appears to be almost unaffected by the presence of the macromolecular system. In other cases, e.g., for the non-covalent adduct between HSA and MS325 (Chart 3), $[GdDTPA(BOM)_3(H_2O)]^-$ (Chart 11) (*115*) or $[GdPCTP[13](H_2O)]^{3-}$ (Chart 8) (*54*) as well as in the host–guest interaction between poly-β-CD and several macrocyclic Gd(III) complexes bearing benzyl-oxy-methylenic (BOM) residues (*56*), the Gd(III) units showed longer τ_M values than the corresponding free chelates. As a general rule, it is likely that the elongation of τ_M upon binding to a macromolecular substrate may be correlated with the occurrence of a significant contribution to the overall relaxivity arising from second sphere exchangeable protons.

V. Responsive Contrast Agents

The term "responsive" (elsewhere indicated as "smart") refers to diagnostic agents whose contrasting properties are sensitive to a given physico-chemical variable that characterizes the microenvironment in which the probe is distributed (*116–117*). Typical parameters of primary diagnostic relevance include pH, temperature, enzymatic activity, redox potential and the concentration of specific ions, and low-weight metabolites.

Unfortunately, their clinical use is still rather far-off, mainly because an accurate measurement of one of the above-mentioned parameters with a given relaxing probe requires a precise knowledge of the local concentration

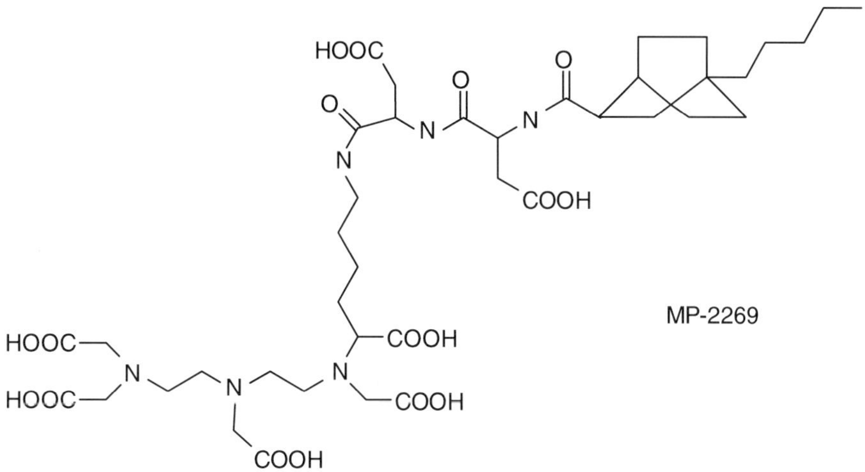

MP-2269

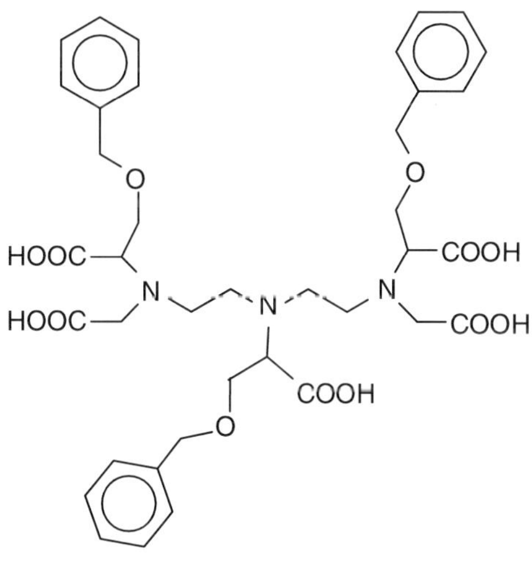

$DTPA(BOM)_3$

CHART 11.

of the contrast medium in the region of interest. Only by this way can the observed change in the relaxation rate of water protons be safely attributed to a change in the parameter to be measured. Recently, a novel class of MRI paramagnetic contrast media, the so-called PARACEST agents, has entered the field with the aim of acting as concentration-independent responsive agents (*118*). These probes do not generate contrast by altering the relaxation rates of water protons, but through a different mechanism based on a transfer of saturated magnetization between exchangeable protons of the probe and water protons. Such a process yields a decrease in the NMR signal of bulk water protons, thus creating a negative contrast in the image.

The saturation transfer is induced by the selective irradiation of the mobile protons of the probe with a proper radiofrequency field. When different pools of exchangeable protons are available, this modality enables the set-up of ratiometric methods in which a responsive PARACEST agent provides a MRI response no longer dependent on its absolute concentration (*119–120*).

As said above, the need to know the concentration of the CA has limited the study of responsive agents mainly to in vitro conditions. Nevertheless, the subject has received much attention, and routes to overcome this drawback have been envisaged. For instance, one may obtain information on the concentration of the responsive agent by acquiring the images with a related, non-responsive agent. Alternatively, one may think of obtaining the required information by measuring the relaxation enhancement of a metabolite non-interacting with any of the structural forms of the responsive agent.

The selected examples will be discussed according to the physico-chemical parameter that has to be measured.

A. pH Sensitive

Many pathological events are associated with alterations of pH homeostasis, including tumors and cardiovascular diseases (*121–124*). The task to develop non-invasive methods for measuring pH in vivo has stimulated many researchers, especially in MR field, in designing probes sensitive to the hydrogen ion concentration (*125–129*).

As far as Gd(III) agents are concerned, several systems have so far been investigated (mostly in vitro). The design of a Gd(III)-based complex whose relaxivity is pH-dependent requires that at least one of the structural or dynamic parameters determining its relaxivity is rendered pH-dependent. In most of the examples so far reported, the pH-dependence of the relaxivity reflects changes in the hydration of the metal complex.

For instance, Lowe *et al.* showed that the relaxivity of a series of macrocyclic Gd(III) complexes bearing β-arylsulfonamide groups is markedly pH-dependent (Fig. 15) on passing from about 8 s^{-1} mM^{-1} at pH < 4 to *ca.* 2.2 s^{-1} mM^{-1} at pH > 8 in one selected case (Chart 12, ligand **2**) (*130*). It has been demonstrated that the observed decrease (about 4-fold) of r_1 is the result of a switch in the number of water molecules coordinated to the Gd(III) ion from 2 (at low pH values) to 0 (at basic pHs). This corresponds to a change in the coordination ability of the β-arylsulfonamide arm that binds the metal ion only when it is in the deprotonated form (Fig. 15).

In some cases the pH dependence of the relaxivity is associated with changes in the structure of the second hydration shell. Two such systems have been reported by Sherry's group. The first case deals with a macrocyclic tetraamide derivative of DOTA (DOTA-4AmP, Chart 12) that possesses an unusual r_1 vs. pH dependence (*131*). In fact, the relaxivity of this complex increases from pH 4 to pH 6, decreases up to pH 8.5, remains constant up to pH 10.5 and then increases again. The authors suggested that this behavior is related to the formation/disruption of the hydrogen bond network between

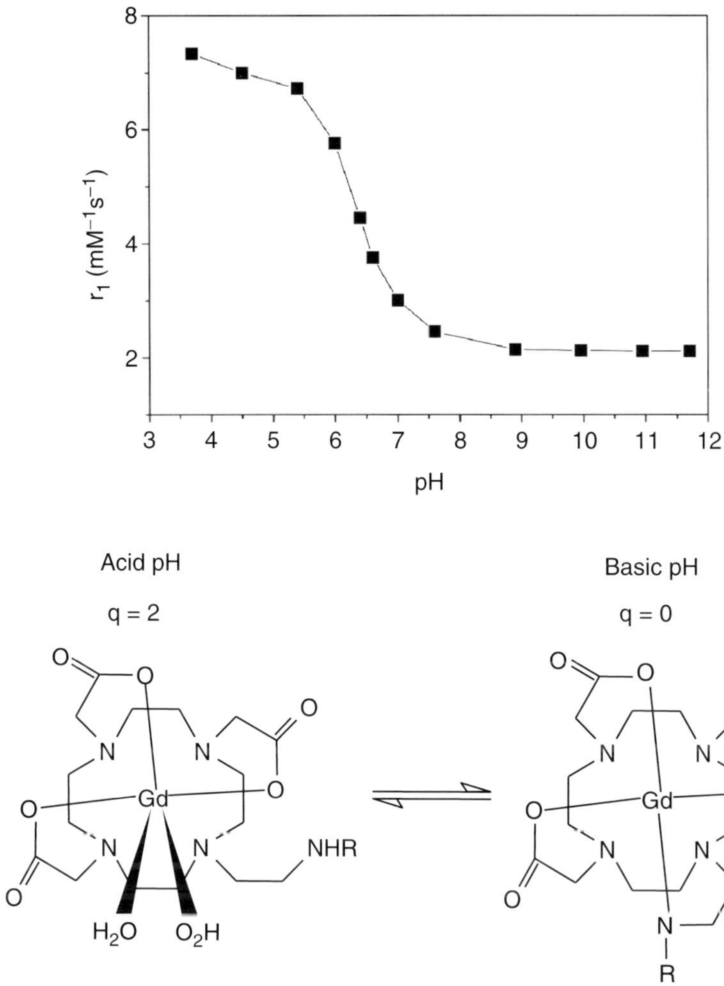

FIG. 15. pH dependence of the relaxivity (20 MHz, 25°C) of a Gd(III) complex bearing an arylsulfonamide group (ligand **2** in Chart 12). The behavior is the result of a switch in the number of water molecules coordinated to the Gd(III) ion from 2 (at low pH values) to 0 (at basic pHs).

the pendant phosphonate groups and the water bound to the Gd(III) ion. The deprotonation of phosphonate occurring at pH $>$ 4 promotes the formation of the hydrogen bond network that slows down the exchange of the metal-bound water protons (analogous to what was already discussed in Section IV.C). On the contrary, the behavior observed at pH $>$ 10.5 was accounted for in terms of a shortening of τ_M catalyzed by OH^- ions. Recently, it has been demonstrated that this complex can be successfully used in vivo for mapping renal and systemic pH, on the assumption that its biodistribution is equal to that of the reference compound $[GdDOTP]^{5-}$ (Chart 5), a complex whose relaxivity is not affected by pH (*132*).

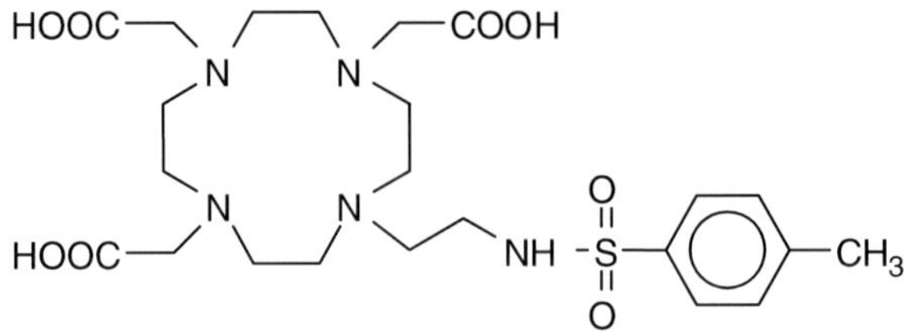

2

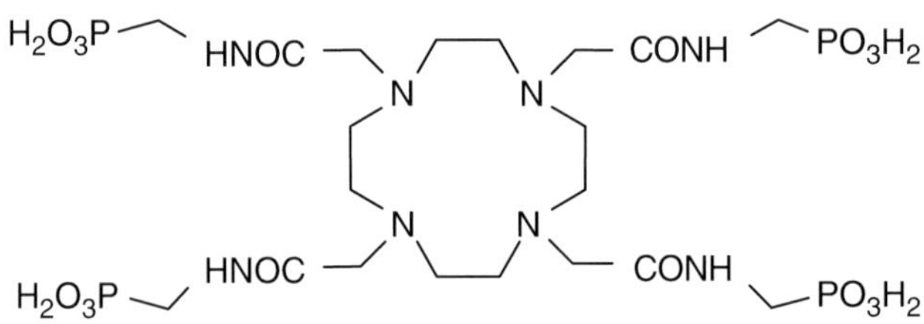

DOTA4-AmP

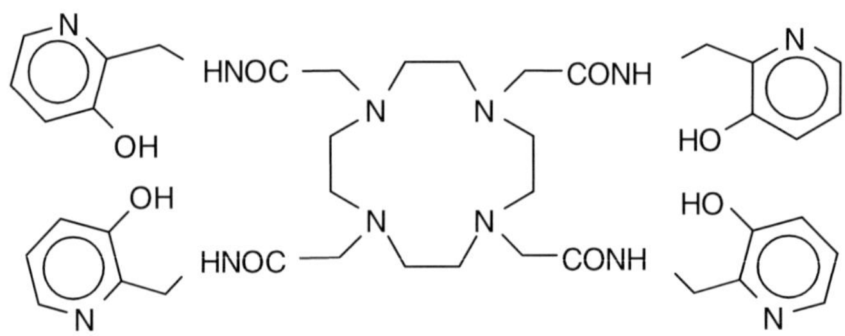

3

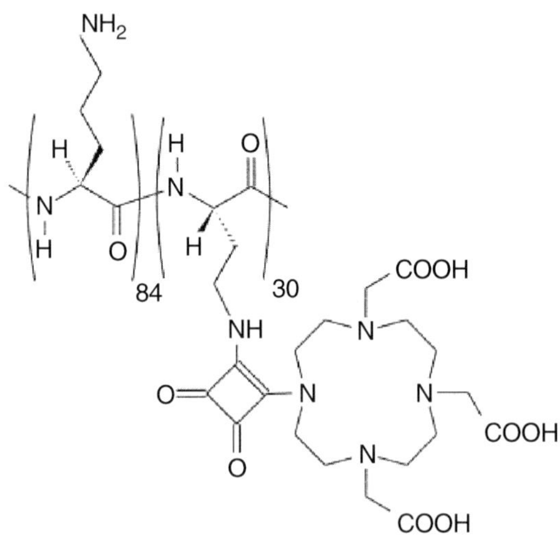

4

CHART 12.

In a very recent paper by Sherry's group, it has been shown that also the second sphere contribution to the relaxivity can be properly modulated by pH. Again, the basic coordination cage is here constituted by the DOTA tetramide motif, but the pendant arms are represented by pyridyl, phenolic, and hydroxypyridyl groups (*37*). Interestingly, whereas the relaxivity of the first two complexes was not significantly affected by pH, the profile for the latter compounds (Chart 12, ligand **3**) was characterized by the presence of two regions of enhanced relaxivity centered around pH 3 and 9, respectively. The modest (25%) enhancement observed at pH 3 was attributed to a decrease in the residence lifetime of the metal-coordinated water protons, whereas the larger enhancement (84%) observed at higher pHs was ascribed to the occurrence of a strong second sphere contribution arising from the structuring effect on the second hydration shell of the metal complex promoted by the deprotonation of the coordinating amide groups.

pH dependent probes can also be obtained when the hydrogen ion concentration is able to affect the rotational tumbling of a slowly moving Gd(III)-based system. A nice example is presented by a macromolecular Gd(III) construct constituted by 30 Gd(III) units covalently linked, by a squaric acid moiety, to a poly-ornithine (114 residues, Chart 12, ligand **4**) (*133*). At acidic pH the unreacted amino groups of the polymer are protonated and, therefore, they tend to be localized as far apart as possible, whereas at basic pH the progressive deprotonation of the $NH_3{}^+$ groups determines an overall rigidification of the polymer structure owing to the formation of intramolecular hydrogen bonds between adjacent peptidic linkages. As expected, the reduced rotational mobility of the polymeric backbone upon increasing pH enhances the relaxivity of the system (Fig. 16). Nevertheless, even if the enhancement is not particularly remarkable (*ca.* 40% in the 3–8 pH interval), it is worth to note that the relaxivity of this system is considerably higher than in the previous examples over the entire pH range, thus allowing, in principle, the detection of pH changes at lower concentration of the responsive probe.

Another interesting route for developing pH sensitive probes, involves the inclusion of Gd(III) complexes, whose relaxivity is not affected by pH, into pH sensitive liposomes (*134*). The main constituent of these liposomes is unsaturated phosphatidylethanolamine that is usually "doped" with negatively charged amphiphilic components in order to stabilize the lamellar phase necessary for the formation of the bilayer structure of liposomes. If the acidic group of the amphiphilic component is protonated, then the liposomes destabilize (transition phase) with the consequent release of their content. On this basis, a relaxometric "off-on" pH switch can be designed by incorporating a Gd(III) complex into intact liposomes endowed with a very low water permeability. In this way, the relaxivity of the system will be very low (about 0.5 s^{-1} $mMGd^{-1}$ at 37°C and 10 MHz), but the protonation of the acidic components determines the release of the paramagnetic content, thus leading to a significant relaxation enhancement (almost 4 s^{-1} $mMGd^{-1}$). This system has been successively tested in human blood where, unfortunately, it showed to be quite unstable owing to the presence of blood components

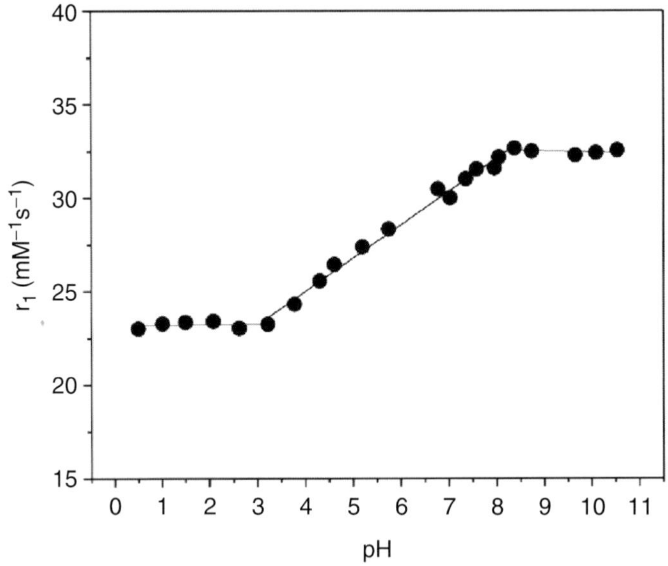

FIG. 16. pH dependence of r_1 at 20 MHz and 25°C for Gd-4 (Chart 12). The relaxivity enhancement upon increasing pH is due to an elongation of τ_R promoted by the progressive deprotonation of the NH_3^+ groups of the polymer structure.

(Na^+, Ca^{2+}, Mg^{2+}, and also HSA) which interfere with the phase transition of the liposomes (*135*).

B. Temperature Sensitive

The need of designing probes for measuring in vivo temperature is primarily dictated by therapeutical purposes. In fact, useful therapies against tumors like hypertermia or thermal ablation are based on localized heating which selectively kills tumor cells. Such therapies require the achievement of well-defined temperatures that, moreover, have to be maintained constant for a given time. Therefore, a continuous temperature monitoring is essential for the success of the therapy (*136*).

In principle, the relaxivity of almost all Gd(III) complexes is affected by temperature as a result of the temperature dependence of the dynamic parameters controlling r_1, namely τ_R, τ_M, τ_V, and D. Nevertheless, the effect of temperature on the relaxivity is usually rather small and, therefore, of little utility for clinical use.

To this purpose, Fossheim *et al.* reported temperature sensitive liposomal Gd(III)-based probes (*137–138*). The composition of the liposomes was chosen in order to tune the temperature of transition between the gel-crystalline, where the liposome is water-impermeable, to the liquid-crystalline state, where water has free access to the interior of the liposome. This means that at temperatures below the transition, the relaxivity of the system is very low (paramagnetic contribution close to zero), whereas at higher temperature the Gd(III) complex ([GdDTPABMA(H_2O)] is the reference) is no longer silent owing to the fast exchange of water molecules through the lipidic bilayer of the liposome. Besides the "on-off" effect, the great advantage of this approach is based on the possibility to modulate the transition temperature through proper modifications of the liposome composition.

C. Agents Sensitive to Redox Potential

A diagnostic MRI probe sensitive to the in vivo redox potential would be very useful for detecting regions with a reduced oxygen partial pressure (PO_2), a typical symptom of several pathologies including strokes.

Very few Gd(III) chelates sensitive to the tissue oxygenation have so far been reported. Our group has investigated the potential ability of $[GdDOTP]^{5-}$ (Chart 5) to act as an allosteric effector of hemoglobin (*139*). In fact, it has been observed that this chelate binds specifically to the T-form of the protein, that is characterized by a lower affinity towards oxygen. The interaction is driven by electrostatic forces and leads to a significant relaxivity enhancement (*ca.* 5-fold) owing to the restricted molecular tumbling of the paramagnetic complex once it is bound to the protein. Although hemoglobin can be considered an excellent indirect target for detecting PO_2, the practical applicability of the method suffers from the inability of $[GdDOTP]^{5-}$ to enter red blood cells.

Another approach deals with the use of liposomes built up with an amphiphilic Gd(III) complex containing a radical-sensitive disulfide bridge (Chart 13, ligand **5**, 140). The relaxivity (at 20 MHz and 25°C) of the liposomal paramagnetic agent is 13.6 s^{-1} mM^{-1}, i.e., 2-fold higher than the free Gd(III) complex (r_1 of 6.5 s^{-1} mM^{-1}). Most likely, the limited relaxivity enhancement is due to the rotational flexibility of the complex bound to the liposome. This system was tested in vitro by inducing the cleavage of the S S bond by chemical (by dithiothreitol) or physical (by γ-rays) means. In both cases the relaxivity decreased from the value of the liposomal agents to that of the free Gd(III) complex.

Although not based on Gd(III) ion, it is worth refering to other paramagnetic probes acting on the relaxation rates of water protons, which have been proposed as responsive agents towards the oxygen concentration. Basically, such systems share the idea of designing probes containing a metal that exists in two redox states endowed with very different relaxing properties.

A first example is represented by the Mn(III)/Mn(II) redox switch. The complexes of Mn(II) and Mn(III) with the water-soluble tetraphenylsulphonate porphyrin (TPPS, Chart 13) display significantly different r_1 values at low magnetic field strength (lower than 1 MHz), but very similar values at the fields currently used in the clinical practice (> 10 MHz) (*141*). However, the longer electronic relaxation rates of the Mn(II) complex makes its relaxivity dependent on the rotational mobility of the chelate. In fact, upon interacting with a poly-β-cyclodextrin, a 4-fold enhancement of the relaxivity of $[Mn(II)\text{-}TPPS(H_2O)_2]^{2-}$ at 20 MHz has been detected, whereas little effect has been observed for the Mn(III)-complex. The ability of the Mn(II)/Mn(III)

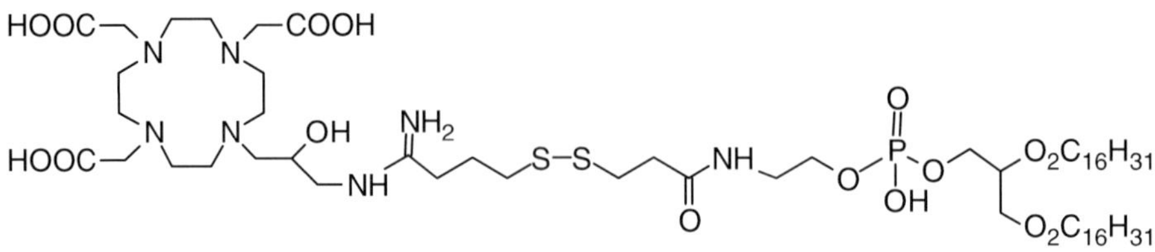

5

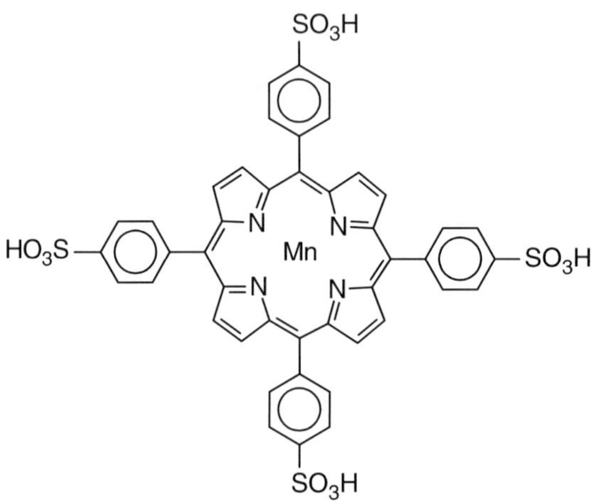

Mn-TPPS

CHART 13.

system to respond to changes in the partial pressure of oxygen has been demonstrated in vitro (Fig. 17).

Another interesting redox switch recently considered, deals with the Eu(III)/Eu(II) couple. Here, Eu(II) has the same electronic ^8S ground state of Gd(III) ion, whereas Eu(III) ion is a poor relaxing probe owing to its very short electronic relaxation times. Unfortunately, Eu(II) is a very unstable cation. Therefore, efforts have been addressed to get more insight into the determinants of thermodynamic and kinetic stability of Eu(II) complexes (*142–143*). Interestingly, it has been shown that the relaxivity of [Eu(II)-DOTA(H_2O)]$^{2-}$ is very similar to that of [Gd(III)DOTA(H_2O)]$^-$, but the residence lifetime of the former is noticeably shorter (τ_M of *ca.* 0.4 ns vs. 200 ns at 25°C) (*144*).

D. ENZYME RESPONSIVE

One possible route to design enzyme responsive agents is to synthesize paramagnetic inhibitors, whose binding to the active site of the protein can be signaled by the consequent relaxivity enhancement. An example of this approach has been provided by Anelli *et al.*, who synthesized a linear Gd(III)

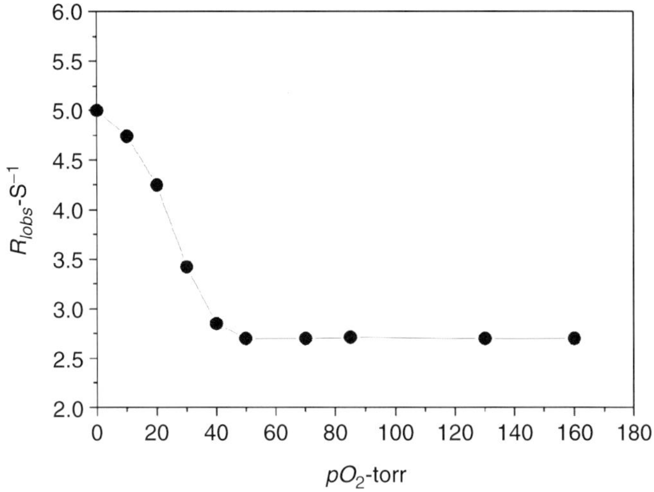

FIG. 17. Plot of the pO_2 dependence of the longitudinal water proton relaxation rate of a 0.125 mM solution of MnTPPS (Chart 13) (20 MHz, 25°C and pH 7).

complex bearing an arylsulfonamide moiety (Chart 14, ligand **6**), that is a well-known inhibitor of carbonic anhydrase (*145*). In vitro experiments demonstrated that this complex binds quite strongly (K_A of about 1.5×10^4) to the enzyme and its relaxivity in the bound form is about 5-fold higher than the free complex (27 s^{-1} mM^{-1} vs. 5 s^{-1} mM^{-1} at 20 MHz). Unfortunately, the attempt to test the validity of this approach in vivo failed, likely owing to the small amount of enzyme circulating in the blood.

An alternative approach is to design Gd(III) complexes acting as substrate for a specific enzyme. Along this direction, an experimental proof of concept has been provided by Lauffer *et al.* who prepared a Gd(III) chelate containing a phosphoric ester sensitive to the attack of the serum alkaline phosphatase (Chart 14, ligand **7**) (*146*). The hydrolysis yields the exposure of a hydrophobic moiety well-suitable to bind to HSA. Upon binding, there is an increase in the relaxivity as a consequence of the lengthening of the molecular reorientational time. This approach was used by the same research group for designing Gd(III)-complexes sensitive to TAFI (thrombin-activatable fibrinolysis inhibitor), a carboxypeptidase B involved in the clot degradation (*147*).

Another example of enzyme responsive agents has been provided by Moats *et al.* who synthesized a GdDOTA-like system containing a galactpyranosyl substituent capping the metal ion (Chart 14, ligand **8**). Hence, the complex shows a low relaxivity (typical $q = 0$ complex). In the presence of β-galactosidase, an enzyme widely used as a marker of gene delivery, the sugar moiety is hydrolyzed, and a water molecule can have free access to the paramagnetic center (*148*). This complex has been successfully used for monitoring the gene expression in vivo during the development of embryos of *Xenopus laevis* (*149*).

Also particles have been considered as enzymatic responsive agents. For this purpose, insoluble Gd(III) chelates have been synthesized by introducing

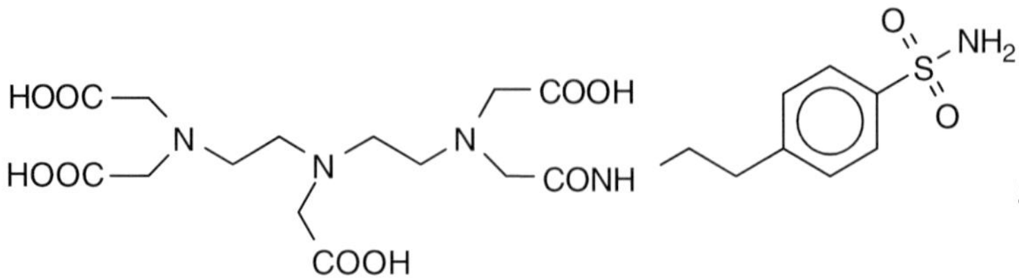

6

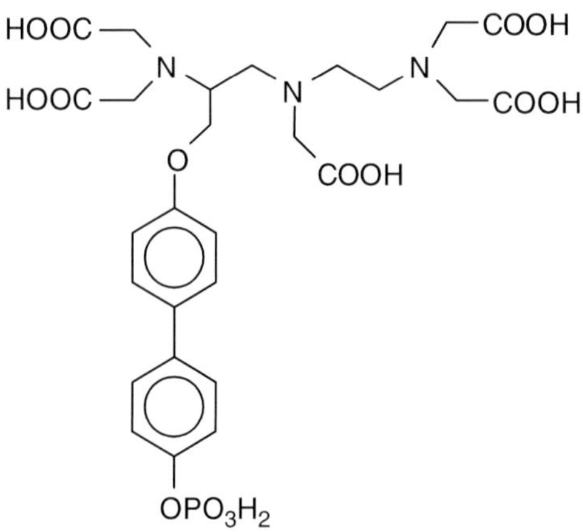

7

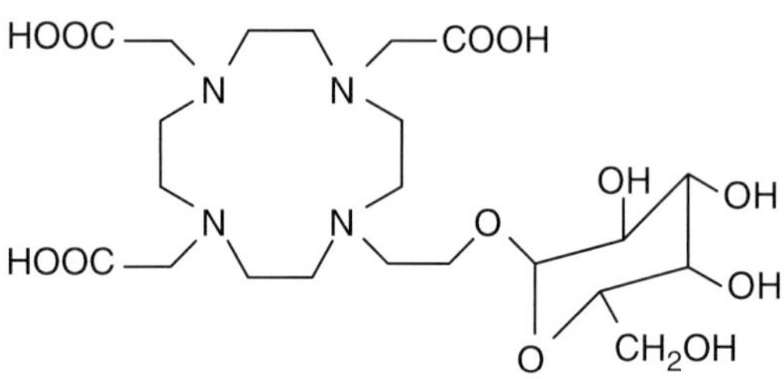

8

CHART 14.

on the ligand surface long aliphatic chains *via* an ester or a peptidic bond. The particles can be internalized into cells having phagocytic activity and then degraded by the action of the proper enzyme which cleaves the bond between the Gd(III) chelate and the insolubilizing moiety (Fig. 18). Thus, the increase in intracellular relaxivity becomes a function of the activity of the enzyme of choice. In principle, such an approach provides a representative example of "responsive particles". Immediately after the internalization, the particles act as negative T_2 agents because they affect the bulk magnetic susceptibility. Then, their dissolution yields soluble Gd(III) chelates which act as positive T_1 agents (*150*).

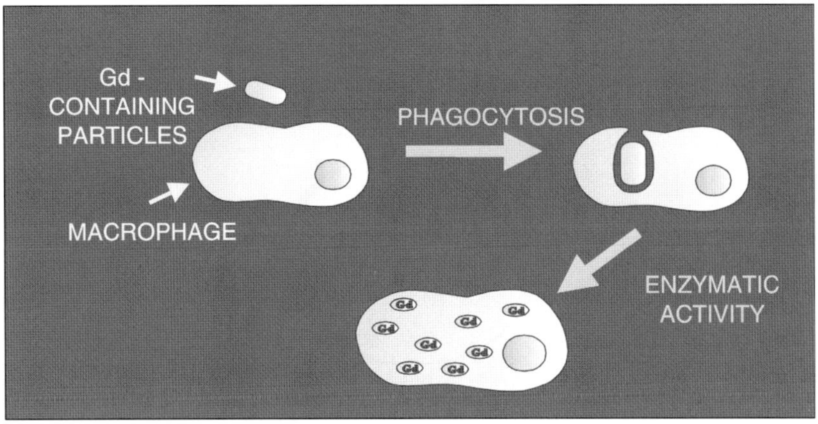

FIG. 18. Schematic representation of the intracellular enzymatic solubilization of a Gd(III)-based insoluble particles phagocytized by a macropahge.

E. AGENTS SENSITIVE TO CHANGES IN THE CONCENTRATION OF IONS AND LOW-MOLECULAR WEIGHT METABOLITES

Several Gd(III)-based agents have been proposed as responsive systems for small-sized metabolites. Most of the attention has been focused on lactate, because this end-product of the anaerobic glycolytic pathway is involved in several pathologies (*151–153*). An over-production of lactate is usually associated with both a decrease in pH and a reduced partial pressure of O_2. Therefore, a MRI agent sensitive to pH or PO_2 may be considered as an indirect responsive probe for lactate.

A route for designing Gd(III) complexes whose relaxivity depends on the presence of lactate, is provided by the ability shown by some hexa- or hepta-coordinate chelates to form ternary complexes with a wide array of anionic species (*154–161*). The interaction between the coordinatively unsatured metal complex and lactate involves the displacement of two water molecules coordinated to Gd(III) ion with the two donor atoms of the substrate, thus leading to a marked decrease in the relaxivity. Lactate is a good ligand for Gd(III) ion because it can form a stable 5-membered ring by using the hydroxo and carboxylic oxygen donor atoms (Fig. 19).

The reference compound of the class of coordinatively unsatured Gd(III) chelates is represented by GdDO3A (Chart 4). It has been reported that lactate binds this chelate with an association constant (pH 7.4 and 25°C) of only 150 (*156*), but the affinity can be significantly enhanced by changing the electric charge of the metal complex (from neutral to positive) or by introducing a substituent on the secondary amine of the macrocycle (*160–161*). As expected, the relaxivities of the ternary complexes are lower than the parent chelates; for instance the relaxivity of $[GdDO3A(H_2O)_2]$ at 10 MHz is 6.95 s^{-1} mM^{-1}, which decreases to 3.1 s^{-1} mM^{-1} upon the binding of lactate. One of the current limitations of the use of these complexes as responsive agents is their poor selectivity. In fact, though the affinity

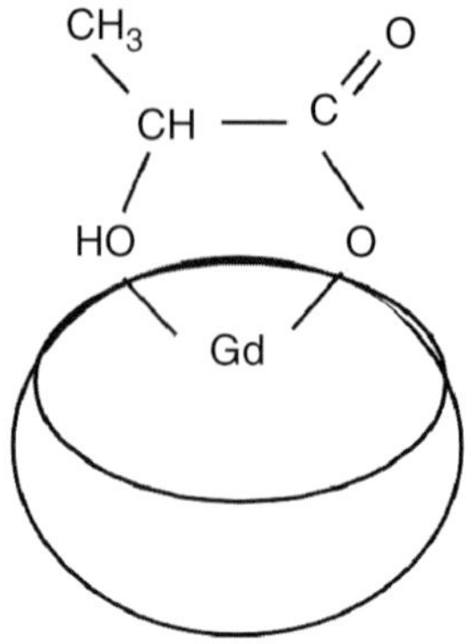

FIG. 19. Schematic representation of the 5-membered ring formed by a coordinatively unsatured Gd(III) complex and lactate.

towards lactate is usually the highest, the DO3A-like complex can interact with other substrates present in biofluids, like phosphates or carbonates.

As far as the design of systems sensitive to specific metal ions is concerned, an interesting example is represented by a dimeric Gd(III) complex ($[Gd_2DOPTA(H_2O)_2]^{4-}$, Chart 15) synthesized by Li and co-workers (*162*). The DOPTA ligand contains two macrocyclic coordination cages for Gd(III) ion bridged by a structure reminiscent of BAPTA (1,2-bis(o-aminophenoxy)ethane-N,N,N′,N′-tetraacetic acid), a ligand known for its affinity towards Ca(II) ions (*163*). In the absence of calcium, the four pendant acetate arms of the linking bridge of $[Gd_2DOPTA(H_2O)_2]^{4-}$ coordinate Gd(III) ions, thus leaving one inner-sphere water molecule for each coordination unit. In the presence of Ca(II) ions, the four acetate moieties, in cooperation with the two oxygens of the ether bridge, become involved in the coordination of the alkaline-earth ion, thus promoting an increased hydration of the two Gd(III) ions. This leads to a relaxivity enhancement of about 80%. It has been shown that this complex is specific towards Ca(II) ions, at least when its relaxometric behavior is investigated as a function of pH or pMg.

Another target ion is zinc(II) ion, an essential component of many enzymes. Recently, Hanaoka *et al.* synthesized a bis-amide derivative of $[GdDTPA(H_2O)]^{2-}$ bearing two TPEN moieties (Chart 15) (TPEN = N,N,N′,N′-tetrakis(2-pyridylmethyl)ethylene-diamine)(*164*), a structure known for chelating selectively Zn(II) ions (*165*). Upon addition of Zn(II), the relaxivity of this complex (measured at 300 MHz, 25°C and pH 8) monotonically decreases from 6 to 4 s^{-1} mM^{-1} (decrease of *ca.* 35 %) when an equimolar amount of Zn(II) is added. It has been suggested that the lower relaxivity in the presence of Zn(II) is the result of the reduced water accessibility at the Gd(III) site caused by the Zn(II) coordination. In fact, in the presence of an excess of Zn(II) (where Zn(II) and GdL form a 2:1 complex) or by using a mono-TPEN derivative of $[GdDTPA(H_2O)]^{2-}$, the relaxivity of the Gd(III) chelate is no longer sensitive to Zn(II) concentration. Interestingly, no effect on the relaxivity of $[GdDTPA(TPEN)_2(H_2O)]$ by H_3O^+, Ca(II) or

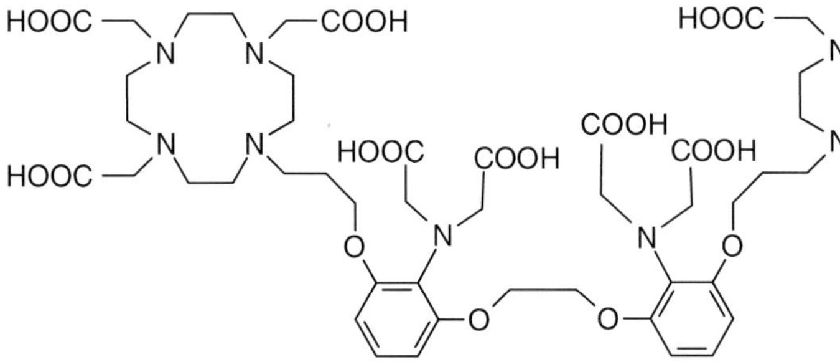

DOPTA

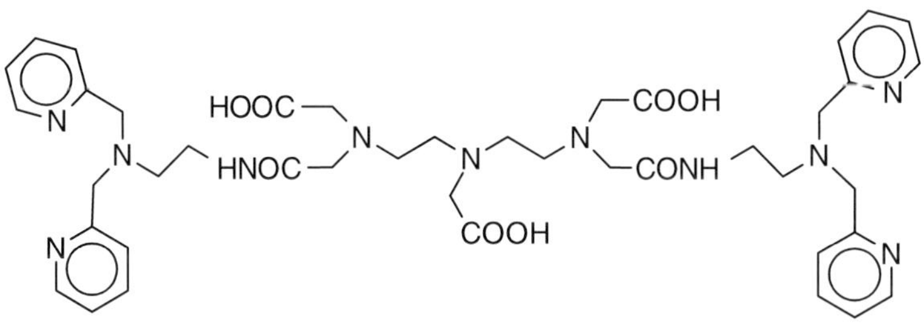

DTPA(TPEN)$_2$

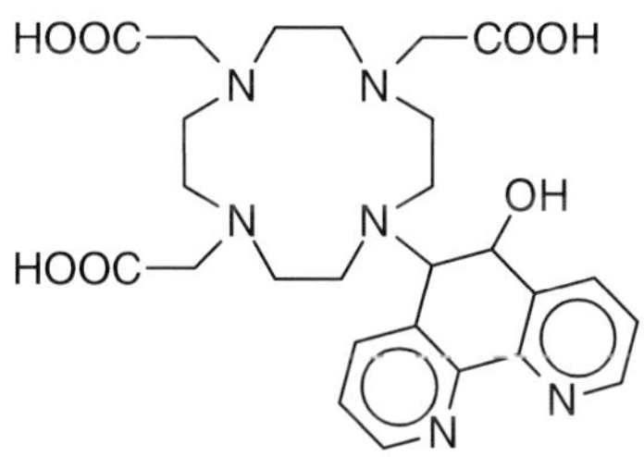

PhenHDO3A

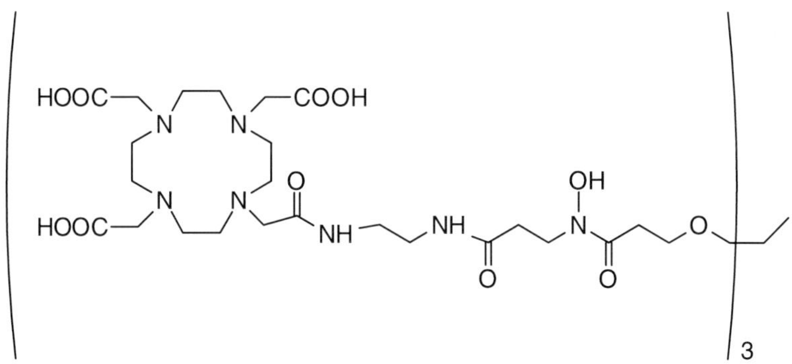

9

CHART 15.

Mg(II) was observed between pH 4 and 10, thus demonstrating the high selectivity of this system towards Zn(II).

Another important metal ion to be monitored is iron. Recently, Desreux *et al.* described a class of agents able to report on the concentration of this transition metal. The first class of compounds, whose prototype is

[Gd-PhenHDO3A(H_2O)] (Chart 15), deals with systems able to self-assemble upon binding to Fe(II). The slower rotational tumbling of such a supramolecular system leads to a relaxation enhancement of about 145% (*97*). An alternative approach is the use of a tripodal Gd(III)-based system characterized by the presence of three hydroxamate groups (Chart 15, ligand **9**). In the presence of Fe(III) ions the "paramagnetic ligand" wraps around Fe(III) ion. At 20 MHz a relaxivity enhancement of 57% has been reported for such system (*116*). Besides the elongation of τ_R, this relaxivity enhancement also receives a contribution from the presence of the paramagnetic Fe(III) ion.

VI. Insights for Molecular Imaging Applications

Molecular Imaging (*166*) deals with in vivo characterization and measurement of biological processes at the cellular and molecular level. With Molecular Imaging, early diagnosis of disease will become possible as the detection of altered biochemical processes largely anticipates the anatomical changes that are the basis of current diagnostic modalities.

Several modalities have contributed to the early stages of this innovative approach, namely PET (Positron Emission Tomography), SPECT (Single Photon Emission Computed Tomography), MRI, and Optical Imaging. MRI-Gd(III) based agents are much less sensitive than Radionuclear and Optical Imaging probes. Therefore, Molecular Imaging based on MRI, invariantly involves the need of accumulating a high number of contrast enhancing units at the site of interest. The basis for the design of a Gd-containing Imaging Probes is first dictated by the concentration and localization (vascular, extracellular matrix, on the cellular membrane, intracellular) of the target molecule (*98*). Of course, the most accessible targets are those present on the surface of endothelial vessels. In principle, they can be visualized by a number of macromolecular conjugates containing many Gd(III) complexes endowed with the proper vector recognizing the given target. A nice example of targeting an endothelial site has been reported by Sipkins *et al.* (*167*) in the targeting of a specific angiogenesis marker, the endothelial integrin $\alpha_v\beta_3$, whose presence has been shown to correlate with tumor grade. The Imaging Probe used in this work is a Gd-containing polymerized liposome. The target is first bound by a biotinylated antibody against $\alpha_v\beta_3$, which is successfully recognized by an avidin moiety on the surface of the liposome. Each liposome has a mean diameter of 300–350 nm, which appears suitable to avoid the uptake by the reticuloendothelial system (Fig. 20). This approach provided enhanced and detailed detection of rabbit carcinoma through the imaging of the angiogenetic vasculature.

Recently, the same $\alpha_v\beta_3$ target has been addressed with lipidic nanoparticles containing a huge number of Gd-chelated units (94,400 Gd/particle characterized by $r_1 = 19.1\ s^{-1}\ mM^{-1}$ (per Gd) $r_1 = 1{,}800{,}000$ per particle). One of the lipidic components is covalently bound to the $\alpha_v\beta_3$ – integrin peptidomimetic antagonist (*168*).

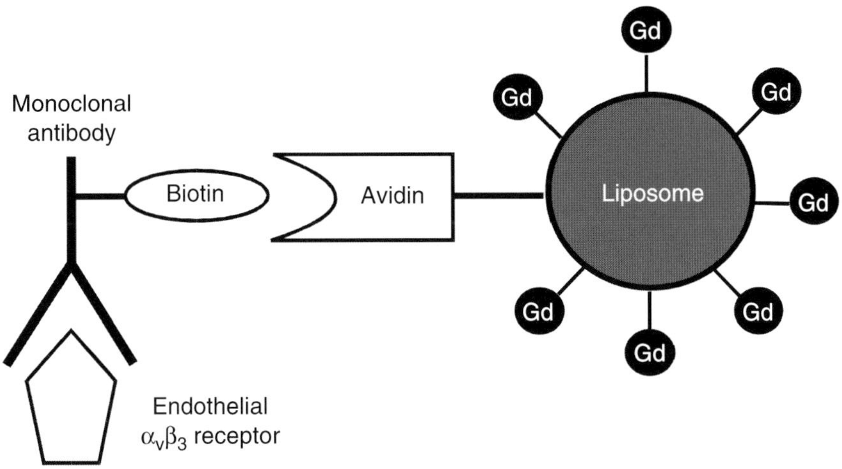

FIG. 20. Targeting of the endothelial integrin $\alpha_v\beta_3$ as a specific angiogenesis marker. The receptor is recognized by a biotinylated antibody that is then bound by an avidin moiety bearing a Gd(III) loaded liposome.

The large molecular size of these constructs limits their delivery to targets on the endothelial walls. To target receptors in solid tissues, other routes have to be followed. Bhujwalla and coworkers (*169*) have recently developed and applied a two-component Gd-based avidin-biotin system for the visualization of HER-2/scan receptors. The latter is a member of the epidermal growth factor family and it is amplified in multiple cancers. Their approach consisted of addressing the extracellular domain of the receptors by means of a biotinylated mAb. After clearance of the unbound mAb, Gd-labeled avidin is administered and it binds, with high affinity, to the biotinylated mAb. The expression level of the receptor was estimated at 7×10^5 receptors/cell and the average number of $[GdDTPA(H_2O)]^{2-}$ units per avidin molecule was 12.5. The method has been succesfully applied to an experimental mouse model of breast carcinoma.

An interesting route to MR signal amplification has been developed by Weissledder and coworkers (*99*). Their approach is based on enzyme-mediated polymerization of paramagnetic substrates into oligomers of higher relaxivity. As substrates they used Gd-chelates functionalized with phenolic substituents which undergo rapid condensation in the presence of H_2O_2 and peroxidase. The increased molecular size of the oligomeric structures causes an increase in the molecular reorientational time which, in turn, results in an increase in the observed relaxivity. This approach has been applied to the imaging of E-selectin-peroxidase conjugate.

As far as the cell internalization of Gd chelates is concerned, several routes have been explored:

(i) *via* Pinocytosis.

The cell may internalize portions of the surrounding fluid by means of the invagination of its membrane and formation of small vesicles ($\leq$150 nm diameter) called endosomes. Therefore, incubation of cells for a sufficiently long time, in a medium containing the Imaging Probe at relatively high concentration, leads to its internalization at amounts that may be sufficient for MRI visualization. Among a number of systems we have considered the neutral, highly hydrophilic [GdHPDO3A(H_2O)] as a good candidate for labeling stem cells by the pinocytotic route (*170*). The in vivo MR visualization of labeled stem cells will allow their monitoring after transplantation. In a typical experiment of uptake *via* pinocytosis, few millions of stem cells are incubated in a culture medium containing [GdHPDO3A(H_2O)] in the mM concentration range (10–50 mM) for a few hours. Upon incubation, no saturation effect is observed and the amount of uptaken Gd is linearly proportional to the concentration of the paramagnetic agent in the incubation medium. Once cell-internalized, the [GdHPDO3A(H_2O)] molecules remain entrapped in endosomic vesicles as can be seen by observing the cells incubated with [EuHPDO3A(H_2O)] in the confocal microscope. In fact, Gd(III) and Eu(III) chelates with the same ligands display the same chemical/biological behavior, and the fluorescent response of [EuHPDO3A(H_2O)] acts as a histological reporter of the localization of [GdHPDO3A(H_2O)] in the cell. We have proved the potential of this approach by observing a mouse model for angiogenesis, on which blood-derived endothelial progenitor cells (EPCs) have been implanted subcutaneously, within a matrigel plug. After few days, the histologic examination showed large capillary structure transposing the gel plugs. MR images parallel histologic findings since hyperintense spots, corresponding to the labeled cells, were clearly detected. In Fig. 21, we report a MR image taken 7 days after implantation. As control, matrigel embedding unlabeled cells implanted under the same conditions are always negative for MRI signal.

The cell labeling procedure described above appears to be of general applicability. We have tested it on several tumor cell lines obtaining invariantly a very efficient uptake with no apparent cytotoxicity. Likely, the entrapment of [GdHPDO3A(H_2O)] into the endosomic vesicles prevents any impact of the paramagnetic agent on relevant cellular processes while maintaining the full accessibility to cytoplasmatic water molecules.

(ii) *via* Phagocytosis.

It is the process of internalization of particles by cells endowed with phagocytic activity. In such a case, this route appears highly efficient for a single step internalization of a large amount of imaging probes. However to be effective on MR images, Gd-chelates must be water soluble. Therefore the particles must be biodegradable in order to release soluble Gd-chelates once internalized into phagocytic cells. One may envisage several ways for the release of the Gd-chelates. For instance, one may think of gel nano-particles of chitosan loaded with negatively charged Gd-chelates. Such particles (200–400 nm diameter) are easily phagocytosed and slowly degraded once internalized into the cells (*171*). Another approach to biodegradable Gd-containing particles has been pursued by designing particles whose insolubility is a property of the Gd-chelates themselves (*158*). This goal is easily achieved by introducing long aliphatic chains on the surface of the ligand. However one can control the fate of these systems by means of the

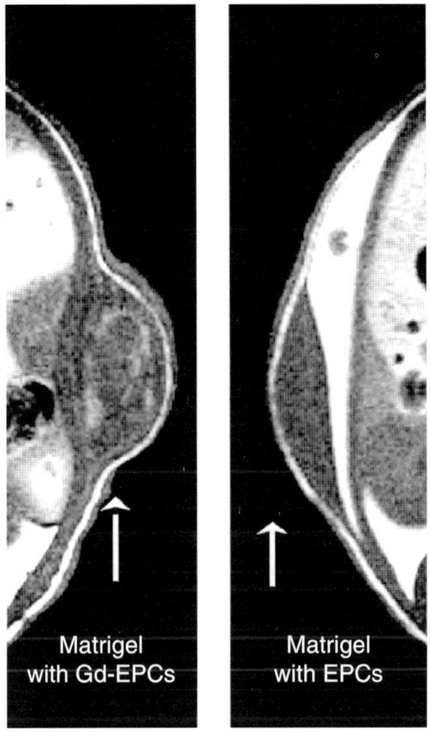

FIG. 21. In vivo T_1 weighted spin echo MR image (7.05 T) of EPCs labeled with [GdHPDO3A(H_2O)] (on the left). The cells are dispersed into a subcutaneous matrigel plug seven days after the implantation. On the right is shown the control image of the same in the absence of the paramagnetic label.

functionality used to link the insolubilizing moiety to the Gd-chelate. In fact, by using ester or peptidic functionalities, the insolubilizing synthon can be displaced from the Gd-chelates by the activity of the proper enzyme as discussed in Section V.D.

(iii) *via* Receptors.

Cell-internalization *via* receptors is the route of choice in a number of Nuclear Medicine assays. For MRI, the design of the Imaging Probe requires the attachment of one or more Gd(III)-chelates to the ligand molecule. Such structural modification may drastically affect the internalization process with respect to the mechanism occurring for the native ligand.

In order to deal with a system whose structural characteristics were unaltered by the loading with Gd(III) chelates, we choose Apoferritin because it allows the Imaging Probes to be entrapped inside its inner cavity (*60*). The exterior of such Gd(III)-loaded Apoferritin is exactly the same as in the parent Ferritin and then, once administered intravenously, it is quickly cleared-up by the proper receptors on hepatocytes (*172*). The process of loading Apoferritin with [GdHPDO3A(H_2O)] consists first of the dissociation of the protein into subunits at pH 2, followed by its reforming at pH 7, thereby trapping the solution components (e.g., [GdHPDO3A(H_2O)])

within its interior. In such a system, water can freely diffuse through the channels formed at the intersection of the protein subunits (10 channels), but the larger [GdHPDO3A(H_2O)] molecules cannot. As discussed in Section IV, the relaxivity shown by each [GdHPDO3A(H_2O)] complex entrapped in the Apoferritin cavity is very high (*ca.* 80 s^{-1} $mMGd^{-1}$ at 20 MHz and 25°C). It has been possible to assess that the Gd-loaded Apoferritin maintains its integrity in the cell internalization process, as the relaxivity observed for the cytoplasmatic extract corresponds to that of the intact system. Finally, the amount of cell-internalized Gd-loaded Apoferritin is similar to that reported for the native Ferritin (6.5×10^6 molecules per cell in 6 hours).

(iv) *via* Receptor mediated endocytosis.

Probably this is the most common route for the internalization of Gd(III) chelates since it is expected to occur any time a given ligand is modified by attaching one or more Gd-chelating units. Thus, the internalization process is no longer the one followed by the native ligand but the binding to the receptor stimulates an endocytotic process which starts with the introflection of a portion of cellular membrane and ends up with the welding of its extremities. Through this process, a number of substances remain entrapped in endosomic vesicles, primarily the molecules bound to the membrane or in close proximity of the portion of the membrane involved in the endosome formation.

Weiner *et al.* (*173*) showed that the uptake of a folate-conjugate dendrimer into tumor cells overexpressing high affinity Folate Receptor (hFR), occurs through this type of endocytotic pathway.

Another example of receptor mediated endocytosis of a large dendrimer has been recently reported (*93*). The macromolecular construct comprised of Avidin and a biotinylated dendrimer bearing 254 GdDTPA chelates (AV-G6Gd) has shown to accumulate in vitro into SHIN3 cells (a cell-line obtained originally from human ovarian cancer) 50-fold greater than $[GdDTPA(H_2O)]^{2-}$. The internalization process is driven by the avidin molecule a glycoprotein that binds to β-D-galactose receptors which are present on these tumor cells.

Although not yet proven, it is likely that receptor-mediated endocytosis occurs with a number of systems in which the imaging Gd-probes are linked either to peptides as targeting vectors that bind to specific receptors or to nutrient or pseudo-nutrient moieties that interact with the proper transporter up-regulated in tumor cells.

(v) *via* transmembrane carrier peptides.

Another route to enter cells with Gd(III) chelates is based on the use of membrane translocation peptides which have been proven useful for the internalization of a number of substrates like proteins, oligonucleotides and plasmid DNA. For instance, Bhorade *et al.* (*174*) showed that GdDTPA bound to 13-merHIV-tat peptide is efficiently internalized.

Another example has been recently reported by Allen and Meade (*175*) who showed that GdDTPA conjugated to polyarginine (8–16 monomer units) is able to permeate all membranes.

Finally, along this line, an interesting development has been recently reported by Heckl *et al.* (*68*). They synthesized an Imaging Probe consisting of a Gd-complex, a PNA (Peptide Nucleic Acid) sequence, and a transmembrane carrier peptide. Although the system enters any type of cell, it accumulates only in tumor cells because of the specific binding of the PNA moiety of the c-myc mRNA whose production is upregulated in those cells.

VII. Final Remarks

Gd(III) chelates have played an important role in the development of clinical applications of MRI technique by adding relevant physiological information to the superb anatomical resolution attainable with this imaging modality.

More is expected with the currently available contrast agents especially in the field of Dynamic Contrast Enhancement protocols reporting on changes in the vascular permeability associated with the staging and therapeutic follow-up of important pathologies. However, the major challenges are in the emerging field of Molecular Imaging where the competition with other Imaging modalities can be very tight. Targeting of thrombi and atherosclerotic plaques by peptides functionalized with Gd(III) chelates appears to be the next goal for industrial research. The possibility of identifying and characterizing vulnerable plaques will certainly represent an important task. Clearly, the need for new ideas in enhancing the attainable relaxivity at higher fields as the 3 T indication for clinical imagers seems to be quite established. Moreover, it is necessary to improve the efficiency of the available delivery systems and, possibly, to exploit suitable amplification procedures in order to reach the sensitivity required for the visualization of target molecules present at low concentrations.

Upon indicating the efforts necessary for further positioning Gd-based probes in the field of Molecular Imaging, it is worth to recall the outstanding feed-back that coordination chemistry has gained from nearly two decades of development of paramagnetic agents for MRI. The search for very safe agents has led to concentrate huge efforts in the design and synthesis of a number of new ligands endowed with remarkably high thermodynamic and kinetic stability. As it was shown that the attainable relaxivity is quenched by the occurrence of slow exchange rates of the coordinated water, detailed studies have addressed the understanding of the determinants of the lifetime of a water molecule bound to a lanthanide(III) ion. Remarkably, it has been found that it is possible to modulate such exchange process over a range of six orders of magnitude by controlling the structure and the electric charge of the coordination cage.

In conclusion, the search for improved Gd(III)-based agents has been highly beneficial for the growth of lanthanide coordination chemistry and has created a very fertile interdisciplinary area with contributions from Chemistry, Biology, Medicine, and Imaging technology. The requisite for more efficient Gd(III)-based probes for more advanced MRI applications will be the driving force for further goals in lanthanide coordination chemistry.

References

1. Young, I. R. "*Methods in Biomedical Magnetic Resonance Imaging and Spectroscopy*"; John Wiley & Sons: Chichester **2000**.
2. Rinck, P. A. "*Magnetic Resonance in Medicine*"; ABW Wissenschaftsverlag GmbH: Berlin **2003**.
3. Merbach, A. E.; Tóth, E. "*The Chemistry of Contrast Agents in Medical Magnetic Resonance Imaging*"; John Wiley & Sons: Chichester **2001**.
4. Rummeny, E. J.; Marchal, G. *Acta Radiol.* **1997** *38*, 626–630.
5. Wang, Y. X. J.; Hussain S. M.; Krestin, G. P. *Eur. Radiol.* **2001** *11*, 2319–2331.
6. Reimer, P.; Balzer, T. *Eur. Radiol.* **2003** *13*, 1266–1276.
7. Caravan, P.; Ellison, J. J.; Mc Murry, T. J.; Lauffer, R. B. *Chem Rev.* **1999** *99*, 2293–2352.
8. Brücher, E.; Sherry, A. D. "*The Chemistry of Contrast Agents in Medical Magnetic Resonance Imaging*"; Eds. Merbach, A. E.; Tóth, E.; John Wiley & Sons: Chichester, **2001**, pp. 243–279.
9. Padhani, A. R. *JMRI* **2002** *16*, 407–422.
10. Lauffer, R. B. *Chem. Rev.* **1987** *87*, 901–927.
11. Aime, S.; Botta M.; Fasano, M.; Terreno, E. *Chem. Soc. Rev.* **1998** *27*, 19–29.
12. Tóth, É.; Helm, L.; Merbach, A. E. "*The Chemistry of Contrast Agents in Medical Magnetic Resonance Imaging*"; Eds. Merbach, A. E.; Tóth, E.; John Wiley & Sons: Chichester, **2001**, pp. 45–119.
13. Banci, L.; Bertini, I.; Luchinat, C. "*Nuclear and Electron Relaxation*"; VCH: Weinheim **1991**.
14. Bertini, I.; Luchinat, C. "*NMR of Paramagnetic Molecules in Biological Systems*"; Benjamin/Cummings: Menlo Park CA, **1986**.
15. Xu, J.; Franklin, S. J.; Whisenhunt, D. W.; Raymond, K. N. *J. Am. Chem. Soc.* **1995**, *1173*, 7245–7246.
16. Hajela, S.; Botta, M.; Girando, S.; Xu, J.; Raymond, K. N.; Aime, S. *J. Am. Chem. Soc.* **2000**, *122*, 11228–11229.
17. Cohen, S. M.; Xu, J.; Radkov, E.; Raymond, K. N.; Botta, M.; Barge, A.; Aime, S. *Inorg. Chem.* **2000** *39*, 5747–5756.
18. Aime, S.; Botta, M.; Fasano, M.; Terreno, E. *Acc. Chem. Res.* **1999** *32*, 941–949.
19. Aime, S.; Botta, M.; Fasano, M.; Paoletti, S.; Anelli, P. L.; Uggeri, F.; Virtuani, M. *Inorg. Chem.* **1994** *33*, 4707–4711.
20. Pubanz, D.; González, G.; Powell, D. H.; Merbach, A. E. *Inorg. Chem.* **1995** *34*, 4447–4453.
21. Aime, S.; Barge, A.; Bruce, J. I.; Botta, M.; Howard, J. A. K.; Moloney, J. M.; Parker, D.; de Sousa, A. S.; Woods, M. *J. Am. Chem. Soc.* **1999**, *121*, 5762–5771.
22. Powell, D. H.; Merbach, A. E.; Gonzalez, G.; Brücher, E.; Micskei, K.; Ottaviani, F.; Köhler, K.; von Zelewsky, A.; Grinberg, O. Y.; Lebedev, Y. S. *Helv. Chim. Acta* **1993** *76*, 2129–2146.
23. Powell, D. H.; Ni Dhubhghaill, O. M.; Pubanz, D.; Helm, L.; Lebedev, Y. S.; Schlaepfer, W.; Merbach, A. E. *J. Am. Chem. Soc.* **1996**, *118*, 9333–9346.
24. O'Sullivan, B.; Doble, D. M. J.; Thompson, M. K.; Siering, C.; Xu, J.; Botta, M.; Aime, S.; Raymond, K. N. *Inorg. Chem.* **2003** *42*, 2577–2583.
25. Freed, J. *J. Chem. Phys.* **1978**, *68*, 4034–4037.
26. Koenig, S. H.; Brown III, R. D. *Prog. NMR Spectrosc.* **1990**, *22*, 487–567.
27. Kim, W. D.; Kiefer, G. E.; Maton, F.; Mc Millan, K.; Muller, R. N.; Sherry, A. D. *Inorg. Chem.* **1995** *34*, 2233–2243.
28. Chen, J. W.; Belford, R. L.; Clarkson, R. B. *J. Phys. Chem.* **1998**, *102*, 2117–2130.
29. Borel, A.; Helm, L.; Merbach, A. E. *Chem. Eur. J.* **2001** *7*, 600–610.
30. Aime, S.; Batsanov, A. S.; Botta, M.; Howard, J. A. K.; Parker, D.; Senanayake, K.; Williams, G. *Inorg. Chem.* **1994** *33*, 4696–4706.

31. Easteal, A. J. "*Measurements of the Transport Properties of Fluids*" vol. III, Eds. Wakeham, W. A.; Nagashima, A.; Sengers, J. V.; Blackwell Scientific Publications: Oxford, **1994**, p. 265.
32. Botta, M.; Ravera, M.; Barge, A.; Bottaro, M.; Osella, D. *Dalton Trans.* **2003**, 1628–1633.
33. Botta, M. *Eur. J. Inorg. Chem.* **2000**, 399–407.
34. Aime, S.; Botta, M.; Parker, D.; Williams, G. J. A. *Dalton Trans.* **1996**, 17–23.
35. Aime, S.; Botta, M.; Crich, S. G.; Giovenzana, G.; Pagliarin, R.; Sisti, M.; Terreno, E. *Magn. Reson. Chem.* **1998**, *36*, S200–S208.
36. Aime, S.; Barge, A.; Botta, M.; Parker, D.; de Sousa, A. S. *J. Am. Chem. Soc.* **1997**, *119*, 4767–4768.
37. Woods, M.; Zhang, S. R.; Ebron, V.; Sherry, A. D. *Chem. Eur. J.* **2003** *9*, 4634–4640.
38. Aime, S.; Botta, M.; Garino, E.; Geninatti Crich, S.; Giovenzana, G.; Pagliarin, R.; Palmisano, G.; Sisti, M. *Chem. Eur. J.* **2000** *6*, 2609–2617.
39. Helm, L.; Tóth, E.; Merbach, A.E. "*Metal Ions in Biological Systems*"; Eds. Sigel, A.; Sigel, H.; Marcel Dekker, Inc.: New York, **2003**, Chapter 15, pp. 589–641.
40. Tóth, E.; Helm, L.; Merbach, A. E. *Top. Curr. Chem.* **2002** *221*, 61–101.
41. Bruce, J. I.; Lowe, M. P.; Parker, D. "*The Chemistry of Contrast Agents in Medical Magnetic Resonance Imaging*"; Eds. Merbach, A. E.; Tóth, E.; John Wiley & Sons: Chichester, **2001**, pp. 437–460.
42. Horrocks, W. deW. Jr, Sudnick, D. R. *Acc. Chem. Res.* **1981** *14*, 384–395.
43. Beeby, A.; Clarkson, I. M.; Dickins, R. S.; Faulkner, S.; Parker, D.; Royle, L.; de Sousa, A. S.; Williams, J. A. G.; Woods, M. *J. Chem. Soc., Perkin Trans 2* **1999**, 493–499.
44. Alpoim, M. C.; Urbano, A. M.; Geraldes, C. F. G. C.; Peters, J. A. *J. Chem. Soc., Dalton Trans.* **1992**, 463–468.
45. Caravan, P.; Astashkin, A. V.; Raitsimring, A. M. *Inorg. Chem.* **2003** *42*, 3972–3974.
46. Aime, S.; Botta, M.; Ermondi, G. *J. Magn. Reson.* **1991**, *92*, 572–580.
47. Lammers, H.; Maton, F.; Pubanz, D.; Van Laren, M. W.; Van Bekkum, H.; Merbach, A. E.; Muller, R. N.; Peters, J. A. *Inorg. Chem.* **1997** *36*, 2527–2532.
48. Dunand, F. A.; Borel, A.; Merbach, A. E. *J. Am. Chem. Soc.* **2002**, *124*, 710–716.
49. Micksei, K.; Helm, L.; Brücher, E.; Merbach, A. E. *Inorg. Chem.* **1993** *32*, 3844–3850.
50. Rast, S.; Fries, P. H.; Belorizky, E. *J. Chem. Phys.* **2000**, *113*, 8724–8735.
51. Rast, S.; Borel, A.; Helm, L.; Belorizky, E.; Fries, P. H.; Merbach, A. E. *J. Am. Chem. Soc.* **2001**, *123*, 2637–2644.
52. Messeri, D.; Lowe, M.P.; Parker, D.; Botta, M. *Chem. Comm.* **2001**, 2742–2743.
53. Terreno, E.; Boniforte, P.; Botta, M.; Fedeli, F.; Milone, L.; Mortillaro, A.; Aime, S. *Eur. J. Inorg. Chem.* **2003** *19*, 3530–3533.
54. Aime, S.; Botta, M.; Crich, S. G.; Giovenzana, G. B.; Pagliarin, R.; Piccinini, M.; Sisti, M.; Terreno, E. *J. Biol. Inorg. Chem.* **1997**, *2*, 470–479.
55. Aime, S.; Botta, M.; Frullano, L.; Geninatti Crich, S.; Giovenzana, G.; Pagliarin, R.; Palmisano, G.; Riccardi Sirtori, F.; Sisti, M. *J. Med. Chem.* **2000**, *43*, 4017–4024.
56. Aime, S.; Botta, M.; Fedeli, F.; Gianolio, E.; Terreno, E.; Anelli, P. L. *Chem. Eur. J.* **2001** *7*, 5262–5269.
57. Okumura, M.; Mikawa, M.; Yokawa, T.; Kanazawa, Y.; Kato, H.; Shinohara, H. *Acad. Radiol.* **2002** *9*, S495–S497.
58. Bolskar, R. D.; Benedetto, A. F.; Husebo, L. O.; Price, R. E.; Jackson, E. F.; Wallace, S.; Wilson, L. J.; Alford, J. M. *J. Am. Chem. Soc.* **2003**, *125*, 5471–5478.
59. Funasaka, H.; Sakurai, K.; Oda, Y.; Yamamoto, K.; Takahashi, T. *Chem. Phys. Lett.* **1995** *232*, 273–277.
60. Aime, S.; Frullano, L.; Geninatti Crich, S. *Angew. Chemie Int. Ed.* **2002** *41*, 1017–1019.
61. Aime, S.; Botta, M.; Fasano, M.; Terreno, E. "*The Chemistry of Contrast Agents in Medical Magnetic Resonance Imaging*" Eds. Merbach, A.E.; Tóth, E.; John Wiley & Sons: Chichester, 2001, pp. 193–242 and references therein.
62. Sieving, P. F.; Watson, A. D.; Rocklage, S. M. *Bioconjugate Chem.* **1990** *1*, 65–70.
63. Schuhmann-Giampieri, G.; Schmitt-Willich, H.; Frenzel, T.; Press, W. R.; Weinmann, H. J. *Invest. Radiol.* **1991** *26*, 969–974.

64. Desser, T.; Rubin, D.; Muller, H.; Qing, F.; Khodor, S.; Zanazzi, Y.; Ladd, D.; Wellons, J.; Kellar, K.; Toner, J.; Snow, R. J. *Magn. Reson. Imaging* **1994** *4*, 467–472.

65. Vexler, V. S.; Clement, O.; Schmitt-Willich, H.; Brasch, R. C. *JMRI* **1994** *4*, 381–388.

66. Aime, S.; Botta, M.; Crich, S. G.; Giovenzana, G.; Palmisano, G.; Sisti, M. *Bioconjugate Chem.* **1999** *10*, 192–199.

67. Lu, Z. R.; Wang, X. H.; Parker, D. L.; Goodrich, K. C.; Buswell, H. R. *Bioconjugate Chem.* **2003** *14*, 715–719.

68. Heckl, S.; Pipkorn, R.; Waldeck, W.; Spring, H.; Jenne, J.; von der Lieth, C. W.; Corban-Wilhelm, H.; Debus, J.; Braun, K. *Cancer Res.* **2003** *63*, 4766–4772.

69. Corot, C.; Schaefer, M.; Beaute, S.; Bourrinet, P.; Zehaf, S.; Benize, V.; Sabatou, M.; Meyer, D. *Acta Radiol.* **1997**, *412* (Suppl.), 91–99.

70. Casali, C.; Janier, M.; Canet, E.; Obadia, J. F.; Benderbous, S.; Corot, C.; Revel, D. *Acad. Radiol.* **1998**, *5* (Suppl. 1), S214–S218.

71. Meyer, D.; Schaefer, M.; Bouillot, A.; Beaute, S.; Chambon, C. *Invest. Radiol.* **1991** *26*, S50–S52.

72. Rebizak, R.; Schaefer, M.; Dellacherie, E. *Eur. J. Pharm. Sci.* **1999** *7*, 243–248.

73. Helbich, T. H.; Gossman, A.; Mareski, P. A.; Raduchel, B.; Roberts, T. P. L.; Shames, D. M.; Muhler, M.; Turetschek, K.; Brasch, R. C. *JMRI* **2000** *11*, 694–701.

74. Sun, G.; Feng, J.; Jing, F.; Pei, F.; Liu, M. *J. Magn. Magn. Mat.* **2003**, *265*, 123–129.

75. Aime, S.; Barbero, L.; Botta, M. *Magn. Reson. Imaging* **1991** *9*, 843–847.

76. Andrè, J. P.; Toth, E.; Maecke, H. R.; Merbach, A. E. *Chem. Eur. J.* **1999** *5*, 2977–2983.

77. Nicolle, G. M.; Toth, E.; Eisenwiener, K. P.; Maecke, H. R.; Merbach, A. E. *J. Biol. Inorg. Chem.* **2002**, *7*, 757–769.

78. Tournier, H.; Hyacinthe, R.; Schneider, M. *Acad. Radiol.* **2002**, *9* (Suppl. 1), S20–S28.

79. Torchilin, V. P. *Adv. Drug Deliv. Rev.* **2002** *54*, 235–252.

80. Glogard, C.; Hovland, R.; Fossheim, S. L.; Aasen, A. J.; Klaveness, J. *J. Chem. Soc. Perkin Trans. 2* **2000**, *5*, 1047–1052.

81. Alhaique, F.; Bertini, I.; Fragai, M.; Carafa, M.; Luchinat, C.; Parigi, G. *Inorg. Chim. Acta* **2002** *331*, 151–157.

82. Glogard, C.; Stensrud, G.; Hovland, R.; Fossheim, S. L.; Klaveness, J. *Int. J. Pharm.* **2002** *233*, 131–140.

83. Suga, K.; Mikawa, M.; Ogasawara, N.; Okazaki, H.; Matsunaga, N. *Invest. Radiol.* **2001** *36*, 136–145.

84. Weissig, V.; Babich, J.; Torchilin, V. *Coll. Surf. Biointerfaces* **2000** *18*, 293–299.

85. Fossheim, S. L.; Fahlvik, A. K.; Klaveness, J.; Muller, R. N. *Magn. Reson. Imaging* **1999** *17*, 83–89.

86. Bryant, L. H.; Brechbiel, M. W.; Wu, C. C.; Bulte, J. W. M.; Herynek, V.; Frank, J. A. *JMRI* **1999** *9*, 348–352.

87. Margerum, L. D.; Campion, B. K.; Koo, M.; Shargill, N.; Lai, J. J.; Marumoto, A.; Sontum, P. C. *J. Alloys Comp.* **1997**, *249*, 185–190.

88. Wiener, E. C.; Brechbiel, M. W.; Brothers, H.; Magin, R. L.; Gansow, O. A.; Tomalia, D. A.; Lauterbur, P. C. *Magn. Reson. Med.* **1994** *31*, 1–8.

89. Tacke, J.; Adam, G.; Classen, H.; Muhler, A.; Prescher, A.; Gunther, R. W. *JMRI* **1997** *7*, 678–682.

90. Roberts, H. C.; Saeed, M.; Roberts, T. P. L.; Muhler, A.; Shames, D. M.; Mann, J. S.; Stiskal, M.; Demsar, F.; Brasch, R. C. *JMRI* **1997** *7*, 331–338.

91. Dong, Q.; Hurst, D. R.; Weinmann, H. J.; Chenevert, T. L.; Londy, F. J.; Prince, M. R. *Invest. Radiol.* **1998** *33*, 699–708.

92. Tóth, E.; Pubanz, D.; Vauthey, S.; Helm, L.; Merbach, A. E. *Chem. Eur. J.* **1996** *2*, 1607–1615.

93. Kobayashi, H.; Brechbiel, M. W. *Mol. Imaging* **2003** *2*, 1–10.

94. Tóth, E.; Helm, L.; Kellar, K. E.; Merbach, A. E. *Chem. Eur. J.* **1999** *5*, 1202–1211.

95. Toth, E.; van Uffelen, I.; Helm, L.; Merbach, A. E.; Ladd, D.; Briley-Saebo, K.; Kellar, K. E. *Magn. Reson. Chem.* **1998** *36*, S125–S134.

96. Duarte, M. G.; Gil, M. H.; Peters, J. A.; Colet, J. M.; Vander Elst, L.; Muller, R. N.; Geraldes, C. F. G. C. *Bioconjugate Chem.* **2001** *12*, 170–177.

97. Comblin, V.; Gilsoul, D.; Hermann, M.; Humblet, V.; Jacques, V.; Mesbahi, M.; Sauvage, C.; Desreux, J. F. *Coord. Chem. Rev.* **1999** *186*, 451–470.
98. Aime, S.; Cabella, C.; Colombatto, S.; Crich, S. G.; Gianolio, E.; Maggioni, F. *JMRI* **2002** *16*, 394–406.
99. Bogdanov, A.; Matuszewski, L.; Bremer, C.; Petrovski, A.; Weissleder, R. *Molec. Imag.* **2002** *1*, 16–23.
100. Port, M.; Meyer, D.; Bonnemain, B.; Corot, C.; Schaefer, M.; Rousseaux, O.; Simonot, C.; Bourrinet, P.; Benderbous, S.; Dencausse, A.; Devoldere, L. *MAGMA* **1999** *8*, 172–176.
101. Vander Elst, L.; Port, M.; Raynal, I.; Simonot, C.; Muller, R.N. *Eur. J. Inorg. Chem.* **2003**, 2495–2501.
102. Nicolle, G. M.; Tóth, E.; Schmitt-Willich, H.; Radüchel, B.; Merbach, A. E. *Chem. Eur. J.* **2002** *8*, 1040–1048.
103. Aime, S.; Gianolio, E.; Terreno, E.; Giovenzana, G. B.; Pagliarin, R.; Sisti, M.; Palmisano, G.; Botta, M.; Lowe, M. P.; Parker, D. *J. Biol. Inorg. Chem.* **2000**, *5*, 488–497.
104. Aime, S.; D'Amelio, N.; Fragai, M.; Lee, Y. M.; Luchinat, C.; Terreno, E.; Valensin, G. *J. Biol. Inorg. Chem.* **2002**, *7*, 617–622.
105. Parker, D.; Dickins, R. S.; Puschmann, H.; Crossland, C.; Howard, J. A. K. *Chem. Rev.* **2002** *102*, 1977–2010.
106. Aime, S.; Barge, A.; Bruce, J. I.; Botta, M.; Howard, J. A. K.; Moloney, J. M.; Parker, D.; de Sousa, A. S.; Woods, M. *J. Am. Chem. Soc.* **1999**, *121*, 5762–5771.
107. Woods, M.; Aime, S.; Botta, M.; Howard, J. A. K.; Moloney, J. M.; Navet, M.; Parker, D.; Port, M.; Rousseaux, O. *J. Am. Chem. Soc.* **2000**, *122*, 9781–9792.
108. Pubanz, D.; Gonzalez, G.; Powell, D. H.; Merbach, A. E. *Inorg. Chem.* **1995** *34*, 4447–4453.
109. Parker, D.; Puschmann, H.; Batsanov, A. S.; Senanayake, K. *Inorg. Chem.* **2003**, *42*, 8646–8651.
110. Aime, S.; Barge, A.; Batsanov, A.S.; Botta, M.; Castelli, D.D.; Fedeli, F.; Mortillaro, A.; Parker, D.; Puschmann, H. *Chem. Comm.* **2002**, 1120–1121.
111. Aime, S.; Barge, A.; Borel, A.; Botta, M.; Chemerisov, S.; Merbach, A. E.; Muller, U.; Pubanz, D. *Inorg. Chem.* **1997** *36*, 5104–5112.
112. Barge, A.; Botta, M.; Parker, D.; Puschmann, H. *Chem. Comm.* **2003**, 1386–1387.
113. Thompson, M. K.; Botta, M.; Nicolle, G.; Helm, L.; Aime, S.; Merbach, A. E.; Raymond, K. N. *J. Am. Chem. Soc.* **2003**, *125*, 14274–14275.
114. Toth, E.; Connac, F.; Helm, L.; Adamzli, K.; Merbach, A. E. *J. Biol. Inorg. Chem.* **1998**, *3*, 606–613.
115. Aime, S.; Chiaussa, M.; Digilio, G.; Gianolio, E.; Terreno, E. *J. Biol. Inorg. Chem.* **1999**, *4*, 766–774.
116. Jacques, V.; Desreux, J. F. *Top. Curr. Chem.* **2002** *221*, 123–164.
117. Meade, T. J.; Taylor, A. K.; Bull, S. R. *Curr. Opin. Neurobiol.* **2003** *13*, 597–602.
118. Zhang, S. R.; Merritt, M.; Woessner, D. E.; Lenkinski, R. E.; Sherry, A. D. *Acc. Chem. Res.* **2003** *36*, 783–790.
119. Aime, S.; Barge, A.; Castelli, D. D.; Fedeli, F.; Mortillaro, A.; Nielsen, F. U.; Terreno, E. *Magn. Reson. Med.* **2002** *47*, 639–648.
120. Aime, S.; Delli Castelli, D.; Terreno, E. *Angew. Chemie Int. Ed.* **2002** *41*, 4334–4336.
121. Raghunand, N.; Altbach, M. I.; Van Sluis, R.; Bagget, B.; Taylor, C. W.; Bhuiwalla, Z. M.; Gillies, R. J. *Biochem. Pharmacol.* **1999** *57*, 309–312.
122. Gerweck, L. E.; Seetharaman, K. *Cancer Res.* **1996** *56*, 1194–1199.
123. Tannock, I. F.; Rotin, D. *Cancer Res.* **1989** *49*, 4373–4384.
124. Naghavi, M.; John, R.; Naguib, S.; Siadaty, M. S.; Grasu, R.; Kurian, K. C.; van Winkle, W. B.; Soller, B.; Litovsky, S.; Madjid, M.; Willerson, J. T.; Casscells, W. *Atherosclerosis* **2002** *164*, 27–35.
125. Aime, S.; Botta, M.; Milone, L.; Terreno, E. *Chem. Comm.* **1996**, 1265–1266.
126. Zhou, J. Y.; Payen, J. F.; Wilson, D. A.; Traystman, R. J.; van Zijl, P. C. M. *Nat. Med.* **2003** *9*, 1085–1090.
127. Pilatus, U.; Aboagye, E.; Artemov, D.; Mori, N.; Ackerstaff, E.; Bhujwalla, Z. M. *Magn. Reson. Med.* **2001** *45*, 749–755.

128. Sun, Y.; Sugawara, M.; Mulkern, R. V.; Hynynen, K.; Mochizuki, S.; Albert, M.; Zuo, C. S. *NMR Biomed.* **2000** *13*, 460–466.
129. Ward, K. M.; Balaban, R. S. *Magn. Reson. Med.* **2000** *44*, 799–802.
130. Lowe, M. P.; Parker, D.; Reany, O.; Aime, S.; Botta, M.; Castellano, G.; Gianolio, E.; Pagliarin, R. *J. Am. Chem. Soc.* **2001**, *123*, 7601–7609.
131. Zhang, S. R.; Wu, K. C.; Sherry, A. D. *Angew. Chemie Int. Ed.* **1999** *38*, 3192–3194.
132. Raghunand, N.; Howison, C.; Sherry, A. D.; Zhang, S. R.; Gillies, R. J. *Magn. Reson. Med.* **2003** *49*, 249–257.
133. Aime, S.; Botta, M.; Crich, S.G.; Giovenzana, G.; Palmisano, G.; Sisti, M. *Chem. Comm.* **1999**, 1577–1578.
134. Lokling, K. E.; Fossheim, S. L.; Skurtveit, R.; Bjornerud, A.; Klaveness, J. *Magn. Reson. Imag.* **2001** *19*, 731–738.
135. Lokling, K. E.; Skurtveit, R.; Fossheim, S. L.; Smistad, G.; Henriksen, I.; Klaveness, J. *Magn. Reson. Imag.* **2003** *21*, 531–540.
136. Quesson, B.; de Zwart, J. A.; Moonen, C. T. W. *JMRI* **2000** *12*, 525–533.
137. Fossheim, S. L.; Il'yasov, K. A.; Hennig, J.; Bjornerud, A. *Acad. Radiol.* **2000** *7*, 1107–1115.
138. Fossheim, S. L.; Fahlvik, A. K.; Klaveness, J.; Muller, R. N. *Magn. Reson. Imag.* **1999** *17*, 83–89.
139. Aime, S.; Ascenzi, P.; Comoglio, E.; Fasano, M.; Paoletti, S. *J. Am. Chem. Soc.* **1995**, *117*, 9365–9366.
140. Glogard, C.; Stensrud, G.; Aime, S. *Magn. Reson. Chem.* **2003** *41*, 585–588.
141. Aime, S.; Botta, M.; Gianolio, E.; Terreno, E. *Angew. Chemie Int. Ed.* **2000** *39*, 747–748.
142. Seibig, S.; Toth, E.; Merbach, A. E. *J. Am. Chem. Soc.* **2000**, *122*, 5822–5830.
143. Burai, L.; Toth, E.; Seibig, S.; Scopelliti, R.; Merbach, A. E. *Chem. Eur. J.* **2000** *6*, 3761–3770.
144. Burai, L.; Toth, E.; Moreau, G.; Sour, A.; Scopelliti, R.; Merbach, A. E. *Chem. Eur. J.* **2003** *9*, 1394–1404.
145. Anelli, P.L.; Bertini, I.; Fragai, M.; Lattuada, L.; Luchinat, C.; Parigi, G. *Eur. J. Inorg. Chem.* **2000**, 625–630.
146. Lauffer, R.B.; Mc Murry, T.J.; Dunham, S.O.; Scott, D.M.; Parmelee, D.J.; Dumas, S. PCT Int. Appl. WO9736619, **1997**.
147. Nivorozhkin, A. L.; Kolodziej, A. F.; Caravan, P.; Greenfield, M. T.; Lauffer, R. B.; McMurry, T. J. *Angew. Chemie Int. Ed.* **2001** *40*, 2903–2906.
148. Moats, R. A.; Fraser, S. E.; Meade, T. J. *Angew. Chemie Int. Ed.* **1997** *36*, 726–728.
149. Louie, A. Y.; Huber, M. M.; Ahrens, E. T.; Rothbacher, U.; Moats, R.; Jacobs, R. E.; Fraser, S. E.; Meade, T. J. *Nat. Biotechnol.* **2000** *18*, 321–325.
150. Aime, S.; Dastru, W.; Crich, S. G.; Gianolio, E.; Mainero, V. *Biopolymers* **2002** *66*, 419–428.
151. Henriksen, O.; Weislander, S.; Gierris, F.; Jensen, K. M. *Acta Radiol.* **1991** *32*, 95–100.
152. Castillo, M.; Kwock, L. "*Proton MR Spectroscopy of Common Brain Tumors in Neuroimaging clinics of North America*" vol. 8, Ed. Castillo M. W. B.; Saunders Company: Philadelphia, **1998**, pp. 733–752.
153. Grodd, W.; Krägeloh-Mann, I.; Klose, U.; Santer, R. *Radiology* **1991** *181*, 173–180.
154. VanderElst, L.; Vanhaverbeke, Y.; Goudemant, J. F.; Muller, R. N. *Magn. Reson. Med.* **1994** *31*, 437–444.
155. Bruce, J. I.; Dickins, R. S.; Govenlock, L. J.; Gunnlaugsson, T.; Lopinski, S.; Lowe, M. P.; Parker, D.; Peacock, R. D.; Perry, J. J. B.; Aime, S.; Botta, M. *J. Am. Chem. Soc.* **2000**, *122*, 9674–9684.
156. Aime, S.; Botta, M.; Mainero, V.; Terreno, E. *Magn. Reson. Med.* **2002** *47*, 10–13.
157. Aime, S.; Delli Castelli, D.; Fedeli, F.; Terreno, E. *J. Am. Chem. Soc.* **2002**, *124*, 9364–9365.
158. Faulkner, S.; Burton-Pye, B.P.; Khan, T.; Martin, L.R.; Wray, S.D.; Skabara, P. J. *Chem. Comm.* **2002**, 1668–1669.
159. Dickins, R. S.; Love, C. S.; Puschmann, H. *Chem. Comm.* **2001** *22*, 2308–2309.
160. Terreno, E.; Botta, M.; Fedeli, F.; Mondino, B.; Milone, L.; Aime, S. *Inorg. Chem.* **2003** *42*, 4891–4897.
161. Botta, M.; Aime, S.; Barge, A.; Bobba, G.; Dickins, R. S.; Parker, D.; Terreno, E. *Chem. Eur. J.* **2003** *9*, 2102–2109.
162. Li, W.; Fraser, S.; Meade, T. J. *J. Am. Chem. Soc.* **1999**, *121*, 1413–1414.

163. Grynkiewicz, G.; Poenie, M.; Tsien, R. Y. *J. Biol. Chem.* **1985**, *260*, 3440–3448.
164. Hanaoka, K.; Kikuchi, K.; Urano, Y.; Nagano, T. *J. Chem. Soc. Perkin Trans. 2* **2001**, 1840–1843.
165. Lawrence, Y.; Ozil, J. P.; Swann, K. *Biochem. J.* **1998** *335*, 335–341.
166. Weissleder, R.; Mahmood, U. *Radiology* **2001** *219*, 316–333.
167. Sipkins, D. A.; Cheresh, D. A.; Kazemi, M. R.; Nevin, L. M.; Bednarski, M. D.; Li, K. C. P. *Nat. Med.* **1998** *4*, 623–626.
168. Winter, P. M.; Caruthers, S. D.; Kassner, A.; Harris, T. D.; Chinen, L. K.; Allen, J. S.; Lacy, E. K.; Zhang, H. Y.; Robertson, J. D.; Wickline, S. A.; Lanza, G. M. *Cancer Res.* **2003** *63*, 5838–5843.
169. Artemov, D.; Mori, N.; Ravi, R.; Bhujwalla, Z. M. *Cancer Res.* **2003** *63*, 2723–2727.
170. Geninatti Crich, S.; Biancone, L.; Cantaluppi, U.; Duò, D.; Esposito, G.; Russo, S.; Camussi, G.; Aime, S. *Magn. Reson. Med.*, accepted for publication.
171. Aime, S.; Cavalli, R.; Cravotto, G.; Gianolio, E. *unpublished results.*
172. Osterloh, K.; Aisen, P. *Biochem. Biophys. Acta* **1996** *1275*, 161–203.
173. Wiener, E. C.; Konda, S.; Shadron, A.; Brechbiel, M.; Gansow, O. *Invest. Radiol.* **1997** *32*, 748–754.
174. Bhorade, R.; Weissleder, R.; Nakakoshi, T.; Moore, A.; Tung, C. H. *Bioconjugate Chem.* **2000** *11*, 301–305.
175. Allen, M. J.; Meade, T. J. *J. Biol. Inorg. Chem.* **2003**, *8*, 746–750.

RELAXATION BY METAL-CONTAINING NANOSYSTEMS

R. N. MULLER[a], L. VANDER ELST[a], A. ROCH[a], J. A. PETERS[b], E. CSAJBOK[b], P. GILLIS[c] and Y. GOSSUIN[c]

[a]NMR and Molecular Imaging Laboratory, Department of Organic and Biomedical Chemistry, University of Mons-Hainaut, Avenue du Champ de Mars 24, 7000 Mons, Belgium

[b]Laboratory of Applied Organic Chemistry and Catalysis, Delft University of Technology, Julianalaan 136, 2628 BL, Delft, The Netherlands

[c]Biological Physics Department, University of Mons-Hainaut. Avenue du Champ de Mars 24, 7000 Mons, Belgium

I. Super-Paramagnetic Particles

A. Introduction

Super-paramagnetic (spm) crystals are used as contrast agents for Magnetic Resonance Imaging (MRI). Improving their efficiency requires a

ADVANCES IN INORGANIC CHEMISTRY
VOLUME 57 ISSN 0898-8838 / DOI 10.1016/S0898-8838(05)57005-3

better knowledge of their physical and morphological properties, which can be deduced from experimental and theoretical studies of proton relaxation in ferrofluids.

Colloidal suspensions of super-paramagnetic nanocrystals are candidates for the development of new intelligent contrast agents, opening an early detection of several pathologies. Firstly, it is because of their very high transverse relaxivity (*1*), which is the parameter defining the efficiency of a negative contrast agent in MRI (*2,3*). Figure 1 shows images of a pathological liver before and after administration of such a super-paramagnetic contrast agent, with obvious differences.

A second argument supports the development of ferrofluid contrast agents is the successful grafting of molecular entities on super-paramagnetic particles targeting very specific cellular receptors, as attested by numerous studies. The pioneering work of Weissleder's team aimed at grafting antigen on super-paramagnetic particles is a remarkable example of this new application (*4,5*): a few thousands of active iron atoms can indeed be fixed onto a specific receptor, which is much more than the unique one being possibly fixed onto a classical molecular paramagnetic contrast agent.

The optimization of the efficiency of these "intelligent" contrast agents requires a good knowledge of the relationship between proton relaxation, and physical and morphological properties of the particle.

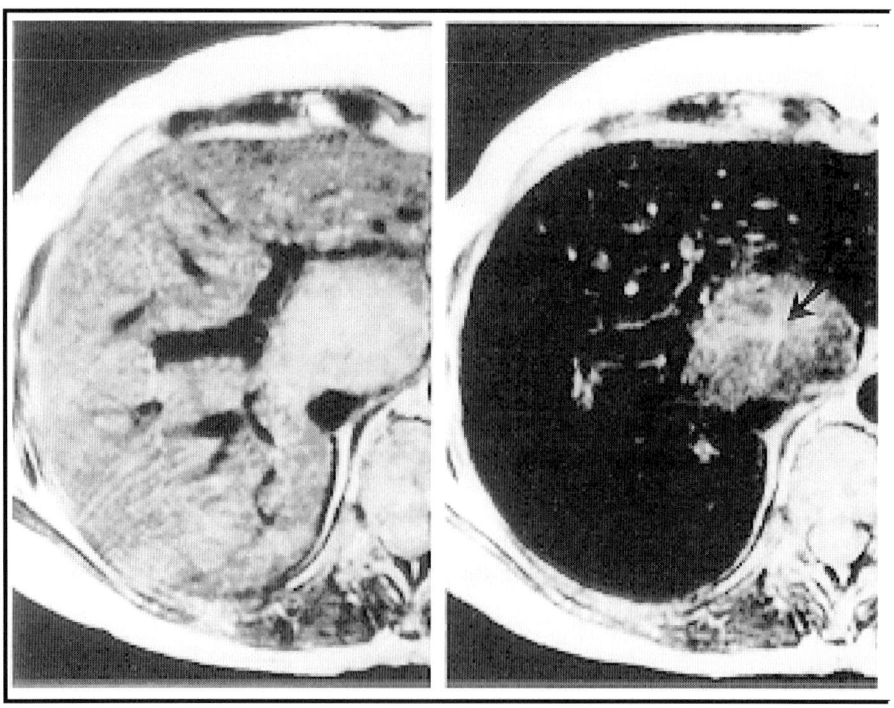

FIG. 1. Liver MR images before and after administration of a super-paramagnetic contrast agent.

The purpose of this section is to review the parameters influencing the proton relaxation of a nanomagnet suspension. It will include an analysis of NMRD profiles, which provide the relaxivity dependence with the external field, expressed in proton Larmor frequency units.

Ferrofluid NMR studies can also be used in order to determine geometrical and physical properties of the super-paramagnetic crystals, like their specific magnetization or radius. They also give valuable information on the aggregation level and on anisotropy.

Ferrofluids are ferro- or ferrimagnetic, i.e., each particle has a spontaneous magnetic moment arising from the alignment of a macroscopic ensemble of electronic magnetic moments. Bulk ferromagnetic materials are organized in micron size, fully magnetized domains. Those domains generally have magnetizations with different directions, so that a bulk ferromagnetic material is not a magnet except if it has previously been submitted to a magnetic field (*6*). In ferrofluids, super-paramagnetic crystals are nanometer sized, i.e., much smaller than ferromagnetic domains. Each crystal is thus a single monodomain and completely magnetized, which explains why it may be called a nanomagnet.

In spm crystals, the aligned spins are not allowed to point arbitrarily towards any direction regarding the crystal structure, which favors some axes against others: aligning onto directions called "easy directions" is less energy consuming than aligning onto other directions.

The anisotropy field arises from four contributions, and it may furthermore be influenced by the extent of crystal agglomeration:

(1) The first one is the bulk magneto crystalline anisotropy field. It depends on the chemical composition and crystallographic structure of the material. For instance, the anisotropy constant is 30 times larger in cobalt ferrite particles than in magnetite particles.
(2) The second one is the demagnetizing field, which is determined by the shape of the crystal. This component of the anisotropy is equal to zero for a sphere and increases with the elongation of the crystal.
(3) The anisotropy constant depends also on the surface structure of the crystal.
(4) For agglomerated structures, the dipolar interaction between two neighboring crystals contributes to the anisotropy energy. This contribution increases when the inter-crystal distance decreases.

Models generally only account for a uni-axial symmetry. Although rather crude if applied to ferrite crystals, this assumption is reasonable for systems of higher symmetry based on cubic models, since some of the contributions mentioned above, namely the demagnetizing field, is often characterized by a uni-axial symmetry. Within that approximation, the energy needed to have the magnetic moment making an angle θ with the unique easy direction may be written as:

$$E_{\mathrm{A}} = K_{\mathrm{A}} V \sin^2\theta,$$

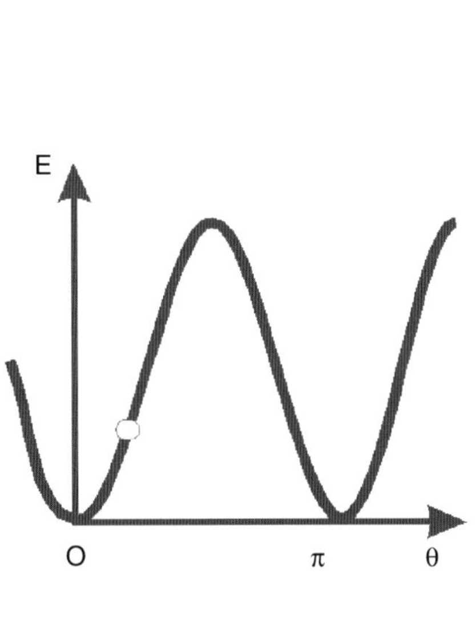

FIG. 2. Magnetic energy of a spm crystal as a function of the tilt angle between the magnetization and the easy axis of uni-axial anisotropy.

where V is the crystal volume (see Fig. 2) (*7*). $K_A V$ is called the anisotropy energy, and K_A, the proportionality constant, is called the anisotropy constant (*8*).

Anisotropy is also associated to a magnetic field, whose modulus is $B_A = 2K_A/M_S$, M_S being the crystal magnetization. The precession frequency of the magnetic moment into this field corresponds to the frequency of the peak characterizing the distribution of the transition frequencies.

B. Phenomenological Approach of the Proton Relaxation in a Ferrofluid

As illustrated in Fig. 3, the proton relaxation in super-paramagnetic colloids occurs because of the fluctuations of the dipolar magnetic coupling between the nanocrystal magnetization and the proton spin. The relaxation rate increases with the fluctuation correlation time and with the magnitude of this fluctuation. Different processes cause the fluctuation of the magnetic interaction.

One is translational diffusion, which produces a variation of the distance between the particle and the water molecule. The translation correlation time is the time for the water molecule to diffuse a crystal radius:

$\tau_D = r^2/D$ where r is the crystal radius and D the water diffusion coefficient.

Another process responsible for a fluctuation of the local magnetic field is Néel relaxation. It corresponds to the flip of the crystal magnetization vector from one easy direction of anisotropy to another. The correlation time of this

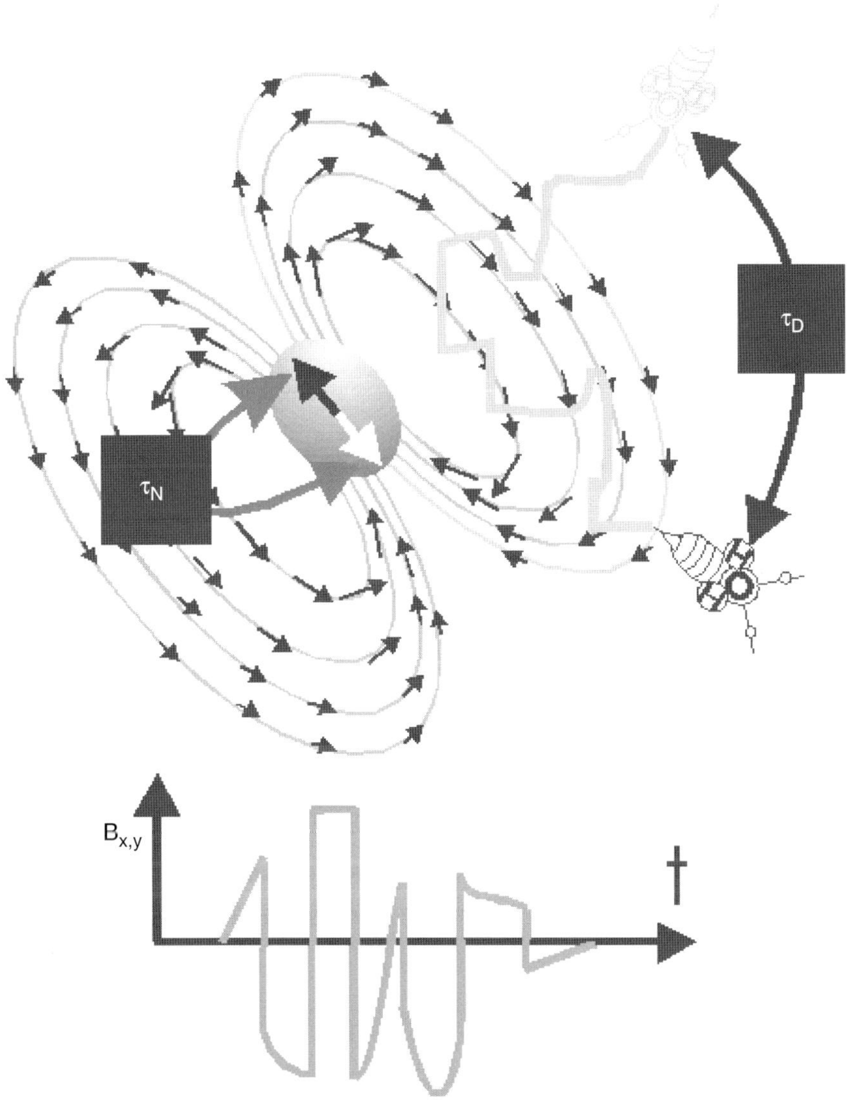

FIG. 3. Illustration of the origin of proton nuclear magnetic relaxation induced by a super-paramagnetic crystal. The water molecule (symbolized by a bee) experiences a magnetic field which fluctuates because of the translational diffusion and because of Néel relaxation. The bottom curve represents a typical time evolution of this field.

fluctuation is called τ_N, the Néel relaxation time of the super-paramagnetic crystal.

Combining both processes results in defining a global correlation time τ_C:

$$1/\tau_C = 1/\tau_D + 1/\tau_N \tag{1}$$

Relaxation is then generally governed by the equations of Freed (*9*). In the special case where the translational diffusion correlation time is much shorter than the Néel relaxation time, τ_C is dominated by diffusion and the equations of Freed reduce to the earlier equations of Ayant (*10*).

The ability of a fluctuation to relax the proton spins depends on whether its correlation time is longer or shorter than the precession period of the spin within the external magnetic field B_0: if τ_C is longer than this period, the fluctuation is averaged by the precession and it is inefficient, while it is efficient in the opposite situation. Furthermore, electron polarization may itself be relaxed or not relaxed by the same fluctuation, depending on how its correlation time is situated (longer or shorter) regarding the electron spin precession period. The equation $\omega_I\tau_C = 1$ thus defines a boundary between two domains, one where the fluctuation characterized by a correlation time τ_C induces relaxation ($\omega_I\tau_C < 1$) and one where it does not ($\omega_I\tau_C > 1$), where ω_I is the angular frequency of the proton precession. The indirect influence of the electron spin relaxation is also visible through the dispersion centered around $\omega_e\tau_C = 1$, where ω_e is the electron spin angular frequency; ω_e being 658 times faster than ω_I, the center of the electron dispersion points towards fluctuations much faster than the center of the proton dispersion, if the same static external magnetic field is applied to the system. If this field is varied, from low to high values, the same fluctuation will first cause an indirect relaxation dispersion, at low field, through the electron relaxation, and then another dispersion at higher field (*11*).

Relaxation induced by super-paramagnetic crystals is moreover complicated by another feature: the influence of the electron magnetic moment is modulated by Néel relaxation, which depends on the crystal anisotropy (see Fig. 4).

For large super-paramagnetic crystals or for crystals with a very high anisotropy constant (*12*), the anisotropy energy is larger than the thermal energy, which maintains the direction of the crystal magnetic moment very

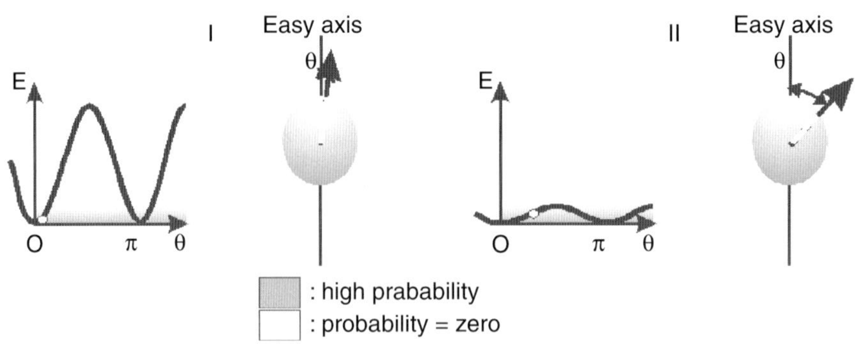

FIG. 4. Constraints exerted by the anisotropy field upon the orientation of the magnetization vector. (I) High anisotropy: magnetization is almost locked onto the easy axis. (II) Low anisotropy: thermal energy is sufficient to move the magnetization vector aside from the easy axis.

close to that of the anisotropy axes. This feature validates a simplified model where the precession of the electron magnetization is forbidden.

On the contrary, in small crystals the anisotropy energy is comparable with the thermal energy, so that the probability that the magnetic moment points towards a direction far from the easy axis is no longer negligible, which allows at least for some electron precession. Explaining the field dependence of the longitudinal relaxation rate (NMRD profile) is in any case based on the so-called Curie relaxation (*13*), which arises from considering separately two time regimes for relaxation: firstly, diffusion into the inhomogeneous non-fluctuating magnetic field created by the mean crystal moment, aligned onto B_0 (strictly speaking, this contribution is precisely termed the Curie relaxation) and secondly, the fluctuations of the electron magnetic moment, or the Néel relaxation. The Curie relaxation accounts essentially for the high field part of the NMRD profiles ($B_0 > 0.02$ T), by considering that the mean magnetization is an increasing function of B_0, given by the Langevin function.

C. High Anisotropy Model

When the anisotropy energy is large enough it prevents any precession of the magnetic moment of super-paramagnetic crystals. The magnetic fluctuations then arise from the jumps of the moment between different easy directions. The precession prohibition is introduced into the Freed equations in order to meet that requirement: every time the electron Larmor precession frequency appears in the equations, it is set to zero (*12*).

The resulting relaxation rates are given by the following expressions:

$$R_1 = (2c\mu^2/3)\left[9L^2(x)J_A\left(\sqrt{2\omega_I\tau_D}\right) + 10\ \Delta L^2(x)J_F(\omega_I, \tau_D, \tau_N)\right] \tag{2}$$

$$R_2 = c\mu^2\left\{L^2(x)\left[3J_A\left(\sqrt{2\omega_I\tau_D}\right) + 4J_A(0)\right] + \Delta L^2(x)[8/3\ J_F(\omega_I, \tau_D, \tau_N) + 4J_F(0, \tau_D, \tau_N)]\right\} \tag{3}$$

where

$$c = 10^3\left(\mu_0^2/135\pi\right)\gamma^2 N_A[M]\tau_D/r^3 \tag{4}$$

In Eqs. (2)–(4), μ_0 is the vacuum magnetic permeability, N_A is Avogadro's number, μ is the modulus of the particle magnetic moment, r is the particle radius, and γ is the proton gyromagnetic ratio. The particle molar concentration $[M]$ is given in moles per liter. The magnetic moment depends on B_0. The mean value of its longitudinal component is given by a Langevin function:

$$\langle\mu_z\rangle = \mu L(x), \tag{5}$$

with $L(x) = \coth(x) - 1/x$, and $x = \mu B_0/(kT)$.

The fluctuating part of the longitudinal magnetic moment was calculated as (*14*):

$$\langle \mu_z^2 \rangle - \langle \mu_z \rangle^2 = \mu^2 \Delta L^2(x),$$
$$\text{with } \Delta L^2(x) = 1 - 2L(x)/x - L^2(x). \tag{6}$$

$L(x)$ increases from 0 at null field up to 1 at high field, while $\Delta L^2(x)$ is 1/3 at null field and decreases towards zero at high field. One can thus conclude that [Eqs. (2) and (3)] are dominated by the terms proportional to $\Delta L^2(x)$ at low field, and by the terms proportional to $L^2(x)$ at high field.

Two spectral density functions determine the relaxation rates in [Eqs. (2) and (3)].

J_F, called Freed function, is defined as:

$$J_F(\omega_I, \tau_D, \tau_N) = Re\left(\frac{1 + \frac{1}{4}\Omega^{1/2}}{1 + \Omega^{1/2} + \frac{4}{9}\Omega + \frac{1}{9}\Omega^{3/2}}\right), \tag{7}$$

with $\Omega = i\omega_I\tau_D + \tau_D/\tau_N$, and accounts for both correlation times (diffusion and Néel relaxation time). Figure 5 shows the dispersion of this density spectral function, centered around $\omega_I = 1/\tau_C$, with the definition of τ_C given by Eq. (1).

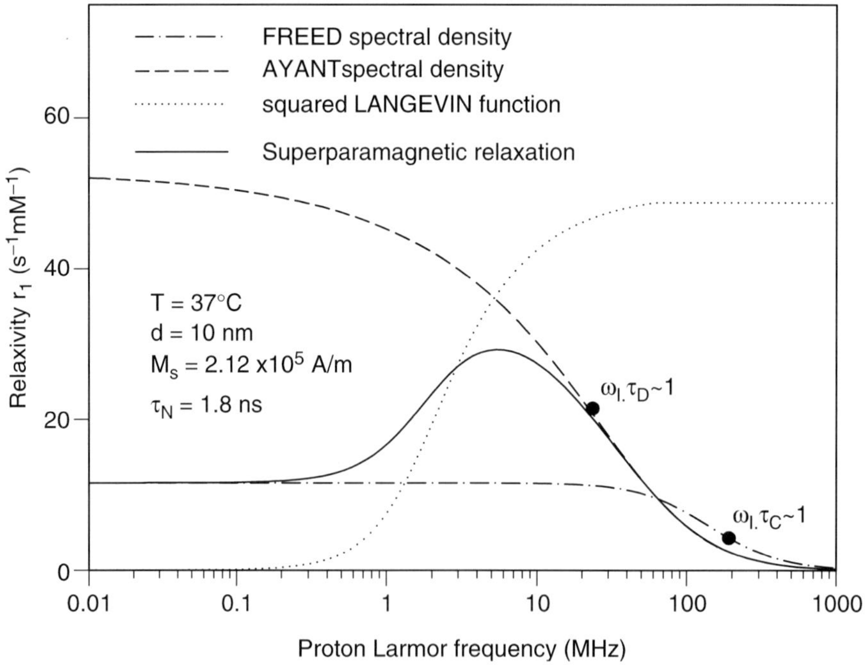

FIG. 5. Contributions to the proton relaxation in the high anisotropy model.

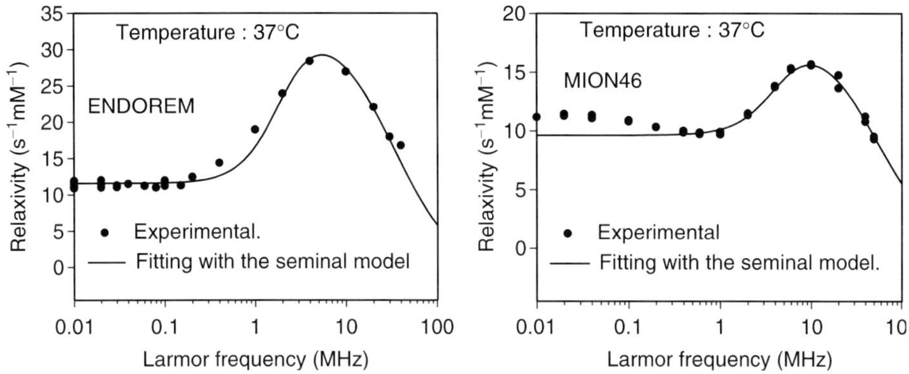

FIG. 6. Fit of the high anisotropy model to the longitudinal NMRD profile of an Endorem solution (SPIO) and of a MION46 solution (USPIO). The fitting parameters for Endorem are: $M_s = 2.12 \times 10^5$ A/m; $r = 6.4$ nm, and $\tau_N = 1.8 \times 10^{-9}$ s; and for MION46: $M_s = 2.23 \times 10^5$ A/m, $r = 4.9$ nm and $\tau_N = 2.6 \times 10^{-9}$ s.

J_A is called Ayant function, and is the limit of the Freed function when $\tau_N \rightarrow \infty$:

$$J_A(z) = \frac{1 + \frac{5z}{8} + \frac{z^2}{8}}{1 + z + \frac{z^2}{2} + \frac{z^3}{6} + \frac{4z^4}{81} + \frac{z^5}{81} + \frac{z^6}{648}} \tag{8}$$

The dispersion of this spectral density occurs about $\omega_I \tau_D = 1$. Its amplitude is always larger than the Freed component. This model matches the experimental relaxation results for large particles containing more than one crystal by coating flake (Small Particles of Iron Oxide, SPIO), but fails to describe the low field part of the NMRD curves of Ultra Small Particles of Iron Oxide (USPIO) containing only one magnetic crystal by particle (Fig. 6).

D. Small Crystal and Low Anisotropy Energy Limit

Equations (2) and (3) arise from an assumption of rigorous locking of the magnetization along the easy axes, assuming an infinite anisotropy energy. This assumption becomes less and less valid when the particle size and consequently the anisotropy energy decrease. Accounting for such a reduced coupling with the anisotropy field requires a new theory (*15*), aimed at introducing the anisotropy energy as a quantitative parameter of the problem, to go beyond the two limits considered so far: Eqs. (2) and (3), where it has been assumed to be infinite, and the classical outer sphere theory, adapted to high susceptibility paramagnetic materials (*16*), where it is neglected.

Exchange energy within super-paramagnetic crystals is sufficiently large to deal with the ensemble of electronic spins as with one large superspin S, submitted to B_0, the external magnetic field, and to the anisotropy field. The coupling of the superspin with the anisotropy field may be written as:

$$H = -(h\nu_a/2S)(S_a)^2 \tag{9}$$

where S_a is the spin component along the easy direction, and ν_a is the electron spin precession frequency at the anisotropy field (*15*).

The transition frequency is a linear function of the component of the magnetization along the easy axis; it reaches its maximum value for a null tilt angle, and it goes to zero when the magnetization becomes perpendicular to the anisotropy axis.

A wide distribution of transition frequencies has thus to be substituted to the unique electron transition frequency typical of the Zeeman interaction. Now, as mentioned above, the condition $\omega\tau_C = 1$ discriminates between fluctuations able to induce relaxation ($\tau_C < 1/\omega$) and fluctuations unable to induce relaxation ($\tau_C > 1/\omega$). Keeping this criterion in mind, the anisotropy coupling strongly differs from the well-known Zeeman coupling. At zero field, the Zeeman transition frequency vanishes, so that all fluctuations are able to induce relaxation, and consequently, their contribution to relaxation disappears for higher frequencies, giving rise to an observable dispersion in the NMRD profile. The anisotropy coupling, at least for small tilt angles, is characterized by non-zero transition frequencies, even at zero field, so that the condition $\tau_C < 1/\omega$ may not be satisfied – it can only be satisfied for magnetization vectors with rather large tilt angles. As a consequence, only the magnetization states with large tilt angles contribute to relaxation at zero field, which implies that only their contribution can disappear for increasing fields, i.e., can be responsible for a dispersion in the NMRD profile. When the anisotropy is very large, those states are essentially unoccupied, which explains intuitively the absence of low field dispersion. We propose to call this relaxation slackening a hidden transition effect.

Another effect of increasing the anisotropy is the lengthening of Néel relaxation time, which generally dominates τ_C. Since the zero-field relaxation rate is proportional to τ_C, this lengthening will increase the low field relaxivity, a consequence opposed to the slackening due to the hidden transitions.

A quantitative evaluation of the relaxivities as a function of the magnetic field B_0 requires extensive numerical calculations because of the presence of two different axes (the anisotropy and the external field axis), resulting in non-zero off-diagonal elements in the Hamiltonian matrix (*15*). Furthermore, the anisotropy energy has to be included in the thermal equilibrium density matrix. Figures 7 and 8 show the attenuation of the low field dispersion of the calculated NMRD profile when either the crystal size or the anisotropy field increases.

Theoretical predictions are clearly different in the low field relaxation profiles observed for small particles (weak dispersion) and larger particles (no dispersion). These predictions are confirmed by the difference between the NMRD profiles of SPIO and USPIO particles. Further experimental

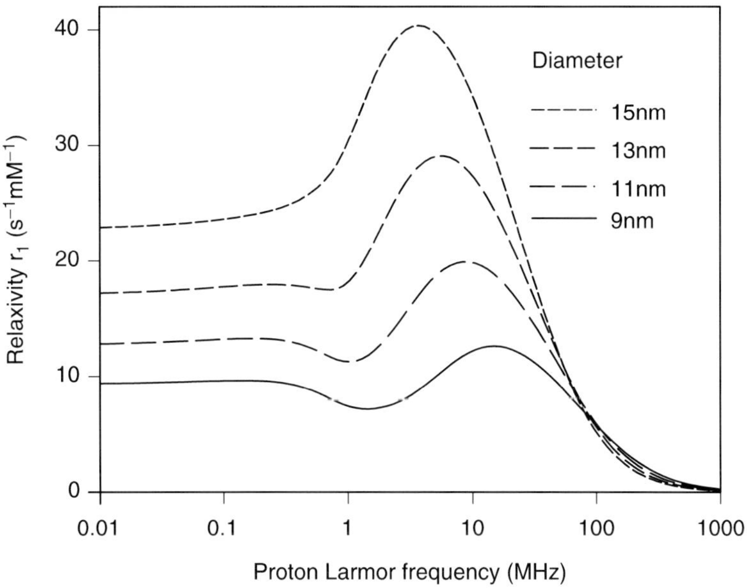

FIG. 7. Longitudinal NMRD profiles for SPM particles with 4 different diameters, $\tau_N = 20$ ns, $M_S = 2.07 \times 10^5$ A/m, $D = 3.4 \times 10^{-9}$ m^2/s, $\nu_A = 1$ GHz at $T = 310$ K (theoretical simulation).

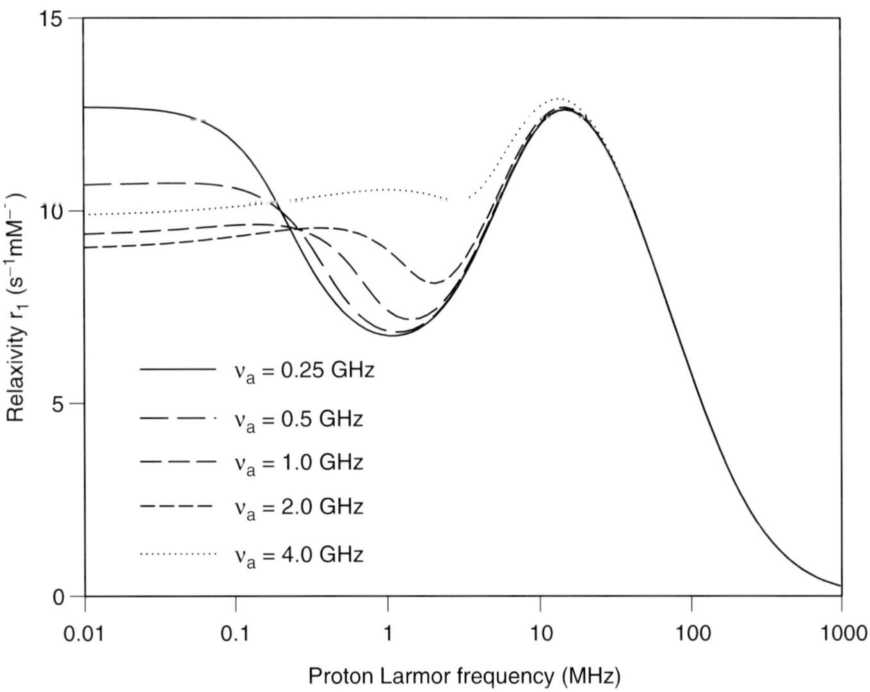

FIG. 8. Longitudinal NMRD profiles for SPM particles with 5 different anisotropy fields, with $R = 4.5$ nm, $\tau_N = 20$ ns, $M_S = 2.07 \times 10^5$ A/m, $D = 3.4 \times 10^{-9}$ m^2/s, at $T = 310$ K (theoretical simulation).

confirmation of our theoretical approach was provided by the NMRD curves of suspensions of colloidal magnetite doped with Cobalt, an element which considerably enhances the energy of anisotropy (*17*). For these small particles, a weak low field dispersion is apparent at very low cobalt content, but it disappears when doping the crystals more heavily.

E. Relaxation of Agglomerated Systems

The aggregation of the nanomagnets has two different consequences on the proton relaxation properties: on one hand, those related to the global structure of the cluster and to the magnetic field distribution around them and, on the other hand, those limited to the inner part of the aggregate (*18*). While the global effect dominantly affects R_2, the inner one influences R_1.

The agglomerate itself may be considered as a large magnetized sphere. Its magnetization increases with the external magnetic field and contributes strongly at high field to the secular part of the transverse relaxivity (see Fig. 9: accounting for the agglomerate obviously brings the theoretical prediction much closer to the experimental measurements).

The longitudinal relaxation rate inversely decreases with the residence time of water molecules inside the agglomerate. This effect was demonstrated thanks to a controlled and chemically induced process of agglomeration amongst ferrite nanomagnets coated by polyelectrolyte polymers. The NMRD profile becomes flatter on increasing agglomeration (Fig. 10).

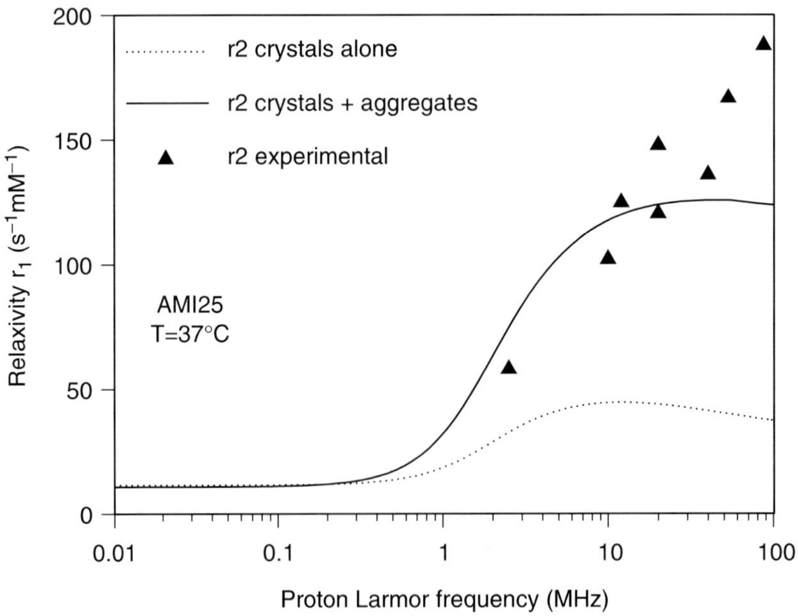

FIG. 9. Transverse NMRD profile of a sinerem sample.

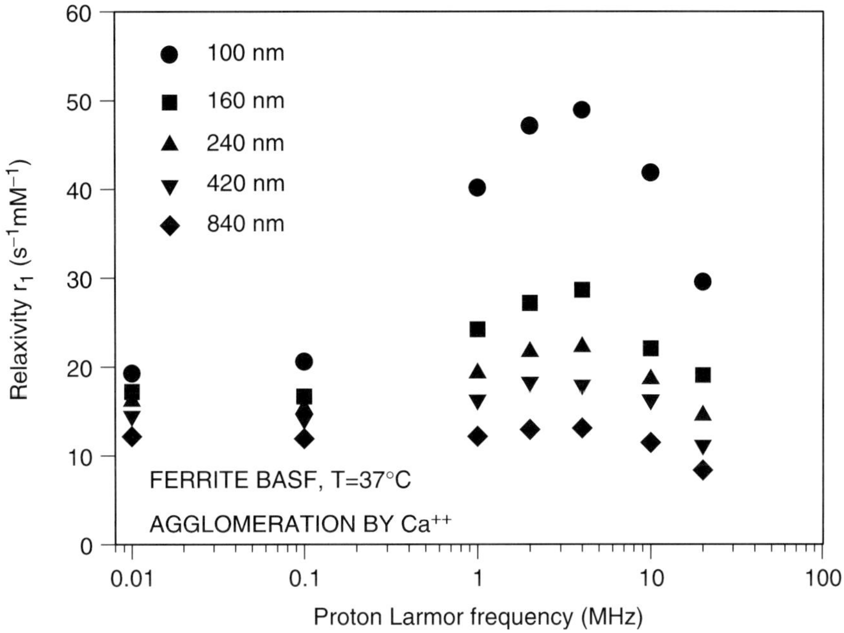

FIG. 10. Evolution of the longitudinal NMRD profile during the agglomeration of ferrite particles coated by a polyelectrolyte.

This experimental observation was explained by a model considering water as made up of two fractions in fast exchange: the water in the inner part of the agglomerate, relaxing rapidly, and the water outside the agglomerate, relaxing slowly. The exchange is produced by the translational diffusion phenomenon (*18*).

F. Large Sphere of Ferromagnetic Material (Strongly Agglomerated Particles)

The theory presented above is valid as long as the Redfield condition is satisfied. In the case of susceptibility-induced relaxation, this condition may be written as $\Delta\omega\tau_D < 3.4$, where $\Delta\omega$ is the equatorial magnetic field at the surface of the particle (expressed in angular frequency units). The Redfield condition defines thus the distinction between "small" and "large" particles.

For magnetite nanoparticles, $\Delta\omega \sim 3 \times 10^8\ s^{-1}$, so that the boundary corresponds to a radius of about 7 nm. When the magnetite cores are diluted in a dextran coating, the equatorial field of the agglomerated cluster is lower and our theoretical approach could remain valid for up to radii of approximately 50 nm.

As expected, small dimension particles return to equilibrium through mono-exponential decays, for longitudinal relaxation as well as for transverse relaxation. The observed R_2 and R_1 are in remarkable agreement with the predictions of the previous paragraph.

For larger magnetized spheres, transverse and longitudinal magnetizations exhibit obvious multi-exponential decays (*19*). In addition, the transverse magnetization decay, measured through CPMG sequences, strongly depends on the echo time.

So far, there is no theoretical work able to match the longitudinal relaxation typical of such large particles. The transverse relaxation is dominated by the secular term, which was first estimated by numerical procedures (*20,21*). These numerical computations simulate the diffusion of elementary magnetic moments (isochromats) through local magnetic fields due to the presence of impenetrable magnetized spheres. The secular contribution is calculated from the phase evolution of the isochromats in the x-y plane and the magnetization decay is obtained by averaging over a large number of isochromats. The 180° pulses are simulated by instantaneous changes of the phase sign.

Figure 11 shows the R_2 dependence on the particle size for systems containing a constant amount of magnetized material, without refocusing pulse (like in a gradient echo sequence), and with 7 different echo times used in CPMG sequences.

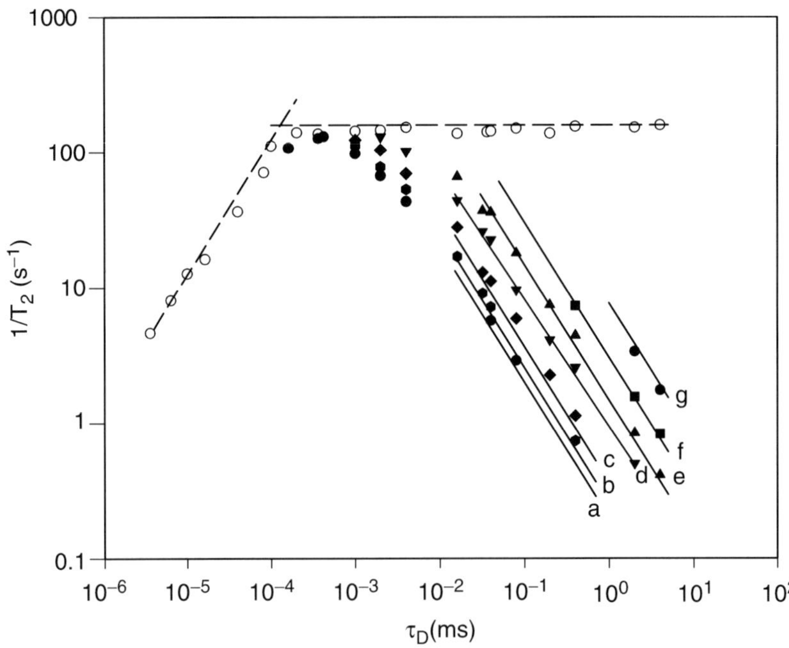

FIG. 11. Computer-generated data for spheres with a volumic fraction $\nu = 5 \times 10^{-6}$ and $\Delta\omega = 2.64 \times 10^7$ rad/s, plotted *vs.* τ_D. The open symbols represent $1/T_2^*$ value, while the filled symbols represent rates obtained, respectively, with $\tau_{CP} = 0.1$ ms (•, and line a), 0.2 ms (•, and line b), 0.5 ms (◆, and line c), 2 ms (▼, and line d), 5 ms (▲, and line e), 10 ms (■, and line f), 20 ms (●, and line g). The short dashed line is the rate predicted by outer sphere theory and the long dashed line is the Static Dephasing Regime model from (*22*). Lines (a–g) are the rates given by Eq. (11).

Without spin refocusing, the simulations show that $R_2{}^*$ reaches a plateau for diameters above 0.02 μm. The corresponding maximum relaxation rate of about 100 s^{-1} is predicted by the Static Dephasing Regime (SDR) model (*22, 23*), which assumes that, as depicted above, transverse relaxation is only attributed to differences in precession frequencies of static spins. The relaxation rate is then given by:

$$\frac{1}{T_2{}^*} = \frac{8\sqrt{3}\pi^2}{27}\mu_0 v\gamma M \tag{10}$$

where ν is the volume fraction occupied by the particles, M their specific magnetization and γ the gyromagnetic ratio of the proton.

A recent theoretical study (*24*) investigated the decay of the transverse magnetization in CPMG sequences, in the slow diffusion limit (the diffusion time must be longer than the time for the signal to be reduced by a factor *e* from its initial value). However, the conditions of slow diffusion are generally not fulfilled for highly magnetized particles, their radii being too small.

A more phenomenological approach (*25*) allows to overcome this limitation. The space around each particle is divided between two regions; the boundary is echo-time dependent. In the first one, near the particle, the gradients are too large for the refocusing pulses to be effective, so that the moments situated in this region will be rapidly dephased. These protons contribute a fast signal decay that is unobservable with MRI techniques.

In the second region, relatively far from the particle, the refocusing pulses are effective, which corresponds to a weak magnetization condition. The relaxation rate can then be analyzed following Majumdar and Gore (*26*). The result of this calculation is in remarkable agreement with the slow rates, obtained from simulated multiexponential decays (*27*):

$$1/T_2 = 1.8\ \nu(\Delta\omega\ \tau_{CP})^{1/3}(1.5 + \nu\ \ \Delta\omega\ \ \tau_{CP})^{5/3}/\tau_D \tag{11}$$

where τ_{CP} is the echo time.

The role of the refocusing pulses is generally understood in the following sense: they lose their efficiency as soon as too many random events take place between two consecutive pulses, creating irreversibility. This idea is translated in a limitation of the allowed diffusion between pulses – the echo time must be shorter than the correlation time characterizing diffusion. However, one question remained and was long debated upon: is this correlation time the diffusion time ($\tau_D = r^2/D$), or the interdiffusion time (the time for diffusing from one particle to another, i.e., $\nu^{-2/3}\tau_D$)?

The answer was also provided by the phenomenological approach leading to Eq. (11), and it depends on the ratio between the echo time and the time to dephase significantly ($1/\Delta\omega$). The SDR regime appears as a ceiling imposed upon the transverse relaxation rate, when the particle radius is reduced starting from large values. When τ_{CP} is on the order of $1/\Delta\omega$, the ceiling is reached for $\tau_D = \tau_{CP}$, the regime described by Eq. (11) being thus valid for

$\tau_D > \tau_{CP}$, which corresponds to the strict slow diffusion condition. The same validity condition becomes $\tau_D \, \nu^{-2/3} > \tau_{CP}$ when $\tau_{CP} > 1/(\nu\Delta\omega)$.

G. Application of the Relaxometric Study of Super-Paramagnetic Particles

The first step in the characterization of a new super-paramagnetic colloid is obviously the evaluation of its relaxometric properties, which determine its potential efficiency for MRI (*27,28*). Relating these valuable relaxometric data to morphological and physical properties of the particles may be carried out thanks to a proton relaxivity theory.

Furthermore, the analysis of the NMRD profiles constitutes an interesting tool to control the reproducibility and to optimize the parameters of nanomagnet synthesis (*29*). The nanomagnet size can be controlled by the $[Fe^{2+}]/[Fe^{3+}]$ ratio (Table I and Fig. 12): the nanomagnet radius and the specific magnetization both increase with this ratio. A widening of the size dispersion can be at the origin of the increasing discrepancy between the mean sizes and the specific magnetization obtained by relaxometry and magnetometry when the $[Fe^{2+}]/[Fe^{3+}]$ ratio is enhanced (*29*).

In conclusion:

For USPIO particles containing only one nanomagnet per particle, the main parameters determining the relaxivity are the crystal radius, the specific magnetization and the anisotropy energy. Indeed, the high field dispersion is determined by the translational correlation time τ_D.

The maximum of relaxivity is proportional to the squared saturation magnetization of the crystal. The low field relaxivity depends on the anisotropy, and the presence of a low field dispersion indicates a low anisotropy energy.

Agglomeration can cause an increase in the transverse relaxation rate and an attenuation of the longitudinal relaxation; therefore the ratio r_2/r_1

TABLE I

Dependence of the Crystal Radius and of the Magnetization Obtained by Relaxometry and Magnetometry with the $[Fe^{2+}]/[Fe^{3+}]$ Ratio

$[Fe^{2+}]/[Fe^{3+}]$	r_{relaxo} (nm)	$M_{S\ relaxo}$ ($\times 10^5$ A.m^{-1})	$r_{magneto}$ (nm)	$M_{S\ magneto}$ ($\times 10^5$ A.m^{-1})
0.5	3.8	2.7	3.1	3.7
1.0	4.2	2.7	2.9	3.9
2.0	4.5	3.1	3.5	4.4
3.0	5.1	2.9	3.6	4.5
3.9	6.4	2.8	4.3	4.7

r_{relaxo} is the radius calculated from NMRD curves.
$M_{s\ relaxo}$ is the specific magnetization calculated from NMRD curves.
$r_{magneto}$ is the radius obtained from magnetometry.
$M_{s\ magneto}$ is the specific magnetization obtained from magnetometry.

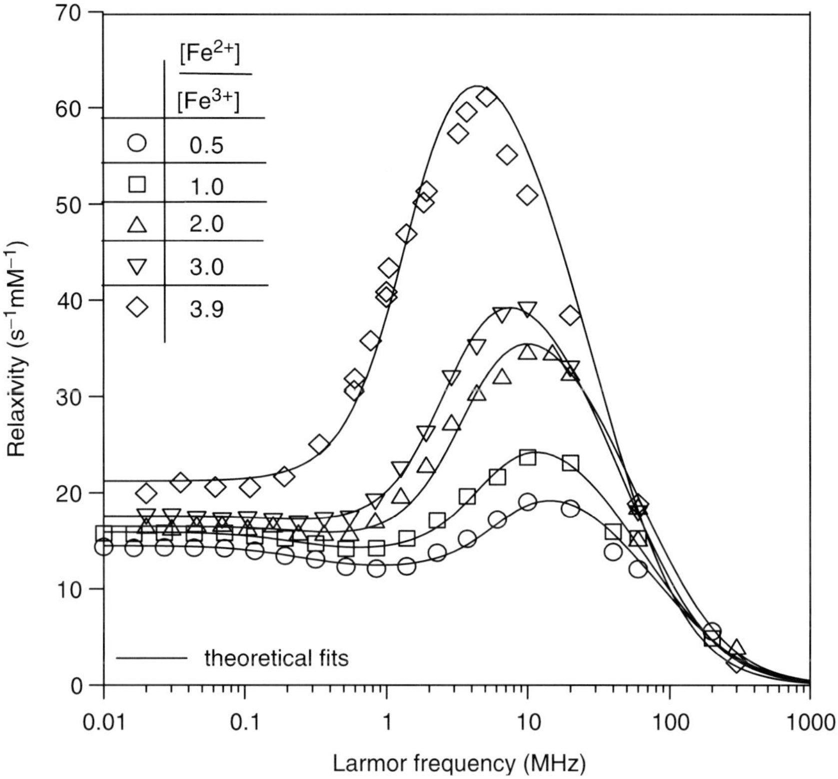

FIG. 12. NMRD profiles of iron oxide colloids for different $[Fe^{2+}]/[Fe^{3+}]$ ratios.

constitutes an indication of the aggregation level of the super-paramagnetic colloid.

The transverse relaxivity induced by large and highly magnetized particles is predicted by phenomenological studies, and by rigorous theoretical studies when diffusion is slow. Finally, no theory is suitable for the proton longitudinal relaxivity of very large magnetized particle ($r > 50$ nm).

II. Ferritin, an Iron-Storage Protein

A. The Iron-Storing Ferritin

A.1. Iron in Biology

Iron is one of the most important elements in biology. In hemoglobin, it allows the binding of molecular oxygen. Iron plays an important role in the active centers of many enzymes. The total quantity of iron in the human body is about 4g. Each day, the body eliminates approximately 1 or 2 mg of iron.

This loss is compensated by the alimentation. 70 % of the body iron is contained in hemoglobin. Transferrin ensures the transport of iron, while ferritin and hemosiderin are used for the storage of iron in a non-toxic form: ferritin is indeed able to transform the highly toxic Fe(II) in to the less toxic Fe(III).

Ferritin is present in numerous living species, from bacteria to elephant. In humans, it is mainly located in the liver, in the spleen, in the bone marrow, and in the extrapyramidal nuclei of the brain (putamen, globus pallidus, caudate and substantia nigra).

A.2. Structure of Ferritin

Ferritin is composed by the arrangement of 24 protein subunits, which results in a hollow shell of 8 nm inner diameter and 13 nm outer diameter (Fig. 13). Ferritin from vertebrates have two types of subunits: heavy (H) and light (L). The subunit composition of human ferritins depends on the origin of the protein: H_2L_{22} for liver ferritin, $H_{20}L_4$ for muscle ferritin, etc. Access channels are formed by the intersection of subunits. The 8 channels located at the intersection of three subunits are hydrophilic while the 6 channels located at the intersection of 4 subunits are hydrophobic. The empty protein is called apoferritin (*30*).

Fe(II) penetrates inside the spherical shell by the hydrophilic channels. After an oxidation on ferroxidase sites, located on H subunits, Fe(III) iron ions migrate to a nucleation site, situated on L subunits, where a crystal of hydrated iron oxide grows. Up to 4500 Fe(III) can be stored inside this mineral phase (*31*). The number of iron atoms contained in the ferritin molecule is called the loading factor (*LF*).

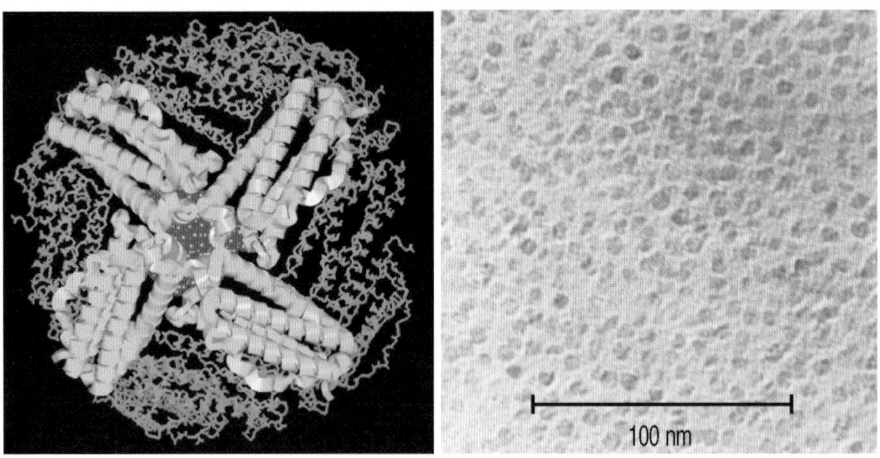

FIG. 13. On the left, representation of the ferritin molecule : the subunits form a spherical shell containing a ferrihydrite crystal. On the right, a Transmission Electron Microscope picture showing the iron core contained inside the proteic shell, appearing as electron-dense dark spots.

The iron core of ferritin is structured as ferrihydrite (5 Fe_2O_3.9 H_2O) of variable crystallinity and contains variable amounts of organic phosphate. Ferrihydrite is a hydrated iron oxide of hexagonal symmetry. It contains iron and oxygen as a classic oxide, but it also includes hydroxyl groups in its structure. Goethite (α-FeOOH), lepidocrocite (γ-FeOOH), feroxyhyte (δ′-FeOOH), and akaganeite (β-FeOOH) belong to the same mineral family (*32*).

A.3. Magnetic Properties of Ferritin

Ferrihydrite is antiferromagnetic: it is composed of two equivalent magnetic sublattices whose magnetic moments are in opposite directions. The resultant magnetic moment of an ideal antiferromagnetic crystal in a null magnetic field is null. However, because of the occurrence of defects in ferrihydrite crystals and because of the canting of the moments from the two sublattices in a non-zero magnetic field, the compensation of the magnetic moments of the two sublattices is not complete: each nanometric particle has a non-null resulting magnetic moment (*33*). The Néel temperature of ferritin, i.e., the temperature of the transition between antiferromagnetism and simple paramagnetism, has been estimated by several groups. Values are ranging from 20 K to 500 K (*34–38*), but the latter temperature, based on a new fitting of the magnetometric data using a more complete model, seems to be the most reliable one (*35*).

Different Mössbauer and magnetometric studies have shown the superparamagnetism of ferritin iron cores. This property is also typical of ferro- and ferrimagnetic materials: when the size of the crystal is small enough, Néel relaxation becomes observable and the resulting magnetic moment of a particle jumps from one direction of easy magnetization to the other. The Néel relaxation time, i.e., the time between two successive jumps – is given by:

$$\tau_N = \tau_0 e^{\frac{KV}{kT}} \tag{12}$$

where τ_0 is the Néel relaxation time at infinite temperature, K the anisotropy energy constant, and V the volume of the crystal (*8,39*). At low temperatures, the Mössbauer spectrum of ferritin is a doublet, typical of a frozen magnetic moment, while at higher temperature it is a sextet typical of magnetic moments undergoing super-paramagnetic relaxation. The transition between the doublet and the sextet occurs at the blocking temperature T_B. For $T > T_B$, τ_N is shorter than the characteristic time of a Mössbauer measurement, called τ_{MES}, so that the super-paramagnetism of the crystal may be observed during this Mössbauer measurement. For $T < T_B$, τ_N is longer than τ_{MES}, and the crystal magnetization appears frozen. Estimating τ_{MES} as equal to 2.5 ns, one obtains a blocking temperature from Mössbauer spectroscopy of about 30–40 K (*40,41*). However, its value depends on the measuring technique: a thermal decay remanence method, or an AC susceptibility measurement will provide other values of T_B. Kilcoyne *et al.* (*40*) deduced the values of τ_0 and K from two different determinations of T_B given by two different experimental

TABLE II

τ_0 AND K VALUES OBTAINED FOR FERRITIN

	Pre-exponential factor τ_0	Anisotropy energy constant K
Kilcoyne *et al.* (*40*)	10^{-11} s	37,000 J/m^3
Dickson *et al.* (*41*)	10^{-12} s	34,000 J/m^3

methods. The same parameters can be calculated from the results of Dickson *et al.* (*41*). Table II shows the corresponding values of τ_0 and K. From these values, one can estimate the Néel relaxation time of a 7 nm diameter ferrihydrite crystal at room temperature: $\tau_N = 0.1$ ns.

Magnetometry curves of a ferritin sample at different temperatures are shown in Fig. 14. Two contributions can be noticed: the first one saturates at high magnetic field, while the second one increases linearly with the field. At room temperature, the magnetometry curve is almost linear with the magnetic field: the magnetization of ferritin shows no saturation up to 5T and remains simply proportional to the applied field, as for a simple

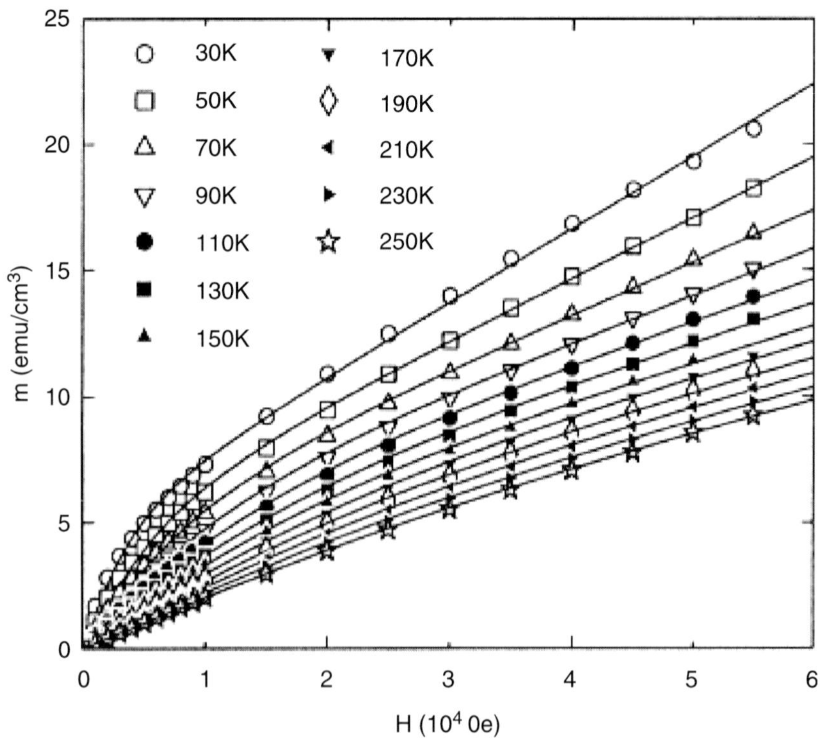

FIG. 14. Magnetization curves of reconstituted horse spleen ferritin. The solid lines are fits to expression (*14*). Reproduced with permission from Ref. (*35*).

paramagnetic system. Different theories have been proposed to describe the saturating contribution of these curves. An early modeling used the Langevin function, but on the one hand, the quality of the fit was rather poor, and on the other hand, Langevin's theory applies to uncoupled spins, allowed to rotate freely, which is not the case for ferrihydrite particles, because of their large anisotropy.

$$M(x) = M_s L(x) + \chi B_0$$
$$L(x) = \coth(x) - \frac{1}{x}$$
$$x = \frac{\mu B_0}{kT} \tag{13}$$

Recently, Gilles *et al.* developed another model, accounting for the anisotropy (*35*), introducing a new function $G(x)$:

$$G(x) = \frac{1}{2}\int_0^{\pi} \sin(\theta)\cos(\theta)\tanh(x\cos(\theta))d\theta = \int_0^1 y\tanh(xy)dy$$
$$M(x) = M_s G(x) + \chi B_0 \tag{14}$$

It can easily be seen that at low field ($x \ll 1$) the first order of the expansion of the tanh function leads to $G(x) \cong x/3$ while the saturation limit ($x \gg 1$), leads to $G(x) \cong 1/2$.

These recent results are thus likely to have put an end to a previously controversial question: ferritin is antiferromagnetic at room temperature (because $T < T_N$) and the size of the ferrihydrite grains is small enough to observe super-paramagnetism with a Néel relaxation time of about 0.1 ns.

B. Proton Relaxation by Ferritin and Similar Nanoparticles

B.1. Why Study Ferritin?

From the very beginning of magnetic resonance imaging (MRI), the presence of ferritin in the liver has been shown to cause a darkening of T_2-weighted images (*42*). To understand this darkening effect, different studies concerning the relaxation caused by ferritin in vivo, in vitro and in solutions have been carried out. This protein contains a nanometric iron core, which makes it similar to a natural contrast agent. Researchers interested in diamagnetic proteins as well as researchers involved in the characterization of contrast agents have thus been working on ferritin-induced relaxation. Moreover, a good understanding of the relaxation mechanism of protons in the presence of ferritin could contribute to develop non-invasive methods for an iron content measurement by MRI (*43*). These methods would be very useful to study the role of ferritin in the brain for Parkinson's and Alzheimer's diseases, and in the liver and spleen for hemochromatosis and thalassemia, Section II.D.

details the role of ferritin in MRI. To provide a basis for comparison with ferritin solutions, studies have been devoted to synthetic hydrated iron oxide suspensions, as for instance akaganeite nanoparticles aqueous solutions, which present similar NMR properties.

B.2. Relaxivity and NMRD Profiles

2a. Relaxivity The relaxivity of ferritin, i.e., the relaxation rate enhancement caused by an increase of 1 mM in the iron concentration of the solution, slightly depends on the loading factor. At 40 MHz and 37°C, for $LF = 2650$, Vymazal *et al.* (*44*) measured relaxivities $r_1 = 0.027\ s^{-1}\ mM^{-1}$ and $r_2 = 0.22\ s^1\ mM^{-1}$. These values are at least 100 times smaller than the relaxivities of magnetite/maghemite particles solutions used as contrast agents (*45*). The corresponding values for akaganeite particles solutions are even smaller: $r_1 = 0.0083\ s^{-1}\ mM^{-1}$ and $r_2 = 0.049\ s^{-1}\ mM^{-1}$ (*46*). This difference between magnetite particles and hydrated iron oxides particles can be partially explained by the difference in susceptibility of both compounds: magnetite and maghemite particles are 100 times more magnetized than ferrihydrite and akaganeite particles. Ferritin and akaganeite particles can thus be considered as poor contrast agents acting primarily on T_2 of water protons.

2b. Longitudinal NMRD The longitudinal NMRD profiles of ferritin and apoferritin aqueous solutions are shown on Fig. 15. They were first measured by Koenig *et al.* (*47*) and by Brooks and co-workers (*44*) at high fields (from 0.05 Tesla). The center of the apoferritin dispersion corresponds to the inverse of the protein rotation time (*47,48*). In order to isolate the contribution from the magnetic core of ferritin, the apoferritin contribution must be subtracted from the global NMRD profile. Both the ferritin iron core profile and the akaganeite particles profile (Fig. 15) show a double dispersion. Correlation times calculated thanks to the Halle method (*49*) are given in Table III. For both particles, the first inflection point is considered as electron-related ($\omega_e\tau_1 = 1$), and the second one as proton-related ($\omega_I\tau_2 = 1$).

Ferritin: The first correlation time is larger at 5°C (1.95 ns) than at 37°C (0.72 ns), as would be a rotational correlation time, a diffusion time or an exchange time. Nevertheless, neither the rotation time (0.19 μs at 37°C) nor the diffusion time (12.8 ns at 37°C) corresponds to this value. The first dispersion was therefore interpreted as arising from a proton exchange. The correlation time of the second dispersion (23 ns at 37°C) also decreases with temperature, and is about 8 times shorter than the rotation time. Nevertheless, it is closer to the value of the diffusion time. The second dispersion was therefore assigned to proton exchange or diffusion.

Akaganeite particles: The first inflection point corresponds to a correlation time larger at 37°C (2.28 ns) than at 5°C (1.76 ns). This temperature dependence is typical of an electronic relaxation time (of Fe^{3+}). The difference between the calculated correlation time and the theoretical relaxation time

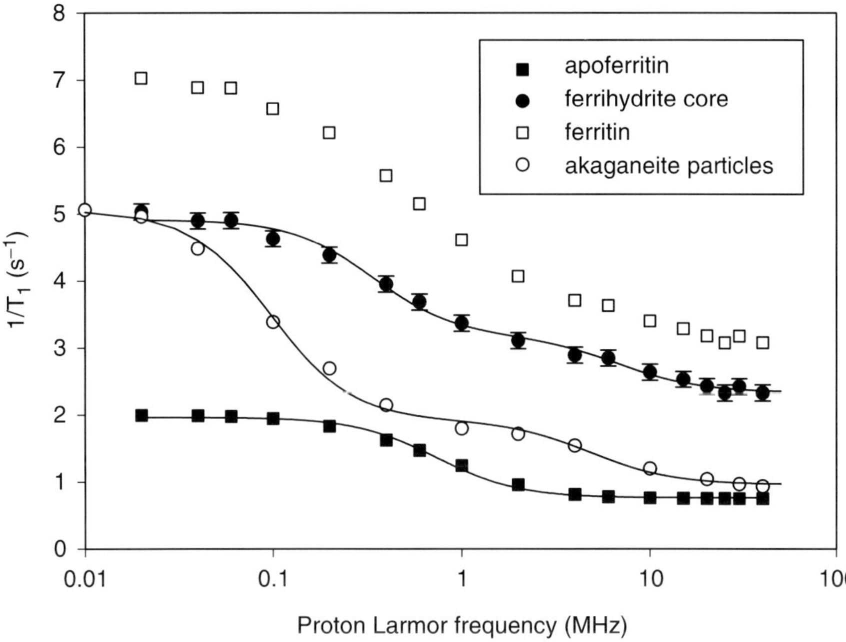

FIG. 15. Longitudinal NMRD profile of ferritin (□) and apoferritin (■) aqueous solutions at 37°C. The contribution of the ferrihydrite core to the relaxation (•) is obtained by the subtraction of the profiles. Ferritin solution has an iron concentration of 100 mM, while the protein concentration of both samples is 0.058 mM. Longitudinal NMRD profile of akaganeite particles (○) with an iron concentration of 100 mM.

TABLE III

CORRELATION TIMES CALCULATED FROM THE FITTINGS OF THE LONGITUDINAL NMRD PROFILES OF APOFERRITIN, FERRITIN, AND AKAGANEITE PARTICLES SOLUTIONS

APOFERRITIN		
1st dispersion	5°C	$\tau_1 = 0.5$ μs
	37°C	$\tau_1 = 0.21$ μs
FERRIHYDRITE CORE		
1st dispersion	5°C	$\tau_1 = 1.95$ ns
	37°C	$\tau_1 = 0.71$ ns
2nd dispersion	5°C	$\tau_2 = 0.039$ μs
	37°C	$\tau_2 = 0.023$ μs
AKAGANEITE NANOPARTICLES		
1st dispersion	5°C	$\tau_1 = 1.76$ ns
	37°C	$\tau_1 = 2.28$ ns
2nd dispersion	5°C	$\tau_2 = 0.0403$ μs
	37°C	$\tau_2 = 0.0266$ μs

of a single Fe^{3+} ion (0.1 ns) could be assigned to the chemical environment of the ion. The inflection point at the center of the second dispersion is almost the same as the one characterizing the second dispersion of the ferritin iron core; it could be associated with diffusion, rotation or proton exchange. The rotation time of a 12 nm radius particle is about 1.2 μs at 37°C, which is much longer than our correlation time (0.027 μs). The value of the diffusion time (43 ns at 37°C) seems more appropriate, and hence we consider the second correlation time as either a diffusion time or a proton exchange time.

2c. Transverse NMRD Different studies have shown that the proton transverse relaxation rate of aqueous solutions of ferritin and akaganeite particles is proportional to the applied magnetic field (*44–46*), which is quite an unexpected result (see Figs. 16 and 17). Indeed, all current theories predict a quadratic dependence of the rate enhancement on the magnetization of the particles, which results in a quadratic dependence against the field if the magnetization is proportional to this field. Furthermore, the transverse relaxation rate measured with a Carr–Purcell–Meiboom–Gill (CPMG)

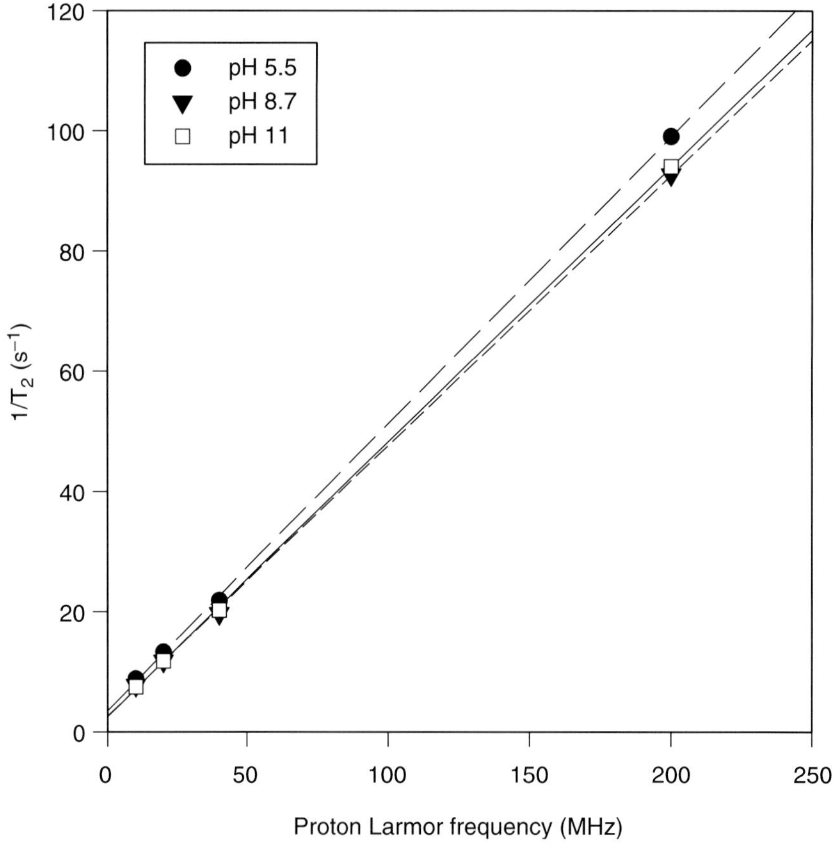

FIG. 16. Transverse NMRD profile of ferritin aqueous solutions at 40°C for different pH values. The iron concentration is 100 mM.

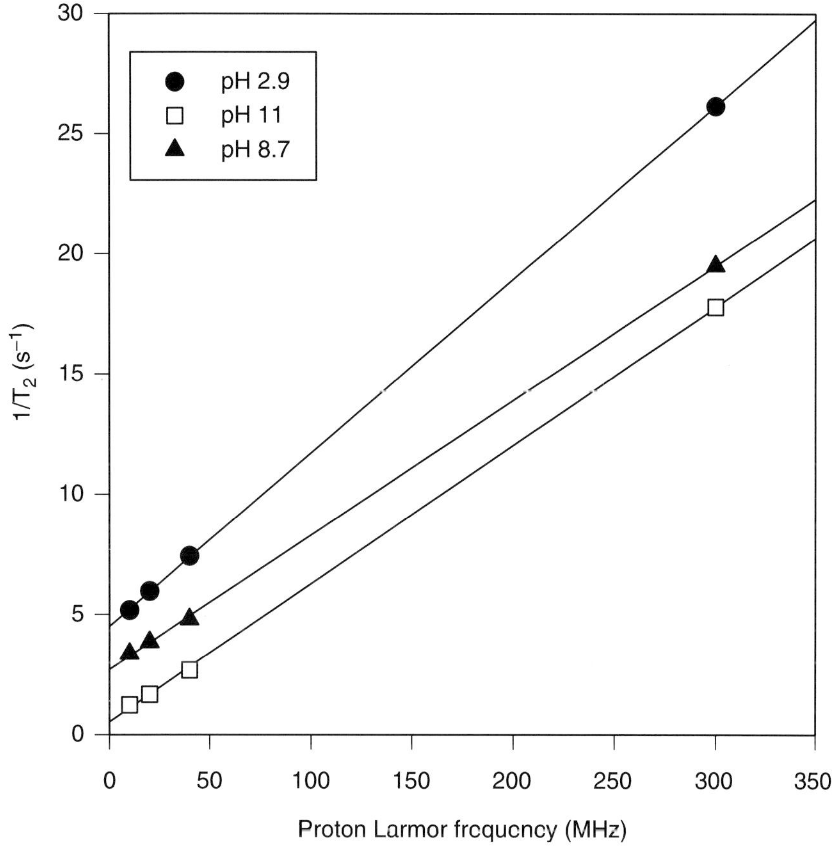

FIG. 17. Transverse NMRD profile of akaganeite particles solutions at 40°C for different pH values. The iron concentration is 100 mM.

TABLE IV

COMPARISON OF α VALUES (CALCULATED FROM THE TRANSVERSE NMRD PROFILE) FOR DIFFERENT SAMPLES AT 40°C

	α (s^{-1} MHz^{-1} mM^{-1})
Ferritin solution (*46*)	0.0039
Akaganeite particles solution (*46*)	0.0011
Ferritin in marmoset liver (*50*)	0.011
Ferritin in human brain (*51*)	0.0082
(*52*)	0.018

sequence was shown to be independent of the echo time τ_{CP} (*44,46*). An important relaxation parameter is α, the slope of the linear NMRD profile normalized by the iron concentration. Table IV shows the value of this parameter α for aqueous solutions of ferritin and akaganeite, and for iron-containing tissues (*50–52*). The value of α is larger for iron-containing tissues than for aqueous solutions of hydrated iron oxide nanoparticles. A possible interpretation of this difference between tissues and solutions is the clustering of ferritin molecules in vivo. However, there is no experimental demonstration on the effect of agglomeration on ferritin relaxation yet.

B.3. Other Relaxation Properties

3a. Effect of pH

Ferritin: The longitudinal relaxation rate of ferritin is not significantly influenced by the pH of the aqueous solution (*47,48*). However, after subtraction of the apoferritin contribution, the effect is more significant (Fig. 18). The transverse relaxation of ferritin solutions is almost pH independent (Fig. 16).

Akaganeite particles: Both T_1 and T_2 are strongly pH-dependent (Figs. 17 and 19). The amplitudes of the longitudinal NMRD profiles drastically decrease when the pH increases from 3.35 to 9.45. The correlation time associated with the first dispersion is only weakly pH dependent, consistent with its former interpretation as an electron relaxation time. However, τ_2, the correlation time characteristic of the second dispersion, increases from 30 ± 8 ns at pH 3.35 to 280 ± 32 ns at pH 9.45, which eliminates its interpretation as a diffusion time: τ_2 can be identified as a proton exchange time.

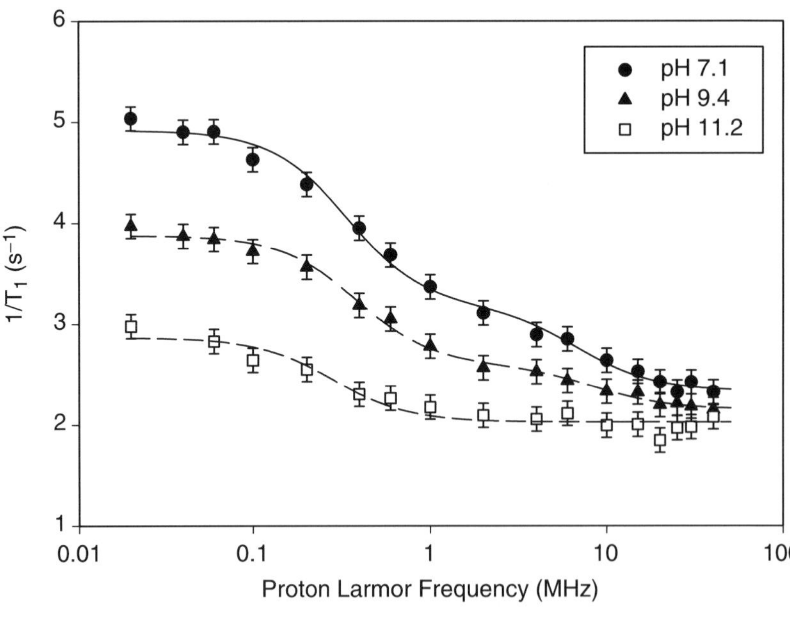

FIG. 18. Contribution of the ferrihydrite core to the longitudinal NMRD profile of ferritin solutions for different pH values at 37°C. The iron concentration is 100 mM.

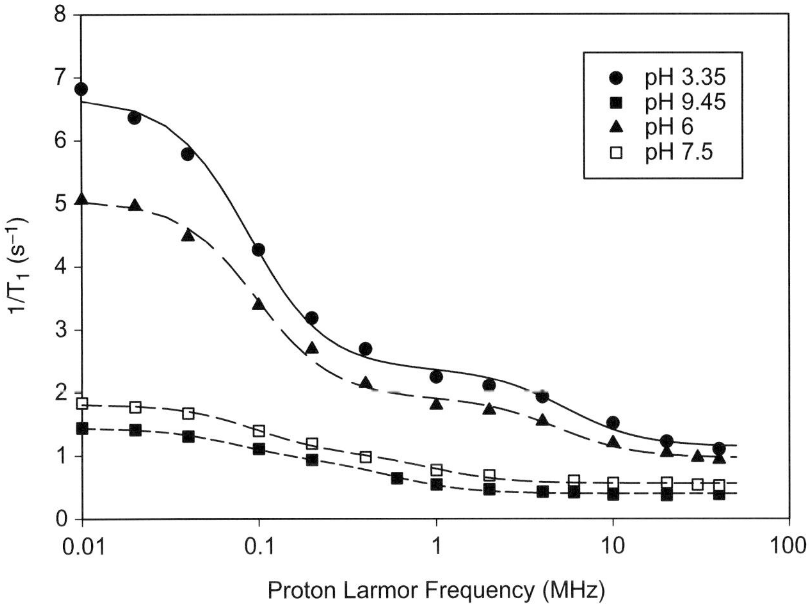

FIG. 19. Longitudinal NMRD profile of akaganeite particles solutions for different pH values at 37°C. The iron concentration is 100 mM.

The linear shape of the transverse NMRD profiles is not altered by changes in pH, but unlike the ferritin solutions, the values of the slope and intercept of the linear regression depend on pH. The intercept decreases when pH increases, while the slope initially decreases towards a minimum which is reached at pH ~ 7.5, and then increases (*48*).

3b. Influence of the Loading Factor Vymazal *et al.* studied the effect of *LF* on the relaxation of ferritin solutions (*38,53,54*). For the same protein concentration, the relaxation induced by ferritin seems to be almost proportional to the loading factor. The measurement of T_2 is thus sufficient to determine the iron concentration in ferritin samples, without regard to the loading factor. This condition is necessary for the development of MRI methods for the in vivo iron content determination.

3c. ^{133}Cs NMR Cesium adsorption on akaganeite particles, on dextran coated ferrihydrite particles and on ferritin in aqueous solutions has been studied by ^{133}Cs Nuclear Magnetic Resonance measurements (*55*). Indeed ^{137}Cs is an important component of nuclear waste which may pollute water. Its migration in natural environments is slowed down by adsorption on minerals. The study of its adsorption on magnetic hydrated iron oxides, very common in clays, is thus of interest.

The longitudinal relaxation time (T_1) of ^{133}Cs in the presence of magnetic particles depends on whether the ions bind to the particle or not. In theory, predictions are that T_1 measured in aqueous solutions containing the same amount of magnetized particles but different cesium concentration will not vary if relaxation is governed by diffusion (if cesium does not bind), while it will depend on cesium concentration if the exchange governs relaxation (if cesium does bind). This method was successfully tested for Tempo, a nitroxide-stable free radical, whose relaxation is due to diffusion. On the contrary, ^{133}Cs relaxation in solutions of ferritin, akaganeite and dextran-coated ferrihydrite particles is due to an exchange of cesium ions between the surface of the particles and bulk ions. This proves the adsorption of cesium on the particles. The effect of pH on ^{133}Cs relaxation in solutions of the studied particles is consistent with the adsorption properties of cations on hydrated iron oxides (*56*): the number of adsorbed cesium ions seems to be more important at alkaline pH than at acidic pH.

B.4. Ferritin-Based Contrast Agents

The spherical shell of apoferritin has been used to synthesize two exotic contrast agents: magnetoferritin and gadolinium-loaded apoferritin. Magnetoferritin is a proteic shell of ferritin filled with a magnetite iron core (*57*). The first step of the synthesis is to remove the ferrihydrite core of a ferritin sample. Then, addition of ferrous ammonium sulfate under strictly anaerobic conditions allows the formation of a magnetite core, identified by electron diffraction, inside the protein shell. The size of the core, determined by transmission electron microscopy, is about 7.3 ± 1.4 nm (*58*). The characteristics of magnetoferritin and AMI-25, a synthetic magnetite contrast agent, are compared in Table V. Their NMR and magnetic properties are rather similar. The NMRD profiles of magnetoferritin are shown in Fig. 20. Magnetoferritin was tested in vivo on rats (*59*), showing that it could be used as a MRI contrast agent for the liver and the spleen.

The gadolinium complex GdHPDO3A can be trapped inside the interior of apoferritin (*60*). The process is based on the dissociation of the apoprotein into subunits at low pH in a concentrated solution of the complex followed by its reforming at pH 7. The complex molecules that are not trapped inside the protein are eliminated by dialysis. The longitudinal relaxivity of the

TABLE V

COMPARISON OF THE CHARACTERISTICS OF MAGNETOFERRITIN AND AMI-25 (*58*)

	Magnetoferritin	AMI-25
Core size	7.3 nm	11 nm
Core crystallinity	Monocrystalline	Oligocrystalline
Surrounding matrix	Protein	Dextran
Total particle size	12 nm	50 nm
r_1	8 $s^{-1} mM^{-1}$	10 $s^{-1} mM^{-1}$
r_2	175 $s^{-1} mM^{-1}$	110 $s^{-1} mM^{-1}$
r_2/r_1	22	11

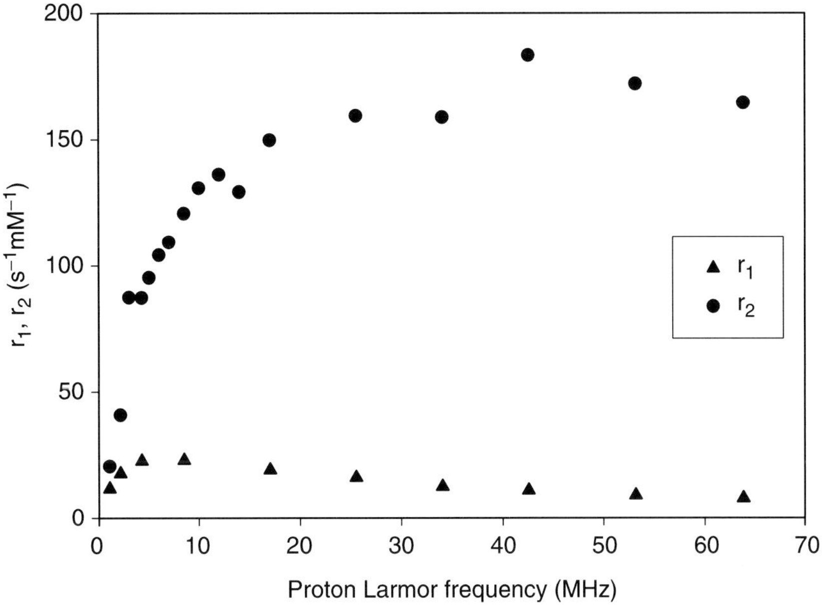

FIG. 20. Longitudinal and transverse NMRD profiles of magnetoferritin at 37°C. Reproduced with permission from Ref. (*58*).

gadolinium loaded apoferritin is 20 times higher than the corresponding value for free GdHPDO3A in water (Fig. 21). This relaxation enhancement is not well understood: it may be related to the presence of exchangeable protons or water molecules inside the protein.

Both compounds may be functionalized to bind specific molecules for the targeting of different pathologies.

C. Theory of Ferritin-Induced Relaxation

C.1. Limitations of the Current Theoretical Models

Outer-sphere model

Relaxation of water in the presence of magnetized nanoparticles is generally well-described by the outer sphere theory (OST), namely for magnetite particles (*61*). This explanation is extended to ferritin solutions in some studies (*38, 47,61*). Outer-sphere relaxation is caused by the diffusion of water protons through the inhomogeneous magnetic field created by the magnetic particles. This model predicts, in high-intensity fields, a quadratic dependence of $1/T_2$ on the magnetic moment of the particles (assumed to be spherical, with a radius R), through $\Delta\omega_r^2$ (*21*):

$$1/T_2 = (4/9)\nu\Delta\omega_r^2\tau_D \qquad (15)$$

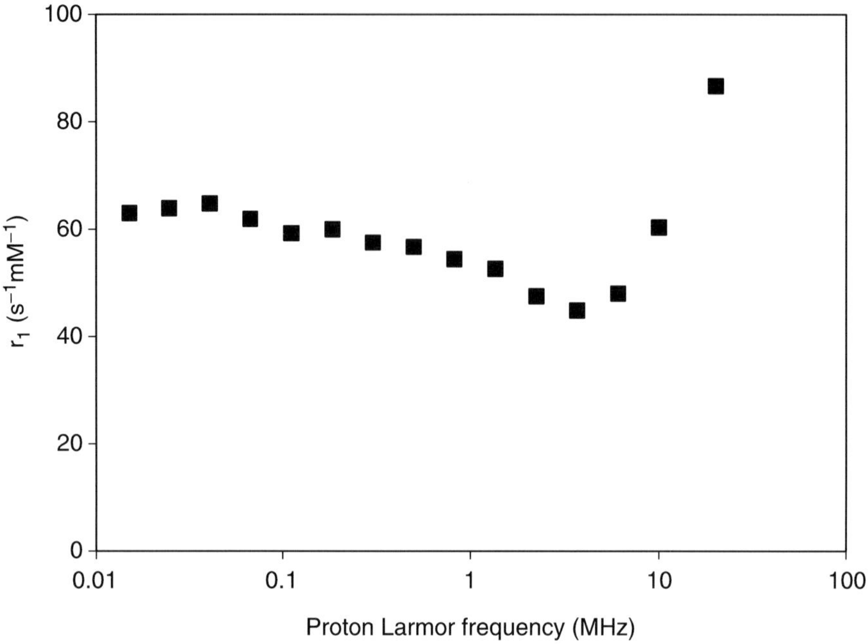

FIG. 21. Longitudinal NMRD profile of gadolinium-loaded apoferritin at pH 7 and 25°C. Reproduced with permission from Ref. (*60*).

where ν is the volume fraction occupied by the particles, $\Delta\omega_r$ is the rms frequency shift at the surface of the particle, and $\tau_D = R^2/D$, with D the water diffusion coefficient.

The relaxation rates calculated from Eq. (15) are smaller than the measured ones at low field, while they are larger at high field. OST is thus obviously unable to match the experimental results. However, water protons actually diffuse around ferrihydrite and akaganeite particles and there is no reason to believe that the contribution to the rate from this diffusion would not be quadratic with the external field. This contribution is not observed, probably because the coefficient of the quadratic dependence with the field is smaller than predicted. This could be explained by an erroneous definition of the correlation length: in OST, this length is the particle radius, whilst the right definition should be the mean distance between random defects of the crystal. This correlation time would then be significantly reduced, hence the contribution to the relaxation rate.

The static dephasing regime

The so-called static dephasing regime (SDR) predicts a linear dependence of $1/T_2$ on the magnetic moment of the particles in solution (*22, 23*). This regime describes the transverse relaxation of homogeneously distributed static protons in the presence of static, randomly distributed point dipoles. The free induction decay rate, $1/T_2^*$, is then proportional to the part of the sample magnetization due to the dipoles (μ), $M = n\mu$, where n is the concentration of the point dipoles. The SDR was later shown to be relevant not only for static protons, but also for protons diffusing slowly amongst spheres with radius R, provided that $\gamma\mu/(RD) > 3.4$, where D is the water diffusion coefficient (*23,25*). For our solutions, this condition is clearly not satisfied (at 1 T, $\gamma\mu/R^3 = 3 \times 10^6$ rad/s and $R^2/D = 2.2$ ns, so $\gamma\mu/(RD) = 0.007$).

Inner sphere model

Hydrated iron oxides can adsorb heavy metals. These adsorption properties arise from the presence of structural hydroxyl groups on their surface, which exhibit amphoterism (*56*):

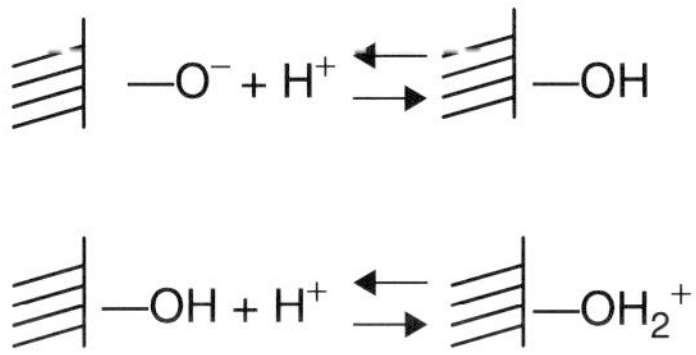

Recent nuclear magnetic resonance work (*48,62*) has confirmed the occurrence of a proton exchange in the relaxation mechanism of solutions of hydrated iron oxide nanoparticles, with an estimated proton exchange time of 33 ns for akaganeite particles at pH 6 and 37°C. However, considering a relaxation mechanism accounting for the exchange does not provide the solution for the problem: the inner sphere theory, aimed at describing relaxation of bound protons, also predicts a quadratic dependence of the rate on B_0 (*21*). Solving this problem requires a new theoretical approach.

C.2. A New Proton Exchange Dephasing Model (PEDM)

The relaxation dynamics

Protons bound to a magnetic particle with adsorption sites characterized by different Larmor angular frequency shifts ($\delta\omega_i$) will progressively become out of phase, and the signal $S(t)$ will decay according to

$$S(t)/S(0) = \int_{-\infty}^{\infty} (\cos\Phi) P_t(\Phi) d\Phi = \mathrm{Re}\left[\widetilde{P}_t(k=1)\right], \tag{16}$$

where $P_t(\Phi)$ is the phase–shift distribution at time t, $\widetilde{P}_t(k)$ is its Fourier transform, and Re stands for "real part of." The decay, and consequently the relaxation rate, is thus determined by the evolution of the phase–shift distribution. If each proton is assumed to jump randomly from one site to another, the accumulated phase shift after l jumps will be $\Phi_l = \sum_{i=1}^{l} \delta\Phi_i$, with $\delta\Phi_i = \delta\omega_i t_i$, and where t_i is the time spent at site i. When a random variable is defined as the sum of several independent random variables, its probability distribution is the convolution product of the distributions of the terms of the sum. The Fourier transform of the sum distribution function is then equal to the algebraic product of the Fourier transforms of the distribution functions of the individual terms of the sum. We thus obtain $\widetilde{P}_t(k) = \prod_{i=1}^{l} \widetilde{P}_{t_i}(k)$, where $t = \sum_{i=1}^{l} t_i$. After spending a time τ at site i in a static situation (between two jumps), the phase–shift distribution is directly deduced from $\Psi(\delta\omega)$, which is the distribution of the Larmor angular frequency shifts of the sites: $P_\tau(\delta\Phi) = \Psi(\delta\Phi/\tau)/\tau$.

Now, the linearity between $1/T_2$ and B_0 predicted by the above SDR is shown in the original article of Brown (*22*) to result from a Lorentzian distribution of phase shifts, which in turn arises directly from a Lorentzian distribution of the longitudinal component of the local magnetic field induced by the magnetic dipoles. Introducing such an assumption into Eq. (16) also leads to a linear dependence: the random jumps do not change the analytical shape of the initial phase–shift distribution, determined by the distribution of the Larmor angular frequencies of the sites: $\Psi(\delta\omega) = K/[\pi(\delta\omega^2 + K^2)]$, where K, the half width at half height of the Lorentzian distribution, is proportional to B_0.

Indeed, remembering that the Fourier transform of a Lorentzian function is an exponential, we obtain

$$\widetilde{P}_t(k) = \prod_{i=1}^{l} e^{-kKt_i} = e^{-kK\sum_{i=1}^{l} t_i} = e^{-kKt}, \qquad (17)$$

which introduced into Eq. (16) obviously yields a relaxation rate K (which, by definition, is proportional to B_0).

This simple model neglects the bulk water (the solvent), focusing only on the adsorbed protons. In fact, between two visits to an adsorption site, the water protons diffuse in the bulk, in the neighborhood of the particle, which enhances their dephasing and produces the OST relaxation, $(1/T_2)_{\text{diff}}$. This diffusion within the bulk water justifies an important assumption of our model: the statistical independence of the phase shifts accumulated during successive periods of adsorption. The estimation of the relaxation rate in the whole system accounts thus for the bulk water only by considering that protons undergo a fast exchange between this bulk and the adsorption sites. This exchange simply results in a weighting of the relaxation rate by the fraction of the adsorbed protons:

$$1/T_2 = (cq/111)K + (1/T_2)_{\text{diff}} \qquad (18)$$

where c is the particle concentration (in moles per liter), q is the number of exchangeable protons per particle, and 111 (moles per liter) is the proton concentration in water.

This model is close to the SDR. However, the SDR only deals with static spins, so that relaxation, more specifically the first-order effect we investigate, is automatically canceled by the refocusing pulses. On the contrary, our model accounts for irreversible dynamics: the randomness of the jumps from one site to another prevents signal restoration as soon as several jumps occur between two refocusing pulses; i.e., if $\tau_{CP} \gg \tau_M$.

How valid are the assumptions of our model? The occurrence of an exchange has been proven, estimates of the residence times are available (*48,62*), and they are within the boundaries ($\tau_M \approx 33$ ns, which is larger than $\tau_D = 2.2$ ns, and smaller than τ_{CP}). However, the Lorentzian shape of the distribution of the adsorption site frequency shifts is not yet verified. We deduced this shape from computer simulations that were aimed at calculating the magnetic field at different positions of the surface of an antiferromagnetic particle. Considering the particle as a homogeneously magnetized entity (e.g., a sphere) is too rough an approximation to obtain any information concerning the distribution of the adsorption site frequency shifts. The computation is therefore based on a more detailed representation of the magnetic structure of ferrihydrite and akaganeite, based on recently published Mössbauer and magnetometric data (*35,39*). The distribution of the adsorption site frequency shifts for ferrihydrite and akaganeite particles was computed from a model accounting for the antiferromagnetic and superparamagnetic nature of the crystal. This feature leads to considering the crystal magnetization as being due to the canting of the two sublattices opposed to each other, and to uncompensated moments resulting from defects of the crystal. The computed distributions are clearly Lorentzian (*63*). The proton exchange dephasing model is therefore validated.

C.3. Application of PEDM

The PEDM is able to explain the anomalous relaxation of solutions of ferritin and akaganeite particles, especially its linear dependence with B_0, the external magnetic field. The model is compatible with the observed dependence of the rate on pH. The relaxation rate predicted by the PEDM is proportional to the number of adsorption sites per particle (q); the values deduced for q from the adjustment of the model to experimental results (from NMR and magnetometry in solutions) are reasonable for hydrated iron oxide nanoparticles (*63*).

However, the application of this model to iron-containing tissues is not straightforward: it does not provide a direct explanation of the difference between ferritin relaxation in tissues and in solutions.

D. Ferritin in MRI

D.1. Introduction

The idea to measure iron by MRI, using the natural darkening caused by the presence of ferritin, has always been very attractive. Such a method could be performed non-invasively and in vivo. The evolution of ferritin concentration during different pathologies could be monitored, even in the brain which is impossible in vivo with current techniques (for example, tissue puncture and subsequent chemical analysis). As iron seems to play a role in Parkinson's and Alzheimer's diseases, the interest on ferritin -induced contrast has been important. Moreover, patients with hemocromatosis and thalassemia accumulate ferritin in the liver and the spleen. MRI detection would be less dangerous than tissue puncture. Many MRI protocols have thus been developed to estimate iron concentration in vivo.

D.2. MRI Methods for Iron-Content Measurement

The first method is to use T_2-weighted images, because the major effect of ferritin is to shorten the transverse relaxation time of water protons. However, the correlation with the iron content is often rather poor. Other groups used T_2^*-weighted imaging or the effect of echo time on the contrast of the images (*64–74*). An interesting study used the increase in $1/T_2$ caused by the increase in magnetic field: the technique consists of recording two images on two imagers working at different magnetic fields (0.5 T and 1.5 T). The difference in relaxation rate should be well-correlated with the ferritin content of the studied tissue (*75,76*). Another attempt with Gradient Recalled Echo sequences resulted in a rather good correlation (*77*).

For all these techniques, the general qualitative correlation between the measured parameter and the iron concentration was established. However, the measurements are too sensitive to the parameters of the experimental procedures and to physiological data beyond the iron content to allow for routine use in hospitals (*43*). Indeed, the relaxivity of ferritin is different in solutions, in the liver, in the spleen, in the brain, etc., which implies that MRI quantification protocols must be developed separately for each organ.

The problems to be solved are listed hereafter: the first problem is the lack of data concerning the exact proportions of low-molecular weight iron complex, ferritin-bound iron and hemosiderin-bound iron (*54*). However, transverse NMRD profiles of iron-loaded tissues indicate that the transverse relaxation is essentially caused by ferritin, because of the linearity between $1/T_2$ and B_0, which is a fingerprint of ferritin-induced relaxation. The relaxation properties of hemosiderin are less understood. If the proportion of ferritin and hemosiderin is not the same in all of the studied tissues, the correlation between the relaxation properties and the total iron concentration could be affected.

The second problem is the possible presence of deoxyhemoglobin in the tissues. Indeed, deoxyhemoglobin could contribute to the transverse relaxation but through a non-linear transverse rate enhancement since for this protein $1/T_2$ increases quadratically with the magnetic field. Such an effect has only been noticed for blood (*51*). As a conclusion, the present state-of-the-art allows to claim that MRI provides a qualitative indication and a quantitative estimation of iron accumulation, but is unable to approach the accuracy of the chemical determination of iron content.

D.3. Perspectives

The use of MRI to quantify iron in vivo is not straightforward: many problems have to be solved before achieving an accurate determination of iron concentration. However, MRI remains an excellent tool for the non-invasive detection of iron overload.

III. Gd(III)-Loaded Porous Systems as MRI Contrast Agents

A. Zeolite Particles as MRI Contrast Agents

Since the 1980s, interest exists in using magnetic resonance imaging (MRI) as an alternative non-invasive technique for the investigation of the gastrointestinal (GI) tract. This has given rise to the need for MRI contrast agents that can be administered orally and that fulfill the requirements for such diagnostic drugs. These are the following criteria: (*78*) (1) good patient acceptance; (2) uniform marking of the GI tract; (3) unchanged characteristics of contrast effects when diluted or concentrated throughout the GI tract; (4) no absorption of contrast material into the systemic circulation or adjacent tissues; (5) complete excretion of the (preferably non-metabolized) contrast agent from the GI tract; (6) no production of artefacts caused by motion or susceptibility; (7) increased sensitivity or specificity for diagnosis of the GI tract; (8) high effectiveness for low dose; (9) acceptable cost.

Microporous metal oxides, as zeolites and clays, modified by ion exchange with Gd(III)-ions have been proposed for this purpose. These crystalline aluminosilicates are chemically and thermally resistant. Zeolites have well-defined pore structures and channel systems of molecular dimensions (Figs. 22a and b) (*79*), while clays have a layered structure. Both of them consist of AlO_4 and SiO_4 tetrahedra, which are linked through oxygen bridges. The cations that balance the negative charge of the AlO_4 units can be exchanged, which allows the incorporation of paramagnetic ions into the structure. Therefore, aluminosilicate frameworks can serve as “carrier matrices”, encapsulating the Gd(III)-ion. This simple model already predicts that the accessibility of the encapsulated Gd(III)-ions for water plays a crucial role in the mechanism of “spreading” the paramagnetic effect to the bulk water.

The earlier works deal with commercially available zeolites and clays with either particle sizes around several micrometers or undefined particle size. The first trials involved zeolites, which were not exchanged with paramagnetic ions (*80*). NMRD profiles (0.01–30 MHz) were measured on 2.5% suspensions in water stabilized by 0.5% agar gel. The contribution of agar to the longitudinal relaxation rate was reproducible. A monotonic decrease (with very small slope) in the relaxation rate with increasing Larmor frequency was obtained in every case. An attempt was made to explain the NMRD profiles with a semi-quantitative model; the zeolites were treated as systems with high internal surface area with a possibility that the diffusion of the water molecules in the cavities is restricted. This should lead to a change in the magnetic field dependence of $1/T_1$. The results of a fitting showed that this restricted diffusion model cannot fully explain the experimental results.

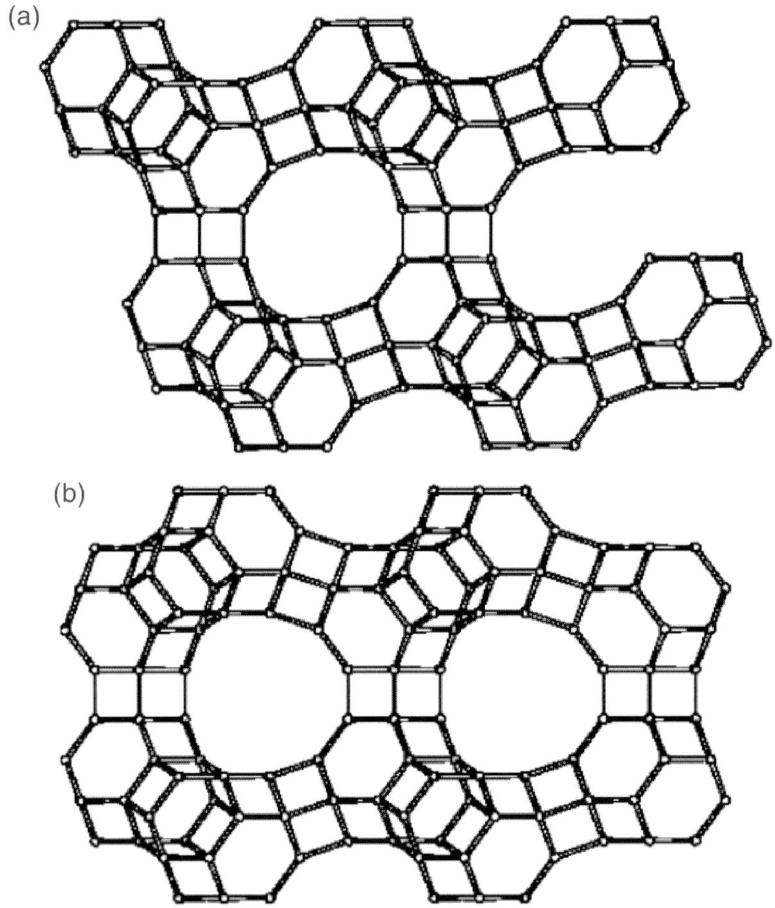

FIG. 22. (a) Connection mode in FAU type zeolite, showing the supercages (*79*); (b) connection mode in EMT type zeolite, showing the hypocages (*79*).

The authors conclude that contamination of the zeolites with paramagnetic ions might also play a role in the relaxation rate enhancement.

Sur *et al.* investigated the effect of direct Gd(III)-exchange in NaY zeolite on the water relaxivity (*81*). The NaY zeolite (faujasite polytype, FAU, Fig. 22a) structurally has a well-defined channel system consisting of 11.8 Å diameter cavities (supercages or α-cages) connected by twelve-membered rings with a diameter of 7.4 Å. Each supercage is surrounded by ten so-called sodalite units (or β-cages) with 6.6 Å internal diameters.

After the partial exchange of the Na^+ counter cations for Gd^{3+}, the specific relaxivity (here defined as the measured water ^1H relaxation rate per gram of material) was measured at Larmor frequencies ranging from 0.01 to 30 MHz. The NMRD profiles obtained were compared with that obtained after complexing the encapsulated Gd(III) with DTPA. After correction for the differences in the Gd(III) content of the samples, the NMRD curves appeared to be superimposable. Based on this result, a semi-quantitative model was applied to describe the mechanism of the relaxation rate enhancement of the solvent protons (see Section III.C).

Balkus *et al.* investigated the commercially available Y zeolites with an unspecified particle size, after Gd(III) exchange (*82,83*). A detailed study was made not only on the magnetic properties, but also on the preparation and the characterization of these materials. Several factors influencing the exchange, including the pH, time, and concentration, were optimized. Procedures to control the Gd(III) loading of the zeolites were developed.

The FT-IR spectra of the resulting Gd(III)-exchanged NaY zeolites did not show any loss of crystallinity or dealumination during the gadolinium exchange. A detailed XRD study also confirmed the retained crystallinity; nevertheless, there are significant changes in the relative intensities of certain reflections.

Two issues concerning the in vivo stability were evaluated: the effect of pH and the possibility of exchange of Gd(III) with cations that may be found in the GI tract. However, it should be noted that the acute toxicity of orally administered $GdCl_3$ is relatively low (LD_{50} mice > 2000 mg/kg).

GdNaY with 8.7 wt % (weight percentage) of gadolinium-loading was shown to be stable at pH > 2.5 with no leaching of Gd(III) or dealumination. At pH 2.0, about 3 % of the incorporated gadolinium was released after 2 h. During the next 6 h, no further leaching was observed, most likely due to the establishment of an exchange equilibrium between H^+ and Gd(III). The framework of the zeolite remains intact under these conditions. However, below pH 2.0, most of the Gd(III) was released within 0.5 h and, additionally, zeolite dealumination and loss of crystallinity occurred.

The stability of the GdNaY zeolites in solutions containing cations (Na^+, K^+, Ca^{2+}, Mg^{2+}) that typically occur in gastric and intestinal fluid was also studied. Using 0.025 M salt solutions (relatively high concentrations compared to the GI conditions), in the case of Na^+ and K^+, no release of gadolinium was detected. However, in the presence of polyvalent cations (Ca^{2+} and Mg^{2+}), leaching of 2–5 % Gd(III) took place.

Relaxivity studies were performed at 0.94 T, which corresponds to a ^1H Larmor frequency of 40 MHz. It was found that the relaxivities of zeolites with low Gd(III)-loadings (15–30 s^{-1} mM^{-1}) were higher than that of a $GdCl_3$ solution (10.3 ± 0.2 s^{-1} mM^{-1}), showing the τ_R effect. Moreover, the relaxivity for GdNaY increases with decreasing Gd(III)-loading. This was explained by the magnetic dilution of the intrazeolite Gd(III) at low Gd(III)-loadings, or the improved water mobility inside the pores. The most probable location of the Gd(III) is in the middle of the 12-ring openings of the supercages and, therefore, it may act as an obstacle for the diffusion of water molecules in and out of the zeolite pores.

Attempts were made to investigate the relaxivity enhancement produced by an MFI-type zeolite with framework-incorporated gadolinium (*84*). A relaxivity less than 3 s^{-1} mM^{-1} was found at room temperature and at 40 MHz. This indicates that the accessibility of gadolinium to water plays a very important role, and that the outer sphere effect might be very small.

Zeolite X and A with Gd(III) complexed by 8-hydroxyquinoline (8-HQ) was prepared either by the introduction of Gd(III)(8-HQ)$_2$Cl complex into the zeolite synthesis gel or by stirring the Gd(III)-exchanged zeolite X in an ethanolic solution of 8-HQ (*85*). In vitro MRI studies showed signal enhancement (T_1 shortening). However, the data cannot be compared with the previous results obtained with Gd(III)-loaded, but uncomplexed zeolites (because of partial complexation of Gd(III), different methods for determining the relaxivity enhancement, etc.)

Synthesis of zeolite Y in the presence of Gd(III) complexes of 18-crown-6 resulted not only in the encapsulation of the complex but the complex also served as a template for EMT polytype zeolite Y (Fig. 22b) (*86*). Feijen *et al.* described how the two different polytypes (the cubic FAU and the hexagonal EMT) can be formed (*87*). In the absence of an organic template, the FAU structure will form. If Na^+-18-crown-6 is present, it can be absorbed on the surface of the growing zeolite layer. This will influence the interconnection of the layers and, therefore, in the presence of this crown ether, the formation of the EMT framework may be favored. The difference between the pore window sizes is that in the EMT there are two different types: 7.3 × 7.3 Å in the [001] direction and 7.5 × 6.5 Å perpendicular to the [001] direction. (The FAU has pore "windows" with 7.4 × 7.4 Å in the [111] direction.)

The structural differences between the two zeolites resulted in a different stability at low pH values. No Gd^{3+} release was detected above pH ~ 2.5 for both materials. However, at lower pH values, the EMT framework appeared to be less stable against acid than the FAU framework.

Relaxivity studies showed increasing relaxivity with decreasing gadolinium-loading, similar to the behavior observed for the FAU type GdNaY zeolite. The measured relaxivities are in the same order of magnitude as those measured for FAU type GdNaY zeolite (*82*). Moreover, the Gd(III)-EMT-zeolite (without the crown-ether) exhibits a relaxivity similar to the Gd(III)-18-crown-6-EMT-zeolite with the same Gd(III)-content. These results are similar to those described for GdNaY and GdDTPANaY (*81*).

B. Clay Minerals

Bryant *et al.* in 1990 proposed the use of aqueous clay suspensions as MRI contrast agents for the abdomen. (*80, 88*). Two different types of clays, kaolin and bentonite were suggested particularly because these materials had been classified "Generally Recognized as Safe" by the Food and Drug Administration. Even the clays containing only diamagnetic ions cause shortening in mainly T_2, and to a lesser extent in T_1, the effects of which can mainly be ascribed to the induced changes in the motion of the water molecules in the anisotropic environment in the interlayer space of clays.

NMRD studies (0.01–30 MHz) on bentonite suspensions showed that the water-proton spin-lattice relaxation rates are dominated by magnetic interactions with paramagnetic centers entrapped in the mineral matrix (*89*). The $1/T_1$ values were linearly dependent on the concentration of the clay mineral (in 1–5 wt % suspensions). The relaxivity increased upon addition of Mn(II) to the clay suspension. Since the $1/T_1$-^{1}H Larmor-frequency profiles did not show inflections at 0.1 and 6 MHz, which were observed for aqueous Mn(II) solutions, the authors concluded that, here, the Mn(II) was fully complexed by the clay suspension. The increase in $1/T_1$ was small (15–30 %), but detectable. NMRD profiles were also measured in bentonite suspensions treated with 50 mM EDTA, which was reported to act not only as a chelating agent, but also as a pillaring agent to hold the clay layers apart. The changes in the longitudinal relaxation rates were in the order of the experimental errors. The conclusion, therefore, was that free metal ions make no significant contribution to the observed relaxation rate. Moreover, the thiocyanate test for detecting iron was negative. The relaxation rates decreased with increasing temperature, which suggests that the translational and rotational motions are important parameters in the relaxation enhancement (τ_R effect). The conclusion that the solvent relaxation in these systems is dominated by the interaction of the paramagnetic centers of the mineral was supported by EPR results.

Gd(III)-exchanged hectorite clay was investigated as a potential MRI contrast agent by Balkus *et al.* (*90*). Detailed studies, concerning the parameters affecting the preparations were conducted. The results showed that the content of the incorporated Gd(III) increases with the initial concentration of the Gd(III) in the exchange solution. The characterization of the Gd(III)-exchanged clay was done by XRD and FT-IR. An increased interlayer spacing was detected, which reflects the larger radius of the Gd(III) cation (3.18 Å) versus that of the original counter cation, Li^+ (0.68 Å).

The stability of a Gd(III)-exchanged hectorite was investigated over a pH range 1–3 at room temperature (26°C) and body temperature (37°C). At pH = 3 no Gd(III) leaching was detected after 4 h. Stirring the clay suspension at pH = 2 for 4 8 h resulted in release of about 15% of the Gd(III). At pH = 1 the structure collapsed (as shown by XRD and FT-IR) and about 90% of the Gd(III) was released. The stability of this clay in acidic medium is relatively low compared to zeolites.

In the presence of Na^+ ions, the Gd(III)-exchanged natural hectorite and the Gd(III)-exchanged synthetic hectorite showed a different behavior. At lower Na^+-concentration (0.13 M), none of the samples showed Gd(III)-release, whereas at higher Na^+-concentration (0.25 M), the natural sample released ~24 % Gd(III) and the synthetic sample did not show any exchange of Na^+ for Gd(III).

Relaxivity studies at 26°C and 40 MHz showed fair reproducibility without a sedimentation effect during the measurement time. An increase in relaxivity with decreasing Gd(III)-loading was observed, similar to the Gd(III)-exchanged zeolite samples. In all cases, the synthetic hectorite exhibited higher relaxivities than the natural hectorite, e.g., for a Gd(III) content of about 1.2 wt % the relaxivities were 36 s^{-1} mM^{-1} and 18 s^{-1} mM^{-1}, respectively. The authors conclude that although the stability of Gd(III)-exchanged clays is lower than that of the zeolite materials, they may be more effective in shortening T_1 than zeolites. In addition, the layered structure can be readily expanded, which facilitates suspension and makes the exchange between the bulk and the entrapped water faster.

A detailed study of the effect of different suspending agents on the relaxivity in Gd(III)-exchanged clay was made (*91*). The adsorption of macromolecules on microporous metal oxides was anticipated to result in enhanced relaxivity. The XRD patterns of the treated clays showed intercalation of poly(ethylene oxide), and an increased interlayer height, but in the presence of other suspending agents poly(ethylene glycol), xanthan gum, cholic acid, α- and β-cyclodextrins), there was no evidence of intercalation. The effect of intercalation of poly(ethylene oxide) on the relaxivity of Gd(III)-hectorite was investigated at 40 MHz and at room temperature for both natural and synthetic hectorite. For both of them, an increase in the room temperature relaxivity induced by the intercalation of poly(ethylene oxide) was observed. The authors ascribe this to: (i) an increase in the rotational correlation time (increase in microviscosity); (ii) an increased interlayer space, which facilitates the water exchange, and (iii) larger water content inside the increased interlayer space, caused by the intercalation of poly(ethylene oxide). A smaller increase in the relaxivity was observed with other surfactants. Here, exclusively the increased viscosity plays a role since intercalation was not observed.

C. Equations Describing the Paramagnetic Relaxivity Enhancement Caused by Gd(III)-Loaded, Porous Systems

Sur *et al.* (*81*) have shown that Gd(III)-loaded porous systems cannot be described by the Solomon–Bloembergen theory. It appeared, for example, that the relaxivity of zeolite-encapsulated Gd(III) is independent of the number of water molecules in the inner sphere of Gd(III), q (*81*). Moreover, the effective exchange between the paramagnetic center and the bulk water is not governed by the first-coordination-sphere dynamics but by other exchange processes, e.g., the diffusion of water into and out of the zeolite cages. Therefore, in a slightly modified model, the residence time of water in the first coordination sphere, τ_M, was replaced by the residence time inside the zeolite, τ_{zeo}; and the relaxation time of water in the first coordination sphere, T_{1M}, by the relaxation time of water inside the cages, T_{1cage}. To take into account the ratio of the paramagnetic cages to the overall number of cages, P_{cage}, the number of the paramagnetic centers, which is proportional to the metal content and inversely proportional to the water content, was included. Therefore, the modified equation is:

$$1/T_1 = 1/T_1^0 + \sum\{P_{cage}/(T_{1cage} + \tau_{zeo})\} \qquad (19)$$

The authors noted that the assumption in the original model that the metal center is dilute, is not appropriate for smaller cages. But the approach can show the approximate dependencies:

$$1/T_{1\text{cage}} = 1/T_1^0 + (q'/w)/(T_{1\text{M}} + \tau_\text{m}) \tag{20}$$

where w is the number of the water molecules inside the cage and q' is the number of water molecules in the first coordination sphere and where $T_{1\text{M}}$ is given by Eq. (21).

$$\frac{1}{T_{1\text{M}}} = \frac{2}{15}\left(\frac{\mu_0}{4\pi}\right)^2 \frac{\hbar^2\gamma_\text{S}^2\gamma_\text{I}^2}{r_\text{GdH}^6} S(S+1)\left(\frac{3\tau_{\text{c}1}}{1+\omega_\text{I}^2\tau_{\text{c}1}^2} + \frac{7\tau_{\text{c}2}}{1+\omega_\text{S}^2\tau_{\text{c}2}^2}\right) \tag{21}$$

Here, r_GdH is the effective distance between the gadolinium electronic spin and the water protons, γ_S (γ_I) is the electron (proton) gyromagnetic ratio and τ_ci is the correlation time, defined by

$$\tau_{ci}^{-1} = \tau_\text{M}^{-1} + \tau_\text{R}^{-1} + \tau_{ie}^{-1} \qquad (i = 1, 2) \tag{22}$$

where τ_R, the rotation correlation time concerns the rotation of the Gd(III)-water proton vector and τ_{ie} is the electron relaxation time.

Platas-Iglesias *et al.* investigated Gd(III)-doped zeolite NaY *nanoparticles* (80–100 nm as determined by XRD and HR-TEM) for their potential application as MRI contrast agent (*92*). These nanoparticles had a relaxivity (37.7 s^{-1} mM^{-1} at 60 MHz, 37°C) substantially higher than that reported by Balkus *et al.* for Gd(III)-doped zeolite NaY particles with undefined size. The exchange of Na^+ by La^{3+} ions was monitored by ^{23}Na NMR and appeared to be quite rapid. The nanoparticles were fully characterized in the solid state by various techniques. A combination of NMRD and EPR measurements was used to determine the parameters, which govern the relaxivity in these systems, and a two-step model was suggested to explain the transmission of the relaxivity from the paramagnetic centers in the zeolite cavities to the bulk water outside the zeolite. The two-step model consists of: (1) the exchange step between the Gd(III)-coordinated water and the zeolitic water inside the zeolite; and (2) the exchange step between the zeolitic water and the bulk water through the zeolite pores (Fig. 23).

For both processes approximate equations were derived from the exact solution of the Bloch equations for the longitudinal relaxation time of a system in which water protons undergo chemical exchange between two magnetically distinct environments A and B;

$$\frac{1}{T_1} = A_1 \pm \left[A_1^2 - \left(\frac{1}{\tau_{1\text{A}}\tau_{1\text{B}}} - \frac{1}{\tau_\text{a}\tau_\text{b}}\right)\right]^{1/2}, A_1 = \frac{1}{2}\left[\frac{1}{\tau_{1\text{A}}} + \frac{1}{\tau_{1\text{B}}}\right], \ \frac{1}{\tau_{1\text{A}}} = \frac{1}{T_{1\text{A}}} + \frac{1}{\tau_\text{a}} \tag{23}$$

where $T_{1\text{A}}$ is the intrinsic relaxation time in the A environment in the absence of exchange and τ_a is the residence time in the A environment, with

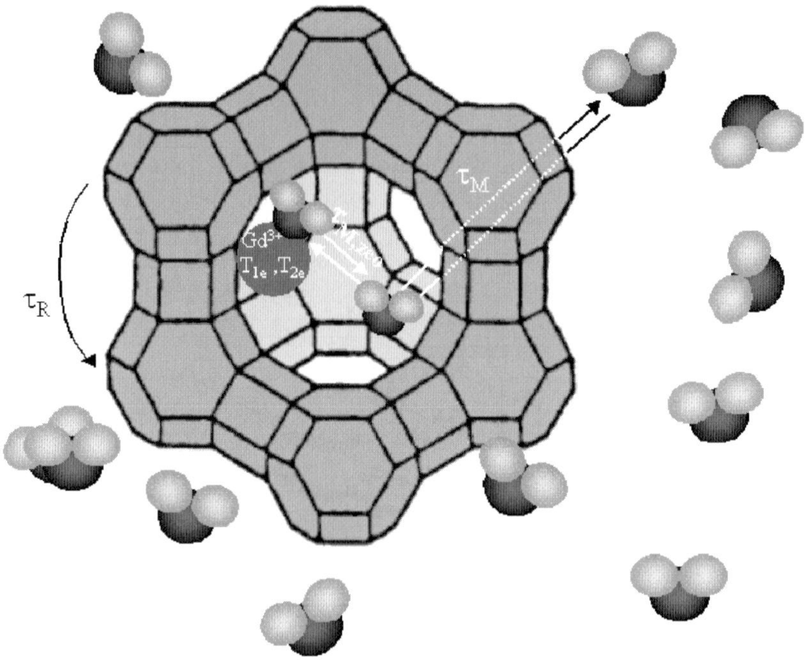

FIG. 23. Schematic representation of the mechanism describing the paramagnetic effect in Gd(III)-doped zeolite suspensions.

analogous definitions for the *B* environment. Considering the *A* system as the "bulk environment" and the *B* system as the water inside the zeolite cavities and assuming that $T_{1A} \gg T_{1B}$ and $\tau_a \gg \tau_b$, the observed proton relaxivity (r_1 in $s^{-1}\,mM^{-1}$) can be expressed as

$$r_1 = \left(\frac{1}{1000\,c}\right)\left(\frac{x+q}{55.5}\right)\left(\frac{1}{T_{1zeo}+\tau_{zeo}}\right) \tag{24}$$

where c is the Gd(III) concentration in mol L^{-1}, x is the number of the free (not Gd(III)-coordinated) water molecules inside the zeolite per Gd(III) ion, q is the number of the inner sphere water molecules coordinating to the Gd(III) ions, τ_{zeo} is the residence time of the water protons inside the zeolite, and T_{1zeo} is the relaxation rate of the water protons inside the zeolite, which can be further evaluated, being determined by the other exchange step undergoing inside the zeolite between the Gd(III)-coordinated water and the zeolitic, non-Gd(III)-coordinated water molecules. However, in this step the condition $\tau_a \gg \tau_b$ is not fulfilled, as it was assumed in *Ref.* (*81*). After the evaluation of this two-step exchange inside the zeolite, the following, surprisingly simple expression was obtained:

$$\frac{1}{T_{1zeo}} = \frac{\frac{q}{x}}{\tau_m + T_{1m}\left(1+\frac{q}{x}\right)} \tag{25}$$

in which τ_m is the mean residence time of water protons in the inner sphere. T_{1m} is the longitudinal relaxation time of the inner sphere protons, given by the Solomon–Bloembergen equation [Eq. (21)].

With this model, an excellent fit of the NMRD profiles of the dispersions containing GdNaY zeolite with different Gd(III)-loading was obtained. The results showed that the Gd(III) ions are immobilized effectively in the interior of the zeolite cavities, which is reflected in a very long rotation correlation time (> 1 ns). The relaxivity is mainly limited by the relatively slow diffusion of water protons through the zeolite channels from the interior of the zeolites to the bulk water. The decrease in relaxivity observed with increased Gd(III)-loading was explained mainly by the decrease in number of the water molecules per Gd(III) ion inside the zeolite cavities. The water exchange rate between the Gd(III)-coordinated water and the zeolitic water is in the same order of magnitude as in the case of the Gd(III) aqua ion and decreases with increasing Gd(III)-loading.

A similar approach could be used to explain the mechanism of the paramagnetic effect caused by other, insoluble particles with encapsulated paramagnetic ions.

D. Other Gd(III)-Loaded, Porous Systems

Particulates consisting of porous, particulate ion exchange resins with various paramagnetic ions (Cu^{2+}, Fe^{2+}, Fe^{3+}, Mn^{2+}, and Gd^{3+}) bound to their surface were investigated by Braybrook *et al.* (*93,94*). The resins used were sulfonated polystyrene, iminodiacetic acid substituted polystyrene and sepharose activated with 1,4-bis(2,3-epoxypropoxy) butane and substituted with iminodiacetic acid.

Some of the exchanged resins were coated with cellulose acetate butyrate or cellulose acetate phthalate. The results showed relaxation rate enhancement in 25% water suspensions containing 2% carboxymethylcellulose (CMC) as a surfactant. The relaxivities, however, were rather low. It should be noted, however, that the measurements were made at high field, 300 MHz, where the relaxivity enhancements are always smaller.

Each of the coated, metallated resins had significantly reduced relaxivities compared to the non-coated samples, which can be explained by the restricted water exchange through the coating.

Acid treatment of the uncoated resins resulted in Gd(III)-loss and, therefore, decrease in the relaxation rates, whereas the coated resins showed increase in the relaxation rate after acid treatment. Apparently, on one hand, the coating prevented the acidic solution to leach large quantities of Gd(III) ions, but on the other hand the acid treatment increased the porosity of the coating and improved the accessibility of water.

A new class of metal-loaded nanoparticles was developed by Reynolds *et al.* (*95*). These materials have a core-shell morphology, where the core is a functionalized polymer with a high affinity to the Gd(III) ions. The core polymer contained monomers with carboxylate pendant arms, such as ethylacrylate, methacrylate, butylacrylate or allylmethacrylate. The shell consisted of a porous hydrophobic polymer that modulates the access of the water molecules to the core. The particle size (120 nm) is small enough to pass through the vasculature. The longitudinal relaxivity measurements at 400 MHz showed significant shortening of T_1 in water suspension ($r_1 \sim 3.2\ s^{-1}\ mM^{-1}$). There was no effect of the Gd-loading on the relaxivity in the range 0.5–4.5 wt % of Gd(III) in the polymer. A slightly higher relaxivity in the core-shell particle as compared to the Gd(III)-containing core-polymer without shell demonstrates that the shell allows good water accessibility. In vivo studies in a rat animal model showed excellent contrast both in the vasculature (intravenous administration) and in the gastrointestinal tract (oral administration). The clearance from the vasculature was renal.

A novel type of colloidal particles with high relaxivity was prepared by Gløgård *et al.* (*96*). The preparation was based on the specific phase organization of amphiphilic Gd(III)-chelates with cholesterol. According to the theory of phase organization, the structure of a certain composition of amphiphiles is determined by the packing parameter, *P*, of the individual molecules. If the *P*-value is close to zero, the bilayer structure is preferred. Combination of a single chain amphiphile, such as Gd(III)-1,4,7-tris(carboxymethyl)-10-(2-hydroxyhexadecyl)-1,4,7,10-tetraazacyclododecane ($P < 1$, conical shape is preferred) with a molecule, such as cholesterol, with the opposite effect ($P > 1$), can form bilayered structures at particular ratios of the two components. The particle size in the various batches varied between 80 and 2000 nm. A cryo-electron micrograph showed mainly disc-shaped particles.

The effect of the composition and the particle size on the stability and on the relaxivity was investigated in detail. The results showed clearly that at least two different structures of the particles were formed depending on the composition. The formulations containing medium to high levels of cholesterol, showed a high T_1-relaxivity dependence on the particle size and on the cholesterol content. Higher relaxivity was obtained at higher cholesterol contents and at correspondingly increased particle sizes. This may be due to the increased rigidity and prolonged τ_R. The overall low relaxivity (12–18 $s^{-1}\ mM^{-1}$ at 20 MHz) in these formulations can be explained by the location of some of the Gd(III) chelates in the interior where the water exchange is slow (smaller τ_M), or by that the lateral motion on the surface is faster (larger τ_R).

The optimal formulations were obtained using a statistical approach (D-optimal design) and the particles obtained with these formulations had high relaxivities (20–25 $s^{-1}\ mM^{-1}$) and small particle sizes (80–100 nm). These formulations appeared to be highly stable in blood, since no change in T_1 relaxivity was observed when they where mixed with whole blood.

A very original application is the use of insoluble, polymeric Gd(III)-chelates as coating materials e.g., on the polypropylene catheters used for medical purposes (*97*). To poly(styrene-maleic acid) copolymer (SMA), HEDTA was conjugated and the Gd(III) ions were complexed by the resulting polymer, forming SMA-HEDTA-Gd(III). As a variation, the citric acid-HEDTA-SMA polymer was synthesized and used for chelating the Gd(III)ions, forming SMA-HEDTA-citric acid-Gd(III). The Gd(III) contents were 3.75 and

3.24%, respectively. The polished and ultrasonically cleaned polypropylene (PP) catheters were coated by immersing the tubes into solutions of the Gd(III)-containing polymer in *N,N*-dimethylformamide, followed by the removal of the solvent. ESCA measurements showed that the surface of the PP catheters coated with SMA-HEDTA-Gd(III) or SMA-HEDTA-citric acid-Gd(III) contained 0.81 and 0.74% Gd(III), respectively.

The water longitudinal relaxation rate measurements (300 MHz) showed $R_{1,\mathrm{obs,dia}} = 0.4\ \mathrm{s}^{-1}$ for water in the vicinity of the uncoated PP catheter's surface, while under the same conditions, R_1 values of 9.1 and 6.4 s^{-1} were measured for SMA- HEDTA-Gd(III) and SMA-HEDTA-citric acid-Gd(III), respectively. Dipping the catheters in BSA solution before the relaxation measurements resulted in a decrease in the water longitudinal relaxation rate. The longer the dip time, the larger was the decrease (50% reduction in 15 min). This might be explained by the aggregation of the BSA around the surface of the Gd(III) complex coating.

E. Conclusion, General Outlook

It is now well-established that paramagnetic ions encapsulated in inorganic or organic porous matrices can be used as MRI contrast agents. These contrast agents can have special use in investigations of the gastrointestinal part, or (in the case of suitable small particles) the vasculature. In order to design such Gd(III)-encapsulating, porous materials, several parameters must be taken into account: (i) good water accessibility with high water-exchange rate between the interior and the bulk (big pores, large surface, small particles), (ii) strong coordination of the Gd(III) into the framework (longer τ_R, no Gd(III)-leaching), and (iii) high water concentration inside the pores and high number of Gd(III)-coordinated water molecules. We also must take into account that the formulation of an MRI contrast agent must show the maximum signal enhancement with the minimum amount of material. It might have special importance in the case of Gd(III)-loaded zeolites and clays, where although the relaxivity ($\mathrm{s}^{-1}\ \mathrm{mM}^{-1}$) decreases with increasing Gd(III)-loading, the specific relaxivity ($\mathrm{s}^{-1}\ \mathrm{g}_{\mathrm{zeolite}}^{-1}$) shows a saturation-type curve as function of the wt % of Gd(III), therefore the optimum is around 4–5 wt % of Gd(III).

IV. Paramagnetic Micelles and Liposomes

Molecular imaging by magnetic resonance requires agents of high relaxivity and specificity. One classical approach to increase the relaxivity is to prolong the dipolar interaction between the paramagnetic metal and the water proton by covalent or non-covalent binding of the Gd-complex to slowly tumbling molecules like dendrimers, linear polymers or proteins. On the other hand, the specificity can be reached by tuning the overall characteristics (size, charge, etc.) of the edifice, or by coupling to a proper vector (peptide, sugar, etc.). The gain in relaxivity obtained by increasing the molecular size of the assembly as well as the vectorization can be obtained with supramolecular nanosystems.

In the context of the development of new contrast agents for medical diagnostic imaging, the nanoparticulate carriers have indeed drawn special attention. These systems allow for an efficient site-specific delivery of targeted contrast agents and, due to their size, provide some relaxivity enhancement with respect to the isolated paramagnetic molecules.

Several types of particulate compounds have been studied so far. Among these are solid nanoparticles containing gadolinium complexes (*98*) or macromolecular systems such as polylysine (*99*).

The earliest and most widely studied particulate carriers for paramagnetic substances are the liposomes, either entrapping hydrophilic contrast agents within the internal aqueous phase, or incorporating lipophilic complexes in their phospholipidic bilayer membrane (*100,101*).

More recently, micelles have also been proposed as contrast agents. They are colloidal particles with a hydrophobic core and a hydrophilic shell, formed by amphiphilic compounds (*102*).

A. Structure of Paramagnetic Micelles

Two types of micellar systems have been described, the first one includes Gd complexes capable of "self-organization" resulting in a supramolecular assembly (*103*), while the other class of micelles, also named "mixed micelles" is made of several components: a lipophilic gadolinium chelate, one or several phospholipid(s) and a non-ionic surfactant containing a polyoxyethylene chain (*104,105*).

One of the major features which made micelles attractive for diagnostic applications is their size, typically smaller than 100 nm, and their flexibility arising from properties (size, charge and surface properties) which are easily controlled and modified by suitable preparation and appropriate choice or combination of their components.

An efficient method for the preparation of mixed micelles has been reported by Tournier *et al.* and yields perfectly monodisperse micelles with a size between 15 and 100 nm (*104*).

The relative concentration of gadolinium can be significantly increased as compared to that achievable with liposomes because the physical stability of the mixed micelles is reached for relatively low amounts of additional lipids and surfactants. It is also worth mentioning that the "gadolinium-heads" of the complexes embedded in micelles are all exposed to the aqueous phase and can interact directly with the water molecules of the bulk, a situation which is usually not met with liposome systems.

B. Structure of Liposomes

Unilamellar liposomes are nanoparticles made of a bilayer, most often phospholipidic, entrapping an internal aqueous core. Formed spontaneously in the presence of an excess of water and above the gel-to-liquid crystal phase transition temperature (T_m) of the lipids, they are heterogeneous in size, number of bilayers and morphology. Well-defined unilamellar liposomes with a narrow size distribution can be obtained by extrusion (*106*).

Depending upon its properties, the paramagnetic material can be encapsulated into the aqueous core of the liposome (if hydrophilic), incorporated or intercalated in the membrane (if lipophilic or amphiphatic), or directly attached by chemical bond to the surface membrane (*107–109*).

For water-soluble agents, the encapsulation efficiency and thus the internal volume of vesicles which determines the amount of encapsulated material, is a very important factor with regard to the in vitro efficacy. It depends on the liposome size, lipid composition and lamellarity. When the number of liposomes delivered has to be known, the enumeration of the suspension can be performed by multinuclear NMR and photon correlation spectroscopy (*110*).

In the case of membrane incorporation, the trapping efficiency depends on the lipophilicity of the material, phospholipid concentration and phase behavior of the membrane. For compounds covalently bound to the membrane, the number of binding sites available is the upper limit of the loading factor.

C. Proton Relaxivity and Water Exchange

As known from classical theories and models, the enhancement of the water proton relaxation rates in the presence of chelated paramagnetic ion results from short distance (inner sphere) and long distance (outer sphere) magnetic interactions between the electronic spin and the hydrogen nuclei. The factors influencing the long distance interactions are the relative diffusion coefficient (D) of the water molecules and of the paramagnetic complex, the distance between them (d), and the electronic relaxation rates of the paramagnetic ion (τ_{S1}^{-1} and τ_{S2}^{-1}). For short distance interactions, a chemical bond between water molecules and the ion is involved. The resulting interaction depends on the distance between the bound nuclei and the paramagnetic ion (r) and on the relative value of the electronic relaxation times, the rotational correlation time of the complex (τ_R) and the residence time of the water molecule(s) bound to the ion (τ_M). The electronic relaxation times are field dependent but at high magnetic fields, only the longitudinal relaxation time still plays a role.

When included in large supramolecular structures, the complexes experience slower translational and rotational mobilities and are thus characterized by smaller translational diffusion coefficient and longer rotational correlation times. The outer sphere relaxation rate thus depends mainly on the water diffusion coefficient and is slightly more efficient. On the contrary, the inner sphere interaction can be largely enhanced due to an increase in the rotational correlation time. In very large structures, the rotational correlation time could even be too large to influence the relaxation rates. For instance, τ_R calculated from the Stokes–Einstein theory is larger than 1 μs for micellar diameters larger than 10 nm. In such a case, the electronic relaxation time will act as the effective correlation time modulating the interaction at low magnetic fields whereas at higher fields, both τ_{S1} and τ_M can be involved. This situation can lead to a very short relaxation time of the bound nuclei (T_{1M}) and hence, to a limitation of the relaxivity by the residence time of the bound water molecule. If $\tau_M \gg T_{1M}$, the proton nuclei are indeed partially or totally relaxed before exchanging with the bulk and the information is lost. Such limitations are encountered in nanosystems like liposomes and micelles.

D. Relaxivity of Micellar Systems

Paramagnetic amphiphilic complexes embedded in micelles have their hydrophilic head and thus the paramagnetic ion in contact with the surrounding water. If the complex has only one hydrophobic chain, the access of water to the ion should be easy but when two hydrophobic chains are involved like in DTPA bisamides complexes, the incorporation of both chains in the micellar structure could reduce either the accessibility of water to the ion or the water exchange rate. It has been reported that micellar Gd-DOTA complexes substituted by one hydrophobic chain are characterized by a water residence time similar to that of Gd-DOTA ($\tau_M^{298} \approx 200$ ns) and that their enhanced relaxivity results mainly from the increased rotational correlation time (*111*). On the contrary, Gd-DTPA bisamide complexes included in mixed micelles were shown to have a limited relaxation rate increase, probably due to prolonged water residence time ($\tau_M^{310} \approx 1.3$–1.5 μs) (Fig. 24 and Table VI) (*112*).

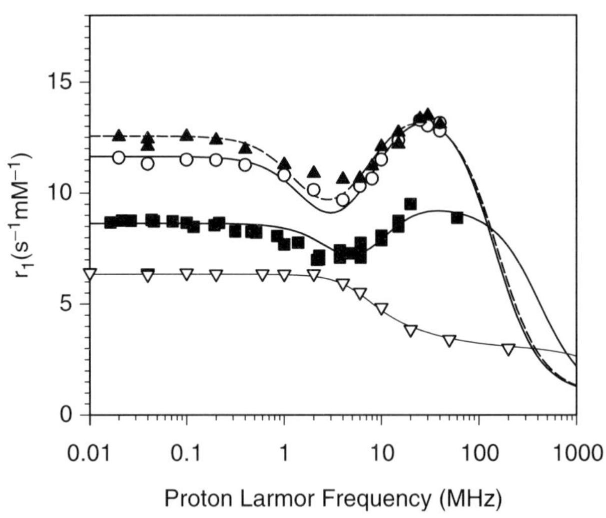

Fig. 24. NMRD profiles of Gd-DTPA-bis-tetradecylamide (circles), Gd-DTPA-bis-hexadecylamide (triangles up), and DTPA-bis-octadecylamide (DTPA-BC18) (squares) incorporated into mixed micelles made of phospholipid (DPPC) and a surfactant (Tween 80) (T = 37°C) (*112*). The mean diameters of the micelles are ranging between 15 and 20 nm. The NMRD profile of Gd-DTPA (triangles down) has been added for comparison.

TABLE VI

PROTON RELAXIVITY OF VARIOUS PARAMAGNETIC MICELLES AND LIPOSOMES

	T (°C)	r_1 at 0.47 T (s^{-1} mM^{-1})
LIPOSOMES		
intraliposomal GdHPDO3A (250 mM)	25	0.23
(DPPC/DPPG liposomes) (*114*)	37	0.60
	45	3.60
intraliposomal GdHPDO3A (250 mM)	25	0.085
(HPC/HPS liposomes) (*114*)	37	0.21
	55	3.00
Gd-bis-SDA-DTPA/OPPC/Biotin-XDHPE (*115*)	37	13.5
Gd-DMPE-DTPA/OPPC/Biotin-X DHPE (*115*)	37	11.5
MICELLES		
Gd-DTPA-(DSPE)2/DPPA/synperonic F108 (*104*)	39	14.3
Gd-DTPA-bis-tetradecylamide/DPPC/Twin80 (*112*)	37	13.2
Gd-DTPA-bis-hexadecylamide/DPPC/Twin80 (*112*)	37	13.3
Gd-DTPA-bis-octadecylamide/DPPC/Twin80 (*112*)	37	7.3
Gd-DOTAC10 (*111*)	25	10.8[a]
Gd-DOTAC12 (*111*)	25	17.9[a]
Gd-DOTAC14 (*111*)	25	22.0[a]
	37	18.8[a]
Gd-DOTASAC18 (*111*)	25	17.0[a]

[a]at 1.41 T

In addition, Nicolle *et al.* (*111*) have shown that the mobility of the embedded Gd-complex can be described by two rotational correlation times using the Lipari–Szabo analysis, i.e., a correlation time describing the overall rotational mobility of the micelle and a correlation time for the internal mobility of the Gd-complex inside the micellar structure.

E. RELAXIVITY OF LIPOSOME SYSTEMS

As mentioned so far, various types of paramagnetic unilamellar liposomes can be considered:

- paramagnetic liposomes containing a hydrophilic paramagnetic complex in their water core
- paramagnetic liposomes with lipophilic paramagnetic complex embedded in the external layer of the membrane
- paramagnetic liposomes with lipophilic paramagnetic complex embedded in the internal layer of the membrane
- paramagnetic liposomes with lipophilic paramagnetic complex embedded in both sides of the bilayer membrane.

Paramagnetic liposomes containing the hydrophilic paramagnetic complex in their water core

For these liposomes and depending on the concentration of the paramagnetic species, the hydrogen nuclei in the water core relax much faster than the hydrogen nuclei of the bulk water. If the water exchange through the lipidic bilayer is very slow when compared to the inner water relaxation rate, the relaxation evolution is bi-exponential with the smallest T_1 characteristic of the internal water and the largest one due to the bulk. If the water exchange through the membrane is faster, the paramagnetic relaxation rate measured on the bulk water peak (R_1^{obs}) in dilute liposomal aqueous systems will be given by Eq. (26):

$$R_1^{obs} = P^{inner}\left(\frac{1}{T_1^{inner} + \tau_{ex}}\right) \tag{26}$$

where P^{inner} is the molar fraction of the water inside the liposome, T_1^{inner} is the relaxation time of the intravesicular water and τ_{ex} is the residence time of water inside the liposome. If the exchange is slow compared to the internal relaxation rate ($\tau_{ex} \gg T_1^{inner}$), the observed relaxation rate is given by P^{inner}/τ_{ex} and is thus quite small. On the contrary, if the exchange is fast ($\tau_{ex} \ll T_1^{inner}$), the observed paramagnetic relaxation rate depends on T_1^{inner} and is thus larger. Such a behavior has been reported during the heating of various types of liposomes entrapping Gd-DTPA or Gd-HPDO3A (Table VI) (113,114) and an increase of the water exchange rate through the membrane on decreasing the size of the liposomes has been reported (*114*).

Paramagnetic liposomes with lipophilic paramagnetic complex embedded in the external layer of the membrane

In such liposomes, the paramagnetic center is in close contact with the external water and its relaxivity takes advantage of the reduced translational and rotational mobilities. This system is quite similar to the paramagnetic micellar structure described above. The relaxation rate increases due to the slower rotation of the complex and is mainly governed by the motion of the molecular segment carrying the paramagnetic head, by the electronic relaxation rates and by the water exchange between coordinated and bulk molecules.

Paramagnetic liposomes with lipophilic paramagnetic complex embedded in the internal layer of the membrane

This situation can be compared with the first one where the water core contains a hydrophilic paramagnetic complex. The vesicular water molecules relax faster than the bulk ones and the overall paramagnetic relaxation rate depends on the exchange rate of the water through the bilayer. The synthesis of such systems is cumbersome and, to the best of our knowledge, has not been reported.

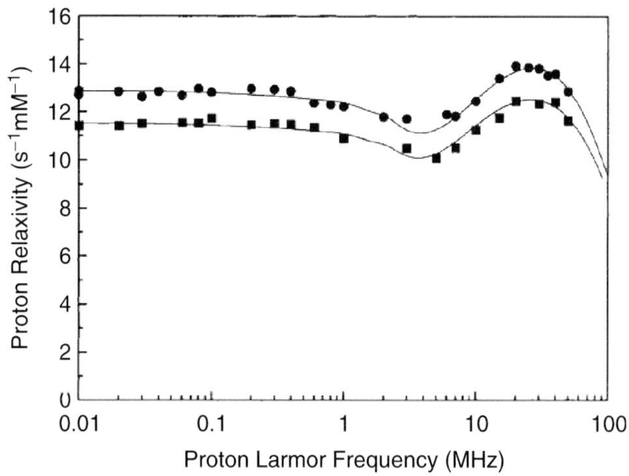

FIG. 25. Water proton NMRD profiles of solutions of paramagnetic liposomes containing Gd-bis-SDA-DTPA, OPPC, and Biotin-X DHPE at 298 K (■) and 308 K (●)(*115*)

Paramagnetic liposomes with lipophilic paramagnetic complex embedded in both sides of the bilayer membrane

This situation is more complex since:

- the number of paramagnetic complex molecules embedded in the inner (C_{in}) and outer (C_{out}) layers could be different with $C_{in} < C_{out}$.
- two contributions arising respectively from the complexes in the inner and in the outer layers have to be taken into account.

If the water exchange through the membrane is very slow, the main effect will be due to the complex of the outer layer. On the contrary, if this water exchange is extremely fast, complexes in the inner and in the outer layers will equally contribute to the observed paramagnetic relaxation rate as shown in Fig. 25 (*115*).

REFERENCES

1. Mendonca Dias, M. H.; Lauterbur, P. C. *Magn. Reson. Med.* **1986**, *3*, 328–330.
2. Stark, D. D.; Weissleder, R.; Elizondo, G.; Hahn, P. F.; Saini, S.; Todd, L. E.; Wittenberg, J.; Ferrucci, J. T. *Radiology* **1988**, *168*, 297–301.
3. Arnold, P.; Ward, J.; Wilson, D.; Guthie, J. A.; Robinson, P. J. *Magn. Reson. Imaging* **2003**, *21*, 695–700.
4. Weissleder, R.; Reimer, P.; Lee, A. S.; Wittenberg, J.; Brady, T. J. *Am. J. Roentgenol.* **1990**, *155*, 1161–1167.
5. Weissleder, R.; Moore, A.; Mahmood, U.; Bhorade, R.; Benveniste, H.; Chiocca, E. A.; Basilion, J. P. *Nature Medicine* **2000**, *6*, 351–355.
6. Grangle, J. "*Solid State Magnetism*"; Edward Arnold: London, 1991, pp. 155–181.
7. Bean, C. P.; Livingston, J. D. *J. Appl. Phys.* **1959**, *30*, 120–129.
8. Dormann, J. L. *Rev. Phys. Appl.* **1981**, *16*, 275–301.
9. Freed, J. H. *J. Chem. Phys.* **1978**, *68*, 4034–4037.
10. Ayant, Y.; Belorizky, E.; Alizon, J.; Gallice, J. *J. Phys. A* **1975**, *36*, 991–1004.

11. Pfeifer, H. *Ann. Phys. (Liepzig)* **1961**, *8*, 1–10.
12. Roch, A.; Muller, R.N.; *Proceedings of the 11th Annual Meeting of the Society of Magnetic Resonance in Medicine, Work in Progress* **1992**, *11*, 1447.
13. Gueron, M. *J. Magn. Reson.* **1975**, *19*, 58–66.
14. Gillis, P.; Roch, A.; Brooks, R. *J. Magn. Reson.* **1999**, *137*, 402–407.
15. Roch, A.; Muller, R. N.; Gillis, P. *J. Chem. Phys.* **1999**, *110*, 5403–5411.
16. Koenig, S. H.; Kellar, K. E. *Magn. Reson. Med.* **1995**, *34*, 227–233.
17. Roch, A.; Gillis, P.; Ouakssim, A.; Muller, R. N. *J. Magn. Magn. Mater.* **1999**, *201*, 77–79.
18. Roch, A.; Moiny, F.; Muller, R. N.; Gillis, P. "*Magnetic Resonance in Colloids and Interface Science*"; Eds. Fraissard, J.; Lapina, O.; Kluwer Academic Publishers, 2002, pp. 383–392.
19. Roch, A.; Bach-Gansmo, T.; Muller, R. N. *MAGMA* **1993**, *1*, 83–88.
20. Weisskoff, R. M.; Zuo, C. S.; Boxerman, J. L.; Rosen, B. R. *Magn. Reson. Med.* **1994**, *31*, 601–610.
21. Brooks, R. A.; Moiny, F.; Gillis, P. *Magn. Reson. Med.* **2001**, *45*, 1014–1020.
22. Brown, R. J. S. *Phys. Rev.* **1961**, *121*, 1379–1382.
23. Yablonskiy, D. Y.; Haacke, E. M. *Magn. Reson. Med.* **1994**, *32*, 749–763.
24. Jensen, J. H.; Chandra, R. *Magn. Reson. Med.* **2002**, *47*, 1131–1138.
25. Gillis, P.; Moiny, F.; Brooks, R. A. *Magn. Reson. Med.* **2002**, *47*, 257–263.
26. Majumdar, S.; Gore, J. C. *J. Magn. Reson.* **1988**, *78*, 41–55.
27. Bulte, J. W. M.; Brooks, R. A.; Moskowitz, B. M.; Bryant, L. H.; Frank, J. A. *Magn. Reson. Med.* **1999**, *42*, 379–384.
28. Kellar, K. E.; Fujii, D. K.; Gunther, W. H. H.; Briley-Saebo, K.; Bjornerud, A.; Spiller, M.; Koenig, S. H. *J. Magn. Reson. Imaging* **2000**, *11*, 488–494.
29. Ouakssim, A.; Fastrez, S.; Roch, A.; Laurent, S.; Gossuin, Y.; Piérart, C.; Vander Elst, L.; Muller, R. N. *J. Magn. Magn. Mater.* accepted for publication.
30. Harrison, P. M.; Arosio, P. *Biochim. Biophys. Acta* **1996**, *1275*, 161–203.
31. Russel, S. M.; Harrison, P. M. *Biochem. J.* **1978**, *175*, 91–104.
32. Cornell, R. M.; Schwertmann, U. "*The Iron Oxides*"; VCH: Weinheim, 1996, pp. 7–33.
33. Néel, L., "*Conférence à l'Ecole de physique théorique*"; Ed. De Witt C.; CNRS: Les Houches. Paris, 1961.
34. Schoffa, G. Z. *Z. Naturforshg.* **1965**, *20*, 167–172.
35. Gilles, C.; Bonville, P.; Wong, K. K. W.; Mann, S. *Eur. Phys. J. B* **2000**, *17*, 417–427.
36. Blaise, A.; Féron, J.; Girardet, J.-L.; Lawrence, J.-J. *C. R. Acad. Sc. Paris* **1967**, *265*, 1077–1080.
37. Bauminger, E. R.; Nowik, I. *Hyperfine Interactions* **1989**, *50*, 484–498.
38. Brooks, R. A.; Vymazal, J.; Goldfarb, R. B.; Bulte, J. W.; Aisen, P. *Magn. Reson. Med.* **1998**, *40*, 227–235.
39. Gilles, C.; Bonville, P.; Rakato, H.; Broto, J. M.; Wong, K. K. W.; Mann, S. *J. Magn. Magn. Mater.* **2002**, *241*, 430–440.
40. Kilcoyne, S. H.; Cywinsky, R. *J. Magn. Magn. Mater.* **1995**, *140–144*, 1466–1467.
41. Dickson, D. P. E. *J. Magn. Magn. Mater.* **1999**, *203*, 46–49.
42. Doyle, F. H.; Pennock, J. M.; Banks, L. M.; Mc Donnel, M. J.; Bydder, G. M.; Steiner, R. E.; Young, I. R.; Clarke, G. J.; Pasmore, T.; Gilderdale, D. J. *Am. J. Roentgenol.* **1982**, *138*, 193–200.
43. Brittenham, G. M.; Badman, D. G. *Blood* **2003**, *101*, 15–19.
44. Vymazal, J.; Brooks, R. A.; Zak, O.; Mc Rill, C.; Shen, C.; Di Chiro, G. *Magn. Reson. Med.* **1992**, *27*, 368–374.
45. Bulte, J. W.; Vymazal, J.; Brooks, R. A.; Pierpaoli, C.; Frank, J. A. *J. Magn. Reson. Imaging* **1993**, *3*, 641–648.
46. Gossuin, Y.; Roch, A.; Muller, R. N.; Gillis, P. *Magn. Reson. Med.* **2000**, *43*, 237–243.
47. Koenig, S. H.; Brown, R. D. III; Gibson, J. F.; Ward, R. A.; Peters, T. J. *Magn. Reson. Med.* **1986**, *3*, 755–767.
48. Gossuin, Y.; Roch, A.; Lo Bue, F.; Muller, R. N.; Gillis, P. *Magn. Reson. Med.* **2001**, *46*, 476–481.
49. Halle, B.; Johannesson, H.; Venu, K. *J. Magn. Reson.* **1998**, *135*, 1–13.

50. Bulte, J. W. M.; Miller, G. F.; Vymazal, J.; Brooks, R. A.; Frank, J. A. *Magn. Reson. Med.* **1997**, *37*, 530–536.
51. Brooks, R. A.; Vymazal, J.; Bulte, J. W. M.; Baumgarner, C. D.; Tran, V. *J. Magn. Reson. Imaging* **1995**, *5*, 446–450.
52. Gelman, N.; Gorell, J. M.; Barker, P. B.; Savage, R. M.; Spickler, E. M.; Windham, J. P.; Knight, R. A. *Radiology* **1999**, *210*, 759–767.
53. Vymazal, J.; Zak, O.; Bulte, J. W.; Aisen, P.; Brooks, R. A. *Magn. Reson. Med.* **1996**, *36*, 61–65.
54. Vymazal, J.; Urgosik, D.; Bulte, J. W. *Cell. Mol. Biol.* **2000**, *46*, 835–842.
55. Gossuin, Y.; Colet, J.-M.; Roch, A.; Muller, R. N.; Gillis, P. *J. Magn. Reson.* **2002**, *157*, 132–136.
56. Dzombak, D. A.; Morel, F. M. "*Surface Complexation Modeling: Hydrous Ferric Oxide*"; John Wiley & Sons: New York, 1990.
57. Meldrum, F. C.; Heywood, B. R.; Mann, S. *Science* **1992**, *257*, 522–523.
58. Bulte, J. W. M.; Douglas, T.; Mann, S.; Frankel, R. B.; Moskowitz, B. M.; Brooks, R. A.; Baumgarner, C. D.; Vymazal, J.; Strub, M.-P.; Frank, J. A. *J. Magn. Reson. Imaging* **1994**, *4*, 497–505.
59. Bulte, J. W. M.; Douglas, T.; Mann, S.; Vymazal, J.; Laughlin, P. G.; Frank, J. A. *Acad. Radiol.* **1995**, *2*, 871–878.
60. Aime, S.; Frullano, L.; Crich, S. G. *Angew Chem Int Edit* **2002**, *41*, 1017–1019.
61. Gillis, P.; Koenig, S. H. *Magn. Reson. Med.* **1987**, *5*, 323–345.
62. Gossuin, Y.; Roch, A.; Muller, R. N.; Gillis, P. *J. Magn. Reson.* **2002**, *158*, 36–42.
63. Gossuin, Y.; Roch, A.; Muller, R. N.; Gillis; Lo Bue, F. *Magn. Reson. Med.* **2002**, *48*, 959–964.
64. Stark, D. D.; Bass, N. M.; Moss, A. A.; Bacon, B. R.; McKerrow, J. H.; Cann, C. E.; Brito, A.; Goldberg, H. I. *Radiology* **1983**, *148*, 743–751.
65. Stark, D. D.; Moseley, M. E.; Bacon, B. R.; Moss, A. A.; Goldberg, H. I.; Bass, N. M.; James, T. L. *Radiology* **1985**, *154*, 137–142.
66. Engelhardt, R.; Langkowski, J. H.; Fischer, R.; Nielsen, P.; Kooijman, H.; Heinrich, H. C.; Bücheler, E. *Magn. Reson. Imaging* **1994**, *12*, 999–1007.
67. Ordidge, R. J.; Gorell, J. M.; Deniau, J. C.; Knight, R. A.; Helpern, J. A. *Magn. Reson. Med.* **1994**, *32*, 335–341.
68. Gomori, J. M.; Horev, G.; Tamary, H.; Zandback, J.; Kornreich, L.; Zaizov, R.; Freud, E.; Krief, O.; Ben-Meir, J.; Rotem, H.; Kuspet, M.; Rosen, P.; Rachmilewitz, E. A.; Loewenthal, E.; Gorodetsky, R. *Radiology* **1991**, *179*, 367–369.
69. Mavrogeni, S. I.; Gotsis, E. D.; Markussis, V.; Tsekos, N.; Politis, C.; Vretou, E.; Kremastinos, D. *MAGMA* **1998**, *6*, 7–12.
70. Papakonstantinou, O. G.; Maris, T. G.; Kostaridou, V.; Gouliamos, A. D.; Koutoulas, G. K.; Kalovidouris, A. E.; Papavassiliou, G. B.; Kordas, G.; Kattamis, C.; Vlahos, L. J.; Papavassiliou, C. G. *J. Magn. Reson. Imaging* **1995**, *13*, 967–977.
71. Fenzi, A.; Bortolazzi, M.; Marzola, P.; Colombari, R. *J. Magn. Reson. Imaging* **2001**, *13*, 392–396.
72. Clark, P. R.; Chua-anusorm, W.; St Pierre, T. G. *Magn. Reson. Med.* **2003**, *49*, 572–575.
73. Ye, F. Q.; Martin, W. R.; Allen, P. S. *Magn. Reson. Med.* **1996**, *36*, 153–158.
74. Ye, F. Q.; Allen, P. S.; Martin, W. R. *Mov. Disord.* **1996**, *11*, 243–249.
75. Bartzokis, G.; Aravagiri, M.; Oldendorf, W. H.; Mintz, J.; Marder, S. R. *Magn. Reson. Med.* **1993**, *29*, 459–464.
76. Bartzokis, G.; Cummings, J. L.; Markham, C. H.; Marmarelis, P. Z.; Treciokas, L. J.; Tishler, T. A.; Marder, S. R.; Mintz, J. *Magn. Reson. Imaging* **1999**, *17*, 213–222.
77. Bonkovsky, H. L.; Rubin, R. B.; Cable, E. E.; Davidoff, A.; Rijcken, T. H. P.; Stark, D. D. *Radiology* **1999**, *212*, 227–234.
78. Giovagnioni, A.; Fabbri, A.; Macconi, F. *Abdominal Imaging* **2002**, *27*, 367–375.
79. van Koningsveld, H. "*Schemes for Building Zeolite Framework Models*" http://www.iza-structure.org/databases Published on behalf of the Structure Commission of the International Zeolite Association; version okt. 2003.
80. Sur, S. K.; Heinsbergen, J. F.; Listinsky, J. J.; Bryant, R. G. *J. Colloid Interface Sci.* **1993**, *157*, 141–146.
81. Sur, S. K.; Heinsbergen, J. F.; Bryant, R. G. *J. Magn. Reson. Series A* **1993**, *103*, 27–33.

82. Bresinska, I.; Balkus, K. J. *J. Phys. Chem.* **1994**, *98*, 12989–12994.

83. Balkus, K. J.; Bresinska, I. *J. Alloys Compd. 1994, 207/208*, 25–28.

84. Bresinska, I.; Balkus, K. J.; Kowalak, S. *Petr. Preprints 38* 494 *in* Balkus, K. J.; Bresinska, I. *J. Alloys Compd.* **1993**, *207/208* 25–28.

85. Balkus, K. J.; Bresinska, I.; Young, S. W. *Proceedings of the 9th International Zeolite Conference* **1993**, 193.

86. Balkus, K. J.; Shi, J. *Microporous Mater.* **1997**, *11*, 325–333.

87. Feijen, E. J. P.; Vadder, K. D.; Bosschaerts, M. H.; Lievens, J. L.; Martens, J. A.; Grobet, P. J.; Jacobs, P. A. *J. Am. Chem. Soc.* **1994**, *116*, 2950–2957.

88. Bryant, R. G.; Listinsky, J. J. U.S. Patent 4,927,624.

89. Sur, S. K.; Heinsbergen, J. F.; Bryant, R. G. *J. Magn. Reson. Series A* **1993**, *103*, 8–12.

90. Balkus, K. J.; Shi, J. *J. Phys. Chem.* **1996**, *100*, 16429–16434.

91. Balkus, K. J.; Shi, J. *Langmuir* **1996**, *12*, 6277–6281.

92. Platas-Iglesias, C.; Elst, L. V.; Zhou, W.; Muller, R. N.; Geraldes, C. F. G. C.; Maschmeyer, T.; Peters, J. A. *Chem. Eur. J.* **2002**, *8*, 5121–5131.

93. Braybrook, J. H.; Hall, L. D. *Carbohydr. Res.* **1989**, *187*, c6–c8.

94. Braybrook, J. H.; Hall, L. D. *Polym. Int.* **1991**, *26*, 251–259.

95. Reynolds, C. H.; Annan, N.; Beshah, K.; Huber, J. H. *J. Am. Chem. Soc.* **2000**, *122*, 8940–8945.

96. Gløgård, C.; Stensrud, G.; Klaveness, J. *Int. J. Pharm.* **2003**, *253*, 39–48.

97. Guo, J.; Jiang, X.; Yang, C.-Z. *J. Appl. Polym. Sci.* **2003**, *87*, 1358–1364.

98. Morel, S.; Terreno, E.; Ugazio, E; Aime, S.; Gasco, M. R. *Eur. J. Pharm. Biopharm.* **1998**, *45*, 157–163.

99. Aime, S.; Botta, M.; Crich, S. G.; Giovenzana, G.; Palmisano, G.; Sisti, M. *Bioconjug. Chem.* **1999**, *10*, 192–199.

100. Unger, E. C.; Winokur, T. W.; MacDougall, P.; Rosenblum, J.; Clair, M.; Gatenby, R.; Tilcock, C. *Radiology* **1989**, *171*, 81–85.

101. Kabalka, G. W.; Davis, M. A.; Moss, T. H.; Buonocore, E.; Hubner, K.; Holmberg, E.; Maruyama, K.; Huang, L. *Magn. Reson. Med.* **1991**, *19*, 406–415.

102. Misselwitz, B.; Platzek, J.; Radüchel, B.; Oellinger, J.; Weinmann, H.-J. *MAGMA* **1999**, *8*, 190–195.

103. André, J. P.; Toth, E.; Fischer, H.; Seelig, A.; Mäcke, H. R.; Merbach, A. E. *Chem. Eur. J.* **1999**, *5*, 2977–2983.

104. Tournier, H.; Hyacinthe, R.; Schneider, M. *Acad. Radiol.* **2002**, *9* (suppl 1), S20–S28.

105. Anelli, P. L.; Lattuada, L.; Lorusso, V.; Schneider, M.; Tournier, H.; Uggeri, F. *MAGMA* **2001**, *12*, 114–120.

106. Mayer, L. D.; Hope, M. J.; Cullis, P. R. *Biochim. Biophys. Acta* **1986**, *858*, 161–168.

107. Fossheim, S. L.; Colet, J. M.; Mansson, S.; Fahlvik, A. K.; Muller, R. N.; Klaveness, J. *Invest. Radiol.* **1999**, *33*, 810–821.

108. Alhaique, F.; Bertini, I.; Fragai, M.; Carafa, M.; Luchinat, C.; Parigi, G. *Inorg. Chim. Acta* **2002**, *331*, 151–157.

109. Tilcock, C.; Ahkong, Q. F.; Koenig, S. H.; Brown, R. D. III; Davis, M.; Kabalka, G. *Magn. Reson. Med.* **1992**, *27*, 44–51.

110. Vander Elst, L.; Piérart, C.; Fossheim, S. L.; Raux, J. C.; Roch, A.; Muller, R. N. *Supramol. Chem.* **2002**, *14(5)*, 411–417.

111. Nicolle, G. M.; Toth, E.; Eisenwiener, K.-P.; Mäcke, H. R.; Merbach, A. E. *J. Biol. Inorg. Chem.* **2002**, *7*, 757–769.

112. Kimpe, K; Parac-Vogt, T.N.; Laurent, S.; Pierart, C.; Vander Elst, L.; Muller, R.N.; Binnemans, K. *Eur. J. Inorg. Chem.* **2003**, 3021–3027.

113. Koenig, S. H.; Ahkong, Q. F.; Brown, R. D. III; Lafleur, M.; Spiller, M.; Unger, E.; Tilcock, C. *Magn. Reson. Med.* **1992**, *23*, 275–286.

114. Fossheim, S. L.; Fahlvik, A. K.; Klaveness, J.; Muller, R. N. *Magn. Reson. Imaging* **1999**, *17*, 83–89.

115. Alhaique, F.; Bertini, I.; Fragai, M.; Carafa, M.; Luchinat, C.; Parigi, G. *Inorg. Chim. Acta* **2002**, *331*, 151–157.

MAGNETIC RELAXATION DISPERSION IN POROUS AND DYNAMICALLY HETEROGENEOUS MATERIALS

JEAN-PIERRE KORB[a] and ROBERT G. BRYANT[b]

[a]Laboratoire de Physique de la Matière Condensée, CNRS UMR 7643, Ecole Polytechnique, 91128 Palaiseau, France

[b]Chemistry Department, University of Virginia, Charlottesville, VA 22904-4319, USA

I. Introduction

Microporous and dynamically heterogeneous materials are common in everyday experience and consist of an open solid matrix that may be filled with a liquid. Many materials fall in this category including biological tissues, gels, food preparations, plasticized polymers, porous rocks, zeolites, among others. For our purposes, microporous systems permit percolation of liquids into and throughout the solid matrix where the dimensions of the liquid spaces are small, on the order of tens of Angstroms to a few microns. Some may consider the lower end of the dimension scale nanoporous, but we will not belabor this distinction. The liquid compartments may be geometrically complex with various levels of compartmental connectivity and different degrees of confinement. Both the degree of confinement and the nature of the connectivity may affect the characteristics of the liquid dynamics, and therefore, nuclear spin relaxation of the liquids in this environment. We distinguish two classes of the solid component for the purposes of discussing magnetic relaxation: proton-poor or inorganic solids, and proton–rich or organic solids. These systems contain in one way or another most aspects of the relaxation problems discussed in other contributions in this volume. The general introduction by Canet, which introduces the relaxation dispersion and the fundamental diamagnetic theories, is a foundation that we build on in the more complex heterogeneous

ADVANCES IN INORGANIC CHEMISTRY
VOLUME 57 ISSN 0898-8838 / DOI 10.1016/S0898-8838(05)57006-5

systems. Because adventitious paramagnetic metals are generally a component of all natural systems, the contributions by Kowalewski, Bertini, Aime, and Desreux are directly relevant although the dynamical regime suffered by the metal ion sites may be different from those discussed here. The contribution by Muller on relaxation by nanosystems has a different emphasis from the present one, but is clearly complimentary. Finally the importance of magnetic as well as atom or material exchange is central to almost all observations in heterogeneous systems, which is addressed in the contribution by Merbach.

In the case of inorganic solids such as natural rocks, clays, glasses, or ceramics, the bulk of the solid does not contain protons and those that are present, for example in surface silanol groups, are magnetically dilute. The magnetically active spins of the solid such as ^{29}Si or ^{11}B are poorly coupled to the liquid proton spins and, with few exceptions, are unimportant in the proton spin relaxation of the liquid. Unless great care is taken to clean high surface area systems, proton spin relaxation of the mobile liquid is usually dominated by the effects of relatively rare paramagnetic centers provided by the solid matrix as radical centers or inorganic ions such as $Fe^{(III)}$ or $Mn^{(II)}$ which are present in remarkably high levels in natural materials. Iron and manganese are also present at the level of parts per million or tens of parts per million in most reagent grade chemicals used for the synthesis of microporous materials, and as a result, may dominate liquid proton spin relaxation in a nominally diamagnetic material. Many of these systems are discussed in the recent proceedings of the International Conference on Magnetic Resonance in Porous Media (*1–3*).

In the case that the solid component is proton–rich as in a biological tissue, a polymer gel, or a plasticized organic engineering polymer, the proton spin system of the solid component may be magnetically dense, which permits efficient proton–proton coupling between the solid protons. Because the proton–proton dipolar coupling is not averaged by rapid rotation in a solid phase, the proton linewidth is generally in the range of tens of kHz which precludes high resolution detection unless high-speed magic angle spinning techniques are employed. The spin–lattice relaxation rate of the solid protons, however, may be slow or fast depending on the dynamics present in the molecules comprising the material. Generally, the efficient proton–proton coupling in the solid spin system causes spin-spin communication to be rapid compared with the spin–lattice relaxation. As a consequence, the solid spin system maintains a common spin temperature throughout a spin–lattice relaxation experiment, and a single spin–lattice relaxation rate constant characterizes the population. However, if a liquid experiences the solid surface and there is magnetic contact between the liquid and the solid spins, which is common, the magnetic relaxation of the liquid and the solid spins are coupled, and the spin–lattice relaxation is generally not described by a single exponential decay. Often the slow relaxation components are characterized for the liquid, but standard liquid relaxation equations are inappropriate for interpretation of the relaxation rate constants because the detected relaxation rate represents a mixture of effects from the liquid and the solid spin populations. In this case, the interpretation of the liquid-spin relaxation may contain important information about the solid component relaxation and vice versa. This situation is common for biological materials where the spin–lattice relaxation of mobile solvent or co-solute spins is dominated by the spin–lattice relaxation of the solid component.

The convenient NMR observables depend on the characteristics of the system studied, but generally the protons of the liquid are readily detected. The apparent NMR linewidths are often determined by the magnetic susceptibility inhomogeneities in the sample and do not directly reflect the dynamics of the liquid. This report will focus largely on theoretical approaches to understand the spin–lattice relaxation rate constants for both classes of microporous materials. The magnetic field dependence of the spin–lattice relaxation rate constant generally provides a useful dynamical characterization of the liquid and often a structural characterization of the confining media.

II. The Magnetic Relaxation Dispersion

The magnetic field dependence of the nuclear spin–lattice relaxation rate constant, $1/T_1$, is often called the Magnetic Relaxation Dispersion, MRD (*4–6*). The advantage of the MRD experiment is that it provides an accurate dynamical characterization at constant temperature. Thus, changes in the structure of the system with temperature do not pose as a complication in experimental characterization of the dynamics. In addition, phase changes may also be avoided, and the extensive magnetic field dependence provides a rigorous test of the theory employed to interpret the data. Nuclear magnetic resonance is experimentally executed in the range of radio frequencies from kilohertz to approximately 1 gigahertz; however, at all of these presently available Larmor frequencies, the relaxation processes must be stimulated because spontaneous relaxation is very inefficient. The nuclear spins couple to the magnetic fluctuations in the sample, which are caused by molecular motions including rotation, translation, atom transfer reactions, and internal molecular structural rearrangements. The fluctuations in the system are generally characterized in the time domain by an autocorrelation function; the nuclear relaxation rate constant is proportional to the spectral density function which is the Fourier transform of the correlation function evaluated usually at the Larmor frequency or twice the Larmor frequency of the nuclear or electron spin involved. Thus, the MRD profile provides a characterization of the molecular motions that drive the spin relaxation; the inflection frequencies in the MRD plot provide the dynamical rate constants directly.

The sensitivity of the NMR experiment is strongly dependent on the magnetic field used for the experiment, which precludes measurements at low field strengths with high sensitivity. A solution to this problem is to make the magnetic field time-dependent either by moving the sample from one field strength to another (*7*) or by changing the current in a magnet rapidly (*5,8–10*). The spins are brought to equilibrium in a relatively high field strength to achieve a large Boltzmann population, then the field is switched to a target value and the spins allowed to relax for a varied time after which the field is switched again to a convenient field strength where the magnetization is sampled. The detection sensitivity is related to the strength of the initial polarization field and the final detection field used, but is approximately independent of the relaxation fields sampled. This strategy provides a robust means for recording the relaxation dispersion profiles over four or more decades in the Larmor frequency at essentially constant signal-to-noise ratio. However, the finite field switching times limit the magnitude of the rate constants that may be observed. The present instrumentation technique permits the measurement of rate constants up to approximately 2000 s^{-1} in favorable samples; however, even at this large rate, it is possible to miss a rapidly decaying component that is lost during the field switching periods. Thus, in cases where magnetic coupling between populations may be important, the relaxation detected is generally the one corresponding to the slow component of the coupled relaxation solution.

III. Relaxation Dispersion of Mobile Liquids in Inorganic Matrices

A. Highly Purified Systems

It is difficult to obtain a high surface area system that confines a liquid in which the paramagnetic impurities are sufficiently dilute that the paramagnetic contribution to proton spin-relaxation is negligible. Nevertheless, several have been reported. There is usually no magnetic field dependence for the spin–lattice relaxation rate constant for an unconfined diamagnetic liquid at the presently achievable magnetic field strengths, except at the lowest field strengths when a chemical exchange event may contribute as in water (*11*), because the motions that modulate the magnetic couplings are more rapid than the nuclear Larmor frequencies of the highest fields. However, when a paramagnetic center is present, the paramagnetic contribution to the relaxation rate has terms involving both the nuclear and the electron Larmor frequency. Because the electron Larmor frequency is at least 658 times larger than the nuclear Larmor frequency, both translational and rotational contributions change with magnetic field strength, and correlation times in the range of tens of picoseconds may be easily measured experimentally as discussed by the other contributions to this volume. Because confinement of a liquid may change the correlation function significantly and introduce low frequency terms, both diamagnetic and paramagnetic effects are important for spin relaxation for confined liquids (*12–16*).

An important feature of liquids at surfaces or interfaces is that generally the molecules at the surface are translationally mobile with translational diffusion constants remarkably similar to those of the bulk liquid. There are several means for measuring translational diffusion constants of liquids at surfaces using NMR, including use of pulsed field gradient methods and relaxation dispersion methods. The experimental issue is to isolate the surface from bulk liquid, which may be done using physical methods such as freezing the bulk water phase or using the strong distance dependence of the dipole–dipole coupling. Pulsed field gradient NMR experiments have been used to characterize diffusion coefficients for liquids at interfaces sometimes taking advantage of the remaining liquid NMR signal in systems below the liquid freezing point. The resulting diffusion constant is generally found to be within a factor of ten for that of the bulk liquid at similar temperatures (*17–26*). This conclusion is supported by MRD experiments in which a stable radical with a long electron spin–lattice relaxation time is affixed to the surface studied (*27–29*). In this case, the relaxation of the liquid is dominated by the relative translational motion of the liquid in the vicinity of the immobile electron spin. The analysis of the relaxation dispersion profiles using translational correlation functions yield translational diffusion constants similar to those deduced from the pulsed field gradient experiments. Thus, the mobility of a liquid molecule in the immediatevicinity of a solid or macromolecular interface is not at all solid-like.

The local rapid rotational and translational motion of liquids at interfaces, and therefore in microporous materials, produces a very significant averaging of the potentially anisotropic magnetic interactions in the liquid spin system. Dipolar splittings in proton spectra are rare in wetted systems, and even quadrupolar splittings in deuteron systems are not common except in systems with significant long-range order such as liquid crystals. Thus, the effects of the surface are not immediately apparent in the readily detected proton spectrum of the liquid. The deuteron nuclear electric quadrupole coupling provides larger intramolecular interaction that requires faster motions to average completely. There are more reports of deuteron splittings, but in most systems, the first order quadrupole splittings are not apparent, and the effects of surface orientation are reflected in the spin relaxation (*30*).

The definition of pore size in a microporous material is, to some extent, a function of the measurement technique used. The temperature dependence of the liquid signal intensity may be used to measure the pore volume, resulting in generally larger pore volumes than those obtained using the mercury-intrusion method (*31–33*). Stapf and Kimmich have proposed a method based on the change in relaxation rates at low magnetic field strength and argue that deuterons provide an advantage because unlike protons, cross-relaxation to surface protons is unimportant (*34*). Pore dimension is often a parameter in more complex models discussed below.

Kimmich and coworkers have studied the magnetic relaxation dispersion of liquids adsorbed on or contained in microporous inorganic materials such as glasses and packed silica (*34–43*) and analyze the relaxation dispersion data using Lévy walk statistics for surface diffusion to build models for surface reorientations mediated by translational diffusion (*30*). The detected spectral line is generally narrow reflecting rapid local rotational and translational diffusion; however, the averaging of the anisotropic interactions may not be complete. The long-time averaging processes then lead to relaxation dispersion in the accessible range of Larmor frequencies. The surface boundaries partly constrain the liquid molecules on the assumption that the rotational reorientation is coupled to translational motion, which may experience a reduced dimensionality caused by the confinement. This class of models does not exclude local rotational reorientation; indeed, there is ample evidence of such motion in the narrow resonance lines detected; however, these local dynamics are incomplete and the averaging of the residual line width is presumed to be a slow process coupled to translational motion on a rough surface. A simple example of such processes not directly relevant to the relaxation model involves averaging of first order quadrupolar splittings by diffusion of deuterium oxide from one micro liquid crystalline domain to another, the sizes of which may be controlled by the mechanical and thermal history of the sample. If the domains are small, a non-Lorentzian line is obtained, but if the domains are large, a quadrupole splitting is resolved. These spectroscopic differences may be readily observed even though the thermodynamic phase is the same for both spectra (*44*).

Figure 1 shows data for the relaxation of liquids in Vycor glasses (*34*) which are representative. The field dependence is described by a power law,

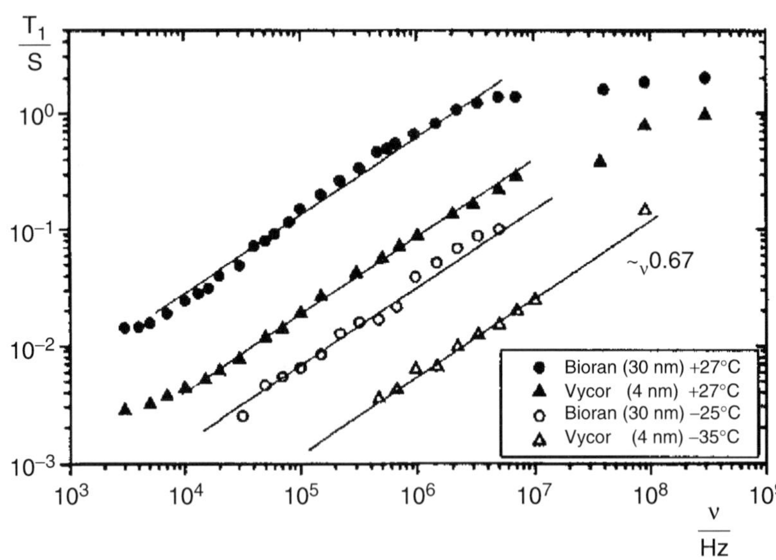

FIG. 1. Magnetic field dependences of the proton spin–lattice relaxation time of water in Bioran B30 and Vycor glasses at temperatures above 27°C and below the temperature where the non-surface water freezes (−25°C and −35°C). The solid lines represent the power law in the Larmor frequency with an exponent of 0.67 (*34*).

$T_1 \propto \nu^{0.67}$. The observed NMR signal generally decays exponentially, which supports the conclusion that there is rapid material exchange between surface and bulk phases of liquid in the pore. The shift on the frequency axis associated with the freezing event is consistent with the blocking of rapid exchange between bulk and surface phases; i.e., elimination of the bulk phase. A similar power law is found for tetradecane in Vycor above the freezing temperature.

Relaxation dispersion data for water on Cab-O-Sil, which is a monodisperse silica fine particulate, are shown in Fig. 2 (*45*). The data are analyzed in terms of the model summarized schematically in Fig. 3. The γ process characterizes the high frequency local motions of the liquid in the surface phase and defines the high field relaxation dispersion. There is little field dependence because the local motions are rapid. The β process defines the power-law region of the relaxation dispersion in this model and characterizes the molecular reorientations mediated by translational displacements on the length scale of the order of the monomer size, or the particle size. The α process represents averaging of molecular orientations by translational displacements on the order of the particle cluster size, which is limited to the long time or low frequency end by exchange with bulk or free water. This model has been discussed in a number of contexts and extended studies have been conducted (*34,41,43*).

Working with different microporous materials, Korb and coworkers built a different model for the dynamics of the liquid. As in other approaches (*46*), the effective dimensionality for the diffusive process that drives relaxation is lower than three, and the model is based on locally planar constraints.

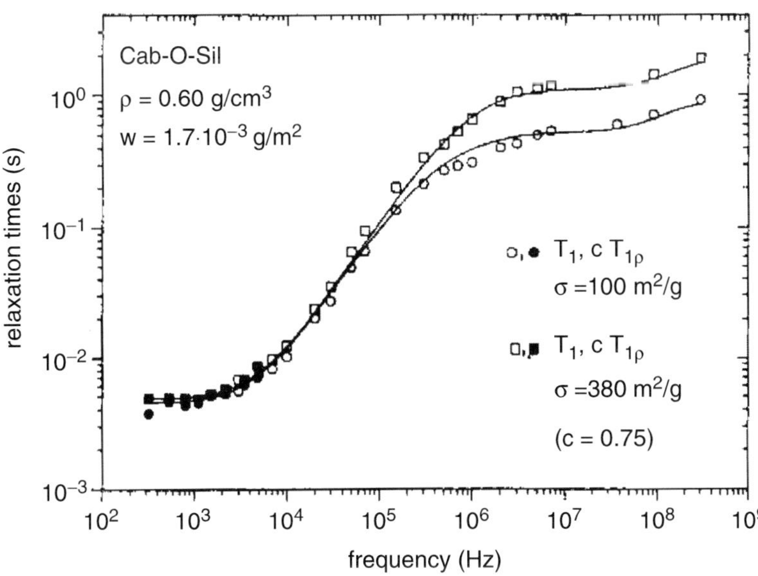

FIG. 2. Water proton spin–lattice relaxation dispersion in Cab-O-Sil M- samples with two different degrees of compression. The solid lines were computed using the model as in Ref. (*45*).

FIG. 3. Schematic representation of the topological space of hydration water in silica fine-particle cluster (*45*). The processes responsible for the water spin–lattice relaxation behavior are restricted rotational diffusion about an axis normal to the local surface (γ process), reorientations mediated by translational displacements on the length scale of a monomer (β process), reorientations mediated by translational displacements in the length scale of the clusters (α process), and exchange with free water as a cutoff limit.

The relaxation equation derived for this case is:

$$\frac{1}{T_1} = \frac{\pi\gamma^4\hbar^2}{80}\frac{\sigma}{d^2\delta^2}\tau_\perp\left\{3\ln\left(1+\frac{1}{(\omega_I\tau_\perp)^2}\right) + 12\ln\left(1+\frac{1}{4(\omega_I\tau_\perp)^2}\right)\right.$$
$$\left. +2\left(1+\frac{d^2}{\delta^2}\right)\left[5+\omega_I\tau_\perp\left(\arctan(\omega_I\tau_\perp)+\arctan(2\omega_I\tau_\perp)-\frac{9\pi}{2}\right)\right]\right\} \quad (1)$$

where d is the average interplane spacing or pore size, δ is a diffusional jump length, which is of the order of the molecular diameter or smaller, $\tau_\perp$ the translational correlation time for diffusion perpendicular to the director normal, $\tau_\perp = \delta^2/D_\perp$, where D is the translational diffusion constant perpendicular to the director normal, σ is the number of spins per unit of surface area, γ the magnetogyric ratio for the nucleus, and ω_I the nuclear Larmor frequency (*47,48*). The geometry is summarized in Fig. 4. The spin–lattice relaxation rate for non-wetting liquids was examined in model porous silica glasses. The relaxation is intermediate between that for two-dimensional and

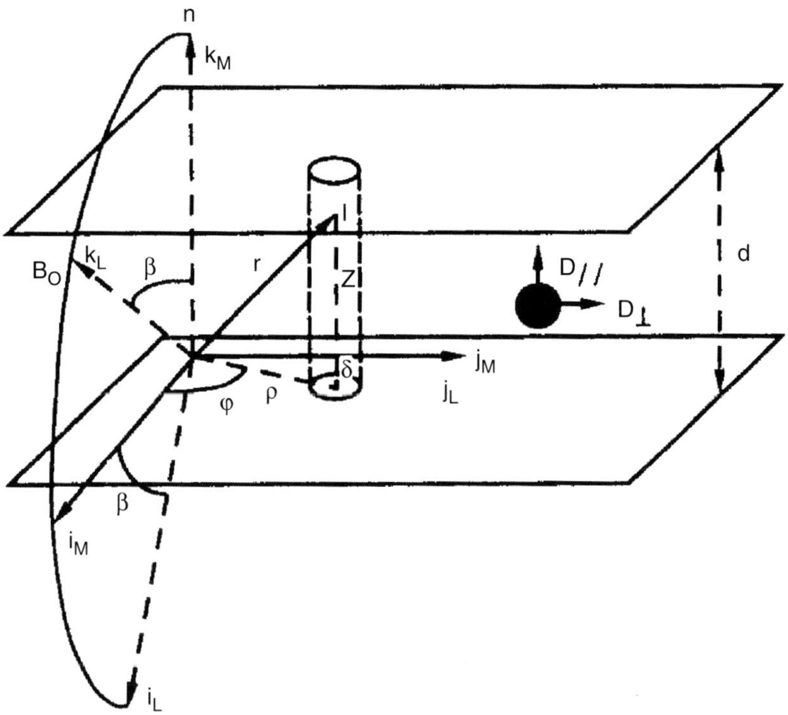

FIG. 4. Schematic diagram of the layered model for a pore (*47*). The two nuclear spins diffuse in an infinite layer of finite thickness d between two flat surfaces. The M axes are fixed in the layer system. The L axes are fixed in the laboratory frame, with $\mathbf{B}_o$ oriented at the angle β from the normal axis **n**. The cylindrical polar relative coordinates ρ, φ, and z are based on the M axis. The smallest value of ρ corresponding to the distance of minimal approach between the two spin bearing molecules is δ.

three-dimensional cases. At low magnetic fields, the relaxation varies logarithmically for small pore sizes and tends to a constant as the pore size increases. For small pore sizes, the relaxation depends on $1/d^2$ provided that the diffusing liquid does not have a significant interaction with the confining surface, which can be chemically adjusted by surface modifications (*49*).

In the more general case where there may be surface interactions, i.e., a chemical exchange with surface sites where more efficient relaxation may occur, a term is added to the relaxation equation that is proportional to $1/d$. The relaxation may generally be written as the sum of contributions:

$$\frac{1}{T_1} = \frac{a}{d^2} + \frac{b}{d} + \frac{1}{T_{1\text{bulk}}} \tag{2}$$

where the first term results from the geometrical confinement and the second from specific surface interactions. Figure 5 shows that for methylcyclohexane in a microporous silica system, the geometrical confinement dominates and the quadratic pore size dependence is observed. However, for pyridine,

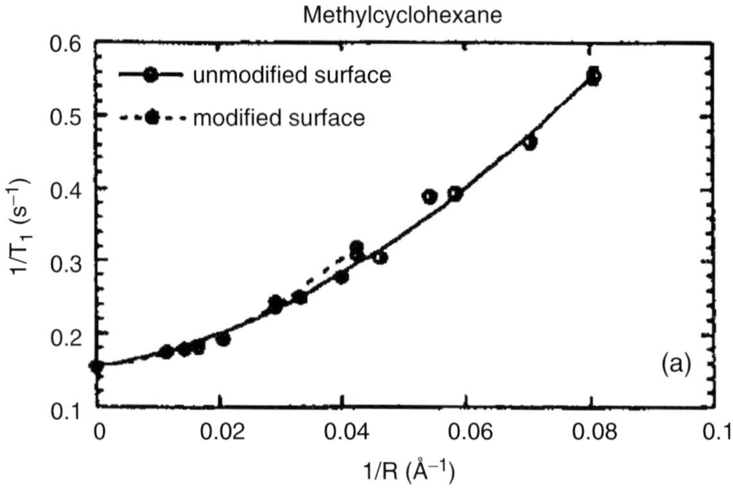

FIG. 5. The ^{1}H spin–lattice relaxation rates of methylcyclohexane protons as a function of the pore radius reciprocal in porous sol-gel glasses at 300 K and a resonance frequency of 180 MHz (*49*).

there is a linear relationship between the spin–lattice relaxation rate and $1/d$, which indicates that the exchange with surface sites dominates. Although the effects of confinement on a completely intramolecular process such as deuterium relaxation does not necessarily follow, it is interesting that for the deuterated pyridine, the spin–lattice relaxation rate depends on $1/d$ as shown in Fig. 6 (*50*).

B. Natural Systems

It is often useful or necessary to study systems that have not been extensively cleaned to remove trace paramagnetic impurities. In these cases, the paramagnetism generally dominates the proton spin relaxation. If the predominant motion driving relaxation is molecular rotation, for example in a paramagnetic metal ion aqua complex, the correlation function describing the electron nuclear coupling is exponential and the relaxation dispersion profile, a Lorentzian function or a sum of Lorentzian functions. Hexaaquairon(III) ion is a classic example as shown in Fig. 7, where the MRD profile is described by the Solomon, Bloembergen, Morgan equations (*51,52*). If the electron–nuclear coupling is modulated by the relative translational motion of the spins, the MRD profile is broadened as shown in Fig. 8 and described by a more complex translational correlation function (*53–55*). Analysis of the MRD profile in this case using theories by Ayant or Freed yields the relative translational diffusion constant in the vicinity of the

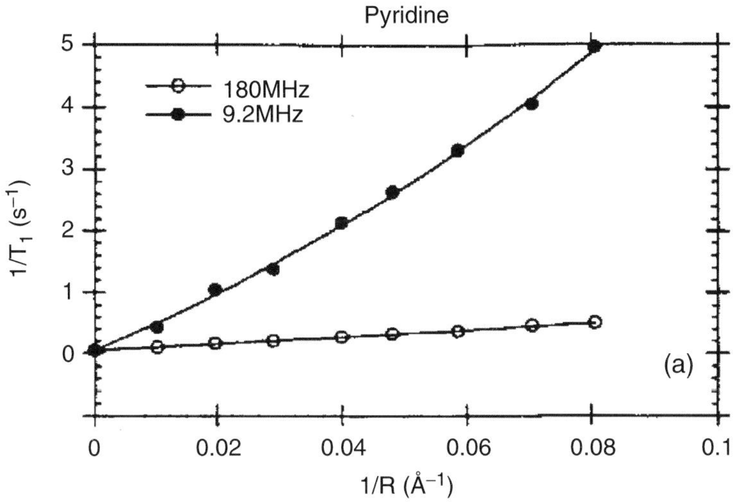

FIG. 6. The ¹H spin–lattice relaxation rates of pyridine as a function of the pore radius reciprocal in porous sol-gel glasses at 300 K at Larmor frequencies of 180 MHz and 9.2 MHz. The solid lines are the best fits to the equation $1/T_1 = a_1/R^2 + b_1/R + 1/T_{1b}$. There is only a slight curvature at 9.2 MHz (*50*).

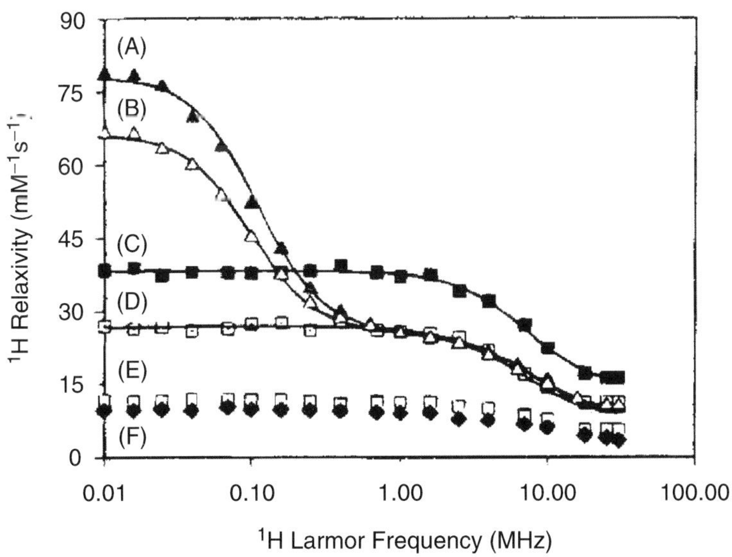

FIG. 7. ¹H water proton relaxivity, i.e., the nuclear spin–lattice relaxation rate per mM of metal, plotted as a function of the magnetic field strength expressed as the proton Larmor frequency for aqueous solutions of manganese(II) and iron(III) ions at 298 K. (A) 0.10 mM manganese(II) chloride in 2.80 M perchloric acid; (B) 0.1 mM aqueous manganese(II) chloride at pH 6.6; (C) 0.5 mM iron(III) perchlorate in 2.80 M perchloric acid; (D) 0.5 mM iron(III) perchlorate in water at pH 3.1; (F) 2.0 mM Fe(III) in 2.0 M ammonium fluoride at pH 7, which causes a distribution of species dominated by $[FeF_6]^{3-}$.

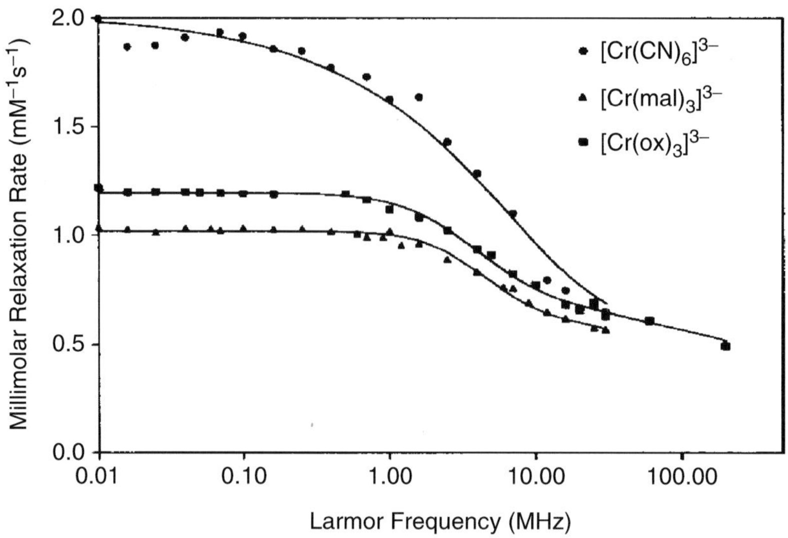

FIG. 8. The water-proton spin–lattice relaxation rates *vs.* magnetic field strength plotted as the Larmor frequency at 282 K for hexacyanochromate(II) ion (●), trioxalatochromate(III) ion (■), and trimalonatochromate(III) ion (▲). The lines were computed using translational diffusion models developed by Freed with and without the inclusion of electron spin relaxation effects *(54,121)*.

paramagnetic center and a distance of closest approach between the electron and nuclear spins. Because the dipole–dipole coupling is so strongly dependent on distance, this measure of the diffusion constant is heavily weighted by the dynamics within a few molecular diameters of the interacting spin pairs and provides one approach for dynamical characterization of liquid dynamics close to surfaces *(27,28,56)*.

If the paramagnetic center is part of a solid matrix, the nature of the fluctuations in the electron–nuclear dipolar coupling change, and the relaxation dispersion profile depends on the nature of the paramagnetic center and the trajectory of the nuclear spin in the vicinity of the paramagnetic center that is permitted by the spatial constraints of the matrix. The paramagnetic contribution to the relaxation equation rate constant may be generally written as

$$\frac{1}{T_{1I}} = \frac{2}{3}\gamma_I^2\gamma_S^2\hbar^2 S(S+1)\left[\frac{1}{3}J_L^{(0)}(\omega_I - \omega_S) + J_L^{(1)}(\omega_I) + 2J_L^{(2)}(\omega_I + \omega_S)\right] \quad (3)$$

where the spectral $J_L^{(m)}$ in the laboratory frame (L) associated with the direction of the magnetic field are the exponential Fourier transforms of the pairwise dipolar correlation functions *(51,52)*. The particular form for the spectral densities depends on the way that the observed liquid spin moves in the vicinity of the paramagnet, which is determined by the nature of the confinement.

There are three limiting cases: (1) If the confinement is minimal and the diffusive exploration three-dimensional, the relaxation rate is described by a Lorentzian function as shown in Fig. 7 (*57*), (2) If the spaces explored by the diffusing spins are essentially planar, or the exploration 2-dimensional, the relaxation rate depends on the logarithm of the nuclear Larmor frequency, and (3) If the spaces are very confined such that the branch points are uncommon compared with the diffusive encounter with a paramagnet, the diffusive exploration is practically one-dimensional and the MRD profile is a power law in the nuclear Larmor frequency. The first case describes both diamagnetic and paramagnetic systems in liquids, which have been the subject of extensive discussion for many years. The paramagnetic contributions to relaxation provide an excellent basis for the characterization of liquid dynamics in the picosecond regime (*58–64*). This discussion will focus on cases 2 and 3.

Figure 9 shows MRD profiles for acetone protons in calibrated chromatographic porous glasses (*65*). The field dependence consists of two logarithmic

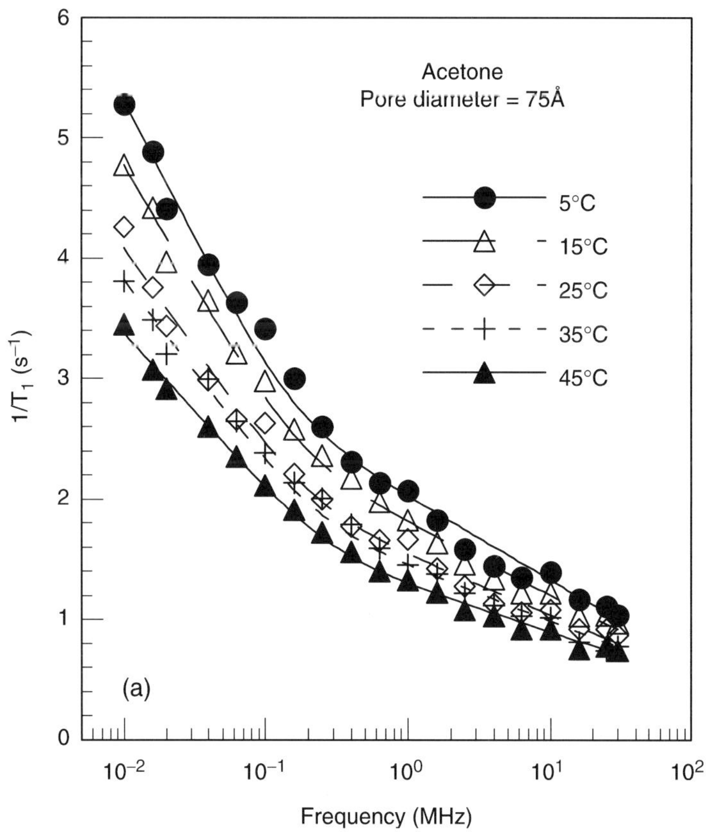

FIG. 9. Magnetic field dependence of ^{1}H spin–lattice relaxation rates of acetone embedded in packed samples of calibrated porous glass beads with pore diameter of 75 Å for several temperatures. The solid lines are best fits to the theory (*66*).

regions with slopes in the ratio of 10/3 is predicted by Eq. (3) when paramagnetic contributions dominate relaxation. From the large magnetic moment of the paramagnetic species S ($\gamma_S = 658.21\gamma_I$), there is no ambiguity about the relaxation mechanism of the diffusing proton spins, I, which is the intermolecular dipolar relaxation process induced by fixed spins S and modulated by the translational diffusion of the mobile spins, I, in close proximity to these surfaces. The MRD profile, when combined with an appropriate theoretical framework (*65,66*), provides a measurement of the effective translational diffusion constant for the liquid in close proximity to the paramagnetic centers at the pore surface and the diffusion constant generally decreases with decreasing pore diameter. The analysis in terms of a locally two-dimensional model for the motion of the nuclear spin in the vicinity of the paramagnetic center is quantitatively successful for several non-protic liquids including acetone, dimethyl sulfoxide, acetonitrile, and dimethyl formamide. The translational diffusion constant for these molecules in the vicinity of the surface is found to be approximately proportional to the reciprocal of the molecular diameter which supports the model. The temperature dependence of the deduced diffusion constants demonstrates that the local motion is a thermally activated process for these liquids with activation energies in the range of 8–12 kJ/mol (*65*).

Figure 10 shows that for the same controlled pore glasses, water magnetic relaxation dispersion is different and shows a power-law dependence on magnetic field strength (*67*). Water is unique in that it is small, has extensive hydrogen-bonding capabilities and may exchange protons with other molecules or surface sites. It may behave as both a Lewis acid or base, and generally coordinates with most metal ions. The MRD profile results from I–S correlations that persist much longer in the surface region than in the bulk, thus leading to a significant increase in $1/T_1$ (*67*). The nuclear spin–lattice relaxation rate at the pore surface is dominated by dynamic processes that appear to be one-dimensional. An interesting feature is that the temperature dependence is opposite to that usually observed for diffusion-induced relaxation, which is found for aprotic liquids. The anomalous temperature dependence is interpreted in terms of a diffusive process at the pore surface that is interrupted by a chemical exchange with the bulk phase. The fundamental difference between water and other solvents in these glasses is the spatial extent of the surface explored by the diffusing protons of the liquid. The possibility that water may coordinate directly to paramagnetic relaxation centers is shown to be of minor importance to the observed relaxation dispersion profiles.

Figure 11 shows the MRD profiles for aqueous suspensions of three commercial clay preparations, Volclay, Polargel, and Magnabrite. The relaxation profiles are linear in the logarithm of the Larmor frequency, the signature of relaxation being a two-dimensional diffusive process (*66*). The up-turn in the relaxation rate apparent for Volclay is expected in the case where the electron relaxation is short and makes a significant but not completely dominant contribution to the effective correlation time for the electron–nuclear coupling. The importance of the paramagnetic species in the nuclear spin

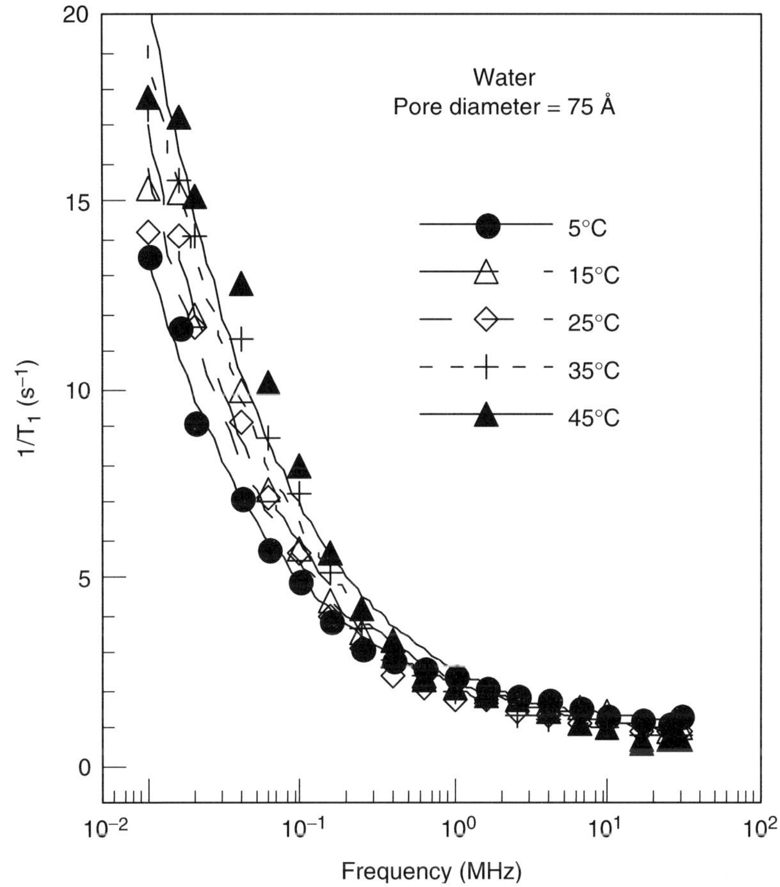

FIG. 10. The water ^{1}H spin–lattice relaxation rates as a function of magnetic field strength represented as the Larmor frequency for packed samples of calibrated porous glass beads with pore diameter of 75 Å at several temperatures. The solid lines are best fits to the theory (*67*).

relaxation is supported by the ESR spectra shown in Fig. 12, where broad components from iron at $g \sim 4.3$ are apparent in all samples. In addition, the characteristic 6-line spectrum associated with manganese is a strong feature of Magnabrite, Polargel, and Fuller's Earth (*68*). The presence of both iron and manganese in natural products is common. It is important to note that even in synthetic systems such as aerogels, the starting materials often contain heavy metals, notably iron and manganese, at the level of several parts per million, which may be sufficient to make significant if not dominant contributions to the proton relaxation dispersion profile.

Figure 13 shows MRD profiles for zeolites and kaolin suspended in aqueous agar. The agar prevents the particulates from settling in the sample tube, and eliminates uninteresting time dependence in the data acquisition. The agar contribution has been subtracted to yield the MRD profiles displayed in

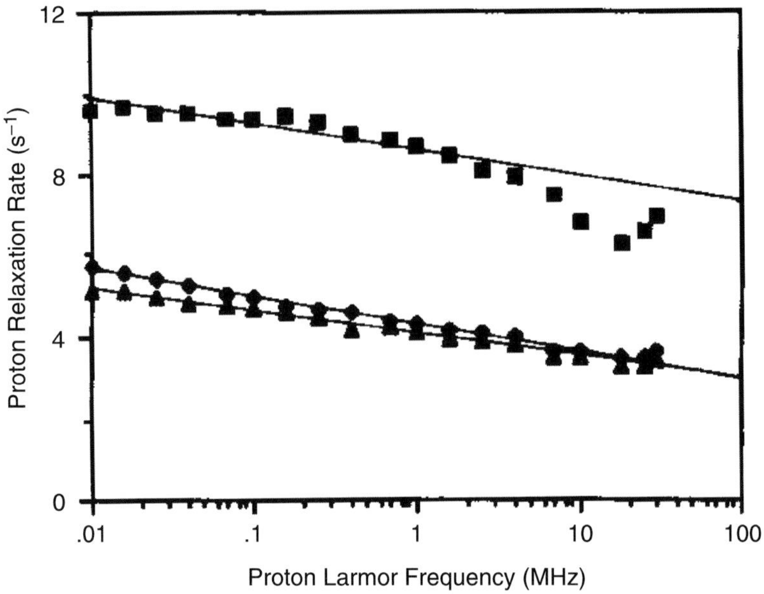

FIG. 11. The nuclear magnetic spin–lattice relaxation rate for water protons as a function of the magnetic field strength reported as the proton Larmor frequency at 298 K for Volclay (■), Polargel (◆), and Magnabrite (▲) (*68*).

Fig. 13. With the exception of kaolin, the relaxation rate depends on the field strength as a power law where the exponent of the Larmor frequency is a function of the sample (*68*). The zeolites provide an interesting system of cages with interconnections of various sizes. If a spin may diffuse throughout the system before encountering a relaxation sink, the paramagnetic contribution will be negligible and the relaxation described by a diamagnetic model will have restricted diffusion. However, when the paramagnetic centers are present, the effective dimensionality of the diffusive exploration of the observed spin may be practically affected by the concentration of the relaxation centers in the matrix. As a result the relaxation may be either linear in the logarithm of the Larmor frequency or a power law, even though the underlying structure of the matrix does not change (*68,69*).

Bentonite is a clay in the montmorillonite class. These planar structures swell on the addition of water and the interlamellar spacing changes with increasing water content. The water proton relaxation rate is linear in the concentration of the mineral, as expected. As shown in Fig. 11, the field dependence is linear in the logarithm of the Larmor frequency at low field strengths reflecting both the locally planar organization of the system and the diffusive exploration to paramagnetic relaxation sinks. At higher magnetic field strengths, the relaxation falls more rapidly, and at the highest field strengths, the relaxation rate may increase, which is the signature of the magnetic field dependence of the electron–spin relaxation rate for the paramagnetic center. A useful feature of these systems is that the magnetic field dependence is weak over a considerable range, so that the montmorillonites

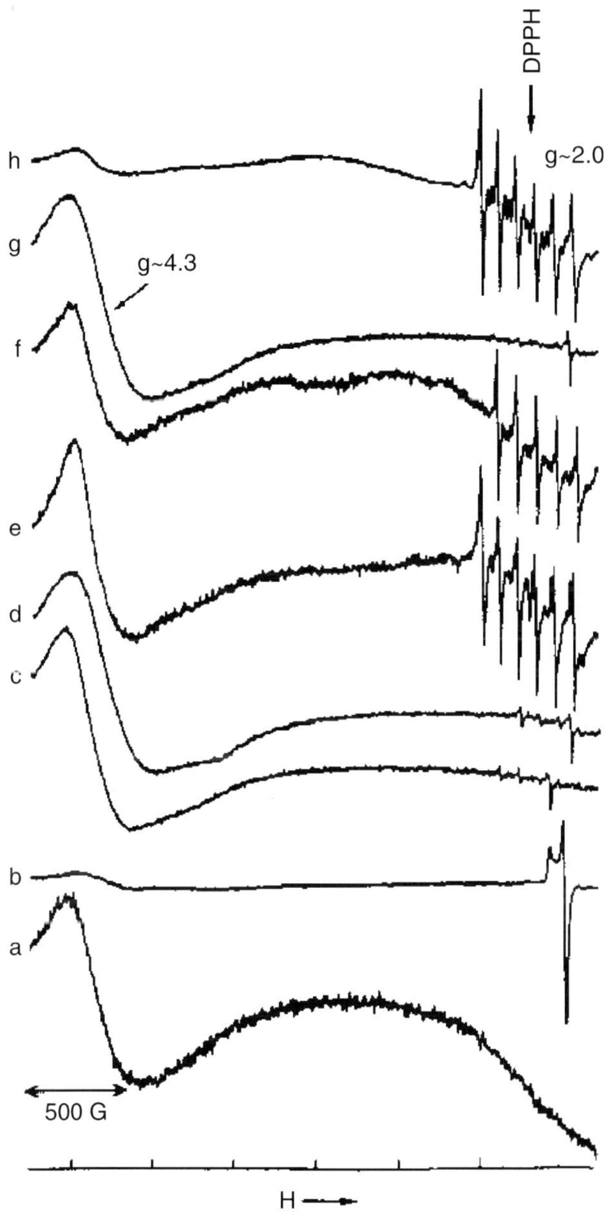

FIG. 12. Electron spin resonance spectra recorded at X-band for powdered clay samples at ambient laboratory temperature: (a) montmorillonite, (b) kaolin, (c) Bentonite (Aldrich Chemical), (d) Bentonite (SSP/NF), (e) Magnabrite, (f) Polargel, (g) Volclay, and (h) Fuller's Earth (*68*).

may be of use as magnetic contrast agents in MRI applications at most clinically available magnetic field strengths (*70*).

Magnetic resonance dispersion is a useful aid in the characterization of liquids in porous rocks from which one may extract oil; however, the method

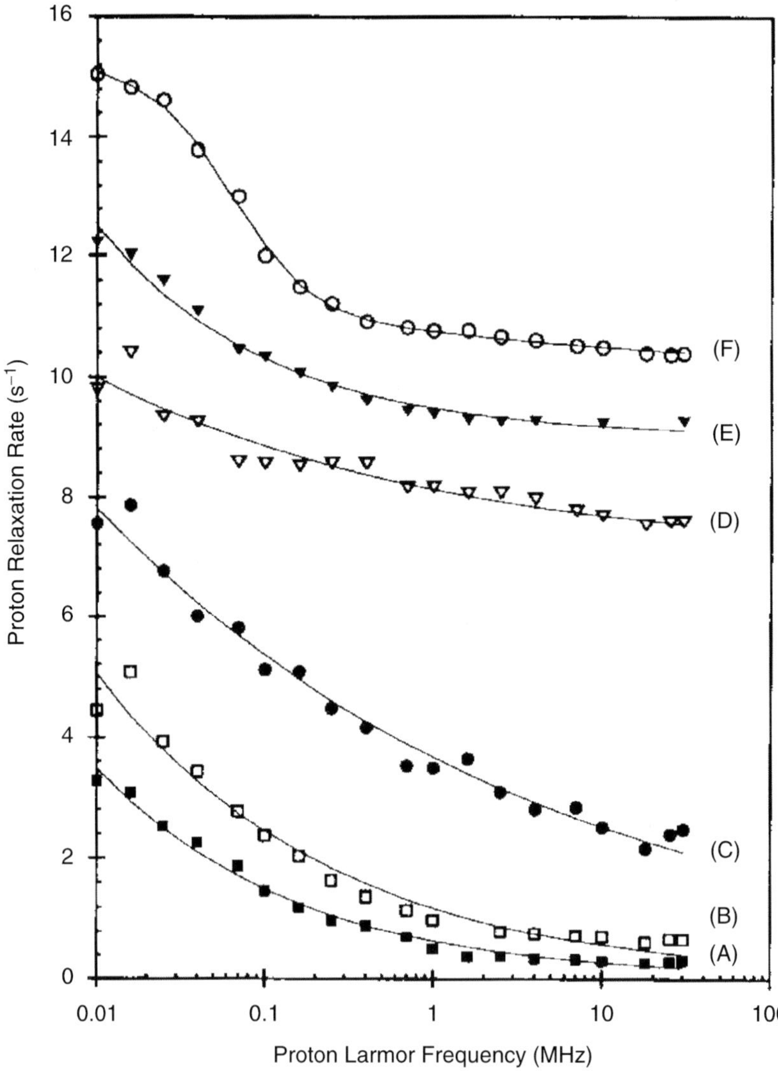

FIG. 13. The nuclear magnetic spin–lattice relaxation rate for water protons as a function of magnetic field strength reported as the proton Larmor frequency at 298 K for 5% suspensions of the particulate stabilized in a 0.5% agar gel presented as the difference plot: (A) Zeolite 3A; (B) Zeolite 13X; (C) Zeolite NaY; (D) kaolin with 7 s^{-1} added to each point to offset the data presentation; (E) Cancrinite with 9 s^{-1} added to each point to offset the data presentation; and (F) 0.5% agar gel profile with 10 s^{-1} added to each point. The solid lines are fits to a power law (*68*).

is not directly a well-logging tool because the apparatus cannot presently be compacted to the small space of a well bore. However, the MRD profiles provide a useful foundation upon which somewhat simpler experiments may be built. Studies of oil and water in rocks or rock models are similar to those on inorganic systems such as porous glasses (*1–3,71–74*). The temperature and magnetic field dependencies of the longitudinal nuclear spin–lattice relaxation rates of water and oil confined in a series of calibrated porous media and various natural rocks have been interpreted with an original model of surface diffusion, taking into account the solid–liquid interactions with the pore surface (*72,74*). Depending on the strength of such interactions, one observes an anomalous temperature dependence of the transverse $1/T_2$ and longitudinal $1/T_1$ nuclear spin relaxation rates for water in silica and various sandstones, but a normal temperature dependence for water-saturated calcium carbonate surfaces. The model also supports the normal temperature dependence observed for oil-saturated porous silica, sandstone, and carbonate surfaces.

Recent applications of relaxation dispersion measurements to concrete or cement-based materials are promising for characterizing reactive nanoporous materials, the structure of which may evolve over time (*75–78*). The MRD profiles have provided, for the first time, a direct means for characterizing the specific surface area, S_p, of a hydrated cement-based material (*79*), without exposing the sample to extremes of temperature or pressure (*80–83*). The interest in such a surface area is to provide information on the microsctructure and its impact on macroscopic or structural properties. The method is based on a clear separation of surface and bulk contributions of the overall ^{1}H nuclear spin–lattice relaxation rate of water confined within the hydrated cement. Representative relaxation dispersion data are shown in Fig. 14. An original model based on solid/liquid cross-relaxation, proton surface diffusion and nuclear paramagnetic relaxation, in presence of progressive hydration, has been proposed (*79*), which permits characterization of the surface area of a very reactive and disordered material from the MRD data as shown in Fig. 15. The method is sufficiently rapid to be applied continuously during the progressive hydration and setting of cement. A reliable measured value of S_p is fundamental for characterizing the highly disordered internal porous microstructure encountered in these materials. This is more important for hardened cement paste, which is the main ingredient in concrete, the most widely manufactured material in the world. Here the value of S_p is directly related to the mechanical performance. However, most of the existing techniques have been shown to be inadequate because of strong perturbations on the sample. MRD provides a non-destructive method that does not require drying or temperature cycling (*79*). Although this non-destructive method has been applied to cement based materials, it is general and may be used to characterize any water-saturated reactive porous material.

IV. Proton Rich Solid Phases

A. Cross-Relaxation

For the present discussion, we define a solid phase as any phase where the rotational mobility of the molecules involved is sufficiently slow so that the

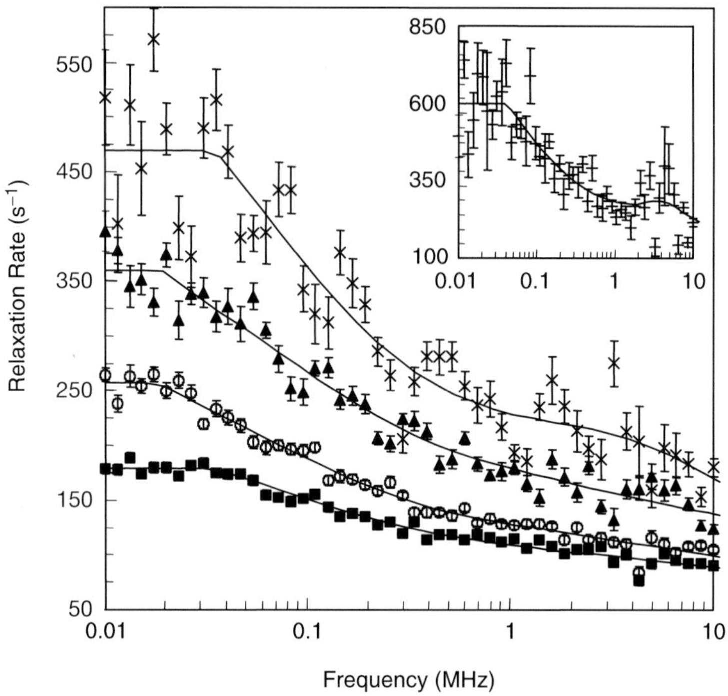

FIG. 14. Measured water ^{1}H spin–lattice relaxation rates of a hydrated mortar at w/c = 0.38 as a function of the proton Larmor frequency, for different duration of hydration: 0h 34 min (■), 7h 27 min (○), 8h 45 min (□), and 9h 40 min (×), upwards. The insert represents the data obtained after a hydration time of 10h 32 (+), the labels for the two axis are equivalent to those of the main figure. The continuous lines correspond to the best fits to the theory.

proton–proton dipolar couplings between spins is not averaged completely by rapid reorientation. Proton–rich solid phases include any organic solid including hydrocarbons, peptides, and polysaccharides where the spatial distribution of the protons in the material is sufficiently dense that the protons are not magnetically isolated; that is, most natural materials. The NMR spectrum of the rotationally immobilized material is typically a broad line with linewidths in the range of tens of kilohertz depending on the local motion within the material. The proton spectrum of the solid may be detected directly with many current commercial spectrometers, although finite receiver recovery time may distort the line shape. For many proton–rich solid phases, the magnetic coupling with a liquid in contact with the interface is sufficient that the spectrum of the solid may be detected indirectly from the response of the liquid signal to off-resonance irradiation of the solid spins (*84,85*). The relaxation of the protons detected by observing either the liquid or the solid spins in such systems is complicated by the magnetic coupling between the two easily distinguished populations of protons. Generally, the liquid spins are detected. The relaxation of the liquid spins is generally not exponential, and the detectable time constants for the

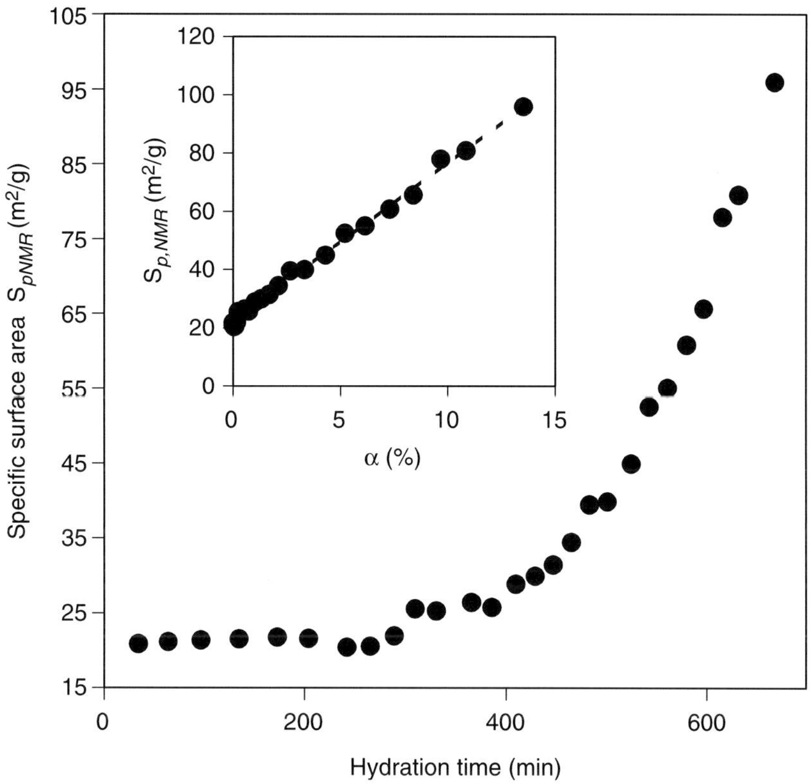

FIG. 15. Variation of the NMR-based specific area of hydrated material as a function of the hydration time (main figure) and the extent, α (%), of chemical reactions (insert).

magnetization growth or decay are mixtures of rate constants associated with solving the coupled differential equations that characterize the time dependence of the magnetization in the presence of communication between the spins. The slow component of the liquid relaxation response is usually detected, and the time constant for this part of the decay may be written as:

$$R_{\mathrm{slow}} = \frac{1}{2}\left[R_L + R_S + R_{SL}\left(1 + \frac{1}{F}\right) - \left[\left(R_S - R_L - R_{SL}\left(1 - \frac{1}{F}\right)\right)^2 + \frac{4R_{SL}^2}{F}\right]^{0.5}\right] \tag{4}$$

where R_{L} and R_{S} are the spin–lattice relaxation rate constants for the water and protein protons, R_{SL} is the cross-relaxation rate constant between the water and the protein, and F is the ratio of the protein–proton population to the water-proton population at equilibrium $(M_S^{\infty}/M_L^{\infty})$ (*84*). The rate constant R_L is generally independent of the magnetic field strength range accessible because the liquid motions are rapid compared with the Larmor (even the highest Larmor) frequencies. The rate constant R_{SL} may be a function of the magnetic field strength as treated by Solomon (*51*). However, in most biological systems, the liquid is generally water and interacts with the solid matrix by binding at specific sites for times of the order of microseconds. While bound, the magnetization transfer rate is rapid because the liquid spins are not magnetically different from the solid spins for which the spin–spin communication rate is given approximately by $1/T_{2\text{solid}}$, which is about $10^5\ \text{s}^{-1}$ (*86*). In this case, which is appropriate for proteins, the magnetic field dependence of the spin–lattice relaxation rate detected in the liquid spins is created by the magnetic field dependence in the solid-protein spin systems.

A clear demonstration of the effects of the magnetic coupling are shown in Fig. 16 where the liquid proton signal intensity is plotted as a function of the frequency offset of a preparation pulse (*84,87*). The essence of the

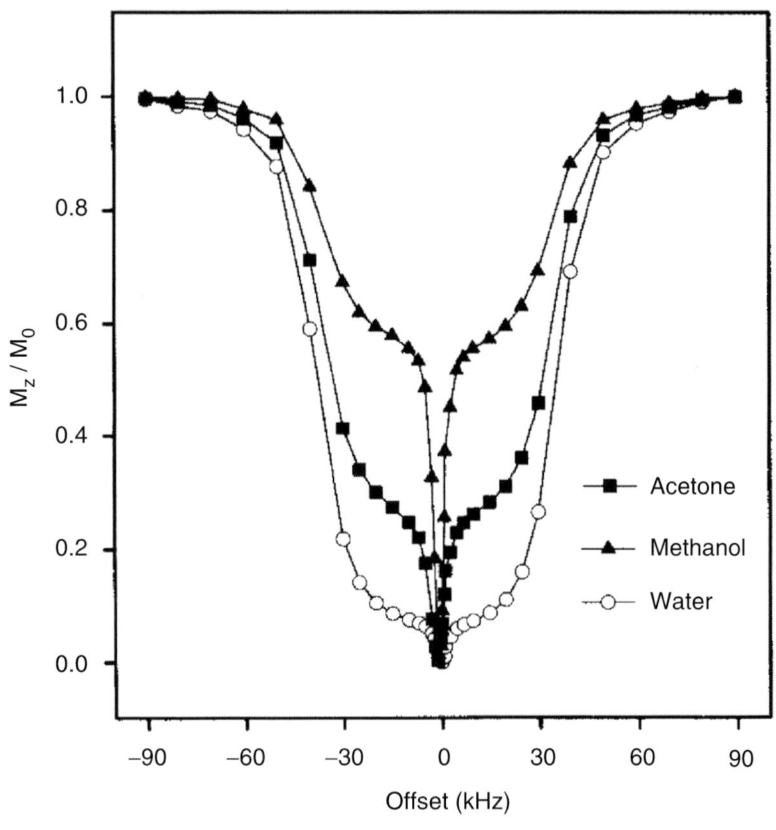

FIG. 16. Cross-relaxation or Z-spectra derived from the solvent proton spectra of cross-linked bovine serum albumin gel supported in a ternary solvent system consisting of 8.8% H_2O, 8.7% acetone, and 8.8% methanol and 63% D_2O. The offset axis represents the frequency offset of the off-resonance preparation pulse which partly saturates the immobilized spin. In this case the 3 s preparation had an amplitude of 880 Hz at a Larmor frequency of 500 MHz (*87*).

experiment is that the preparative rf pulse partly saturates the solid component spin system, i.e., the population changes from the equilibrium value. Then, the liquid protons that are magnetically coupled to the solid spin population in transiently bound environments are saturated as well by magnetization transfer or cross-relaxation in the bound environment. The effect is carried to the liquid population of spins by chemical exchange from the bound to the bulk liquid environment. This spin coupling between the liquid and solid spin system provides an important class of experiments that permit interrogation of the solid spin properties by detection of the liquid spin without suffering severe dynamic range problems usually associated with attempts to detect the broad solid spectrum in the presence of a relatively narrow liquid spin signal. This experiment demonstrates the effects of coupling spin population responses clearly and provides a number of useful applications, particularly in the context of magnetic imaging (*88*).

The importance of the magnetic coupling is easily seen in Fig. 17 which shows two water proton MRD profiles for serum albumin solutions at the same composition (*89*). The approximately Lorentzian dispersion is obtained for the solution, and reports the effective rotational correlation time for the protein. The magnetic coupling between the protein and the water protons carries the information on the slow reorientation of the protein to the water spins by chemical exchange of the water molecules and protons between the protein and the bulk solution. When the protein is cross-linked with itself at the same total concentration of protein, the rotational motion of the protein

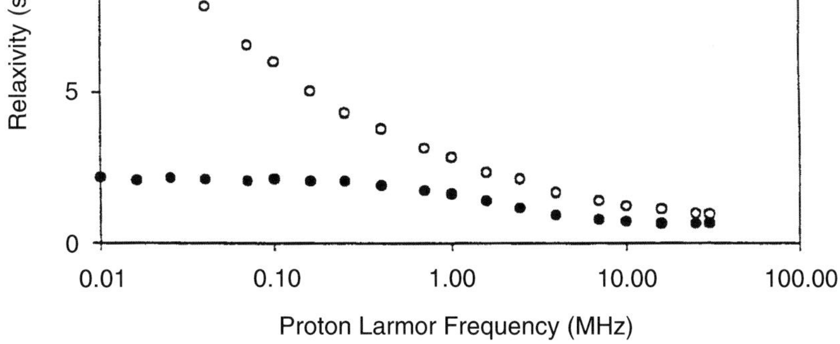

FIG. 17. The water proton spin–lattice relaxation rates as a function of magnetic field strength reported as the proton Larmor frequency in aqueous 1.8 mM samples of bovine serum albumin. The lower data set was taken on the solution, the open circles taken after the sample had been cross-linked with glutaraldehyde to stop rotational motion (*89*).

stops and the solution becomes a gel. The relaxation dispersion profile of the water protons changes profoundly from a Lorentzian to a power law. The relaxation profile is dominated by the magnetic field dependence of the protein protons, which is a power law.

B. The Solid Proton Component Relaxtion

There are a variety of models for the magnetic field dependence for polymers and variously ordered phases coming largely from the groups of Kimmich and Noack. We focus here on the protein system that provides unique structural and physical characteristics. Recent advances in the speed of current-switched field-cycling instruments have provided a direct measure of the MRD profile of the protein protons as shown in Fig. 18. The relaxation

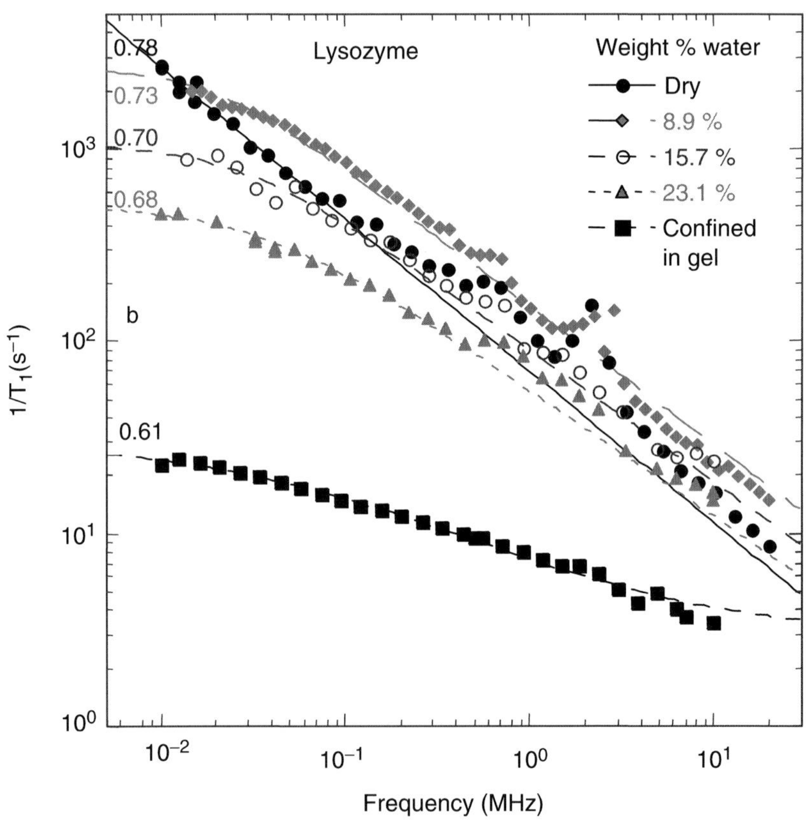

FIG. 18. The proton spin–lattice relaxation rate recorded as a function of the magnetic field strength plotted as the proton Larmor frequency for lysozyme samples. Dry (•), hydrated to 8.9% (◆), 15.7% (○). 23.1% (▲), and cross-linked in a gel (■). The solid lines were computed from the theory. The solid lines are fits to the data using Eq. (4) with R_S given by Eq. (6). The two parameters adjusted are R_{SL} and b (*97*). The small peaks most apparent in the dry samples are caused by cross-relaxation to the peptide nitrogen spin (*90,122*).

dispersion over the field range studied is described by a power law,

$$\frac{1}{T_1(\omega)} = A\omega^{-b} + C \tag{5}$$

where b is 0.78 for dry proteins. Kimmich and collaborators, who in elegant experiments have shown that the field dependence of the proton relaxation rates is not dependent on the dynamics of the side chains in the polypeptide (*19,90–93*), but is caused by the dynamics of the main chain itself, have reported similar data. The basic idea is that the motions of the chain make the proton–proton dipolar couplings time dependent which then drives proton spin relaxation.

The origin of the power law in the protein case derives from the chain motion, which is caused by thermal fluctuations. However, Abragam (*94*) demonstrated that spin-phonon coupling in a three-dimensional crystalline solid was completely unimportant because the correlation was lost rapidly in three dimensions. However, in a polypeptide and perhaps other long chain macromolecules, a disturbance does not propagate uniformly in a three-dimensional space. The firm connections are along the main chain of the polymer; the mechanical connectivity between chains, even in a folded protein, is weak by comparison. Thus, a disturbance tends to propagate along the chain and the effective dimensionality of the resulting spin–phonon interaction is reduced and the importance of spin–phonon coupling is increased dramatically. The theoretical details have been recently developed for a quantitative model that relates the spin–lattice relaxation rate constant to the strength of the proton–proton dipolar coupling, ω_{dip}, the effective fractal dimensionality, d_f, of the proton distribution in space, i.e., within the structure of the protein, and a spectral dimension, d_s that is approximately 4/3, that replaces the Euclidean geometric dimensionality in the expression for the low frequency distribution of vibrational states (*95–97*).

$$\frac{1}{T_1} = \frac{9\pi\beta d_s}{5}\left(\frac{k_B T}{\hbar}\right)\left\{\left(\frac{3}{4}\right)\left(1+\frac{1}{2^b}\right)\left(\frac{\hbar\omega_{dip}}{E^0_{\nu\parallel}}\right)^2\left(\frac{\hbar\omega_0}{E^0_{\nu\parallel}}\right)^{-b} + \left(\frac{1}{6}\right)\left(\frac{7}{2}+\frac{1}{2^b}\right)\left(\frac{\hbar\omega_{dip}}{E^0_{\nu\perp}}\right)^2\left(\frac{\hbar\omega_0}{E^0_{\nu\perp}}\right)^{-b}\right\} \tag{6}$$

where $E^0_{\nu\parallel}$ and $E^0_{\nu\perp}$ are the cut-off frequencies for the parallel and perpendicular amide I modes and β is a numerical factor that may be obtained from the second moment of the proton line or the linewidth (*98*).

The exponent, b, is then given by

$$b = 3 - \frac{2d_s}{d_f} - d_s \tag{7}$$

which corresponds to the exponent in Eq. (5).

Data for the dry protein are shown in Fig. 18. The exponent for the experimental power law is 0.78, which corresponds to a value of d_f equal to 3. Physically, this implies that the proton distribution in space for the dry lyophilized protein is essentially uniform. A low frequency plateau is expected for solid systems in general if the magnetic field switches are adiabatic. If the impressed magnetic field strength drops to zero, the protons remain in the effective local dipolar field and relax at this constant but small field strength, corresponding to approximately 11 kHz in the protein. Observation of the low-field plateau requires careful elimination of stray magnetic fields including the earth field (*99*). These data are fit well by the first term in Eq. (6) corresponding to the longitudinal modes of the chain. The perpendicular modes of the chain are not required to account for the data quantitatively; thus, these low frequency transverse modes appear to be quenched in the dry and partially denatured protein.

Three characteristics of the MRD profile change when the protein is hydrated with either H_2O or D_2O. Both terms of Eq. (6) are required to provide an accurate fit to the data. The second or perpendicular term dominates once the transverse modes become important. The power law for the MRD profile is retained, but the exponent takes values between 0.78 and 0.5 depending on the degree of hydration. A low frequency plateau is apparent for samples containing H_2O which derives from two sources: the field limitation of the local proton dipolar field as mentioned above, and from limitations in the magnetization transfer rates that may be a bottleneck in bringing the liquid spins into equilibrium with the solid spins.

The value of b or d_f changes slowly with increasing hydration; no sudden changes are detected at hydration levels that correspond to solvating charged groups, for example. The value of d_f is a single parameter measure of the protein structure as reflected in the spatial distribution of protons within the structure. The smooth change in d_f with hydration levels of the protein implies that the protein structure changes slowly and incrementally from the dry lyophilized state to the native fully hydrated state. That the value of d_f is not equal to 3 in the folded protein simply means that the protons in this structure are not uniformly packed in the solid, which is consistent with the protein structures derived from X-ray diffraction studies. This conclusion may be verified computationally based on the proton distribution in space for proteins of known structure. Calculations of the fractal dimensionality of the proton distribution in ribonuclease A and lysozyme from the X-ray coordinates in the protein data bank yield a value of approximately 2.5, which is consistent with the experimental value obtained from Eq. (6) for hydrated systems. It should be noted that a protein crystal has a high water content, typically in the range of 40%, thus the protein is essentially fully hydrated in most protein crystals.

The need to include the transverse terms in Eq. (6) is consistent with the water-molecule dipole moments solvating charges and local protein component dipole moments, which may then unlock low frequency modes in the protein structure that, in the absence of a high dielectric medium may lock down the structure by relatively strong intramolecular cross-links. The transverse modes assume dominant importance because of the much smaller value of the vibrational cut-off frequency that enters Eq. (6); i.e., the relaxation equation results from integration over the accessible vibrational density of states up to a high frequency cut-off, which we take as the frequency of the amide modes of a peptide bond. The longitudinal mode is of a much higher frequency than the softer transverse modes, and the lower frequency cut-off increases the amplitude of the contribution to the spectral density profile as expected.

The low-field plateau is dominated by the effects of magnetization coupling between the immobilized protein spins and the mobile water spins. Although the plateau results in part from the local dipolar field limitation to adjustment of the field strength, the dipolar field effects are the same or smaller in the hydrated case than in the dry case, and they are small in the dry case. Thus, the plateau in the hydrated case is dominated by the effects of relaxation coupling between the two proton populations. The effects of the relaxation coupling are easily handled by incorporating the magnetic field dependence of the solid component relaxation directly in Eq. (4), which provides an excellent description of the experiment.

The low field relaxation limitation is mostly caused by the magnetization transfer rate between the solid and the liquid spins. It is important to note that there is no significant pH dependence of the relaxation dispersion profile in either cross-linked protein gels or in hydrated protein powders. This observation demonstrates that proton exchange is not the dominant mechanism for magnetization transfer between the protein and the water spins. In the gel case where pH is well defined, one would expect that the catalysis of the proton exchange by hydroxide ion would be kinetically first order in hydroxide ion; however, the pH dependence is very weak over 4 decades in hydrogen ion concentration from 4–8 so that the magnetization coupling between the solid protein and the water spins is not carried predominately by proton transfer. Instead, the magnetic coupling is carried by the exchange of water molecules between specific sites on the protein and the bulk water population. The proton relaxation rate at the plateau provides an indication of the effective magnetization transfer rate that may in turn be limited by water molecule residence time in protein sites. For the proteins studied, the mean lifetime is in the range of a microsecond or shorter; however, it is anticipated that the lifetime of water molecules in such environments will depend on the detailed structure and dynamics of the protein.

The temperature dependence of the MRD profile for the protein-water systems where the protein is magnetically a solid, is remarkably weak. The relaxation rate is proportional to $1/T$, which is consistent with Eq. (4) that was derived on the assumption that the relaxation process is a direct spin–phonon coupling rather than an indirect or Raman process. If it were a Raman process, there would be no magnetic field dependence of the relaxation rate; therefore, the temperature dependence provides good evidence in support of the theoretical foundations of Eq. (6).

C. Porous Protonic Solids

Many natural materials are porous but also proton–rich such as wood or other plant products. Relaxation of liquids in these materials has features in common with both inorganic matrices and the protein systems discussed above. The class of porous polysaccharide materials used for size exclusion chromatography provides an example; one commercial product is Sephadex. The material swells on solvation to form a controlled pore gel. The main application involves excess liquid, generally water, which flows through the gel bed carrying solutes of various size. The large solutes are excluded from the pore interior and elute rapidly while the smaller ones equilibrate with the pore interior and elute later. The solvent generally samples the pore interior as well as the bulk phase.

MRD profiles are shown in Fig. 19 for Sephadex gels at different pH values (*100*). The curves appear at first glance to be approximately Lorentzian, but are more accurately represented by a sum of a plateau region and a power law higher field portion. The relaxation rates increase with increasing pH value, which may be expected if proton exchange is an important

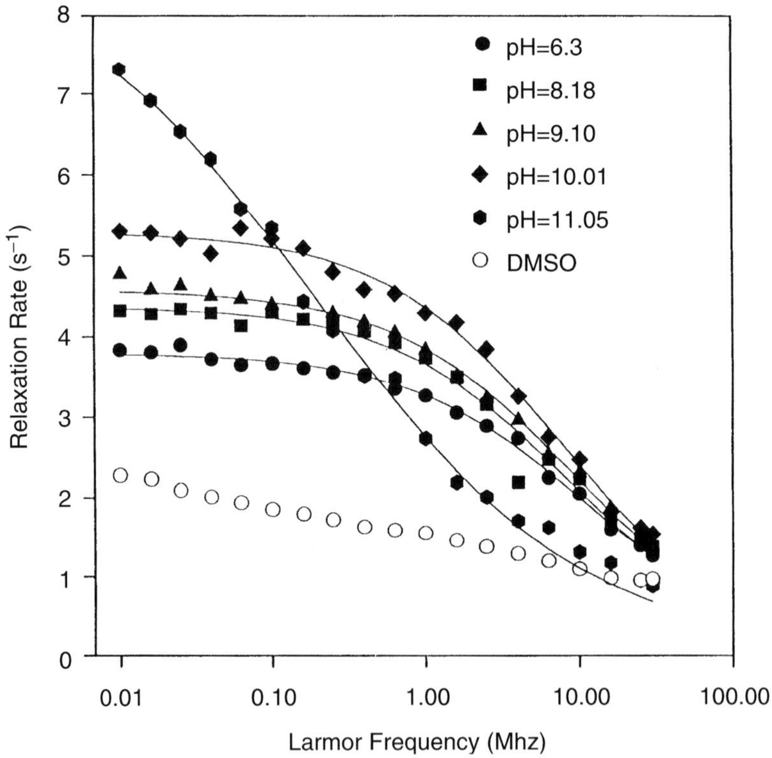

FIG. 19. The magnetic relaxation dispersion for water proton in a Sephadex G-25 sample swollen to equilibrium at different values of pH at 298 K. The open circles are the relaxation rates for the methyl protons of dimethyl sulfoxide. The solid lines were computed from a two-stage exchange model (*100*).

contribution to relaxation. However, if simple water proton exchange with sugar hydroxyl groups were the dominant exchange event controlling relaxation, then the exchange rate would generally be first order in hydroxide concentration. For the data shown in Fig. 19, the hydroxide ion concentration changes by several orders of magnitude; however, the relaxation rate does not. Therefore, the proton exchange between sugar hydroxyl and water cannot be the dominant mechanism in this case. The power law at high field is consistent with a mechanism involving magnetization transfer from the Sephadex protons to the water protons as in the protein case. What is different in this case is that there is a macroscopic microporous structure in the gel that creates pores. If the exchange between the pore interior and the bulk phase is rapid, then the field dependence of the solid protons is observed, and it is a power law. The relaxation equation may be written as:

$$\frac{1}{T_1} = \frac{1}{T_{10}} + \frac{P_P}{T_{1P} + \tau_{ex}} \tag{8}$$

where T_{10} is the relaxation time constant outside the pore, and P_P is the probability that the liquid proton is inside a pore characterized by a relaxation time T_{1P} and a mean lifetime τ_{ex}. The relaxation time inside the pore may itself be determined by exchange events between the surface and the liquid in the pore, but that process may not be apparent. In the Sephadex case, the exchange rate limits the low field spin relaxation rate. The spin–lattice relaxation of the plateau increases with increasing temperature with an activation barrier of 4.2 kcal/mol, which is consistent with water transport. Thus, the picture that emerges from these data is that the spin–lattice relaxation rate is determined by exchange events between the pore interior and the bulk, and that relaxation within the pore is determined by a combination of chemical and magnetic exchange events that map the power law dependence of the solid Sephadex proton onto the liquid spins. One contributing mechanism to the interior exchange coupling may involve binding of water molecules in the Sephadex matrix, where efficient magnetization transfer occurs, similar to the protein gel case. However, significant contributions inside the pore from proton exchange with sugar hydroxyl groups are not ruled out. The overall process is then a two-step exchange involving equilibration within the pore followed by exchange of water between the pore interior and the bulk. Although the power law is qualitatively consistent with the model presented in Eq. (6), there is not sufficient structural data available to examine it in detail and more work is required to examine the implications of possibly greater or lesser interchain connectivity.

V. Rotational Dynamics of Liquids at Surfaces

If intramolecular pathways dominate the nuclear spin relaxation mechanism, a separation between translational and rotational contributions is possible that permits definition of rotational dynamics at surfaces. The simplest approach is to study nuclei with nuclear electric quadrupole moments so that relaxation is dominated by reorientation of the electric field gradient. Schauer and coworkers (*101*) have exploited the deuteron in water to characterize the dynamics of water in the hydration domain of proteins and propose a model that involves coupling between translational motion on the surface and rotational reorientation that accounts well for the deuteron MRD (*19,90,91,101–103*). Halle and coworkers (*111*) have published the majority of the work directed at characterizing water–protein interactions using the magnetic field dependence of deuterons and ^{17}O relaxation rates (*104–116*).

An interesting application in porous silica glass employs deuterium spin–lattice relaxation rates in selectively deuterated pyridine (*50*). The pore-size dependence shown in Fig. 20 is linear in the reciprocal of the pore radius for both the para and ortho positions, but the significant difference in the relaxation at these two positions precludes analysis by an isotropic

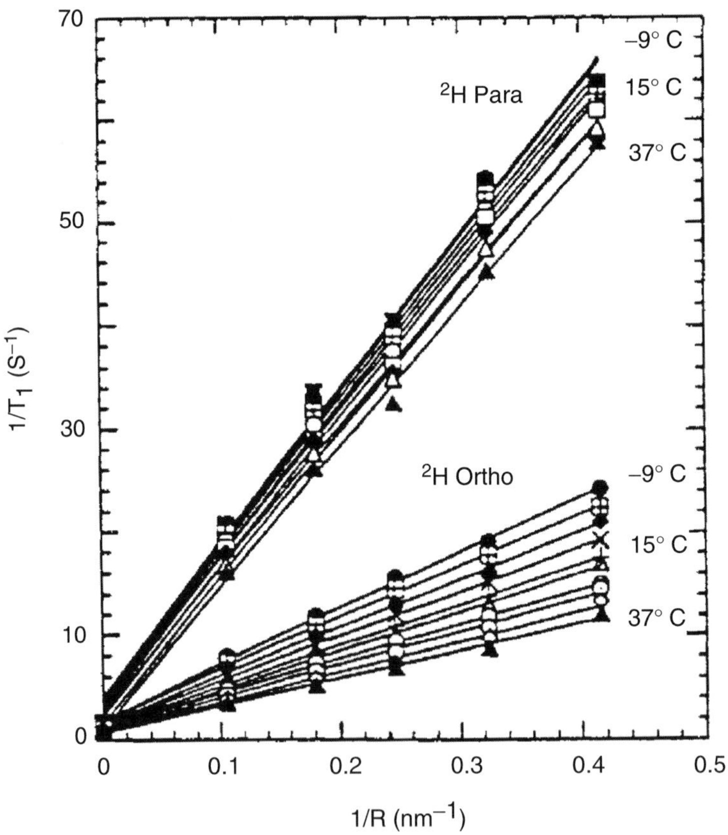

FIG. 20. The ^2H spin–lattice relaxation rates of liquid pyridine deuterated in ortho and para positions as a function of the pore radius reciprocal in porous sol-gel silica glasses (*50*).

reorientation model. The surface relaxation rates extracted using a two-site exchange model are about an order of magnitude larger than the bulk rates, consistent with slower molecular reorientation at the surface. However, the anisotropy of the relaxation is also reflected in the temperature dependence as shown in Fig. 21. The apparent activation energy for the para position is only 2.1 kJ mol^{-1}, which is of the order of RT and thus does not involve a thermally activated reorientation. This result also shows that the exchange from the surface does not dominate the observed relaxation otherwise the barrier would be larger. The activation barrier for reorientation sensed at the ortho positions is much more and approximately that of a hydrogen bond. The relaxation was successfully modeled with pyridine ordered by hydrogen bonding to the surface SiOH groups and reorienting rapidly about the C2 axis. Reorientation of the C2-axis was activated and was found to be much slower. This basic picture is supported by other NMR and spectroscopic studies (*117*). This class of experiments holds considerable promise for defining both structural and dynamical characteristics of

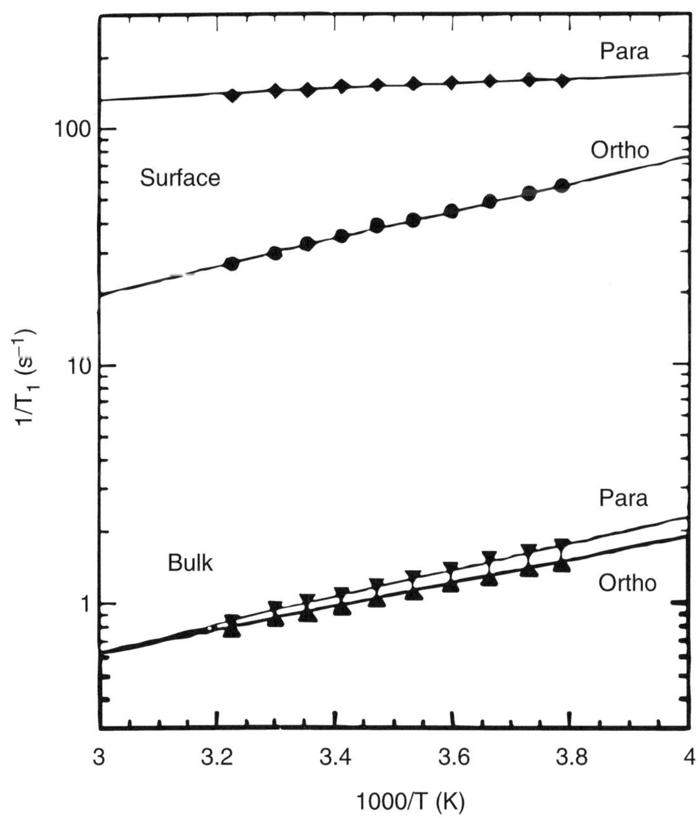

FIG. 21. Semi-logarithmic plots of the surface ^2H spin–lattice relaxation rates for ortho and para positions of pyridine as a function of the reciprocal temperature in porous sol-gel silica glasses. The surface rates are shown on the top and the rates in bulk pyridine are shown at the bottom (*50*).

molecular interactions with surfaces that should be helpful in defining components of catalytic mechanisms and lubrication (*118–120*).

References

1. Various *Magn. Reson. Imaging* **2003**, *21*, 159–450.
2. Various *Concepts Magn. Reson.* **2001**, *13*, 340–411.
3. Various *Magn. Reson. Imaging* **2001**, *19*, 291–593.
4. Koenig, S. H.; Schillinger, W. E. *J. Biol. Chem.* **1969**, *244*, 3283–3289.
5. Redfield, A. G.; Fite, W. I.; Bleich, H. E. *Rev. Sci. Instrum.* **1968**, *39*, 710.
6. Noack, F. *Progr. Nucl. Magn. Reson. Spectrosc.* **1986**, *18*, 171–276.
7. Wagner, S.; Dinesen, T. R. J.; Rayner, T.; Bryant, R. G. *J. Magn. Reson.* **1999**, *140*, 172–178.
8. Schauer, G.; Nusser, W.; Blanz, M.; Kimmich, R. *J. Phys. [E]* **1987**, *20*, 43–46.
9. Grossl, C.; Winter, F.; Kimmich, R. *J. Phys. [E]* **1985**, *18*, 358–360.
10. Rommel, E.; Mischker, K.; Osswald, G.; Schweikert, K. H.; Noack, F. *J. Magn. Reson.* **1986**, *70*, 219–234.
11. Graf, V.; Noack, F. *J. Chem. Phys.* **1980**, *72*, 861–863.
12. Korb, J. P.; Torney, D. C.; McConnell, H. M. *J. Chem. Phys.* **1983**, *78*, 5782–5789.
13. Korb, J. P.; Winterhalter, M.; McConnell, H. M. *J. Chem. Phys.* **1984**, *80*, 1059–1068.
14. Korb, J. P. *J. Chem. Phys.* **1985**, *82*, 1061–1062.
15. Korb, J. P.; Ahadi, M.; Zientara, G. P.; Freed, J. H. *J. Chem. Phys.* **1987**, *86*, 1125–1130.
16. Tabony, J.; Korb, J. P. *Mol. Phys.* **1985**, *56*, 1281–1305.
17. Kimmich, R.; Unrath, W.; Schnur, G.; Rommel, E. *J. Magn. Reson.* **1991**, *91*, 136–140.
18. Kimmich, R.; Klammler, F.; Skirda, V. D.; Serebrennikova, I. A.; Maklakov, A. I.; Fatkullin, N. *Appl. Magn. Reson.* **1993**, *4*, 425–440.
19. Kimmich, R.; Gneiting, T.; Kotitschke, K.; Schnur, G. *Biophys. J.* **1990**, *58*, 1183–1197.
20. Karger, J. *Mol. Phys.* **1981**, *43*, 1189–1191.
21. Heink, W.; Karger, J.; Walter, A. *Exp. Tech. Physik* **1978**, *26*, 161–173.
22. Karger, J. *Surf. Sci.* **1976**, *57*, 749–754.
23. Karger, J.; Herrmann, P. *Ann. Physik (Leipzig)* **1974**, *31*, 277–286.
24. Karger, J. *Z. Phys. Chem.* **1973**, *252*, 299–304.
25. D'Orazio, F.; Bhattacharja, S.; Halperin, W. P.; Eguchi, K.; Mizusaki, T. *Phys. Rev., B Condens. Matter* **1990**, *42*, 9810–9818.
26. D'Orazio, F.; Bhattacharja, S.; Halperin, W. P.; Gerhardt, R. *Phys. Rev., B Condens. Matter* **1990**, *42*, 6503–6508.
27. Polnaszek, C. F.; Bryant, R. G. *J. Chem. Phys.* **1984**, *81*, 4038–4045.
28. Polnaszek, C. F.; Hanggi, D.; Carr, P. W.; Bryant, R. G. *Anal. Chem. Acta* **1987**, *194*, 311–316.
29. Hodges, M. W.; Cafiso, D. S.; Polnaszek, C. F.; Lester, C. C.; Bryant, R. G. *Biophys. J.* **1997**, *73*, 2575–2579.
30. Kimmich, R.; Weber, H. W. *Phys. Rev., B Condens. Matter* **1993**, *47*, 11788–11794.
31. Overloop, K.; VanGerven, L. *J. Magn. Reson.* **1993**, *Series A 101*, 179–187.
32. Strange, J. H.; Rahman, M.; Smith, E. G. *Phys. Rev. Lett.* **1993**, *71*, 358.
33. Winslow, D. N. *J. Colloid Interface Sci.* **1978**, *67*, 42–47.
34. Stapf, S.; Kimmich, R. *J. Chem. Phys.* **1995**, *103*, 2247–2250.
35. Zavada, T.; Stapf, S.; Beginn, U.; Kimmich, R. *Magn. Reson. Imaging* **1998**, *16*, 711–713.
36. Zavada, T.; Stapf, S.; Kimmich, R. *Magn. Reson. Imaging* **1998**, *16*, 695–697.
37. Zavada, T.; Stapf, S.; Beginn, U.; Kimmich, R. *Magn. Reson. Imaging* **1998**, *16*, 711–713.
38. Ylihautala, M.; Jokisaari, J.; Fischer, E.; Kimmich, R. *Phys. Rev. E Stat. Phys. Plasmas Fluids Relat. Interdiscip. Topics* **1998**, *57*, 6844–6850.
39. Valiullin, R.; Kimmich, R.; Fatkullin, N. *Phys. Rev. E Stat. Phys. Plasmas Fluids Relat. Interdiscip. Topics* **1997**, *56*, 4371–4375.
40. Stapf, S.; Kimmich, R. *Chem. Phys. Lett.* **1997**, *275*, 261–268.
41. Stapf, S.; Kimmich, R.; Seitter, R. O. *Magn. Reson. Imaging* **1996**, *14*, 841–846.
42. Kimmich, R.; Stapf, S.; Maklakov, A. I.; Skirda, V. D.; Khozina, E. V. *Magn. Reson. Imaging* **1996**, *14*, 793–797.

43. Stapf, S.; Kimmich, R.; Seitter, R. *Phys. Rev. Lett.* **1995**, *75*, 2855–2858.
44. Blum, F. D.; Francses, E. I.; Rose, K. D.; Bryant, R. G.; Miller, W. G. *Langmuir* **1987**, *3*, 448–452.
45. Stapf, S.; Kimmich, R.; Niess, J. *J. Appl. Phys.* **1994**, *75*, 529–537.
46. Klafter, J.; Blumen, A.; Shlesinger, M. F. *Phys. Rev. A* **1987**, *35*, 3081–3085.
47. Korb, J. P.; Xu, S.; Jonas, J. *J. Chem. Phys.* **1993**, *98*, 2411–2422.
48. Korb, J. P.; Delville, A.; Xu, S.; Jonas, J. *Magn. Reson. Imaging* **1994**, *12*, 179–181.
49. Korb, J. P.; Delville, A.; Xu, S.; Demeulenaere, G.; Costa, P.; Jonas, J. *J. Chem. Phys.* **1994**, *101*, 7074–7081.
50. Korb, J. P.; Malier, L.; Cros, F.; Xu, S.; Jonas, J. *Phys. Rev. Lett.* **1996**, *77*, 2312–2316. S0031-9007(2396)01130–01131.
51. Solomon, I. *Phys. Rev.* **1955**, *99*, 559–565.
52. Bloembergen, N.; Morgan, L. O. *J. Chem. Phys.* **1961**, *34*, 842–850.
53. Hwang, L.-P.; Freed, J. H. *J. Chem. Phys.* **1975**, *63*, 4017–4025.
54. Freed, J. H. *J. Chem. Phys.* **1978**, *94*, 2843–2847.
55. Ayant, Y.; Belorizky, E.; Alizon, J.; Gallice, J. *J. Phys. I (Paris)* **1975**, *36*, 991–1004.
56. Polnaszek, C. F.; Bryant, R. G. *Magn. Reson. Med.* **1985**, *2*, 296–297.
57. Sur, S. K.; Bryant, R. G. *J. Phys. Chem.* **1995**, *99*, 6301–6308.
58. Bertini, I.; Luchinat, C. "*NMR of Paramagnetic Molecules in Biological Systems*"; Benjamin/ Cummings: Menlo Park, CA, 1986.
59. Nilsson, T.; Kowalewski, J. *J. Magn. Reson.* **2000**, *146*, 345–358.
60. Nilsson, T.; Svoboda, J.; Westlund, P. O.; Kowalewski, J. *J. Chem. Phys.* **1998**, *109*, 6364–6375.
61. Larsson, T.; Westlund, P. O.; Kowalewski, J.; Koenig, S. H. *J. Chem. Phys.* **1994**, *101*, 1116–1128.
62. Dinesen, T. R. J.; Wagner, S.; Bryant, R. G. *J. Am. Chem. Soc.* **1998**, *120*, 7004–7009.
63. Dinesen, T. R. J.; Bryant, R. G. *J. Magn. Reson.* **1998**, *132*, 19–24.
64. Dinesen, T. R. J.; Seymour, J.; McGowan, L.; Wagner, S.; Bryant, R. G. *J. Phys. Chem.* **1999**, *103*, 782–786.
65. Korb, J. P.; Whaley-Hodges, M.; Bryant, R. G. *Phys. Rev. E Stat. Phys. Plasmas Fluids Relat. Interdiscip. Topics* **1997**, *56*, 1934–1945.
66. Korb, J. P.; Hodges, M. W.; Bryant, R. *Magn. Reson. Imaging* **1998**, *16*, 575–578.
67. Korb, J. P.; Hodges, M. W.; Gobron, T.; Bryant, R. G. *Phys. Rev. E Stat. Phys. Plasmas Fluids Relat. Interdiscip. Topics* **1999**, *60*, 3097–3106.
68. Sur, S. K.; Heinsbergen, J. F.; Listinsky, J. J.; Bryant, R. G. *J. Colloid Interface Sci.* **1993**, *157*, 141–146.
69. Sur, S. K.; Heinsbergen, J. F.; Bryant, R. G. *J. Magn. Reson.* **1993**, *103*, 27–33.
70. Listinsky, J. J.; Bryant, R. G. *Magn. Reson. Med.* **1988**, *8*, 285–292.
71. Kleinberg, R. L. *Encyclopedia of nuclear magnetic resonance* **1996**, *6*, 4960.
72. Godefroy, S.; Fleury, M.; Deflandre, F.; Korb, J. P. *J. Phys. Chem. B* **2002**, *106*, 11183–11190.
73. Godefroy, S.; Korb, J. P.; Fleury, M.; Bryant, R. G. *Phys. Rev. E Stat. Phys. Plasmas Fluids Relat. Interdiscip. Topics* **2001**, *64*, 021605.
74. Godefroy, S.; Korb, J. P.; Fleury, M.; Bryant, R. G. *Magn. Reson. Imaging* **2001**, *19*, 517–519.
75. Nestle, N.; Galvosas, P.; Geier, O.; Zimmermann, C.; Dakkouri, M.; Karger, J. *J. Appl. Phys.* **2001**, *89*, 8061–8065.
76. Nestle, N.; Galvosas, P.; Karger, J. *J. Appl. Phys.* **2001**, *90*, 518–520.
77. Nestle, N.; Zimmermann, C.; Dakkouri, M.; Karger, J. *J. Phys. D Appl. Phys.* **2002**, *35*, 166–171.
78. Nestle, N.; Dakkouri, M.; Geier, O.; Freude, D.; Karger, J. *J. Appl. Phys.* **2000**, *88*, 4269–4273.
79. Barberon, E.; Korb, J. P.; Petit, D.; Morin, V.; Bermeo, E. *Phys. Rev. Lett.* **2003**, *90*, 116103 (1–4).
80. Valckenborg, R. M. E.; Huinink, H. P.; Kopinga, K. *J. Chem. Phys.* **2003**, *118*, 3243–3251.
81. Valckenborg, R. M. E.; Huinink, H. P.; Sande, J.; Kopinga, K. *Phys. Rev. E Stat. Nonlin. Soft Matter Phys.* **2002**, *65*, 021306/021301–021308.
82. Valckenborg, R. M. E.; Pel, L.; Kopinga, K. *J. Magn. Reson.* **2001**, *151*, 291–297.
83. Valckenborg, R. M. E.; Pel, L.; Kopinga, K. *J. Phys. D Appl. Phys.* **2002**, *35*, 249–256.
84. Grad, J.; Bryant, R. G. *J. Magn. Reson.* **1990**, *90*, 1–8.

85. Wolff, S. D.; Balaban, R. S. *Magn. Reson. Med.* **1989**, *10*, 135–144.
86. Bryant, R. G. *Annu. Rev. Biophys. Biomol. Struct.* **1996**, *25*, 29–53.
87. Hinton, D. P.; Bryant, R. G. *J. Phys. Chem.* **1994**, *98*, 7939–7941.
88. Henkelman, R. M.; Stanisz, G. J.; Graham, S. J. *NMR Biomed.* **2001**, *14*, 57–64.
89. Bryant, R. G.; Mendelson, D.; Lester, C. C. *Magn. Reson. Med.* **1991**, *21*, 117–126.
90. Winter, F.; Kimmich, R. *Biophys. J.* **1985**, *48*, 331–335.
91. Nusser, W.; Kimmich, R. *J. Phys. Chem.* **1990**, *94*, 5637–5639.
92. Kimmich, R.; Winter, F.; Nusser, W.; Spohn, K. H. *J. Magn. Reson.* **1986**, *68*, 263–282.
93. Kimmich, R.; Nusser, W.; Winter, F. *Phys. Med. Biol.* **1984**, *29*, 593–596.
94. Abragam, A. "*The Priniciples of Nuclear Magnetism*"; Oxford University Press: Oxford, The Clarendon Press, 1961, pp. 401–411.
95. Korb, J.-P.; Bryant, R. G. *Magn. Reson. Med.* **2002**, *48*, 21–26.
96. Korb, J. P.; Van-Quynh, A.; Bryant, R. G. *Chem. Phys. Lett.* **2001**, *339*, 77–82.
97. Korb, J. P.; Bryant, R. G. *J. Chem. Phys.* **2001**, *115*, 10964–10974.
98. Korb, J.-P.; Bryant, R. G. C. R. *Physique* **2004**, *5*, 349–359.
99. Anoarado, E.; Ferrante, G. M. *Appl. Magn. Reson.* **2003**, *24*, 85–96.
100. Whaley, M.; Lawence, A. J.; Korb, J. P.; Bryant, R. G. *Solid State Nucl. Magn. Reson.* **1996**, *7*, 247–252.
101. Schauer, G.; Kimmich, R.; Nusser, W. *Biophys. J.* **1988**, *53*, 397–404.
102. Winter, F.; Kimmich, R. *Biophys. J.* **1985**, *48*, 331–335.
103. Seitter, R. O.; Link, T.; Kimmich, R.; Kobelkov, A.; Wolfangel, P.; Muller, K. *J. Chem. Phys.* **2000**, *112*, 8715–8722.
104. Denisov, V. P.; Halle, B. *J. Mol. Biol.* **1995**, *245*, 682–697.
105. Denisov, V. P.; Halle, B. *J. Mol. Biol.* **1995**, *245*, 698–709.
106. Denisov, V. P.; Halle, B.; Peters, J.; Horlein, H. D. *Biochemistry* **1995**, *34*, 9046–9051.
107. Denisov, V. P.; Peters, J.; Horlein, H. D.; Halle, B. *Nat. Struct. Biol.* **1996**, *3*, 505–509.
108. Denisov, V. P.; Halle, B. *Faraday Discuss* **1996**, *103*, 227–244.
109. Denisov, V. P.; Halle, B. *Biochemistry* **1998**, *37*, 9595–9604.
110. Denisov, V. P.; Jonsson, B. H.; Halle, B. *Nat. Struct. Biol.* **1999**, *6*, 253–260.
111. Halle, B.; Denisov, V. P. *Biophys. J.* **1995**, *69*, 242–249.
112. Viaznikova, M.; Denisov, V. P.; Nikolaeva, S. S.; Petrusevich Iu, M. *Biofizika* **1993**, *38*, 492–499.
113. Wiesner, S.; Kurian, E.; Prendergast, F. G.; Halle, B. *J. Mol. Biol.* **1999**, *286*, 233–246.
114. Otting, G.; Liepinsh, E.; Halle, B.; Frey, U. *Nat. Struct. Biol.* **1997**, *4*, 396–404.
115. Langhorst, U.; Loris, R.; Denisov, V. P.; Doumen, J.; Roose, P.; Maes, D.; Halle, B.; Steyaert, J. *Protein Sci.* **1999**, *8*, 722–730.
116. Johannesson, H.; Denisov, V. P.; Halle, B. *Protein Sci.* **1997**, *6*, 1756–1763.
117. Yi, J.; Jonas, J. *J. Phys. Chem.* **1996**, *100*, 16789–16793.
118. Zhang, J.; Jonas, J. *J. Phys. Chem.* **1994**, *98*, 6835–6840.
119. Adamy, S. T.; Grandinetti, P. J.; Masuda, Y.; Campbell, D.; Jonas, J. *J. Chem. Phys.* **1991**, *94*, 3568–3576.
120. Sullivan, V. S.; Kim, Y. J.; Xu, S.; Jonas, J.; Korb, J. *Langmuir* **1999**, *15*, 4664–4669.
121. Lester, C. C.; Bryant, R. G. *J. Phys. Chem.* **1990**, *94*, 2843–2847.
122. Winter, F.; Kimmich, R. *Mol. Phys.* **1982**, *45*, 33–49.

WATER AND PROTON EXCHANGE PROCESSES ON METAL IONS

LOTHAR HELM, GAËLLE M. NICOLLE and ANDRÉ E. MERBACH

Laboratoire de chimie inorganique et bioinorganique, Ecole polytechnique fédérale de Lausanne EPFL-BCH, CH-1015 Lausanne, Switzerland

I. Introduction

The simplest reaction on a metal ion in aqueous solution is the exchange of a water molecule between the first and second coordination shells. This reaction is fundamental in understanding not only the reactivity of metal ions in chemical and biological systems but also the metal–water interaction. The replacement of a water molecule from the first coordination shell represents an important step in complex-formation reactions of metal cations and in many redox processes (*1*).

$$[M(H_2O)_n]^{z-} + n\ H_2O^* \underset{}{\overset{k_{ex}}{\rightleftarrows}} [M(H_2O^*)_n]^{z+} + n\ H_2O \qquad (1)$$

ADVANCES IN INORGANIC CHEMISTRY
VOLUME 57 ISSN 0898-8838 / DOI 10.1016/S0898-8838(05)57007-7

In solvent exchange reactions there is no net reaction and the Gibbs free energy change, ΔG^0, of the reaction is zero because the reactant and the product are identical. The measured life times of a water molecule in the first coordination shell of a metal cation cover 19 orders of magnitude (Fig. 1). A water molecule stays at average nearly 300 years (9×10^9 s) in the first coordination shell of $[Ir(H_2O)_6]^{3+}$ (*2*) before it is replaced by another one coming from the bulk solvent. The mean life time of H_2O bound to $[Eu(H_2O)_7]^{2+}$ is however only about 200 ps (2×10^{-10} s) (*3,4*). Astonishingly, both exchange processes follow an associative activation mode which means that for both hydration complexes the transition state or the intermediate encountered during the exchange reaction has an increased coordination number.

The rates for water exchange depend strongly on the nature of the metal ion as observed in Fig. 1. The dependence on solvent is however much less pronounced as shown in different reviews (*5,6*). The discussion of solvent exchange and more specifically of water exchange is therefore, conveniently divided into categories of metal ions. A first group is represented by cations formed by main group elements. These ions have filled electron shells and they differ mainly in electric charge and ionic radius. The number of water molecules in the first shell around the ion, called coordination number CN, ranges from 4 up to 10 (*7*). A second category is formed by the *d*-transition metal ions, which are all hexa-hydrated, with

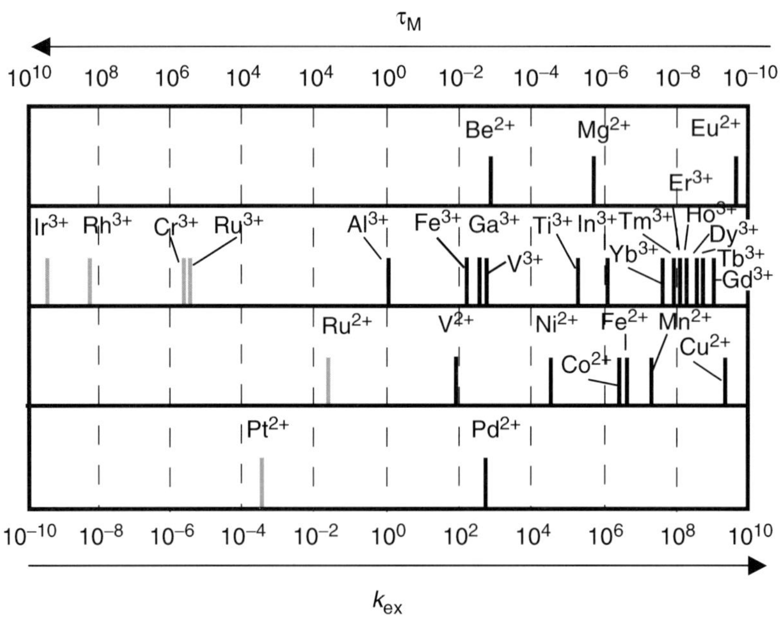

FIG. 1. Rate constants, k_{ex}, for the exchange of water molecules from the first coordination sphere of a given metal ion and the corresponding mean life times, $\tau_M = 1/k_{ex}$, of a particular water molecule measured by ^{17}O NMR. The gray bars indicate values determined by NMR isotope exchange technique (except Cr^{3+}).

the exception of Pd^{2+} and Pt^{2+}, which are four-coordinated (square-planar), Sc^{3+} which is suggested to be hepta-coordinated (*8*) and Cu^{2+} where there is some evidence that it is five-coordinated (*9*). The water exchange rate constants of *d*-transition metal ions are strongly influenced by the occupancy of the *d*-orbitals. Considering the ionic radii, r_M, alone, the first row transition metal ions should have k_{ex} of the same order of magnitude as Zn^{2+} for the divalent ions and as Ga^{3+} for the trivalent ions. The measured exchange rates vary however by 7 (divalent) and 15 (trivalent) orders of magnitude, depending largely on the electronic configuration. A third category involves the lanthanide ions which have eight or nine water molecules in the first coordination shell. Their kinetic behavior is mainly influenced by the decrease in ionic radius along the series and by the coordination equilibrium observed in the middle of the series.

Replacing some of the first shell water molecules by one or more ligands, which are kinetically inert can have a strong influence on the rate and the mechanism of exchange of the remaining water molecule(s). Metal ion complexes still having one or more H_2O directly bound are important in catalysis (*10*), as active centers in bio-molecules (*11*) and in the special case of gadolinium(III), as contrast agents in medical magnetic resonance imaging (*12*).

In section II we will first summarize the concepts of solvent exchange reactions and how exchange reaction mechanisms can be assigned from variable pressure experiments. In section III, we will describe the oxygen-17 NMR techniques available to measure water exchange rates over 18 orders of magnitude. Water exchange rate constants and activation parameters measured directly by ^{17}O NMR on aqua-ions, and on some metal complexes with an inert ligand and water molecules in the first shell, are discussed in the following sections. We would like to mention that solvent exchange reactions including other solvents have been recently reviewed in Advances in Inorganic Chemistry by Dunand, Helm and Merbach (*6*).

II. Concepts of Solvent Exchange Reactions

A. Classification of Mechanisms

Langford and Gray proposed in 1965 (*13*) a mechanistic classification for ligand substitution reactions, which is now generally accepted and summarized here for convenience. In their classification they divided ligand substitution reactions into three categories of stoichiometric mechanisms: associative (**A**) where an intermediate of increased coordination number can be detected, dissociative (**D**) where an intermediate of reduced coordination number can be detected, and interchange (**I**) where there is no kinetically detectable intermediate [Eqs. (2)–(4)]. In Eqs. (2)–(4), $\{MX_{n-1}\}$ and $\{MX_nY\}$

represent intermediates of reduced and increased coordination number, respectively, and $MX_nX \cdots Y$ an outer sphere complex.

$$MX_n \underset{+X}{\overset{-X}{\rightleftarrows}} \{MX_{n-1}\} \underset{-Y}{\overset{+Y}{\rightleftarrows}} MX_{n-1}Y \quad (\mathbf{D} = \text{dissociative}) \tag{2}$$

$$MX_n \underset{-Y}{\overset{+Y}{\rightleftarrows}} \{MX_nY\} \underset{+X}{\overset{-X}{\rightleftarrows}} MX_{n-1}Y \quad (\mathbf{A} = \text{associative}) \tag{3}$$

$$MX_n \cdots Y \rightleftarrows MX_{n-1}\mathrm{Y} \cdots X \quad (\mathbf{I} = \text{interchange}) \tag{4}$$

Langford and Gray furthermore distinguished two categories of intimate mechanisms: those with an associative activation mode (**a**) and those with a dissociative activation mode (**d**). In the first case the reaction rate is sensitive to the nature of the entering group whereas in the second case the reaction rate is not sensitive to the variation of the entering group (Fig. 2).

All **A** mechanisms must be associatively activated and all **D** mechanisms must be dissociatively activated. The interchange mechanisms (**I**) include a continuous spectrum of transition states where the degree of bond-making between the entering ligand and the complex ranges from very substantial ($\mathbf{I_a}$ mechanism) to negligible ($\mathbf{I_d}$ mechanism) and inversely for bond-breaking (*14,15*). For a solvent exchange reaction, the forward and backward reaction coordinates must be symmetrical. For an $\mathbf{I_a}$ mechanism both the entering and the leaving solvent molecule must have considerable bonding to the

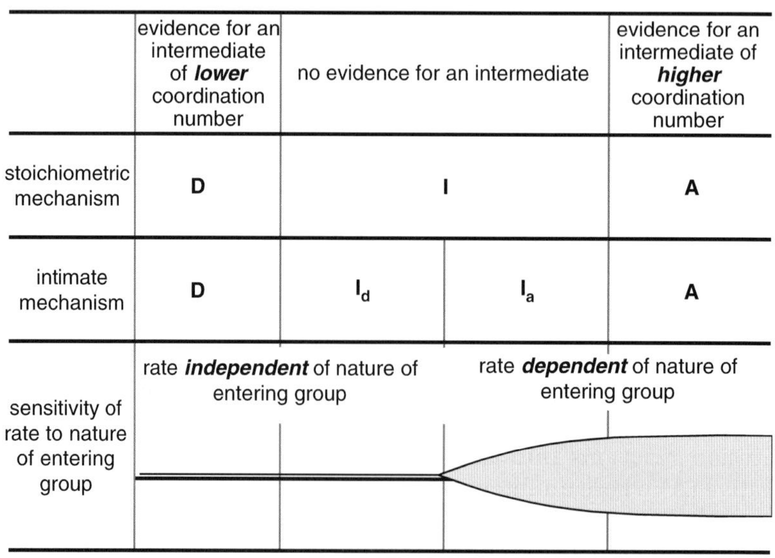

FIG. 2. Classification of substitution reactions.

metal. Inversely, for an $\mathbf{I_d}$ mechanism, if the entering solvent molecule has weak bond-making properties, the leaving one is also weakly bound.

The assignment of a reaction mechanism is mainly based on the response of the reaction rate to the variation of chemical and physical parameters like the nature and concentration of the entering ligand, temperature and pressure. The rate law for solvent exchange reactions can be determined using an inert diluent (*16*). Unfortunately such a diluent does not exist for water. The variation of the enthalpy and entropy of activation, obtained from variable temperature experiments, within a series of similar ligand substitution systems can be a guide to a mechanistic change. Thus **d**-activated reactions tend to have greater $\Delta H^{\ddagger}$ values than do **a**-activated reactions, and $\Delta S^{\ddagger}$ tends to be positive for **d**- and negative for **a**-activated reactions. However, the magnitudes of the contributions of these two parameters arising from interactions that occur beyond the 1st coordination shell can be uncertain and the determination of $\Delta S^{\ddagger}$ is often prone to systematic errors, especially for paramagnetic ions.

The pressure dependence of the exchange rate constant leads to the activation volume, $\Delta V^{\ddagger}$, which has become the major tool for the experimental determination of solvent exchange mechanisms (*15,17,18*). This is mainly due to the direct connection between the sign of $\Delta V^{\ddagger}$ and the intimate mechanism for solvent exchange.

B. The Volume of Activation

The volume of activation, $\Delta V^{\ddagger}$, is defined as the difference between the partial molar volumes of the transition state for the reaction and the reactants and is related to the pressure variation of the rate constant at a constant temperature T by Eq. (5) (*19,20*):

$$\left(\frac{\partial \ln(k)}{\partial P}\right)_T = -\frac{\Delta V^{\ddagger}}{RT} \tag{5}$$

Normally the pressure dependence of $\Delta V^{\ddagger}$ is small which leads to the approximate Eq. (6),

$$\ln\left(\frac{k_P}{k_0}\right) = -\frac{\Delta V_0^{\ddagger} P}{RT} + \frac{\Delta\beta^{\ddagger} P^2}{2RT} \cong -\frac{\Delta V_0^{\ddagger} P}{RT} \tag{6}$$

where k_P and k_0 are the rate constants at pressures P and 0, respectively, $\Delta V_0^{\ddagger}$ is the activation volume at $P = 0$ and $\Delta\beta^{\ddagger}$ is the compressibility coefficient of activation (*20*). For exchange reactions of water the quadratic term in Eq. (6) is small compared to the linear one for pressures generally applied in kinetic studies (typically 0–300 MPa) and therefore $\Delta V^{\ddagger} \cong \Delta V_0^{\ddagger}$ (*21*).

The interpretation of the activation volume for solvent exchange reactions on metal ions is based on the transition state theory (*19*). The experimentally determined activation volume is considered to be the sum of an intrinsic, $\Delta V^{\ddagger}_{\rm int}$, and an electrostriction contribution, $\Delta V^{\ddagger}_{\rm elec}$ (*20,21*). Electrostriction is the reduction of the solvent's molar volume due to very strong electrostatic forces of attraction exerted by the ions on surrounding solvent molecules. The $\Delta V^{\ddagger}_{\rm int}$ results from changes in internuclear distances and angles within the reactants during the formation of the transition state and can therefore be used in the assignment of reaction mechanisms. The term $\Delta V^{\ddagger}_{\rm elec}$ arises from the difference in the electrostriction between the transition state and the reactants, which can be important for substitution reactions when ions or electric dipoles are formed or neutralized at the transition state. For solvent exchange processes, where the charge of the complex does not change, $\Delta V^{\ddagger}_{\rm elec}$ is small. Therefore, $\Delta V^{\ddagger} \cong \Delta V^{\ddagger}_{\rm int}$ and the measured activation volume is a direct measure of the degree of bond-formation and bond-breakage in going to the transition state.

Figure 3 visualizes the pressure-induced changes in the measured exchange rate constants and the underlying water exchange reaction mechanism. In a **d**-activated exchange process the approach to the transition state is characterized mainly by *bond-breaking* and results in an increase in volume, which means a decrease in the observed reaction rate constant, k_P (Eq. 6). In an **a**-activated exchange process, however, the approach to the transition state is characterized mainly by *bond-formation* and results in a decrease in volume which means an increase in k_P. Briefly, $\Delta V^{\ddagger} > 0$ is indicative of dissociatively activated processes and $\Delta V^{\ddagger} < 0$ is indicative of associatively activated processes. However, it is less easy to decide on the basis of activation volumes if the mechanism is a limiting associative, **A,** or

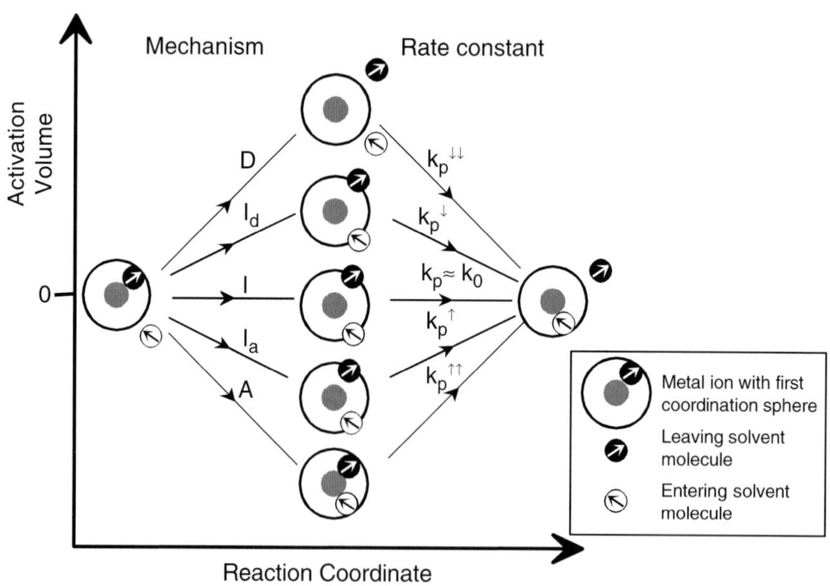

FIG. 3. Water exchange processes: volume profiles connected to the transition states.

dissociative, **D**, or an interchange, **I** with a relatively strong contribution of the entering solvent molecule (**I$_a$**) or with a negligible contribution of the entering solvent molecule (**I$_d$**), respectively. The absolute value of the activation volume, $|\Delta V^{\ddagger}|$, can serve as a guideline: the larger $|\Delta V^{\ddagger}|$, the closer the mechanism is to a limiting **A** or **D**. Swaddle (*22,23*) developed a semi-empirical model for the limiting volumes for water by linking the partial molar volumes of hydrated metal ions to its ionic radius, r_M. Following Shannon (*24*), r_M – values for a specific metal depend on the charge of the ion and on its coordination number. Swaddle's model gives similar limiting values of $|\Delta V^{\ddagger}| = 13.5\ \text{cm}^3\ \text{mol}^{-1}$ for **A** and **D** processes on +2 and +3 octahedral 3*d* transition metal ions. Exceptions to this guideline are however observed and therefore, each case has to be considered individually.

III. Water Exchange Rate Constants Measured by Oxygen-17 NMR

A. Experimental NMR Techniques – General Remarks

Only a small number of experimental techniques allow the direct determination of solvent exchange rate constants. The measurement of water exchange rates in solution today, is almost exclusively performed by NMR (^{18}O mass spectroscopy in the past), which itself offers a variety of different methods like line shape analysis, methods based on relaxation time measurements, or methods based on isotopic labeling (*25*). In principle, both ^{1}H-NMR and ^{17}O-NMR can be used to study the exchange of water molecules. To measure the exchange rate constant of a water molecule, ^{17}O, the only stable oxygen isotope bearing a nuclear spin, is normally used in NMR experiments. [This is in spite of its very low natural abundance (~0.037 %) and as a consequence, its very low sensitivity]. There are mainly two reasons for the use of ^{17}O-NMR: firstly, exchange of H_2O protons can be faster than the exchange of the whole water molecule and secondly, NMR chemical shift differences of ^{17}O between bound and free water are in general much larger than for ^{1}H.

In the case of H_2O exchange on metal ions, line shape analysis technique is not of practical importance because the NMR spectrum consists in the best case of two resonances – a very intense one from bulk water and a small one from water in the first coordination shell of the ion, often designed as bound water. But worse, even under slow-exchange condition, in most cases the resonance due to bound water cannot be observed. In general, the mole fraction of bound water is much smaller than that of bulk water, often well below 0.1. In the case of diamagnetic metal ions the difference in chemical shift between bound and free water is often small and the resonance from the bound water is hidden under the huge resonance from bulk water. In the case of slowly exchanging water bound to paramagnetic metal ions chemical shifts are large, but the resonance of bound water is often extremely broad and disappears in the base line of the spectrum. Therefore, different experimental NMR techniques have to be applied, depending on how fast the exchange between the first coordination shell and the bulk of the solution takes place and depending on the nature of the metal ion also. In the following we will describe the different techniques used in ^{17}O-NMR to study water exchange on metal ions in aqueous solution.

B. Slow Exchange Studied by Isotopic Labeling Technique

To study water exchange on aqua metal ions with very slow exchange of water molecules, an isotopic labeling technique using oxygen-17 can be used. A necessary condition for the applicability of this technique is that the life time, τ_M, of a water molecule in the first coordination shell of the ion is much longer that the time needed to acquire the ^{17}O-NMR spectrum. With modern NMR spectrometers and using enrichments up to 40% in ^{17}O, the acquisition time can be as short as 1 s.

If the aqua complex is very inert, as for example in the case of $[Rh(H_2O)_6]^{3+}$ or $[Ir(H_2O)_6]^{3+}$, the experiment is best performed by dissolving the complex enriched in ^{17}O-water, $[M(H_2{}^{17}O)_6]^{n+}$, in non-enriched water (*2,26*). Preparing the initial condition in this way leads to a relatively intense signal for bound water. The exchange rate constant can then be measured by observing the decrease in the bound water signal with time. The decrease in the mole fraction of labeled water coordinated to the metal, x, is described by Eq. (7):

$$x = x_\infty + (x_0 - x_\infty)\exp\left[\frac{-k_{ex}t}{1 - x_\infty}\right] \tag{7}$$

The mole fractions of labeled water at $t = 0$ and at equilibrium are noted as x_0 and x_∞, respectively (Fig. 4). In the end, the signal of bound water becomes small and difficult to quantify. But, this does not influence the quality of the measured rate constant because the mole fraction at equilibrium, x_∞, is known from the concentration of the metal ion and the coordination number. These experiments can be performed at variable temperature and at variable pressure to obtain activation enthalpies and entropies as well as activation volumes.

If the life time τ_M is below 15 min, the reaction is best started by mixing two thermostated solutions directly inside the NMR magnet using a fast injection device (*27*). A stock solution of the metal ions at a relatively high concentration is rapidly diluted with ^{17}O-enriched water. The exchange reaction is studied after the fast mixing by recording the increasing signal of the bound water molecule [Eq. (8), Fig. 5] (*28,29*). To be able to observe the NMR

$$x = x_\infty\left(1 - \exp\left[\frac{-k_{ex}t}{1 - x_\infty}\right]\right) \tag{8}$$

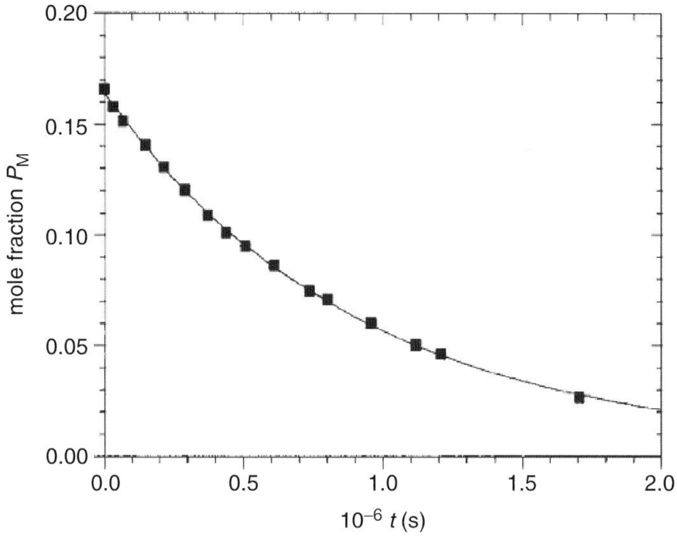

FIG. 4. Mole fraction of bound ^{17}O-water as a function of time (monitored by ^{17}O NMR) for a solution of 0.03 m $[Ir(H_2O)_6](CF_3SO_3)_3$ (3% ^{17}O-water) in 1.0 m CF_3SO_3H at $T = 358.2$ K ($\mu = 5.1$ m $NaCF_3CO_3$).

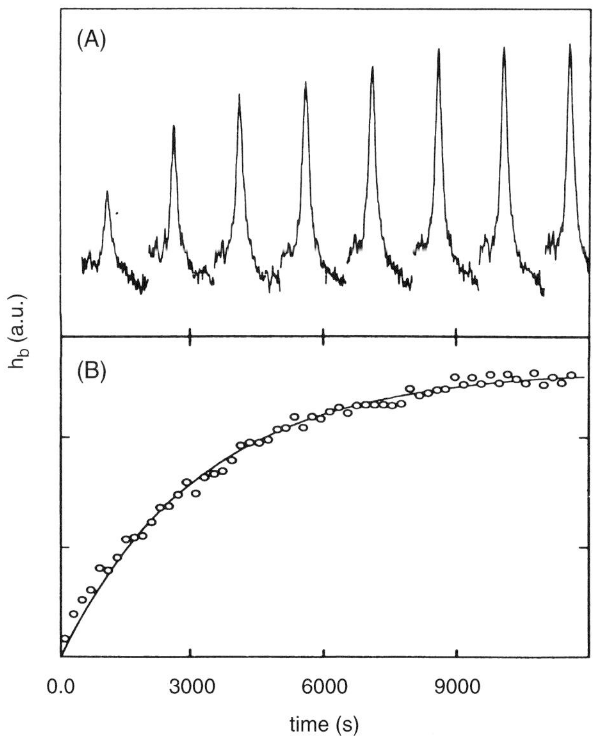

FIG. 5. Spectra (A) and peak heights (B) as a function of time of the ^{17}O-NMR signal of coordinated H_2O after addition of $[Al(H_2O)_6]^{3+}$ to ^{17}O enriched H_2O ($T = 254.5$ K, $[Al^{3+}] = 0.41$ m, $[Mn^{2+}] = 0.44$ m, $[H^+] = 2.57$ m, $k_{ex} = 0.00293 \pm 0.00007$ s^{-1}).

signal of bound water, it may be necessary to add either a paramagnetic shift reagent like $[Co(H_2O)_6]^{2+}$ or a relaxation agent like $[Mn(H_2O)_6]^{2+}$ (*28,30*). It should be mentioned that isotopic labeling technique cannot be performed using ^2H as labeling isotope due to hydrogen exchange which is much faster than water exchange in these cases.

C. Swift and Connick Approach

Water exchange on most metal ions is too fast to be studied by isotopic labeling technique. The measurement of NMR relaxation rates has been shown to be the most widely applicable technique to measure large exchange rate constants. For the case of a simple two-site exchange between bound and bulk water, analytical solutions of the Bloch equations modified by McConnell to include chemical exchange (*31*) exist for both, longitudinal and transverse relaxation (*32*). From these equations approximate solutions for slow and fast exchange have been derived (*33*). These approximations are valid for all two-site exchanges, and not only for the special case where one site is very dilute and fast-relaxing.

In a classical paper, Swift and Connick (*34,35*) derived solutions of the equation for transverse relaxation, $1/T_2$, and chemical shift, $\Delta\omega$, in the case of dilute solutions of paramagnetic ions. Equation (9) gives the increase in transverse relaxation of the bulk water signal, $1/T_{2r}$, due to exchange with water bound to a paramagnetic ion and normalized by the mole fraction of bound water, P_M.

$$\frac{1}{T_{2r}} = \frac{1}{P_M}\left[\frac{1}{T_2} - \frac{1}{T_{2A}}\right] = \frac{1}{\tau_M}\frac{T_{2M}^{-2} + \tau_M^{-1}T_{2M}^{-1} + \Delta\omega_M^2}{\left(\tau_M^{-1} + T_{2M}^{-1}\right)^2 + \Delta\omega_M^2} + \frac{1}{T_{2os}} \tag{9}$$

$1/T_{2A}$ is the relaxation rate of bulk water in absence of exchange, $\Delta\omega_M$ is the chemical shift difference (in rad s^{-1}) between bound and bulk water in absence of exchange, and τ_M is the life time of a water molecule in the first coordination shell of the metal ion M. The reciprocal of τ_M is the exchange rate constant k_{ex} for the exchange of a water molecule. The direct influence of the paramagnetic center on outer sphere molecules is included in a term $1/T_{2os}$.

The chemical shift measured on the bulk water signal and normalized by the mole fraction of bound water, P_M, is given by Eq. (10), where ω and ω_A are the observed chemical shifts with and without the paramagnetic compound and ω_{os} is the shift originating from water molecules outside the first coordination shell.

$$\Delta\omega_r = \frac{1}{P_M}[\omega - \omega_A] = \frac{\tau_M^{-2}\Delta\omega_M^2}{\left(\tau_M^{-1} + T_{2M}^{-1}\right)^2 + \Delta\omega_M^2} + \Delta\omega_{os} \tag{10}$$

The Eqs. (9) and (10) describe the relaxation enhancement and the chemical shift difference measured on the intense bulk water signal, the small and broad ^{17}O-NMR signal of bound water has not to be observed. Transverse relaxation of ^{17}O is either measured by Carr–Purcell–Meiboom–Gill technique (*36*) or more conveniently by measuring the width of the resonance line. Slightly acidified neat water has a line width of about 50 Hz (at ambient temperature), and because its resonance frequency is more than 7 times lower than that of ^{1}H, line broadening due to magnetic field inhomogeneity can often be neglected.

The contributions to $1/T_{2r}$ and $\Delta\omega_r$ due to relaxation, chemical shift and exchange can be separated by measuring their temperature dependences. The temperature variations of the relaxation and chemical shift are shown in Fig. 6 for three magnetic field strengths. Figure 6a shows also the measured transverse relaxation rate of bulk water $1/T_{2A}$ as a function of inverse temperature. Figures 6b and 6c show the reduced relaxation rates $1/T_{2r}$ and chemical shifts $\Delta\omega_r$. Four temperature regions can be distinguished: at the low temperature region, I, $1/T_{2r}$ is given by the outer shell contribution $1/T_{2os}$; at the slow exchange region, II, the observed relaxation increase is proportional to the exchange rate constant $k_{ex} = 1/\tau_M$; at the fast exchange region, III, $1/T_{2r}$ depends normally strongly on the magnetic field strength and is proportional to $\Delta\omega_M{}^2/k_{ex}$; at the highest temperatures in region IV transverse relaxation enhancement is given by $1/T_{2M}$, the relaxation of bound water oxygen. The reduced chemical shift at low temperature is given by $\Delta\omega_{os}$, the outer sphere contribution, which is normally close to zero. In the slow and reasonably fast exchange region $|\Delta\omega_r|$ increases strongly. The inflection points of the $\Delta\omega_r$-curves are at the same temperature as the maxima of $1/T_{2r}$. At fast exchange $\Delta\omega_r$ is given by the chemical shift difference between bound and bulk water, $\Delta\omega_M$. The temperature variation of the exchange rate constant is normally assumed to follow Eyring's equation given in Eq. (11):

$$\frac{1}{\tau_M} = k_{ex} = \frac{k_B T}{h}\exp\left\{\frac{\Delta S^\ddagger}{R} - \frac{\Delta H^\ddagger}{RT}\right\} = k_{ex}^{298}\frac{T}{298.15}\exp\left\{\frac{\Delta H^\ddagger}{RT}\left(\frac{1}{298.15} - \frac{1}{T}\right)\right\} \tag{11}$$

$\Delta H^\ddagger$ and $\Delta S^\ddagger$ are the enthalpy and the entropy of activation for the exchange reaction, respectively.

In most cases only a part of the full curves, as shown in Fig. 6, can be observed experimentally because the temperature range is restricted by the freezing and evaporation of the water solvent. Nevertheless, very often, exchange rate constants can be calculated. In the four temperature regions, the Swift and Connick equations can be approximated by the following expressions:

$$\text{Region I:}\quad \text{very slow exchange}\quad \frac{1}{T_{2r}} \cong \frac{1}{T_{2os}} \quad\text{and}\quad \Delta\omega_r \cong \Delta\omega_{os} \tag{12}$$

$$\text{Region II:}\quad \text{slow exchange}\quad \frac{1}{T_{2r}} \cong \frac{1}{\tau_M} \tag{13}$$

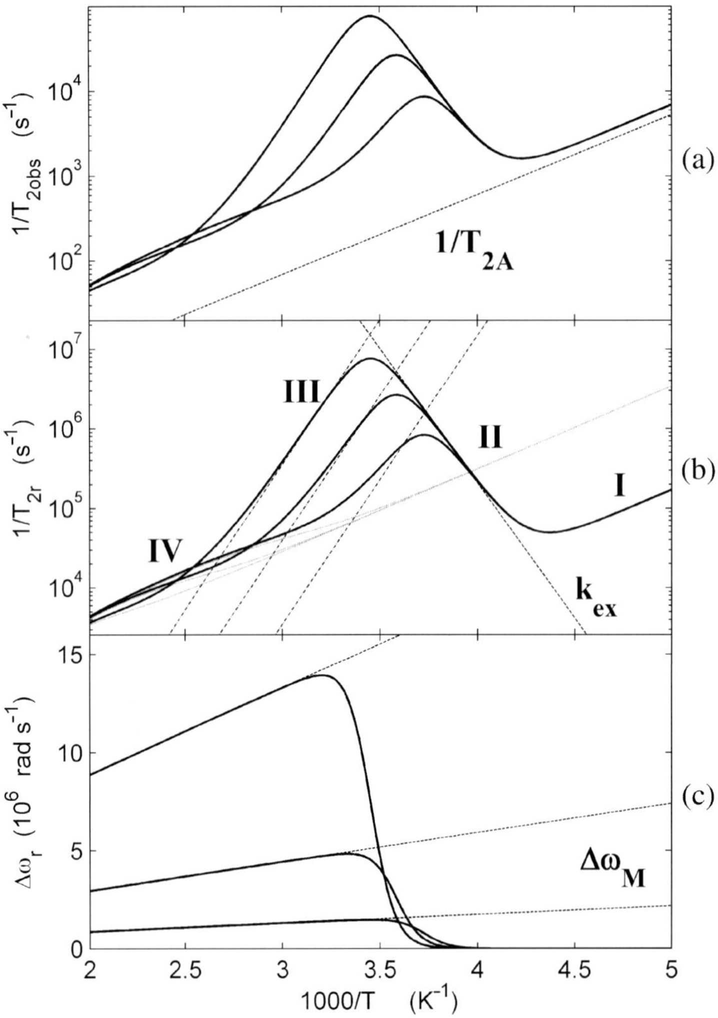

FIG. 6. Temperature variations of observed ^{17}O transverse relaxation rates (a), reduced transverse relaxation rates (b) and reduced chemical shifts (c) calculated from Swift and Connick equations for $B_0 = 14.1$ T (largest effect), 4.7 T, and 1.4 T (smallest effect).

$$\text{Region III}: \quad \text{fast exchange} \quad \frac{1}{T_{2r}} \cong \tau_M \, \Delta\omega_M^2 \tag{14}$$

$$\text{Region IV}: \quad \text{very fast exchange} \quad \frac{1}{T_{2r}} \cong \frac{1}{T_{2M}} \quad \text{and} \quad \Delta\omega_r \cong \Delta\omega_M \tag{15}$$

From these approximate equations, it follows that exchange rate constants $k_{ex} = 1/\tau_M$ are easily obtained in region II and, if the chemical shift difference is large, in region III also. In the very fast exchange, region IV information on water exchange can also be obtained if the interaction between the paramagnetic metal ion and the oxygen-17 spin is dominated by the scalar interaction. The transverse scalar relaxation and the chemical shift are then given by Eqs (16) and (17) (*37*),

$$\frac{1}{T_{2sc}} = \frac{S(S+1)}{3}\left(\frac{A}{\hbar}\right)^2 \tau_{s1} \quad \text{with} \quad \frac{1}{\tau_{s1}} = \frac{1}{\tau_M} + \frac{1}{T_{1e}} \tag{16}$$

$$\Delta\omega_M = \frac{g_L \mu_B S(S+1) B_0}{3 k_B T} \frac{A}{\hbar} \tag{17}$$

where S is the electron spin quantum number, $A/\hbar$ is the scalar coupling constant (in rad s^{-1}) which entirely determines the chemical shift $\Delta\omega_M$, T_{1e} is the longitudinal electron spin relaxation time, g_L is the Landé g-factor, μ_B is the Bohr magneton and k_B the Boltzmann constant. If the exchange rate constant is much larger than the electron spin relaxation rate, then $1/T_{2M} \approx 1/T_{2sc} \propto \tau_M$. Under these conditions the highest water exchange rates directly measured by ^{17}O-NMR have been observed (*3,38*).

D. Water Exchange on Diamagnetic Ions

The measurement of exchange rate constants by relaxation measurements on diamagnetic ions is much more restricted when compared to paramagnetic ions and normally only data on slowly exchanging ions can be measured by the ^{17}O-NMR relaxation technique. From the approximate solutions of the two-site exchange equations (*33*) we obtain under slow exchange conditions for dilute solutions ($\Delta\omega_M \gg 1/\tau_M$ and $1/\tau_M \ll |1/T_{2A} - 1/T_{2M}|$)

$$\frac{1}{T_2} = \frac{1}{T_{2A}} + \frac{1}{\tau_A} \quad \text{and} \quad \frac{1}{T_2} = \frac{1}{T_{2B}} + \frac{1}{\tau_B} \tag{18}$$

If a relaxation agent like $[Mn(H_2O)_6]^{2+}$ is added to the solution, the range of "slow exchange" can be extended considerably as has been shown in examples like $[Al(H_2O)_6]^{3+}$ (Fig. 7) and $[Ga(H_2O)_6]^{3+}$ (*28,39*).

The limit of measurable exchange rate constants on diamagnetic metal ions can be extended to faster reactions by adding a strong chemical shift reagent like the lanthanide ion Tb^{3+} (*40*). The life time of H_2O at $[Tb(H_2O)_8]^{3+}$ is very short (*41*) and the electron spin of Tb^{3+} relaxes extremely fast (*42,43*). As a consequence the resonance signal of bulk water is only slightly broadened but very much shifted towards lower frequency. The exchange between bulk and first shell of a diamagnetic ion like Mg^{2+} broadens the resonance of bulk water further as given by Eq. (19) (*40*)

$$\Delta\frac{1}{T_2} = P_{Mg} \frac{k_{Mg}(P_{Tb}\Delta\omega_{Tb})^2}{k_{Mg}^2 + (P_{Tb}\Delta\omega_{Tb})^2} \tag{19}$$

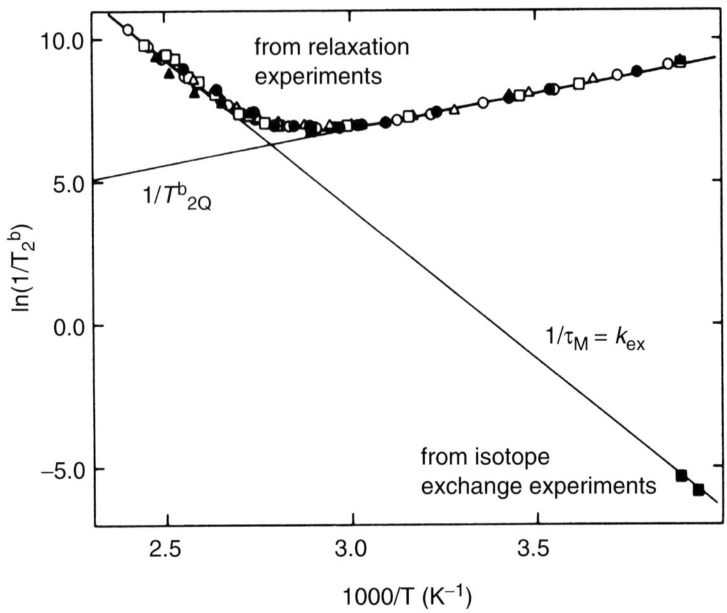

FIG. 7. Temperature dependence of transverse relaxation rates of bound H_2O in $[Al(H_2O)_6]^{3+}$ with various concentrations of $Al(ClO_4)_3$, $Mn(ClO_4)_2$, and $HClO_4$ and exchange rates obtained from isotope exchange experiments (■).

where k_{Mg} is the exchange rate constant from the first shell of the diamagnetic ion Mg^{2+}, $\Delta\omega_{Tb}$ is the chemical shift of water in the first shell of Tb^{3+} and P_{Mg} and P_{Tb} are the mole fractions of Mg^{2+} and Tb^{3+}, respectively (Fig. 8).

Using this technique the limit of exchange rate constants which can be measured on diamagnetic ions depends on the magnetic field available. Using a magnetic field strength of 14.1 T an exchange rate constant of $\sim 10^6$ s^{-1} (at 25°C) could be observed (*40,44*) (Fig. 9).

IV. Water Exchange on Main Group and *d*-Transition Metal Ions

A. Main Group Metal Ions

There are only few main group metal ion hydrates open to detailed mechanistic study of water exchange by NMR: Be^{2+}, Mg^{2+}, Al^{3+}, Ga^{3+} and to a less extent, In^{3+}. They provide the opportunity to study the influence of size and charge on exchange rate constant and mechanism without the complicating effects of the variation of the electronic occupancy of the *d*-orbitals. All of the alkali ions as well as Ca^{2+}, Sr^{2+}, and Ba^{2+} are very labile as a consequence of their relatively low surface charge density. However, indications on water exchange on Sr^{2+} can be obtained from

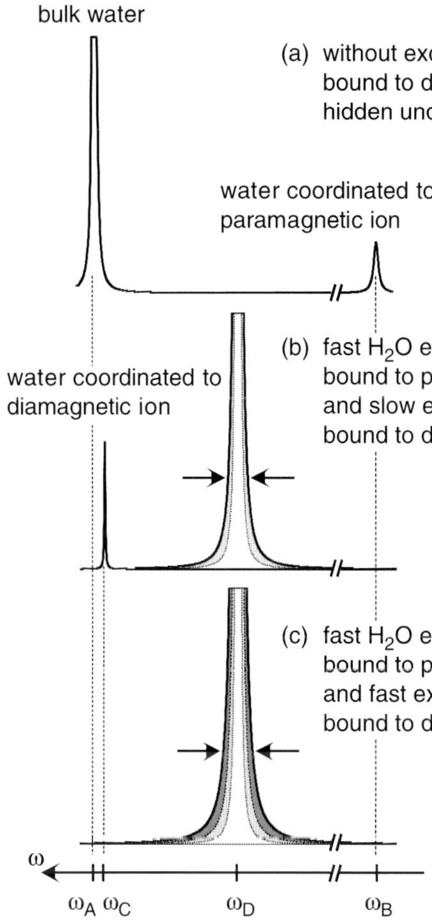

FIG. 8. Schematic representations of NMR spectra with three-site exchange.

studies on the paramagnetic Eu^{2+} which has the same size as the diamagnetic ion (see Section V. C).

Beryllium(II) has a very small ionic radius (Table I) and is tetrahedrally coordinated by water molecules in aqueous solution (*16,45–48*). The water exchange reaction on Be^{2+} is characterized by the most negative activation volume observed for a water exchange process on a hydrated ion ($\Delta V^{\ddagger} = -13.6$ cm^3 mol^{-1}), which is close to a limiting $\Delta V^{\ddagger} = -12.9$ cm^3 mol^{-1} calculated for an **A** mechanism using Swaddle's empirical formula for octahedral 3*d* aqua ions (*16,22,49*). A limiting **A** mechanism was therefore assigned for water exchange on $[Be(H_2O)_4]^{2+}$.

The larger Mg^{2+} ion is hexa-coordinated. The water exchange rate for $[Mg(H_2O)_6]^{2+}$ (*40*) lies between those of the *d*-transition metals $[Co(H_2O)_6]^{2+}$ and $[Ni(H_2O)_6]^{2+}$ (Tables I and Table II), and reflects the order of ionic radii of these three ions. The measured activation volume is

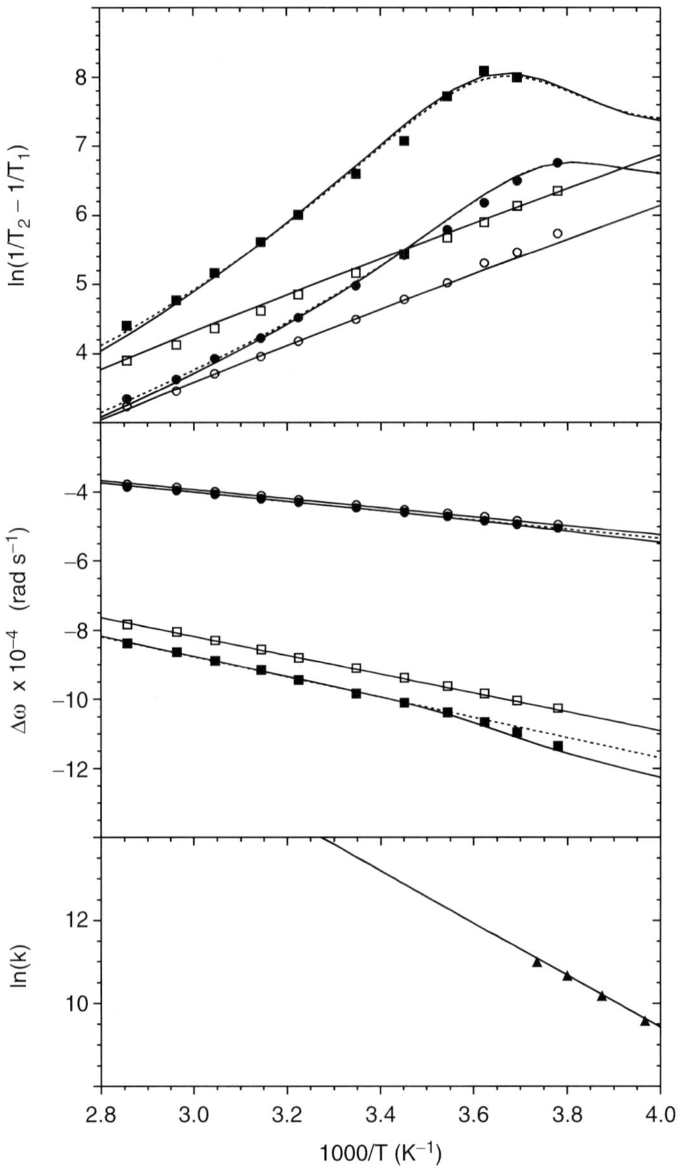

FIG. 9. Variable temperature ^{17}O NMR results ($B_0 = 14.1$ T) for solutions containing $Tb(ClO_4)_3$ (○,□), $Mg(ClO_4)_2$ and $Tb(ClO_4)_3$ (●,■), and $Mg(ClO_4)_2$ and $Mn(ClO_4)_2$ (relaxation agent) (▲). Full lines result from non-linear fitting using Kubo-Sack formalism and short-dashed lines were calculated by an approximate 3-site Swift and Connick method [Ref. (40)].

positive and also intermediate to those obtained for Co^{2+} and Ni^{2+}: a decrease in ionic radii correlates with an increase in activation volumes (Co^{2+} $r_M = 74/\Delta V^{\ddagger} = +6.1$; Mg^{2+} $r_M = 72/\Delta V^{\ddagger} = +6.7$; Ni^{2+} $r_M = 69/\Delta V^{\ddagger} = +7.2$; radii in pm (*24*), volumes in $cm^3\ mol^{-1}$). From the similarity in

TABLE I

RATE CONSTANTS AND ACTIVATION PARAMETERS FOR WATER EXCHANGE ON MAIN GROUP IONS

Ion	r_M[a] (pm)	k (298 K) (s^{-1})	$\Delta H^{\ddagger}$ (kJ mol^{-1})	$\Delta S^{\ddagger}$ (J K^{-1} mol^{-1})	$\Delta V^{\ddagger}$ (cm^3 mol^{-1})	Mechanism	Reference
$[Be(H_2O)_4]^{2+}$	27	730	59.2	+8.4	−13.6	**A**	*(16)*
$[Mg(H_2O)_6]^{2+}$	72	6.7×10^5	49.1	+31.1	+6.7	**D, I_d**	*(40)*
		5.3×10^5	42.7	+8			*(30)*
$[Al(H_2O)_6]^{3+}$	54	1.29	84.7	+41.6	+5.7	**I_d**	*(28)*
		2.0	73	+6.0			*(50)*
			85.4[b]		*+5.6*[b]	*D*[b]	*(51)*
$[Al(OH)(H_2O)_5]^{2+}$		3.1×10^4	36.4	−36.4			*(52)*
$[AlF(H_2O)_5]^{2+}$		1.1×10^2	79	+60			*(52)*
$[Ga(H_2O)_6]^{3+}$	62	400	67.1	+30.1	+5.0	**I_d**	*(39)*
			69.1[b]		*+4.8*[b]	*D*[b]	*(51)*
$[Ga(OH)(H_2O)_5]^{2+}$		1×10^5	58.9		+6.2	**I_d, D**	*(39)*
$[In(H_2O)_6]^{3+}$	80	$\geq 4 \times 10^4$	19.2	−96		**A, I_a**	*(53)*
		$> 1 \times 10^7$	*29.1*[b]		*−5.2*[b]	*A*[b]	*(51)*

[a]From Ref. *(24)*.
[b]From *ab initio* calculations.

activation parameters between Mg^{2+} and these ions, an $\mathbf{I_d}$ or **D** mechanism for H_2O exchange is assumed. Furthermore, the rates of complex formation of $[Mg(H_2O)_6]^{2+}$ (*54*) are close to the water exchange rate which is indicative of metal–water bond rupture as the rate determining step.

The trivalent cations of the main group elements aluminum, gallium and indium differ strongly in their ionic radii (Table I). In aqueous solution, coordination numbers are six as measured by X-ray diffraction for Al^{3+} (55–57) and In^{3+} (*58,59*), and by NMR measurements for Al^{3+} (*45,60,61*) Ga^{3+} (*61,62*) and In^{3+} (*61*). Rate constants and activation parameters for water exchange on Al^{3+} (*28*) and Ga^{3+} (*39*) have been determined from ^{17}O-NMR experiments (Table I). Water exchange processes on $[Al(H_2O)_6]^{3+}$ and $[Ga(H_2O)_6]^{3+}$ are characterized by positive activation volumes which are however substantially smaller than anticipated for a **D** mechanism (*22*). From the high pressure NMR data an $\mathbf{I_d}$ mechanism was therefore proposed for the water exchange reaction on Al^{3+} and Ga^{3+}. The water exchange rate on $[In(H_2O)_6]^{3+}$ was too fast to be measured even by using the chemical shift enhancement technique (*40*) described in Section III.D, and thus precluded the measurement of an activation volume (*51*). Newer limits for water exchange rate constants from these measurements show, however, that older estimates (*53*) of water exchange for In^{3+} are about 3 orders of magnitude slower.

Kowall *et al.* (*51*) modeled different mechanisms for water exchange processes on Al^{3+}, Ga^{3+}, and In^{3+} using *ab initio* calculations at the Hartree–Fock level on clusters with one ion and seven water molecules. For a limiting **D** mechanism, transition states and intermediates have been obtained for all three clusters. The energies of the transition state for Al^{3+} and Ga^{3+} are in excellent agreement with experimental activation enthalpies (Table I). The computation of transition states for an associative exchange mechanism, **A**, was only successful for In^{3+}. The associative pathway calculated for this ion requires a much lower activation energy ($\Delta E^{\ddagger} = 29.1$ kJ mol^{-1}) than the dissociative water exchange pathway ($\Delta E^{\ddagger} = 75.2$ kJ mol^{-1}). The calculated activation energy of 29.1 kJ mol^{-1} is in acceptable agreement with the limit, $\Delta G_{298}{}^{\ddagger} \leq 33.1$ kJ mol^{-1}, obtained experimentally (*51*). For Al^{3+} and Ga^{3+} activation volumes estimated from *ab initio* calculations are very close to experimental ones (Table I). From the associative pathway calculated for In^{3+} a negative $\Delta V^{\ddagger}$ is obtained. These calculations suggest a change from a **d**-activation mode to an **a**-activation mode with increasing ionic radius. From the calculations, limiting mechanisms **D** ($[Al(H_2O)_6]^{3+}$), **D** ($[Ga(H_2O)_6]^{3+}$), and **A** ($[In(H_2O)_6]^{3+}$) are suggested (*51*). Activation volumes determined from pressure variation of the exchange rate constant reflect the volume change between the reactants, here ($[M(H_2O)_6]^{3+}$) and a second shell water molecule, and the *transition state.* The volume change between the reactants and an *intermediate* with higher or lower coordination number is different from the activation volume, $\Delta V^{\ddagger}$. From the *ab initio* calculations volume differences have been calculated: for the dissociative intermediate +7.1 ($[Al(H_2O)_5\cdot(H_2O)_2]^{3+}$) and +6.9 ($[Ga(H_2O)_5\cdot(H_2O)_2]^{3+}$) cm^{3}mol^{-1}, and for the associative

TABLE II

RATE CONSTANTS AND ACTIVATION PARAMETERS FOR WATER EXCHANGE ON DIVALENT FIRST ROW TRANSITION METAL IONS $[M(H_2O)_6]^{2+}$

	3d^3	**3d^5**	**3d^6**	**3d^7**	**3d^8**	**3d^9**
M^{2+}	**V^{2+}**	**Mn^{2+}**	**Fe^{2+}**	**Co^{2+}**	**Ni^{2+}**	**Cu^{2+}** [a]
r_M (pm)	79	83	78	74	69	73
k (298 K) (s^{-1})	8.7×10^1	2.1×10^7	4.4×10^6	3.2×10^6	3.2×10^4	5.7×10^9
$\Delta H^{\ddagger}$ (kJ mol^{-1})	61.8	32.9	41.4	46.9	56.9	11.6
$\Delta S^{\ddagger}$ (J K^{-1} mol^{-1})	−0.4	+5.7	+21.2	+37.2	+32.0	−19.2
$\Delta V^{\ddagger}$ (cm^3 mol^{-1})	−4.1	−5.4	+3.8	+6.1	+7.2	+2.0
Mechanism	$\mathbf{I_a}$	$\mathbf{I_a}$	$\mathbf{I_d}$	$\mathbf{I_d}$	$\mathbf{I_d}$	**I**
Ref.	*(69)*	*(70)*	*(70)*	*(70)*	*(70)*	*(9,71,72)*

[a] CN = 5 [see Ref. *(9)*].

intermediate −7.1 ($[In(H_2O)_7]^{3+}$) cm^3 mol^{-1} *(51)*. The values calculated for the corresponding transition states are given in Table I.

The deprotonation of a water molecule on $[M(H_2O)_6]^{3+}$ (M = Al, Ga) to form $[M(OH)(H_2O)_5]^{2+}$ increases the lability of the other water ligands substantially. This increase in lability is stronger by two orders of magnitude for $[Al(OH)(H_2O)_5]^{2+}$ when compared to $[Ga(OH)(H_2O)_5]^{2+}$ (Table I). For aluminum(III) the effect is among the largest when measured for trivalent metal ions and even much larger than for the isoelectronic $[AlF(H_2O)_5]^{2+}$ (Table I) *(52)*. For water exchange on $[Ga(OH)(H_2O)_5]^{2+}$, an activation volume could be determined which is more positive than for the hexa-aqua ion *(39)* indicating a more dissociative character of the water exchange reaction. For H_2O exchange on $[Al(OH)(H_2O)_5]^{2+}$ there are no variable pressure data. But, despite the negative $\Delta S^{\ddagger}$ observed from variable temperature measurements *(52)*, a more dissociative character of the exchange process, when compared to the hexa-aqua ion, is also expected.

B. FIRST-ROW d-TRANSITION METALS

Water exchange on first row transition metal ions has been studied extensively as a function of temperature and pressure, and has been reviewed several times *(5,14,63)*. Experimental first order rate constants and activation parameters are summarized in Table II for divalent and in Table III for trivalent ions. From the data in these tables it can be seen that the water exchange rate constants do not correlate with ionic radii and that they are largely independent of the reaction mechanism. The rate constants for water exchange are however sensitive to the electronic configuration of the metal ions. Simple calculations of ligand field activation energies, LFAE, were already performed thirty years ago for **D** and **A** exchange mechanisms *(6,15,64)*. Correlations were found between LFAE and exchange rate constants but not for exchange mechanisms. For example, water exchange on

TABLE III

RATE CONSTANTS AND ACTIVATION PARAMETERS FOR WATER EXCHANGE ON TRIVALENT FIRST ROW TRANSITION METAL IONS $[M(H_2O)_6]^{3+}$

	3d¹	**3d²**	**3d³**	**3d⁵**	**3d¹⁰**
M^{3+}	Ti^{3+}	V^{3+}	Cr^{3+}	Fe^{3+}	Ga^{3+}
r_M (pm)	67	64	61	64	62
k(298 K) (s^{-1})	1.8×10^5	5.0×10^2	2.4×10^{-6}	1.6×10^2	4.0×10^2
$\Delta H^‡$ (kJ mol^{-1})	43.4	49.0	108.6	64.0	67.1
$\Delta S^‡$ (J K^{-1} mol^{-1})	+1.2	−27.8	+11.5	+12.1	+30.1
$\Delta V^‡$ (cm^3 mol^{-1})	−12.1	−8.9	−9.6	−5.4	+5.0
Mechanism	**A, I_a**	**I_a**	**I_a**	**I_a**	**I_d, D**
Ref.	*(77)*	*(78)*	*(79)*	*(80, 81)*	*(39)*

V^{2+} and Ni^{2+} is relatively slow and LFAE values calculated for transition states with increased (V^{2+}) and decreased (Ni^{2+}) coordination numbers are larger than for the other divalent ions. Another example is the trivalent d^3 ion Cr^{3+}. It has the slowest k_{ex}^{298} of all first row transition metal ions and it also shows the largest LFAE calculated for an associative exchange mechanism (> 80 kJ mol^{-1}). The important influence of the ligand field activation energies is furthermore demonstrated from the labilities of V^{2+} and V^{3+}. The smaller LFAE of $[V(H_2O)_6]^{3+}$ as compared to $[V(H_2O)_6]^{2+}$ explains why water exchange is about six times faster on the higher charged trivalent vanadium. Both exchange processes are **a**-activated with negative $\Delta V^‡$ values (Tables II and III). To be able to predict reaction mechanisms for water exchange on transition metals, full quantum mechanical calculations have to be performed, including in some cases the second coordination sphere also.

For the divalent first row transition metal ions a progressive change in exchange mechanism from $\mathbf{I_a}$ to $\mathbf{I_d}$ has been observed as the number of d-electrons increases and the ionic radius decreases. This change is most clearly demonstrated by the change in sign of the activation volume. The observed mechanistic changeover is mainly a consequence of the electronic configurations of the aqua complexes. For σ-bonded octahedral complexes t_{2g} orbitals are non-bonding whereas e_g* orbitals are anti-bonding. The t_{2g} orbitals are spread out between the ligands and therefore their gradual filling will electrostatically disfavor the approach of a seventh water molecule towards a face of the octahedron. The possibility of bond-making is decreased. The e_g* orbitals are directed towards the ligands and their increased occupancy will enhance the tendency of bond-breaking. Both effects, combined with steric effects due to a decrease in the ionic radius, explain the change from an **a**-mechanism to a **d**-mechanism by filling of the orbitals. The mechanistic changeover has been confirmed by quantum mechanical calculations *(65–68,73)*. These calculations predict $\mathbf{I_a}$ mechanisms for V^{2+} and Mn^{2+}, and **D** or $\mathbf{I_d}$ mechanism for Fe^{2+}, Cu^{2+}, and Zn^{2+}.

The hydration of the divalent copper ion is a special case. In the solid state $[Cu(H_2O)_6]^{2+}$ shows Jahn–Teller distortion due to the d^9 configuration of Cu^{2+} *(74–76)*. The metal–oxygen bonds of the two axial water molecules are longer than those of the four equatorial waters and they are therefore expected to be more labile. In solution, this Jahn–Teller distortion is expected to be dynamic and because the distortion axis jumps rapidly in time, all water molecules are axial for a certain time. This dynamic process was used to explain the very high lability of copper solvates. A complementary experimental and computational study performed by double difference neutron diffraction and by first-principles molecular dynamics, respectively, showed evidence that the coordination number of Cu^{2+} is five in aqueous solution *(9)*. From the Car–Parrinello molecular dynamics study, an equilibrium between square pyramidal and trigonal bipyramidal structures was proposed with a rapid exchange between them. Oxygen-17 experimental data showed the presence of two time constants, a fast one of ~5 ps and a slower one of ~230 ps, which were in the original article interpreted as the time constant for the rapid reorientation of the Jahn–Teller distortion axis and the life time of a water molecule in the first coordination shell of Cu^{2+}, respectively *(71)*. In a five-coordinated complex the experimental data could be reinterpreted with the time constants $\tau_i \sim 5$ ps for the mean life time of square pyramidal and trigonal bipyramidal structures and $\tau_M = 175$ ps for the life time of a water molecule in the first shell *(9)*. The small and positive activation volume measured is compatible with an interchange process for water exchange on $[Cu(H_2O)_5]^{2+}$ *(72)*.

For the trivalent first row transition metal ions a similar mechanistic behavior has been shown as for the divalent ones. The activation volumes, $\Delta V^{\ddagger}$, become less negative from -12.1 cm^3 mol^{-1} for $[Ti(H_2O)_6]^{3+}$ to -5.4 cm^3 mol^{-1} for $[Fe(H_2O)_6]^{3+}$. The value $\Delta V^{\ddagger}$ of Ti^{3+} is close to -13.5 cm^3 mol^{-1}, which is the value of a limiting **A** mechanism from Swaddle's semi-empirical model *(22)*. With the filling of the t_{2g} orbitals the electronic repulsion towards the entering seventh ligand increases and bond-making becomes less favorable. The associative character of the reaction mechanism becomes less pronounced. The main group ion Ga^{3+} which can be included in this series of trivalent ions has filled anti-bonding e_g* orbitals and bond-breaking is favored. Its size is about the same as that Cr^{3+} but its activation volume is positive and the activation mode for the water exchange process is dissociative. As for the divalent transition metal ions *ab initio* calculations confirmed the changeover of mechanism: hexa-coordinate Sc^{3+}, Ti^{3+}, and V^{3+} react *via* an associative **A** mechanism *(67,82)* and Ga^{3+} *via* a dissociative **D** mechanism *(51)*.

C. Second and Third Row d-Transition Metals

Experimental data for water exchange on aqua ions of second and third row *d*-transition metals are limited to the octahedral species $[Ru(H_2O)_6]^{2+/3+}$ *(83)*, $[Rh(H_2O)_6]^{3+}$ *(26)*, and $[Ir(H_2O)_6]^{3+}$ *(2)*, and to the square planar complexes $[Pd(H_2O)_4]^{2+}$ *(84)* and $[Pt(H_2O)_4]^{2+}$ *(29)*.

Comparison of the exchange rate constants measured on the low spin t_{2g}^6 Ru^{2+} with the analog high-spin t_{2g}^4, e_g^2 Fe^{2+} (Tables II and IV) shows that the low-spin complex is eight orders of magnitude less labile. This is a consequence of the smaller ionic radius for Ru^{2+} and mainly of the larger ligand field, which generate larger LFAE (ligand field activation energies). The activation enthalpy, $\Delta H^{\ddagger}$, is therefore about twice as much for water exchange on $[Ru(H_2O)_6]^{2+}$ than on $[Fe(H_2O)_6]^{2+}$.

For the low spin t_{2g}^6 aqua ions the approach of a seventh water molecule towards an edge or a face of the coordination octahedron is electrostatically disfavored by the filled t_{2g} orbitals spread out between the ligands. One would therefore *a priori* predict a dissociative activation mode for water exchange on $[Ru(H_2O)_6]^{2+}$, $[Rh(H_2O)_6]^{3+}$, and $[Ir(H_2O)_6]^{3+}$. In a first study the similarity of the measured rate constants for anation reactions by halide anions on $[Ru(H_2O)_6]^{2+}$ (*85*) indicated that the rate determining step to reach the transition state is very similar too, namely the dissociation of a water molecule from the first coordination shell. Later this interpretation was enforced by the results of a study of mono-complex formation with the ligands CH_3CN, DMSO, 1,4-thioxane and NMP (*86*): again the complex formation rate was found to be independent of the nature of the entering ligand and an $\mathbf{I_d}$ mechanism was proposed. However, the activation volume measured for water exchange on $[Ru(H_2O)_6]^{2+}$ is close to zero (Table IV), and an interchange **I** mechanism without predominant **a** and **d** character was assigned (*83*). Even more interesting negative $\Delta V^{\ddagger}$ values have been determined for $[Rh(H_2O)_6]^{3+}$, which is isoelectronic with the divalent ruthenium ion, and $[Ir(H_2O)_6]^{3+}$, which is the third row analog (*2,26*). An $\mathbf{I_a}$ mechanism was therefore assigned for H_2O exchange on these ions.

A better insight into the mechanistics of water exchange on the 2nd and 3rd row *d*-transition metal ions came from quantum chemical calculations (*87*). For the rhodium(III) ion two transition states have been obtained, one for the dissociative and one for the interchange mechanisms, whereas for the

TABLE IV

RATE CONSTANTS AND ACTIVATION PARAMETERS FOR WATER EXCHANGE ON SECOND AND THIRD ROW TRANSITION METAL IONS

	4d⁵	**4d⁶**	**4d⁶**	**5d⁴**	**4d⁸**	**5d⁸**
$\mathbf{M^{n+}}$	$\mathbf{Ru^{3+}}$	$\mathbf{Ru^{2+}}$	$\mathbf{Rh^{3+}}$	$\mathbf{Ir^{3+}}$	$\mathbf{Pd^{2+}}$	$\mathbf{Pt^{2+}}$
geometry	octahedral	octahedral	octahedral	octahedral	square planar	square planar
r_M (pm)	68	73	66.5	68	64	60
k (298 K) (s^{-1})	3.5×10^{-6}	1.8×10^{-2}	2.2×10^{-9}	1.2×10^{-9}	5.6×10^{2}	3.9×10^{-4}
$\Delta H^{\ddagger}$ (kJ mol^{-1})	89.8	87.8	131.2	130.5	49.5	89.7
$\Delta S^{\ddagger}$ (J K^{-1} mol^{-1})	−48.2	+16.1	+29.3	+2.1	−26.0	−9.0
$\Delta V^{\ddagger}$ (cm^3 mol^{-1})	−8.3	−0.4	−4.1	−5.7	−2.2	−4.6
Mechanism	$\mathbf{I_a}$	$\mathbf{I_d}$, **D**	$\mathbf{I_a}$	$\mathbf{I_a}$	$\mathbf{I_a}$	$\mathbf{I_a}$
Ref.	(*83*)	(*83*)	(*26*)	(*2*)	(*84*)	(*29*)

isoelectronic ruthenium(II) only the dissociative pathway could be calculated (Fig. 10). For water exchange on $[Rh(H_2O)_6]^{3+}$, the experimental $\Delta G^{\ddagger}$ (298 K) = 122.3 kJ mol^{-1} is intermediate between the $\Delta E^{\ddagger}$ values calculated for the **D** pathway (136.6 kJ mol^{-1}) and for the **I** pathway (114.8 kcal mol^{-1}), but closer to the latter. The calculated change in the sum of all metal–oxygen bond distances, $\Delta\Sigma$, is negative for the interchange mechanism and supports together with the negative volume of activation, an $\mathbf{I_a}$ mechanism. All attempts to calculate a transition state for an interchange mechanism failed for $[Ru(H_2O)_6]^{2+}$. The calculated $\Delta E^{\ddagger}$ (71.9 kJ mol^{-1}) for the **D** mechanism is sufficiently close to both experimental values, $\Delta G^{\ddagger}$ (83.0 kJ mol^{-1}) and $\Delta H^{\ddagger}$ (87.8 kJ mol^{-1}) (*83*) to attribute a **d**-activation mode for water exchange on this complex. The change in mechanism for the isoelectronic ions Rh^{3+} and Ru^{2+} can clearly be linked to the difference in charge. The strong Rh–O bonds forces water exchange on $[Rh(H_2O)_6]^{3+}$ to proceed *via* an $\mathbf{I_a}$ mechanism, whereas the weaker Ru–O bonds favor water exchange on $[Ru(H_2O)_6]^{2+}$, *via* an $\mathbf{I_d}$ or **D** mechanism.

The low-spin $t_{2g}{}^5$ $[Ru(H_2O)_6]^{3+}$ is by four orders of magnitude less labile than the $t_{2g}{}^6$ $[Ru(H_2O)_6]^{2+}$ (*83*). An $\mathbf{I_a}$ mechanism was attributed from the negative $\Delta V^{\ddagger}$. Because the water exchange on both ruthenium complexes is slow, electron exchange could be measured directly by ^{17}O NMR on the $[Ru(H_2O)_6]^{2+/3+}$ couple (*27*). It has been shown that the self-exchange on the

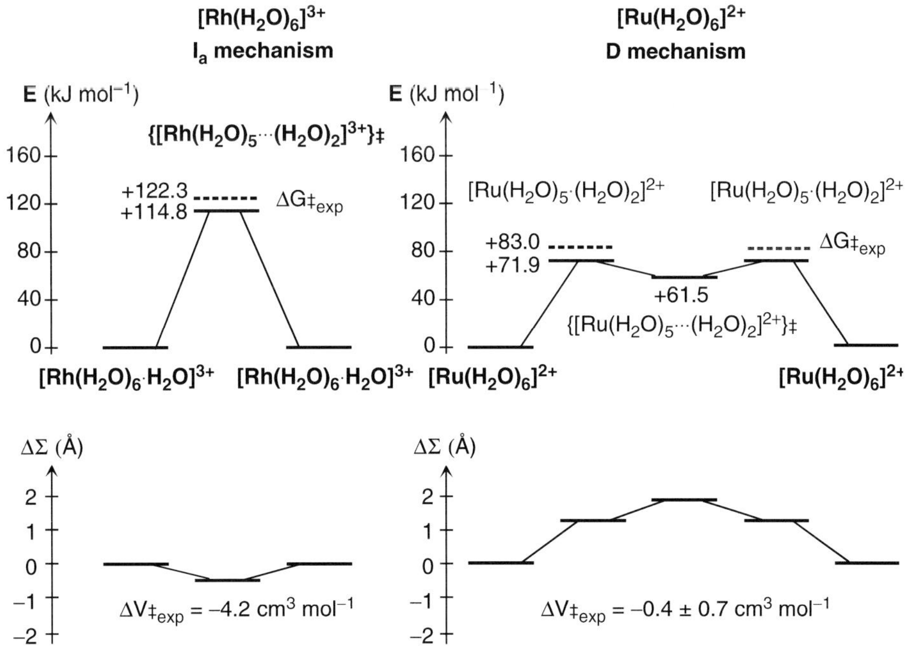

FIG. 10. Water exchange on isoelectronic $[Rh(H_2O)_6]^{3+}$ and $Ru(H_2O)_6]^{2+}$: results of quantum mechanical calculations using a polarizable continuum model; energy profiles and changes of the sum of all M–O bond distances ($\Delta\Sigma$).

2+/3+ ruthenium couple can be considered as a prototype of an outer-sphere electron transfer reaction.

The only water exchange rate constants measured on square-planar aqua ions are on $[Pd(H_2O)_4]^{2+}$ (*84*) and $[Pt(H_2O)_4]^{2+}$ (*29*). The lability of these complexes follows the general trend: palladium(II) complexes are much more labile than those of platinum(II). In the case of aqua complexes this difference is more than six orders of magnitude (Table IV). Because in aqueous solution the rate laws cannot be established from measurements in an inert diluent, mechanistic assignment relies mainly on activation volumes. The four-coordinated d^8 transition metal complexes are coordinatively unsaturated with 16 outer-shell electrons. Therefore the vast majority of solvent exchange reactions on square-planar complexes undergo an **a**-activated mechanism, whereas in the five-coordinate transition state or intermediate, a noble gas configuration is achieved (*88*). The interpretation of activation volumes measured on square-planar complexes is not trivial, as these solvates undergo large changes in coordination geometry on going from the reactant to the transition state. The square-planar species may have their two axial sites occupied by loosely bound water molecules. The simultaneous loss of one of the axially solvent molecules with formation of a new bond with the second solvent molecules in going to a five-coordinate transition state or intermediate may mean that through compensation, the overall volume change is small (*89*). Activation volumes will therefore not allow a straightforward mechanistic assignment.

Deeth *et al.* used density functional theory to model water exchange on square-planar $[Pd(H_2O)_4]^{2+}$ and $[Pt(H_2O)_4]^{2+}$ (*90*). The calculations strongly support that water exchange proceeds *via* an **a**-activation mode on these complexes and the trigonal bipyramidal structures calculated for $[Pd(H_2O)_5]^{2+}$ and $[Pt(H_2O)_5]^{2+}$ were very similar. There is a good agreement between experimental and calculated activation enthalpies for an $\mathbf{I_a}$ mechanism, whereas for an $\mathbf{I_d}$ mechanism, a difference of more than 100 kJ mol^{-1} is observed.

D. Hydrolyzed Trivalent Aqua-Ions

A common observation arising with trivalent metal ions in aqueous solution is the occurrence of the conjugate base species, $[M(OH)(H_2O)_5]^{2+}$ (*1*). At low pH values the water exchange rates, k_{ex}, observed experimentally by ^{17}O NMR, depend linearly on the reciprocal concentration of H^+ as shown by Eq. (20),

$$k_{ex} = k_1 + k_2/[H^+] \tag{20}$$

where k_1 is the rate constant for water exchange on $[M(H_2O)_6]^{3+}$ and $k_2 = k_{OH}K_a$, the product of the rate constant for the monohydroxy species, k_{OH}, and the first hydrolysis constant, K_a, for $[M(H_2O)_6]^{3+}$. It is usually assumed that all 6 oxygens in $[M(OH)(H_2O)_5]^{2+}$ contribute equally to the observed k_2 value (*26,81*). Activation volumes can be obtained for k_1 and k_2 by measuring k_{ex} at variable pressure and at variable $[H^+]$. $\Delta V^{\ddagger}$ values for water exchange on the monohydrolyzed species can be calculated by difference if ΔV_a^0, the thermodynamic volume change for the hydrolysis reaction [Eq. (21)], is known.

$$[M(H_2O)_6]^{3+} \rightleftarrows [M(OH)(H_2O)_5]^{2+} + H^+ \tag{21}$$

Table V summarizes rate constants, and activation volumes for water exchange on $[M(OH)(H_2O)_5]^{2+}$ together with those of $[M(H_2O)_6]^{3+}$. Accuracy of exchange rate constants k_{OH} and its activation parameters $\Delta H^{\ddagger}$ and $\Delta S^{\ddagger}$ relies on the knowledge of K_a, $\Delta H_a{}^0$, and $\Delta S_a{}^0$ (*39,52*). The hydrolysis of aqueous M^{3+} ions is complicated by oligomerization and ranges of hydrolysis constants have been reported for example for Al^{3+} (*91,92*) and Ga^{3+} (*93–95*). As a general trend a strong increase in the lability of the coordinated water molecules is observed for the hydrolyzed species when compared to the hexa-aqua ions. Even at very low pH, where the mole fraction of the hydrolyzed species is extremely small, the water exchange observed by ^{17}O NMR on the bulk water can be dominated by the fast exchange on $[M(OH)(H_2O)_5]^{2+}$ and not by the much slower exchange on $[M(H_2O)_6]^{3+}$.

Activation volumes for water exchange on $[M(OH)(H_2O)_5]^{2+}$ (Table V) are all more positive than those measured on the corresponding hexa-aqua ions indicating a more dissociative character for the water-exchange reaction. The decrease in the positive charge at the metal center loosens the metal–water bonds and facilitates rupture of the M–O bond.

E. Proton Exchange on Metal Aqua Ions

Experimental data on proton exchange rates on aqua-ions in solution is very limited, and there are several reasons for that. One reason is that, if water exchange on $[M(H_2O)_x]^{n+}$ is fast the water molecule exchanges "as a whole" and the "proton exchange rate" is identical to the water exchange rate. If water exchange is slow then protons exchange faster than water oxygen, and the exchange rate constant observed depends on acidity (pH) and pK_a of the aqua ion. Another reason is that a variety of possible pathways have been discussed in literature (*96,97*). Which pathway is the most active one depends on the pK_a values of the aqua ion, the $[H^+]$ concentration and eventually the acid used to adjust the pH. For proton exchange on $[Rh(H_2O)_6]^{3+}$ three possible pathways have been proposed as written in Eqs. (22)–(24) (*97*). (The rate constants correspond to the exchange of any of the protons from the first coordination shell with the bulk.)

$$[Rh(H_2O)_6]^{3+} + {}^*H_2O + \overset{k_{H_2O}}{\rightleftarrows} [Rh({}^*H_2O)_6]^{3+} + H_2O \tag{22}$$

TABLE V

RATE CONSTANTS AND ACTIVATION VOLUMES FOR WATER EXCHANGE ON TRIVALENT HEXA-AQUA TRANSITION METAL IONS and THEIR CONJUGATE BASE SPECIES

		Al^{3+}	Cr^{3+}	Fe^{3+}	Ru^{3+}	Rh^{3+}	Ir^{3+}	Ga^{3+}
$[M(H_2O)_6]^{3+}$	r_M (pm)	54	61	64	68	66.5	68	62
	k_1 (298 K) (s^{-1})	1.29	2.4×10^{-6}	1.6×10^{2}	3.5×10^{-6}	2.2×10^{-9}	1.1×10^{-10}	4.0×10^{2}
	$\Delta V_1^{\ddagger}$ ($cm^3\ mol^{-1}$)	+5.7	−9.6	−5.4	−8.3	−4.1	−5.7	+5.0
	Mechanism	$\mathbf{I_d}$, **D**	$\mathbf{I_a}$	$\mathbf{I_a}$	$\mathbf{I_a}$	$\mathbf{I_a}$	$\mathbf{I_a}$	$\mathbf{I_d}$, **D**
$[M(H_2O)_5OH]^{2+}$	k_2 (298 K) [a] ($M^{-1}\ s^{-1}$)	7.2×10^{-2}	1.4×10^{-8}	1.2×10^{3}	1.1×10^{-6}	1.5×10^{-8}	1.4×10^{-11}	1.4×10^{1}
	$\Delta V_2^{\ddagger}$ ($cm^3\ mol^{-1}$)	–	−1.1	+7.8	−2.1	+1.2	−0.2	+7.8
	pK_a	5.64	4.1	2.9	2.7	3.5	4.45	3.9
	k_{OH} (298 K) (s^{-1})	3.1×10^{4}	1.8×10^{-4}	1.2×10^{5}	5.9×10^{-4}	4.2×10^{-5}	5.6×10^{-7}	1.1×10^{5}
	ΔV_a^{0} ($cm^3\ mol^{-1}$)	–	−3.8	+0.8	−3.0	−0.2	−1.5	+1.5
	$\Delta V_{OH}^{\ddagger}$ ($cm^3\ mol^{-1}$)	–	+2.7	+7.0	+0.9	+1.5	+1.3	+6.2
	Mechanism		**I**	$\mathbf{I_d}$	**I**	**I**	**I**	$\mathbf{I_d}$
	Ref.	*(28,52)*	*(79)*	*(80,81)*	*(83)*	*(26)*	*(2)*	*(39)*

[a] $k_{ex} = k_1 + k_2/[H^+] = k_1 + k_{OH}K_a/[H^+]$.

$$[Rh(OH)(H_2O)_5]^{2+} + {}^*H_2O \underset{}{\overset{k_{OH}}{\rightleftarrows}} [Rh(OH)({}^*H_2O)_5]^{2+} + H_2O \quad (23)$$

$$[Rh(H_2O)_6]^{3+} + {}^*H_3O^+ \underset{}{\overset{k_{H_3O^+}}{\rightleftarrows}} [Rh({}^*H_2O)_6]^{3+} + H_3O^+ \quad (24)$$

The overall rate law is given by the sum of the terms of all parallel pathways, each having a different $[H^+]$ dependence. A minimum in the rate of proton exchange on Rh^{3+} is observed at $pH \approx 3$. The rate of proton exchange increases with increasing pH when $pH > 3$ indicating a reaction path involving exchange between $[Rh(H_20)_50H]^{2+}$ and bulk water protons (reaction 23). When $pH < 3$, the rate of ^1H exchange increases asymptotically with decreasing pH. The pH dependence in this acidic region is explained by a mechanism for which the rate determining step is the exchange of a proton from $[H_30]^+$ in the second sphere of Rh^{3+} with one of the bulk (reaction 24). At very low pH ($<$ 1) the proton exchange rate approaches a constant value where a rate-determining step involving transfer of a proton from $[Rh(H_20)_6]^{3+}$ to the second sphere of hydrogen-bonded water molecules is proposed. The rate constant for this process is $6.0 \times 10^4\ s^{-1}$. The direct exchange between first sphere water protons in $[Rh(H_20)_6]^{3+}$ and bulk water protons (reaction 22) is too slow to be detected.

F. Water Exchange on First-Row Transition Metal Complexes

Replacing one or more water molecules from the first coordination shell of a di- or a trivalent transition metal ion by a kinetically inert mono- or multi-dentate ligand can have a strong effect on the exchange rate constant of the remaining water molecules. In general the remaining water molecules become more labile (Tables VII and VIII): the acceleration can

TABLE VI

Exchange Rate Constants for Proton Exchange from the First Coordination Sphere of Aqua Ions

M^{n+}	pH range	$[M(H_2O)_6{}^{n+}]$ (M)	T(K)	$k_1{}^a$ (s^{-1})	$k_2{}^a$ ($M^{-1}\ s^{-1}$)	pK_a	Ref.
Al^{3+}	0.5–3.3	0.01	303	9.2×10^3	1.1×10^1	4.95	*(98)*
Cr^{3+}	−0.3–2.8	0.06	298	7×10^4	5×10^4	4.13[b]	*(99)*
	0.0		298	$1.3–1.5 \times 10^5$			*(100)*
Rh^{3+}	−0.1–3.7	0.106	298	6×10^4		3.6	*(97)*
Ni^{2+}	0.1–5.8	0.564	298	9×10^4	1.3×10^6		*(99)*

[a]Assumed rate law: $k_{ex} = 12/\tau_H = k_1 + k_2/[H^+]$.
[b]From Ref. *(79)*.

be quite moderate as for manganese(II) or it can be quite dramatic as for nickel(II).

In relation with the ongoing discussion, if Cu^{2+} in aqueous solution has five or six water molecules in its first coordination shell (*9,116,117*), it is interesting to compare water exchange rates measured on five-coordinate copper complexes. Rates of water exchange on five-coordinate complexes of copper(II) are drastically reduced from the rate of exchange on aqua ions of copper(II) (Table VII) (*113*). The mechanism of water exchange is of associative character in all examples studied to date with the exception of $[Cu(tpy)(H_2O)_2]^{2+}$. For that complex the water exchange is very rapid compared to the other complexes and the mechanism is a $\mathbf{I_d}$.

Water exchange rates on $[Fe(L)(H_2O)_x]^{n-}$ chelates of trivalent iron have been studied in the group of Rudi van Eldik. With the exception of $EDDS^{4-}$ -ligand, activation volumes are small and positive, indicating an $\mathbf{I_d}$ mechanism for water exchange on these complexes which are supposed to be seven-coordinate. $[Fe(EDDS)(H_2O)]^-$ is probably an octahedral complex and the

TABLE VII

EFFECT OF INERT LIGANDS ON WATER EXCHANGE RATES ON SOME DIVALENT d-TRANSITION-METAL IONS

$[M(L)_x(H_2O)_y]^{n+}$	k_{ex} (298K) (s^{-1})	$\Delta V^‡$ ($cm^3\ mol^{-1}$)	Ref.
$\mathbf{[Mn(H_2O)_6]^{2+}}$	2.1×10^7	−5.4	(*70*)
$[Mn(EDTA)(H_2O)]^{2-}$	4.4×10^8		(*102*)
$[Mn(EDTA\text{-}BOM)(H_2O)]^{2+}$	9.3×10^7		(*103*)
$[Mn(EDTA\text{-}BOM_2)(H_2O)]^{2+}$	1.3×10^8		(*103*)
$[Mn(PhDTA)(H_2O)]^{2-}$	3.5×10^8		(*104*)
$[Mn(NTA)(H_2O)]^-$	1.5×10^9		(*102*)
$\mathbf{[Ni(H_2O)_6]^{2+}}$	3.2×10^4	+7.2	(*70*)
$[Ni(en)(H_2O)_4]^{2+}$	4.4×10^5		(*105*)
$[Ni(en)_2(H_2O)_2]^{2+}$	5.6×10^6		(*105*)
$[Ni(dien)(H_2O)_3]^{2+}$	120×10^4		(*106*)
$[Ni(trien)(H_2O)_2]^{2+}$	2.9×10^6		(*106*)
$[Ni(tren)(H_2O)_2]^{2+}$	$8.2/90 \times 10^5$		(*107*)
$[Ni(cyclen)(H_2O)_2]^{2+}$	2.1×10^7		(*108*)
$[Ni(Me_4cyclam)(D_2O)_2]^{2+}$	1.6×10^7		(*109*)
$[Ni(Me_4cyclam)(D_2O)]^{2+}$	1.6×10^8		(*109*)
$[Ni(bpy)(H_2O)_4]^{2+}$	4.9×10^4		(*110*)
$[Ni(bpy)_2(H_2O)_2]^{2+}$	6.6×10^4		(*110*)
$[Ni(tpy)(H_2O)_3]^{2+}$	5.2×10^4		(*111*)
$[Ni(EDTA)(H_2O)_2]^{2-}$	7.0×10^5		(*112*)
$[Ni(HEDTA)(H_2O)_3]^-$	2.0×10^5		(*112*)
$\mathbf{[Cu(H_2O)_5]^{2+}}$	5.0×10^9	+2.0	(*9*)
$[Cu(TMPA)(H_2O)]^{2+}$	8.6×10^6	−3.0	(*113*)
$[Cu(tren)(H_2O)]^{2+}$	2.5×10^5	−4.7	(*114*)
$[Cu(tpy)(H_2O)_2]^{2+}$	6.6×10^8	0.0	(*115*)
$[Cu(fz)_2(H_2O)]^{2+}$	3.5×10^5	−4.7	(*113*)

TABLE VIII

EFFECT OF INERT LIGANDS ON WATER EXCHANGE RATES ON TRIVALENT IRON COMPLEXES

$[M(L)_x(H_2O)_y]^{n+}$	k_{ex} (298K) (s^{-1})	$\Delta V^\ddagger$ ($cm^3\,mol^{-1}$)	Mechanism	Ref.
$\mathbf{[Fe(H_2O)_6]^{3+}}$	1.6×10^2	−5.4	$\mathbf{I_a}$	(*81*)
$[Fe(TPPS)(H_2O)_2]^{3-}$	2×10^6	+7.9	$\mathbf{I_d}$	(*118*)
$[Fe(TMPyP)(H_2O)_2]^{3-}$	4.5×10^5	+7.4	$\mathbf{I_d}$	(*118*)
$[Fe(TMPS)(H_2O)_2]^{5+}$	2.1×10^7	+11.9	**D**	(*118*)
$[Fe(CDTA)(H_2O)]^-$	1.3×10^7	+4.0	$\mathbf{I_d}$	(*119*)
$[Fe(EDTA)(H_2O)]^-$	7.2×10^7	+1.6/+2.2	$\mathbf{I_d}$	(*119*)
$[Fe(HEDTA)(H_2O)]$	$7.8/3.8 \times 10^7$	+2.1	$\mathbf{I_d}$	(*119*)
$[Fe(PhDTA)(H_2O)]^-$	1.2×10^7	+4.6	$\mathbf{I_d}$	(*119*)
$Fe(\alpha\text{-}EDDADP)(H_2O)]^-$	1.2×10^7	+3.0	$\mathbf{I_d}$	(*119*)
$[Fe(EDDS)(H_2O)]^-$	4.3×10^5	−14.4	$\mathbf{I_a}$, **A**	(*119*)

water exchange is therefore associative. The water-exchange rate of Fe^{3+} porphyrins are significantly affected by the charge on the porphine and to a lesser degree by steric compression (*118*). The $\Delta V^\ddagger$ measured are positive and $\mathbf{I_d}$ and **D** mechanisms have been assigned. These results are in excellent agreement with the mechanistic interpretation for the complex formation reactions with NO ($\Delta V^\ddagger = +8.3$ for $TPPS^{4-}$, $\Delta V^\ddagger = +13.1$ for $TMPS^{4-}$) (*120*).

V. Water Exchange on *f*-Transition Metal Ions

A. WATER EXCHANGE ON TRIVALENT AQUA-IONS OF LANTHANIDES

The trivalent lanthanides represent the most extended series of chemically similar metal ions. In the order of increasing atomic number they are La^{3+}, Ce^{3+}, Pr^{3+}, Nd^{3+}, Pm^{3+}, Sm^{3+}, Eu^{3+}, Gd^{3+}, Tb^{3+}, Dy^{3+}, Ho^{3+}, Er^{3+}, Tm^{3+}, Yb^{3+} and Lu^{3+}, and are commonly abbreviated as Ln^{3+}. Two factors are of great importance concerning the water exchange on Ln^{3+} (i) Due to the poor shielding ability of the 4*f* electrons, the effective nuclear charge increases across the series. As a consequence, the atomic radii decrease from $r_M = 121.6$ or 116 pm for La^{3+} (CN = 8 or 9) to $r_M = 103.3$ or 97.7 pm for Lu^{3+} (CN = 8 or 9) (*24*). (ii) The shielding of the 4*f* electrons by the 5*s* and 5*p* electrons does not make them readily available for covalent interactions with ligands. As a consequence, the geometry around Ln^{3+} ions is controlled by steric factors rather than electronic ones. Thus the lanthanide contraction, in addition to the absence of significant crystal field effects, allowed to study the influence of the ionic radius on the rate and the mechanism of substitution reactions on the metal center. That is unique for the trivalent lanthanides compared to the *d*-transition metal ions mentioned earlier.

The hydration number of the Ln^{3+} ions has been extensively addressed by various techniques like neutron scattering (*121,122*), X-ray scattering

(*123–125*), extended X-ray absorption fine structures (EXAFS) (*126*), density (*127,128*), and by spectrophotometric means (*129*). It is now well-established that the lighter Ln^{3+} ions, from La^{3+} to Nd^{3+}, are predominantly nine-coordinate, the first-mid of the series, from Pm^{3+} to Eu^{3+}, exist in equilibrium between eight- and nine-coordinate states, and the heavier ions, from Gd^{3+} to Lu^{3+}, are predominantly eight-coordinate. This change of hydration number is a consequence of the lanthanide contraction and can be directly reflected in the evolution of the absolute partial molar volumes, V^0_{abs}, measured (as well as calculated) in aqueous Ln^{3+} solutions, along the lanthanide series, and presented in Fig. 11 (*127,128*).

This change in hydration number across the series has significant consequences for the kinetics of water exchange between the first coordination sphere of the Ln^{3+} ions and the bulk. Equation 14 expresses the ^{17}O transverse relaxation in the fast exchange limit (which is the case for aqueous Ln^{3+} ions) and can be considered as the "kinetic effect". For the heavy ions Tb^{3+} to Yb^{3+}, this kinetic effect is sufficient to be observed at 9.4 T and allows the determination of the water exchange rate constant, k_{ex}, by oxygen-17 NMR (Table IX). The case of Gd^{3+} is special. In presence of $[Gd(H_2O)_8]^{3+}$, the reduced ^{17}O transverse relaxation of bulk water, $1/T_{2r}$, is described by the approximated Swift and Connick equations for region IV (see Section II.C, Eq. (15)) and the exchange rate constant can be obtained because $k_{ex} \ll 1/T_{1e}$ for the aqua ion of Gd^{3+} [Eq. (16)] (*38*). As shown in Fig. 11, following the systematic decrease in the ionic radius from Gd^{3+} to Yb^{3+}, k_{ex} decreases as well (*38,41,130*). Both $\Delta V^{\ddagger}$ and $\Delta S^{\ddagger}$ are negative from Gd^{3+} to Yb^{3+}. The values of $\Delta V^{\ddagger}$ are however far from the limiting values of $-12.9\ cm^3mol^{-1}$ calculated with Swaddle's semi-empirical model for the reaction volume for CN8 $\rightarrow$ CN9 on lanthanides (*22*) (see the limiting dotted lines on Fig. 11) and of the value $\Delta V_0 = -11\ cm^3mol^{-1}$ measured by variable pressure UV-vis experiments on the equilibrium $[Ce(H_2O)_8]^{3+} + H_2O \rightleftarrows [Ce(H_2O)_9]^{3+}$ (*131*). An $\mathbf{I_a}$ mechanism was therefore assigned for the

TABLE IX

RATE CONSTANTS AND ACTIVATION PARAMETERS FOR WATER EXCHANGE ON LANTHANIDE AQUA IONS

Complex	k_{ex} (298K) ($10^7\ s^{-1}$)	$\Delta H^{\ddagger}$ ($kJ\ mol^{-1}$)	$\Delta S^{\ddagger}$ ($J\ K^{-1}\ mol^{-1}$)	$\Delta V^{\ddagger}$ ($cm^3\ mol^{-1}$)	Mechanism	Ref.
$[Eu(H_2O)_7]^{2+}$	500[a]	15.7	−7.0	−11.3	$\mathbf{I_a}$, **A**	(*3,4,135*)
$[Gd(H_2O)_8]^{3+}$	83.0	14.9	−24.1	−3.3	$\mathbf{I_a}$	(*38*)
$[Tb(H_2O)_8]^{3+}$	55.8	12.1	−36.9	−5.7	$\mathbf{I_a}$	(*41,130*)
$[Dy(H_2O)_8]^{3+}$	43.4	16.6	−24.0	−6.0	$\mathbf{I_a}$	(*41,130*)
$[Ho(H_2O)_8]^{3+}$	21.4	16.4	−30.5	−6.6	$\mathbf{I_a}$	(*41,130*)
$[Er(H_2O)_8]^{3+}$	13.3	18.4	−27.8	−6.9	$\mathbf{I_a}$	(*41,130*)
$[Tm(H_2O)_8]^{3+}$	9.1	22.7	−16.4	−6.0	$\mathbf{I_a}$	(*41,130*)
$[Yb(H_2O)_8]^{3+}$	4.7	23.3	−21.0			(*41,130*)

[a]The value originally published for a CN = 8 in Ref. (*3,135*) was corrected for CN = 7 in Ref. (*4*).

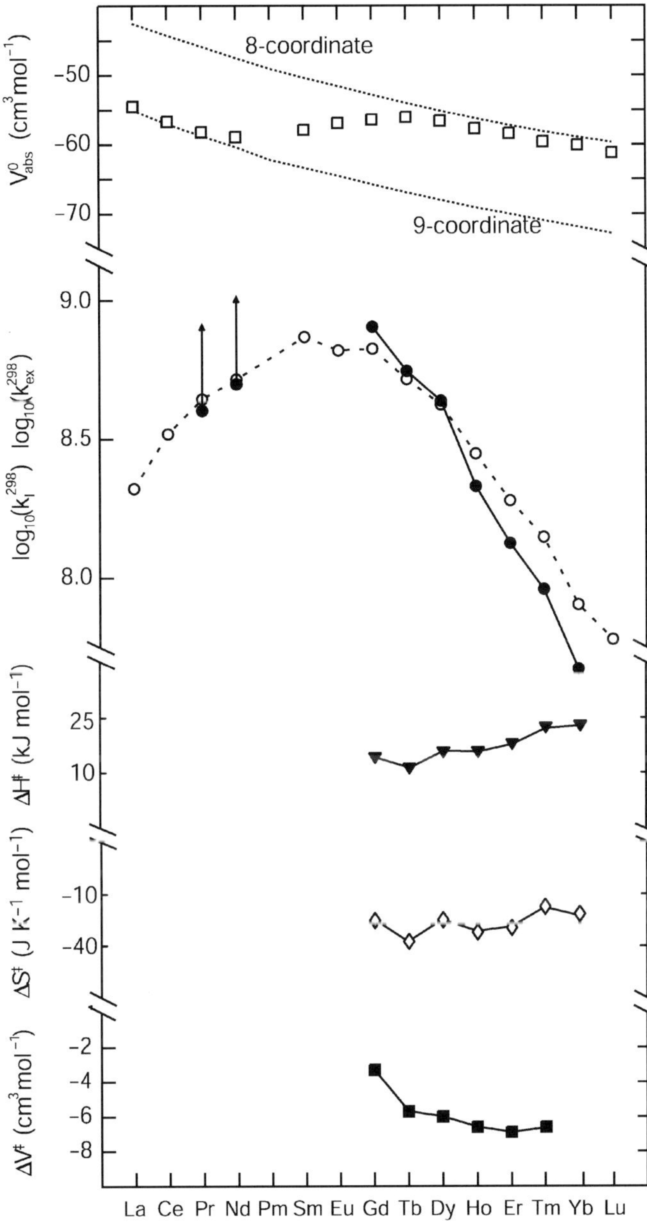

FIG. 11. Absolute partial molar volumes, V^0_{abs}, of $[Ln(H_2O)_n]^{3+}$ is aqueous $LnCl_3$ solutions (□), compared with the calculated V^0_{abs} values for $[Ln(H_2O)_8]^{3+}$ and $[Ln(H_2O)_9]^{3+}$ indicated by the upper and lower dotted curves, respectively. Interchange rate constants, k_i^{298}, for the substitution of SO_4^{2-} on $[Ln(H_2O)_n]^{3+}$ are shown as ○, and water exchange rate constants, k_{ex}^{298}, for $[Ln(H_2O)_8]^{3+}$ are shown as ●. Activation enthalpies, $\Delta H^{\ddagger}$, entropies, $\Delta S^{\ddagger}$, and volumes, $\Delta V^{\ddagger}$, are shown as ▼, ◇, and ■, respectively.

water exchange on $[Ln(H_2O)_8]^{3+}$. The increase in the k_{ex} values up to Gd^{3+} is understandable: according to an $\mathbf{I_a}$ mechanism, the transition state for water exchange is close to a ennea-coordinate Ln^{3+}, which is more and more favored for the larger ions. The same increase is observed for the interchange rate constants, k_i, between an inner sphere water molecule and a SO_4^{2-} ion from the outer sphere (*132*). For the lighter lanthanides the kinetic effect is not sufficient to enable the determination of k_{ex} by ^{17}O transverse NMR relaxation rate measurements at 9.4 T. For $[Pr(H_2O)_9]^{3+}$ and $[Nd(H_2O)_9]^{3+}$, which have the most favorable chemical shifts in the beginning of the lanthanide series, measurements at 14.1 T allowed the determination of lower limits of k_{ex} (see the arrows in Fig. 11) (*133*). The limit of e.g., $k_{ex}^{298} \geq 5 \times 10^8\ s^{-1}$ for $[Nd(H_2O)_9]^{3+}$ is in agreement with the upper limit of $k_{ex}^{298} < 10^{10}\ s^{-1}$ obtained by quasi-elastic neutron scattering (*134*). For the three other lighter paramagnetic lanthanides ions, Ce^{3+}, Pm^{3+}, and Sm^{3+}, the kinetic effect and the chemical shifts are too small to enable the determination of the water exchange rate by ^{17}O NMR relaxation rate measurements.

Concerning the solution structures of both nine and eight coordinates, Ln^{3+}: $[Ln(H_2O)_8]^{3+}$ and $[Ln(H_2O)_9]^{3+}$ adopt a square antiprism geometry (SAP) and a tricapped trigonal prism geometry (TTP), respectively. Consequently, the transition state, $[Ln(H_2O)_9]^{3+}$, in an $\mathbf{I_a}$ water exchange mechanism on $[Ln(H_2O)_8]^{3+}$ is proposed to have a TTP geometry, similar to the one observed in the solid state for the whole series. This leads to the possible mechanistic path for water exchange on $[Ln(H_2O)_8]^{3+}$ illustrated in Fig. 12. It can be well understood by taking the reverse case of an exchange on $[Ln(H_2O)_9]^{3+}$ (case of the early lanthanides) that the transition state should have a metal center with lower CN, like $[Ln(H_2O)_8]^{3+}$. In this latter case the water exchange should proceed through an $\mathbf{I_d}$ mechanism with a transition state that adopts a SAP geometry. This change in mechanism has been comforted by a computational study using classical molecular dynamics simulation (*136*).

B. Water Exchange on Complexes of Trivalent Lanthanides

A large number of water exchange studies have been devoted to the particular case of Gd^{3+}, which is the trivalent ion of the lanthanide series, the most paramagnetic one, and at the same time the one with the slowest electron spin relaxation. In fact since the mid-1980s, Gd^{3+} complexes of poly-aminocarboxylate ligands have been used for medical Magnetic Resonance Imaging (MRI) as drugs to increase the contrast of the image. Among others, one critical parameter for the efficiency of these drugs (which is called relaxivity) is the water exchange rate between an inner sphere water molecule and one from the bulk: that is the way to propagate the paramagnetic effect and therefore to increase the relaxivity of water protons. To achieve optimum water proton relaxation enhancement, the water exchange rate constant has to be in the range of 10^7–$10^8\ s^{-1}$ (*137*), slightly depending on the magnetic field strength used for MRI.

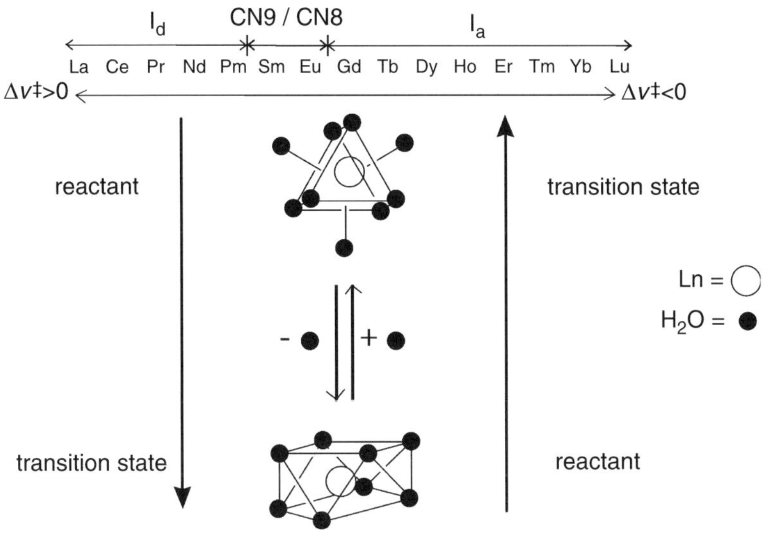

FIG. 12. Mechanistic pathways for water exchange on ennea and octa aqua ions across the lanthanide(III) series; equilibrium between ennea and octa coordinated ions for Sm^{3+} and Eu^{3+}.

Surprisingly, water exchange on the monohydrated Gd^{3+} polyaminocarboxylate chelates is much slower than on $[Gd(H_2O)_8]^{3+}$ (see Table X). The exchange reaction proceeds generally through a **d**-activated mechanism leading to a positive activation volume (*4–6,38,138–143*). The mostly observed nine-coordinate geometry does not allow the transition state to accommodate a ten-coordinate geometry due to steric constraints: the coordination cage formed by the polyaminocarboxylate ligands renders the first shell of Gd^{3+} very rigid (*5,140–142*). This is not in favor of rearrangements around the metal center in going to the transition state unlike the way it is possible for the aqua ion, for example. Since there is no possibility for an incoming water molecule to help the breakage of the bond between the leaving water molecule and the trivalent metal center, the activation energy barrier is high and cuts back the water exchange rate constant of polyaminocarboxylate complexes compared to the one on $[Gd(H_2O)_8]^{3+}$ (see Table X) by several orders of magnitude.

As for aqua lanthanide ions, water exchange on polyaminocarboxylate complexes varies across the lanthanide series. Increasing water exchange rates, from $k_{ex}^{298} = 0.5$ to 5.9×10^6 s^{-1}, have been reported for [Ln(DTPA-BMA)(H_2O)] from Ln = Nd to Ho (*144,145*). The measured activation volumes revealed a change in water exchange mechanism from an associative interchange $\mathbf{I_a}$ to a dissociative $\mathbf{I_d}$ mechanism. A strong change in water exchange rates along the lanthanide series was also observed for complexes of DOTA-tetraamide ligands (*146*). Recent crystallographic work on $DOTA^{4-}$ complexes of Ln^{3+} ions has shown that the size of

TABLE X

EFFECT OF NON-LEAVING LIGAND ON THE RATE CONSTANTS AND ACTIVATION PARAMETERS FOR WATER EXCHANGE ON SELECTED Gd^{3+} AND Eu^{2+} COMPLEXES

Complex	k_{ex} (298K) (10^6 s^{-1})	$\Delta H^‡$ (kJ mol^{-1})	$\Delta S^‡$ (J K^{-1} mol^{-1})	$\Delta V^‡$ (cm^3 mol^{-1})	Ref.
$[Gd(H_2O)_8]^{3+}$	830	14.9	−24.1	−3.3	(*38*)
$[Gd(DTPA)(H_2O)]^{2-}$	4.1[a]	52.0	+56.2	+12.5	(*143*)
	4[b]	52			(*154*)
$[Gd(DTPA\text{-}BMA)(H_2O)]$	0.43	46.6	+18.9	+7.3	(*144*)
$[Gd(EPTPA)(H_2O)]^{2-}$	330	27.7	+11	+6.6	(*140*)
$[Gd(DTTA\text{-}prop)(H_2O)]^{2-}$	31.0	30.8	+2		(*140*)
$[Gd(EGTA)(H_2O)]^{-}$	31	42.7	+42	+10.5	(*155*)
$[Gd(PDTA)(H_2O)_2]^{-}$	102	11.0	−54.6	−1.5	(*38*)
$[Gd(DOTA)(H_2O)]^{-}$	4.8[a]	48.8	+46.6	+10.5	(*143*)
	5.4 (*M*) / 500 (*m*)[b]	55 / 50			(*154*)
$[Gd(DO3A)(H_2O)_2]$	1.1	33.6	+2.4		(*156*)
$[Gd(DO2A)(H_2O)_3]^{+}$	1.0	21.3	−39		(*157*)
$[Gd(TRITA)(H_2O)]^{-}$	270	17.5	−24	−1.5	(*140*)
$[Gd(DOTAM)(H_2O)]^{3+}$	0.053				(*158*)
$[Gd(TREN\text{-}bis(6\text{-}Me\text{-}HOPO)\text{-}(TAM\text{-}TRI)(H_2O)_2]$	53	25.9		−5	(*159*)
$[Gd_2(OHEC)(H_2O)_2]^{2-}$	0.4	30.9	−34	+7.3	(*160*)
$[Gd_3(H_{-3}taci)_2(H_2O)_6]^{3+}$	11.0	59.8	−89	−12.7	(*161*)
$[Eu(DOTAM)(H_2O)]^{3+}$	0.008 (*M*) / 0.33 (*m*)[c]	53.1 / 44.2	+8.4 / +8.8	+4.9 / -	(*150*)
$[Eu(H_2O)_7]^{2+}$	5′000[d]	15.7	−7.0	−11.3	(*3,4,135*)
$[Eu(ODDA)(H_2O)]$	430	22.5	−4.0	−3.9	(*162*)
$[Eu(DTPA)(H_2O)]^{3-}$	1′300	26.3	+18.3	+4.5	(*163*)
$[Eu(DOTA)(H_2O)]^{2-}$	2′460	21.4	+6.9	+0.1	(*164*)
$[Eu(2.2.2)(H_2O)_2]^{2+}$	310	30.6	+20.5	+1.0	(*165*)

[a] Analysis of ^{17}O NMR data only.
[b] Combined analysis of EPR, ^1H, and ^{17}O NMR data.
[c] Independent analysis of the bound water NMR signal of each species.
[d] The value originally published for a CN = 8 in Ref. (*3,135*) was corrected for CN = 7 in Ref. (*4*).

the metal ion determines the structural arrangement of the chelating $DOTA^{4-}$ ligand in the solid state (*147*). For the early lanthanides (Ce) the complex adopts a twisted square antiprismatic geometry (TSA) and the X-ray structure gives Ce–O_{water} = 2.59 Å. From Pr^{3+} to Ho^{3+}, the adopted geometry is a square antiprism (SA) and Ln–O_{water} = 2.53–2.44 Å (with cavity size in the order TSA > SA). At the end of the series, there is no more water molecule in the solid state (*147*). That is an important observation since the elongation of the Ln–O_{water} bond (lower bond enthalpy) has, as a consequence, an increase in the water exchange rate constant (*146,147*).

The importance of the constraint on the inner sphere water molecule for water exchange has been supported by a classical molecular dynamics study (*148*). The value of the solid angle, Ψ, which quantifies the available space for a coordinated water molecule, has been calculated for various Gd^{3+} polyaminocarboxylate complexes with cyclic or linear ligands. This angle is centered on the Gd^{3+} ion and is bordered by the neighboring coordination sites. The water exchange rates increase for decreasing values of Ψ up to a limit where there is not enough space for an inner sphere water molecule. The two extreme cases studied are the binuclear monohydrated Gd^{3+} complex of OHEC for which $k_{ex}^{298} = 0.4 \times 10^6\ s^{-1}$ and $\Psi = 4.9$ steradian, and $[Gd(TETA)]^-$ which has no inner sphere water molecule and $\Psi = 2$ steradian (see Fig. 13).

The use of the steric crowding to tune the water exchange rate on polyaminocarboxylate complexes of Gd^{3+} to its optimum value for MRI contrast

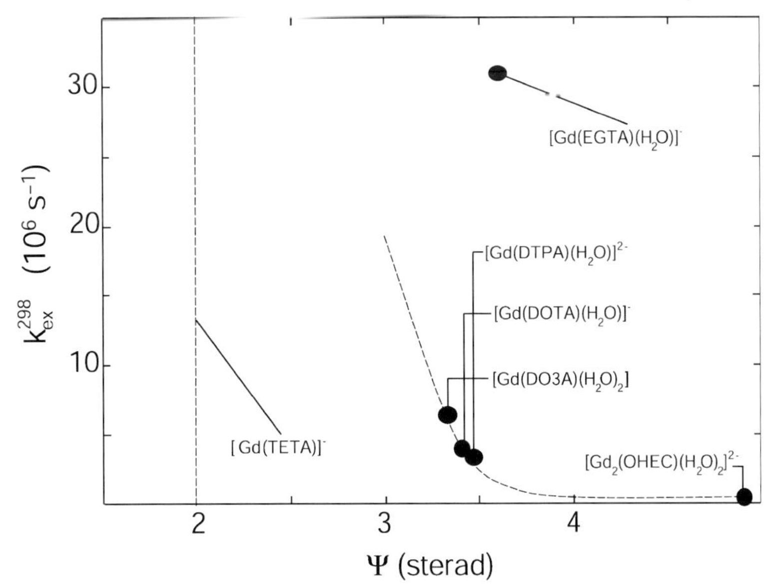

FIG. 13. Influence of the steric constraint on the water molecule represented by the solid angle Ψ, on the water exchange rate (at 298 K) of different Gd^{3+} polyaminocarboxylate complexes.

agents has been recently demonstrated by modifications in the commercial DOTA and DTPA ligands. Addition of one more CH_2 group, in the macrocyclic or linear amine backbone, to give respectively the TRITA or the EPTPA, generates Gd^{3+} complexes with significantly faster water exchange (*140*). For the cyclic TRITA the steric compression also increases the distance between the plane of the carboxylate oxygens and the metal (83 pm in $[Gd(TRITA)(H_2O)]^-$ versus 70 pm in $[Gd(DOTA)(H_2O)]^-$). Since the Gd–O_{water} distance is similar in the two complexes (248 pm in TRITA, and 245 pm in DOTA), the final result is that the bound water molecule in $[Gd(TRITA)(H_2O)]^-$ is much closer to the negatively charged carboxylates which also facilitates its departure (*140*). The water exchange mechanisms assigned from the measured activation volumes are dissociative interchange $\mathbf{I_d}$ for $[Gd(EPTPA)(H_2O)]^{2-}$ and interchange **I** for $[Gd(TRITA)(H_2O)]^-$. For the latter there is a strong participation of the incoming water in the rate-determining step which, in addition to the repulsive effect of the close carboxylates on the bound water, contributes to the increase in the water exchange rate. The DTTA-prop^{5-}, which has an additional CH_2-group on the central acetate has also been studied. Water exchange on $[Gd(DTTA\text{-}prop)(H_2O)]^{2-}$ is one order of magnitude faster than on $[Gd(DTPA)(H_2O)]^{2-}$ but one order of magnitude slower than for $[Gd(EPTPA)(H_2O)]^{2-}$: this implies that elongation of the amine backbone is more effective in producing a crowding effect on the metal than the elongation of the acetate arms. Adding two CH_2-groups on the backbone of DTPA^{5-}, leading to the ligand DPTPA^{5-}, results in the elimination of the water molecule from the inner sphere (see Fig. 14).

Chelates with macrocyclic ligands derived from cyclen, exist as two isomers, a minor one called *m* (twisted antiprismatic geometry) and a major one, *M* (antiprismatic geometry) (*149–152*). Water exchange on the tripositive complexes of DOTA-tetraamides is very slow and allowed for a combined ^{17}O and ^1H NMR study on their Eu^{3+} complexes in acetonitrile–water solvent (*150*). The water exchange rate constants were found to be much larger for the *m* isomer and independent of the ligand within the family of DOTA-tetraamides studied (*153*). However, the *m*/*M* ratio depends on the ligand and therefore determines the overall observed water exchange rate. Favoring the fast exchanging *m*-isomer over the slow exchanging *M*-isomer by modification of the DOTA-type ligand could be a clue for more efficient MRI contrast agents.

The efficiency of Gd-complexes as MRI contrast agents is directly proportional to the number of water molecules in the first coordination shell of the metal ion. Until now all commercial Gd-based agents have only one H_2O directly bound to Gd^{3+} because of the very high metal-ligand stability constant required for such drugs. Scarce examples for Gd^{3+} chelates with two inner sphere water molecules are reported so far in literature . Examples for complexes with two bound water molecules per Gd-ion are $[Gd(PDTA)(H_2O)_2]^-$ (*38*) and the trinuclear complex, $[Gd_3(H_{-3}taci)_2(H_2O)_6]^{3+}$ (*161*). Water exchange was studied for both complexes and the activation volumes measured are both negative. $[Gd(PDTA)(H_2O)_2]^-$ has

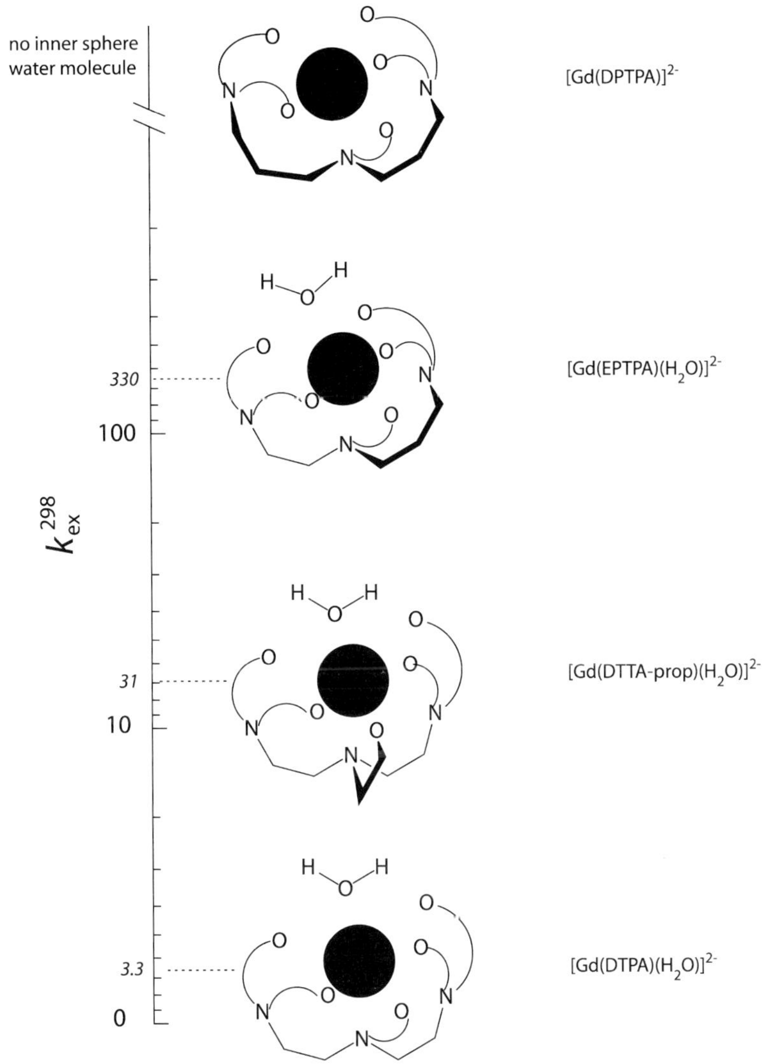

FIG. 14. Schematic representation of the influence on water exchange rate (at 298.15 K) of the crowding effect around the metal due to the elongation of the acetate arm or the amine backbone of $DTPA^{5-}$.

a water exchange which proceeds *via* an associative interchange $\mathbf{I_a}$ mechanism and which is two orders of magnitude faster than on $[Gd(DTPA)(H_2O)]^{2-}$. The trimer has a slower water exchange but the mechanism has a stronger associative character (limiting **A** mechanism) (see for comparison the values of their respective activation volumes in Table X). The difference in water exchange rate was assigned to the more rigid structure of the trimer in comparison with $[Gd(PDTA)(H_2O)_2]^-$ leading to a higher activation energy barrier. Elimination of one acetate arm of $DOTA^{4-}$ results in $DO3A^{3-}$. A UV-vis study on $[Eu(DO3A)(H_2O)_n]$ revealed an equilibrium between a monohydrated ($n = 1$) and a dihydrated ($n = 2$) complex, with a predominance of the second: $K_{1-2}^{298} = 7.7$ (*166*). By analogy with the results on aqua ions of the lanthanides which show a strong increase in lability when an equilibrium between eight and nine-coordinate complexes exist in solution, a low-energy barrier for water exchange would have been expected for $[Gd(DO3A)(H_2O)_{1.9}]$. However, the rigid structure of the macrocyclic $DO3A^{3-}$ ligand compared to the linear $PDTA^{4-}$ one, results in a relatively small water exchange rate constant for $[Gd(DO3A)(H_2O)_{1.9}]$. The elimination of a second acetate arm of $DOTA^{4-}$ to obtain the ligand $DO2A^{2-}$ allowed the Gd^{3+} complex to accommodate a third inner sphere water molecule. The equilibrium constant obtained by UV-vis on $[Eu(DO2A)(H_2O)_n]^+$ is $K_{2-3}^{298} = 4.0$. The only slightly faster water exchange on $[Gd(DO2A)(H_2O)_n]^+$ ($n = 2, 3$) than on $[Gd(DOTA)(H_2O)]^-$ has been interpreted as a consequence of an unfavorable interplay of charge and hydration equilibrium (*167*).

Since a few years, the development of hexadentate or heptadentate tripodal ligands has supplied a new class of polyhydrated Gd^{3+} chelates with fast water exchange. A first example is the Gd^{3+} complex of the heptadentate tripodal ligand $TPAA^{3-}$ (*168*). The low solubility of the complex did not allow common ^{17}O NMR measurements to study the water exchange reaction. Variable temperature ^1H relaxivity measurements conducted at 60 MHz showed that water exchange does not limit the relaxivity. Evaluation of the ^1H NMRD profiles measured at 278 and 298 K gave some evidence for a possible equilibrium between an ennea and a deca coordinate complex. Another tripodal class with catecholate coordination sites results in complexes with two bound water molecules and fast water exchange. A recent ^{17}O NMR study revealed that eight-coordinate $[Gd(TREN\text{-}bis(6\text{-}Me\text{-}HOPO)\text{-}(TAM\text{-}TRI)(H_2O)_2]$ undergoes water exchange *via* an associative interchange $\mathbf{I_a}$ mechanism (Table X) (*169*).

Water exchange on cationic lanthanide chelates can also be influenced by the nature of the counter-anions (*170,171*). Anions like halides, sulfate, nitrate, acetate, and fluoroacetate impose different order on the second coordination shell around the chelate by influencing the hydrogen–bond network. Anions with a high charge density like Cl^- and SO_4^{2-} can break up the hydrogen bond network between water molecules around the metal center and by that, slow down the water exchange rate of the inner shell water molecule (*171*).

MRI contrast agents increase the contrast of the image by influencing the relaxivity of bulk water protons. Besides the exchange of water molecules as a whole, exchange of water protons from the paramagnetic center can also increase the relaxation of bulk H_2O. In practical applications, the pH is around 7 and the H^+ exchange is slow and water exchange determines the relaxivity. However, increasing acidity or basicity of the solution can accelerate the exchange rate of protons considerably (Fig. 15) (*172–174*). Aime *et al.* showed that it is possible to control the rate of the protopic

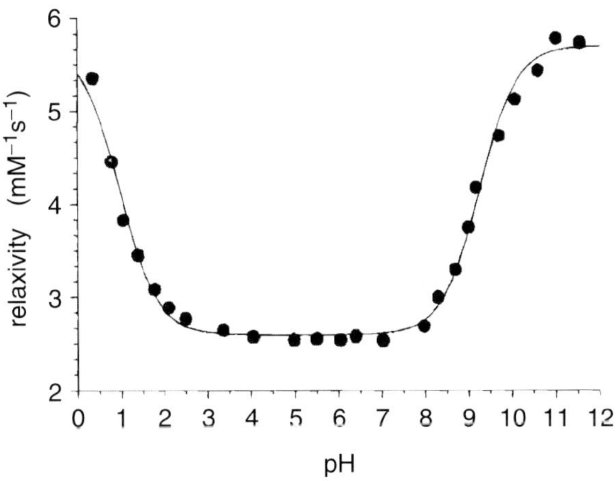

FIG. 15. pH variation of water proton relaxivity for an aqueous solution of $[Gd(DTMA(H_2O)]^{3+}$ as an indication for the increase of proton exchange at low (< 3) and high (> 8) pH (298 K, 20 MHz) (*172*).

exchange of the ligand coordination cage of a Gd^{3+} ion by introducing suitable substituents on the surface of the ligand (*175*). By changing substituents on bis-amide DTPA derivatives residual negative charges are removed, having as a consequence, an increase in the base catalysis of the protopic exchange (Fig. 16).

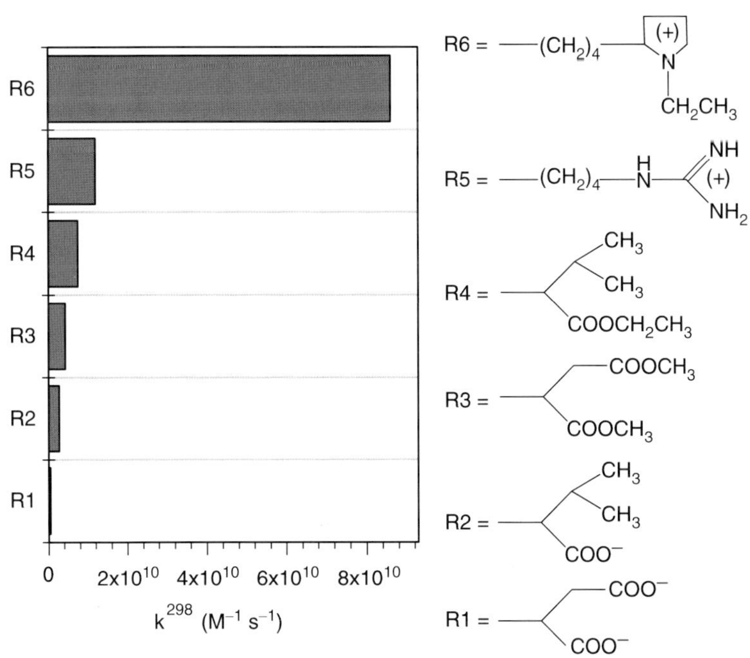

FIG. 16. Prototropic exchange rates at the coordinated water molecule in Gd^{3+} complexes of a series of bis-amide DTPA ligands (*175*).

C. Water Exchange on the Divalent Europium Ion

So far Eu^{2+} is the only divalent lanthanide aqua ion studied kinetically. It has been addressed for research because of its potential use as a "smart" contrast agent for MRI. It is isoelectronic with Gd^{3+}, and is relatively stable against oxidation. The hope is to obtain Eu^{2+} complexes which are stable thermodynamically in the divalent as well as trivalent states in order to have contrast agents responsive to p_{O2}, for example. Since the Eu^{2+} complexes are particularly difficult to crystallize, the X-ray structure of their Sr^{2+} (which has similar ionic radius as Eu^{2+}) analogs is generally used to investigate the coordination geometry of Eu^{2+} complexes (*176*). To measure the coordination number in solution of both metal complexes, extended X-ray absorption fine structures (EXAFS) have been employed. Aqua complexes of Eu^{2+} and Sr^{2+} have been shown to be 7.2 and 8.0 coordinated in solution, respectively (*4*). Water exchange on $[Eu(H_2O)_7]^{2+}$ is faster than on $[Gd(H_2O)_8]^{3+}$: it is the fastest ever directly measured by ^{17}O NMR. Furthermore, the activation volume measured on $[Eu(H_2O)_7]^{2+}$ is more negative (Table IX) than those determined on trivalent lanthanide aqua ions, clearly indicating an **a**-activated exchange process with most probably a limiting **A** mechanism (*3,177*). The coordination number of 7.2 measured by EXAFS indicates an equilibrium between a predominant $[Eu(H_2O)_7]^{2+}$ and a minor $[Eu(H_2O)_8]^{2+}$ species (*4*). The presence of such an equilibrium indicates that the energy barrier for the water exchange reaction on the major seven coordinate species is low and the water exchange reaction is very fast.

Complexes of Eu^{2+} are relatively stable against oxidation, however, most of them are less redox stable than the aqua ion. From measurements of the redox potentials it is clear that the macrocyclic $[Eu(ODDA)(H_2O)]$, $[Eu(ODDM)(H_2O)]^{2-}$, and $[Eu(DOTA)(H_2O)]^{2-}$ complexes are more redox stable than the acyclic $[Eu(DTPA)(H_2O)]^{3-}$ (*176*). The Eu^{2+} complexes of the macrobicyclic cryptand (2.2.2) is even more stable than the aqua ion (*178*). As a consequence of the lower charge and the larger ionic radius of Eu^{2+} compared to Gd^{3+}, the water exchange rate constants on Eu^{2+} polyaminocarboxylate complexes are of the same order of magnitude as that of $[Eu(H_2O)_7]^{2+}$. Thus for $[Eu(DTPA)(H_2O)]^{3-}$ the k_{ex}-value at 298 K is only reduced by a factor of four (*179*). Water exchange rate on $[Eu(DOTA)(H_2O)]^{2-}$ is the highest among all the Eu^{2+} complexes and is just about half of the water exchange rate on $[Eu(H_2O)_7]^{2+}$ (*180*).

The activation mode for water exchange on Eu^{2+} chelates varies as a function of the macrocyclic ligand. From the smaller activation volume it was concluded that water exchange on $[Eu(DTPA)(H_2O)]^{3-}$ ($\Delta V^{\ddagger} = +4.5$ $cm^3\ mol^{-1}$) undergoes an $\mathbf{I_d}$ mechanism as compared to a **D** mechanism for $[Gd(DTPA)(H_2O)]^{2-}$ ($\Delta V^{\ddagger} = +12.5\ cm^3\ mol^{-1}$). This reveals a less steric crowding around the larger and divalent Eu^{2+}. The value of $\Delta V^{\ddagger}$ measured for water exchange on the 18-membered macrocyclic $[Eu(ODDA)(H_2O)]$ (*181*) is slightly negative and an $\mathbf{I_a}$ mechanism was attributed. The $[Eu(ODDA)(H_2O)]$ complex is special in the sense that it is assumed to be analogous to the crystal structure of the corresponding Sr^{2+} complex, that coordination of the water molecule and the caboxylate groups occur at opposite sides of the molecule (*181*). The water coordinating site, located at the hydrophobic side, is much more open in the $ODDA^{2-}$ than in the $DTPA^{5-}$ complex for example. It can therefore accommodate a second water molecule in the inner shell without the preceding departure of the leaving water molecule.

In the cryptate $[Eu(2.2.2)(H_2O)_2]^{2+}$ the encapsulated metal ion is ten-coordinate and has two inner shell water molecules (*165,182*). The exchange rate constant is the lowest of all Eu^{2+} chelates measured so far which is probably due to the positive charge. The value of $\Delta V^{\ddagger}$ is close to zero and therefore an **I** mechanism was assigned for the water exchange process (Table X).

D. Water Exchange on Actinides

Compared to water exchange studies on the lanthanide series, equivalent studies on actinide ions are very scarce. The most studied cation which is also the most easy to manipulate is UO_2^{2+}. Water exchange on $[UO_2(H_2O)_5]^{2+}$ (D_{5h} geometry) was studied by ^{17}O NMR using Tb^{3+} as chemical shift reagent (*44*) (Table XI). From comparison of experimental data with quantum chemical an $\mathbf{I_d}$ mechanism was proposed for the water exchange process. From pure quantum mechanical results Vallet *et al.* attributed however an **I** or **A** mechanism for water exchange on $[UO_2(H_2O)_5]^{2+}$ (*183*). In the same paper, an **A** mechanism was attributed to water exchange on $[UO_2(C_2O_4)_2(H_2O)]^{2-}$. The geometries calculated for **A** and **I** mechanisms on

TABLE XI

Rate Constants and Activation Parameters for Water Exchange on Actinides or Actinyl Aqua Ions

Complex	k_{ex} (298K) (s^{-1})	$\Delta H^{\ddagger}$ (kJ mol^{-1})	$\Delta S^{\ddagger}$ (J K^{-1} mol^{-1})	Mechanism	Ref.
$[Th(H_2O)_{10}]^{4+}$	$> 5 \times 10^7$				(*186*)
$[U(H_2O)_{10}]^{4+}$	5.4×10^6	34	−16		(*186*)
$[U(F)(H_2O)_9]^{4+}$	5.5×10^6	36	+3		(*186*)
$[UO_2(H_2O)_5]^{2+}$	1.3×10^6	26.1	−40	$\mathbf{I_d}$	(*44*)
		21 or *19*[a]		***I*** or ***A***[a]	(*184*)
$[UO_2(H_2O)_5]^{2+}$	12×10^{3} [b]	32	−60		(*185*)
	460×10^{3} [b]	31.7	−30		(*185*)
$[UO_2(C_2O_4)_2(H_2O)]^{2-}$		*12*[a]		***A***[a]	(*184*)
$[UO_2(C_2O_4)(F)(H_2O)_2]^{-}$	1.8×10^4	45	−11		(*44*)
$[NpO_2(H_2O)_5]^{2+}$	5.3×10^{6} [b]	20.2	−72		(*185*)
$[PuO_2(H_2O)_n]^{2+}$	5.7×10^{4} [b]	12	−115		(*185*)

[a]By quantum chemical methods.
[b]Extrapolated from low temperature data obtained in H_2O/acetone mixtures.

$[UO_2(H_2O)_5]^{2+}$ and on $[UO_2(C_2O_4)_2(H_2O)]^{2-}$ indicate that the entering and the leaving water molecules are located outside the pentagonal plane formed by the spectator ligands (*183*). Water exchange on $[UO_2(H_2O)_5]^{2+}$, $[NpO_2(H_2O)_5]^{2+}$, and on $[PuO_2(H_2O)_5]^{2+}$ was also measured at a low temperature in H_2O/acetone mixtures (*184*) and exchange rate constants extrapolated to 298 K are given in Table XI. As the $\Delta S^{\ddagger}$ values become more negative with increasing atomic number, a mechanism with more pronounced associative character was proposed for the compounds with heavier metal centers.

For the tetravalent U^{4+} and Th^{4+} the first experimental kinetic studies on water exchange have been conducted recently (*44,185,186*). The coordination numbers for these two complexes were determined by EXAFS as $CN = 10 \pm 1$ (*187*). Based on the high coordination number (there are no complexes known with unidentate ligands and coordination numbers larger than 10) a limiting associative mechanism (**A**) is unlikely and a **d**-activated mechanism is probable. Surprisingly, the coordination of one fluoride or one hydroxide to U(IV) had no detectable effect on the water exchange rate.

VI. Appendix: Ligand Abbreviations, Formulae, and Structures

- (2.2.2) = 4,7,13,16,21,24-hexaoxa-1,10-diazabicyclo-[8.8.8]hexacosane =

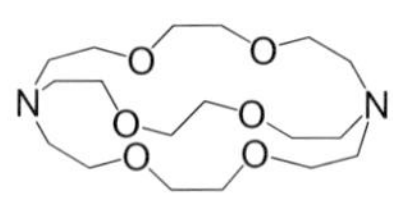

- bpy = 2,2′-bipyridine =

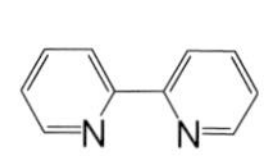

- $C_2O_4^{2-}$ = oxalate = $^{-}$OOCCOO^{-}
- $CDTA^{4-}$ = *trans*-1,2-diaminocyclohexanetetraacetate =

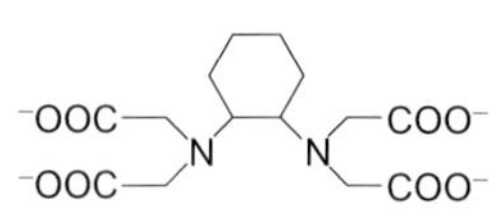

- cyclen = 1,4,7,10-tetraazacyclododecane =

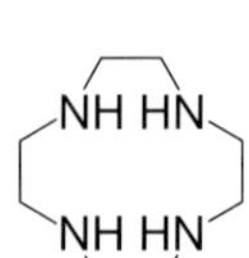

- dien = diethylenetriamine =

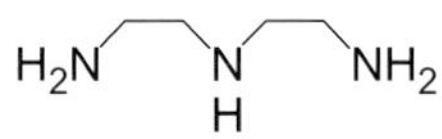

- DMSO = dimethylsulfoxide = $OSMe_2$
- $DO2A^{2-}$ = 1,4,7,10-tetraaza-1,7-bis-(carboxymethyl)cyclododecane =

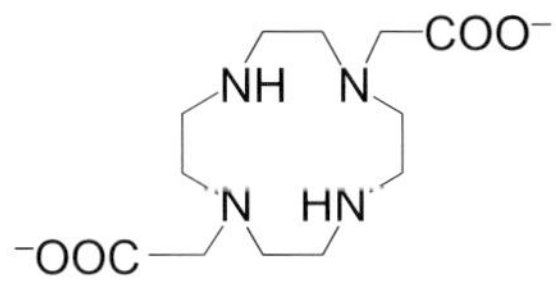

- $DO3A^{3-}$ = 1,4,7,10-tetraaza-1,4,7-tris-(carboxymethyl)cyclododecane =

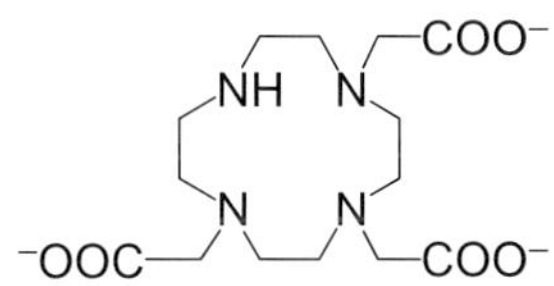

- $DOTA^{4-}$ = 1,4,7,10-tetraaza-1,4,7,10-tetrakis-(carboxymethyl)cyclododecane =

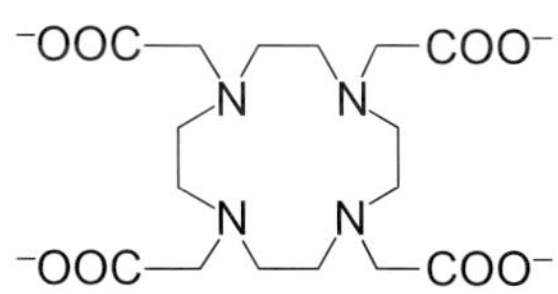

- DOTAM = 1,4,7,10-tetraaza-1,4,7,10-tetrakis-(carbamoylmethyl)cyclododecane =

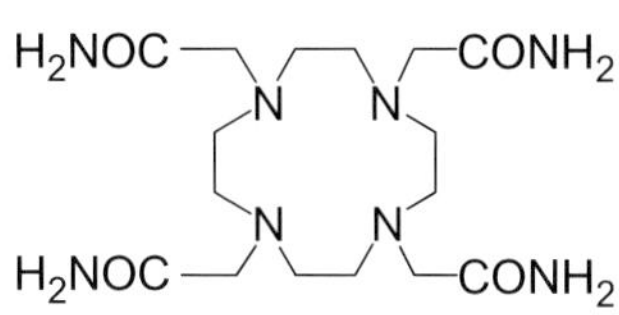

- $DTPA^{5-}$ = 1,1,4,7,7-pentakis(carboxymethyl)-1,4,7-triazaheptane =

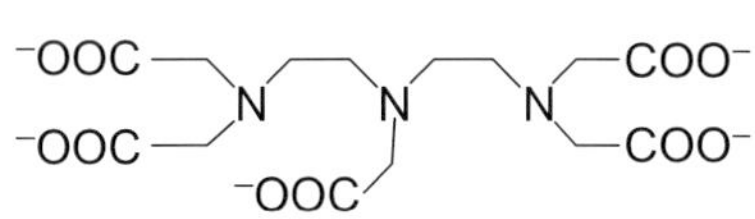

- DTPA-BMA^{3-} = 1,4,7-tri(carboxymethyl)-1,7-bis[(N-methylcarbamoyl)-methyl]-1,4,7-triazaheptane =

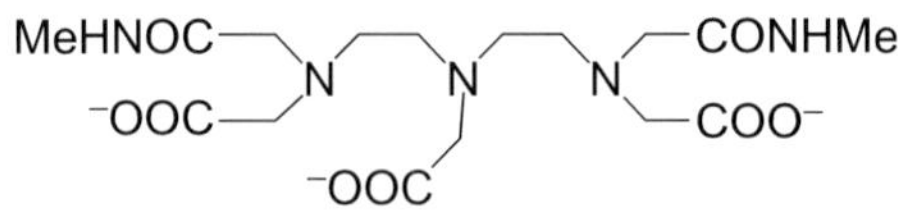

- DTTA-prop^{5-} = diethylenetriamine-1,1,7,7–tetraacetate-4-propionate =

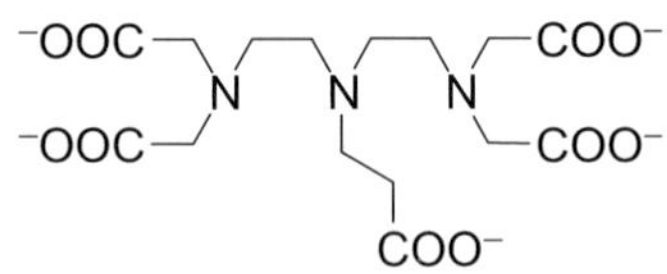

- α-EDDADP^{4-} = α-ethylenediaminediacetatedipropionate =

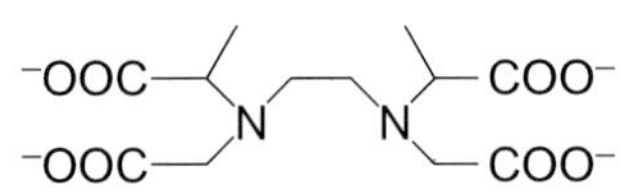

- EDDS^{4-} = s,s-ethylendiaminesisuccinate =

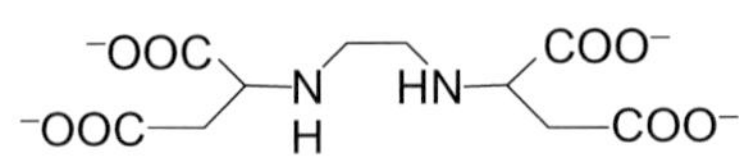

- EDTA^{4-} = ethylendiaminetetraacetate =

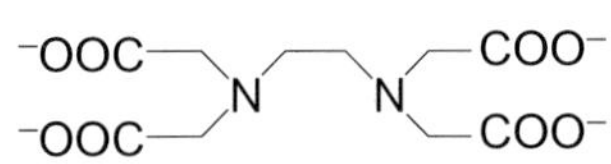

- EDTA-BOM^{4-} : R = H, R′ = BOM

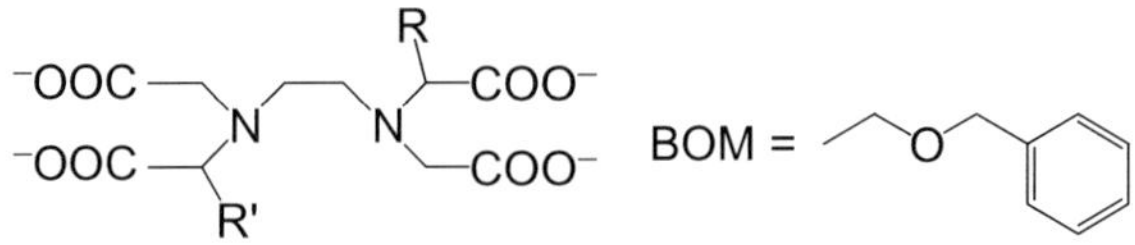

- EDTA-BOM$_2{}^{4-}$: R = R′ = BOM

- $EGTA^{4-}$ = ethylene glycol bis(2-aminoethylether)-1,1,10,10-tetraacetate =

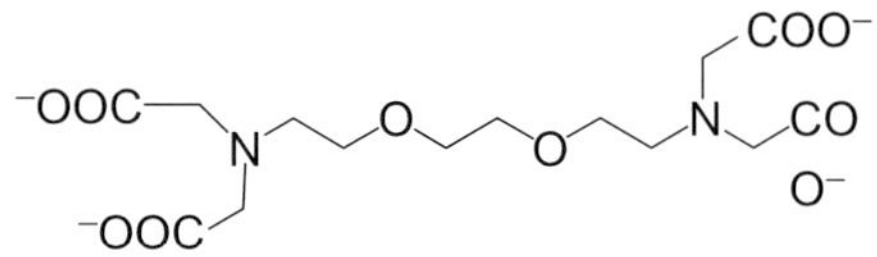

- en = 1,2-diaminoethane =

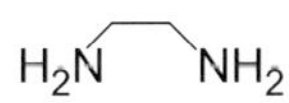

- $EPTPA^{5-}$ = ethylenepropylenetriaminepentaacetate =

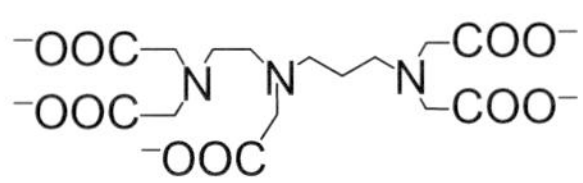

- fz = ferrozine = 3-(2-pyridyl)-5,6-bis(4-phenylsulfonic acid)-1,2,4-triazine =

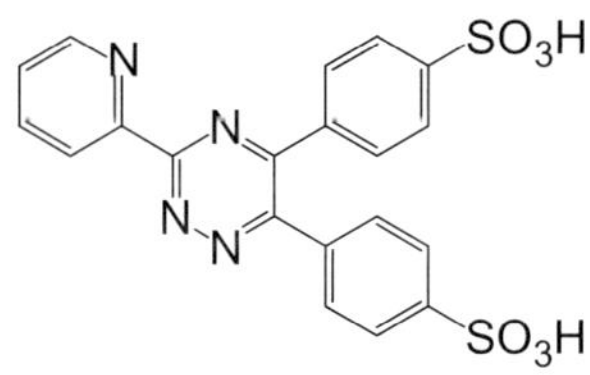

- MeCN = acetonitrile = CH_3CN
- Me_4cyclam = *N,N′,N″,N‴*-tetramethyl-1,4,8,11-tetraazacyclotetradecane =

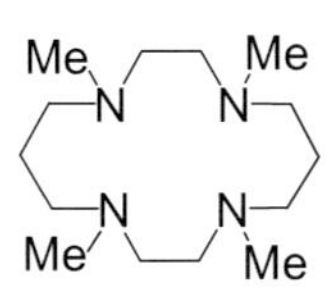

- NMP = N-Methyl-α-pyrrolidinone =

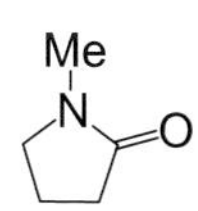

- NTA^{3-} = Nitrilo-2,2′,2″-triacetate =

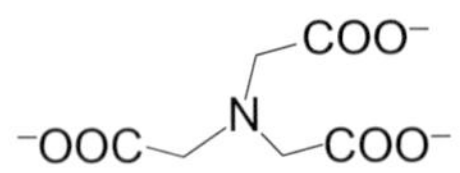

- $ODDA^{2-}$ = 1,4,10,13-tetraoxa-7,16-diazacyclooctadecane-7,16-diacetate =

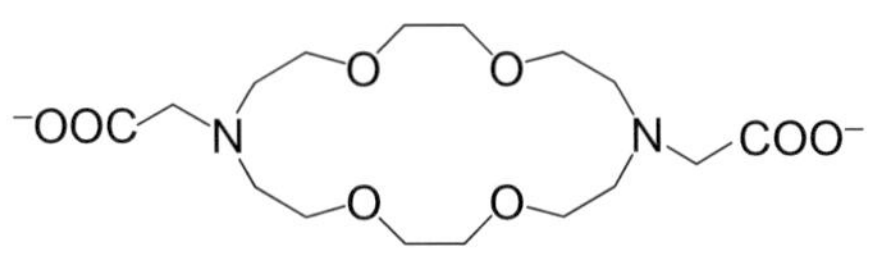

- $OHEC^{8-}$ = octaazacyclohexacosane-1,4,7,10,14,17,20,23-octaacetate =

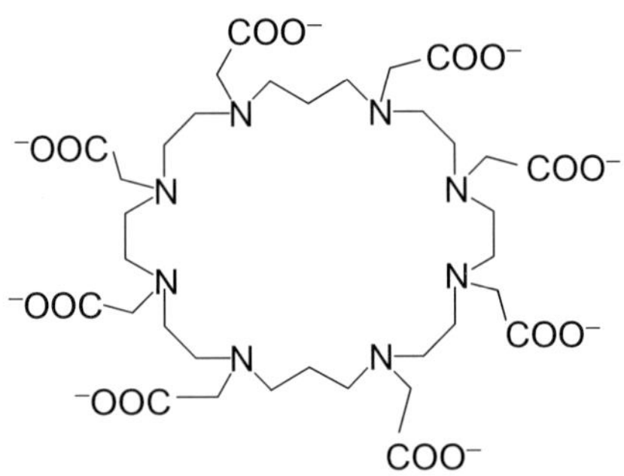

- $PDTA^{4-}$ = 1,3-propylenediaminetetraacetate =

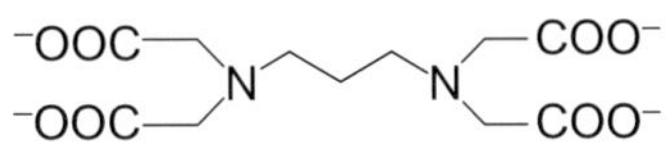

- $PhDTA^{4-}$ = o-phenylenediamine*N,N,N′,N′*-tetraacetate =

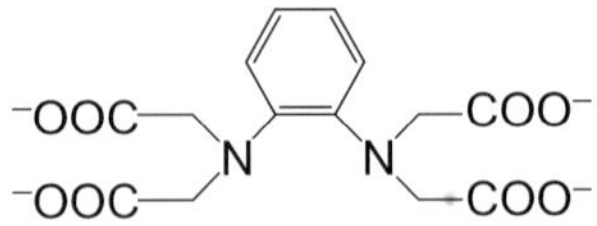

- py = pyridine

- TACI = 1,3,5-triamino-1,3,5-trideoxy-cis-inositol =

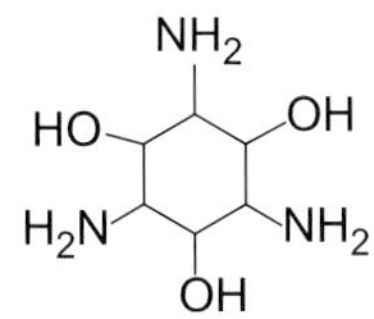

- $TETA^{4-}$ = 1,4,8,11-tetraazacyclotetradecane-1,4,8,11-tetraacetate

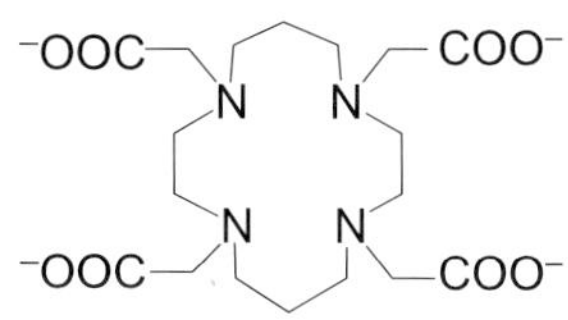

- 1,4-thioxane = 1-Oxa-4-thiacyclohexane =

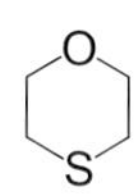

- TMPA = tris(2-pyridylmethyl)amine =

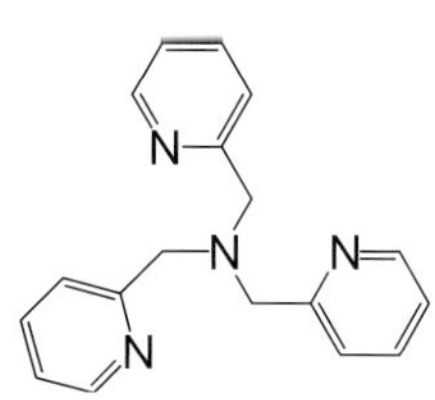

- $TPAA^{3-}$ = α,α',α''nitrilotri(6-methyl-2-pyridinecarboxylate) =

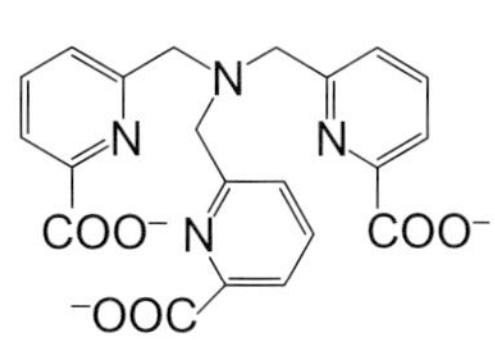

- TPPS = *meso*-tetrakis(*p*-sulfonatophenyl)porphine
- TMPyP = *meso*-tetrakis(*N*-methyl-4pyridyl)porphine

- TMPS = *meso*-tetrakis(sulfonatomesityl)porphine =

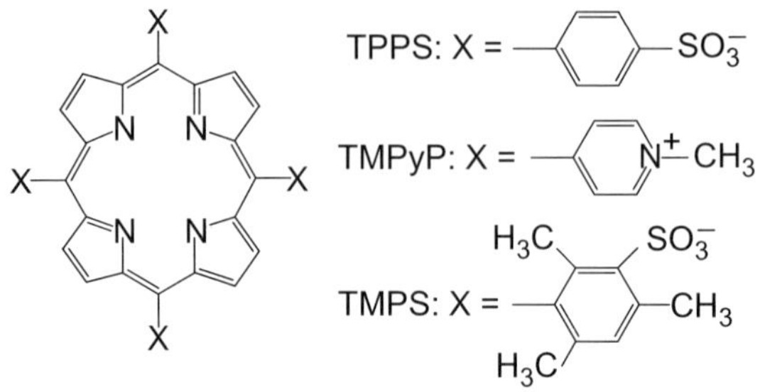

- tpy = 2,2′,2″-terpyridine =

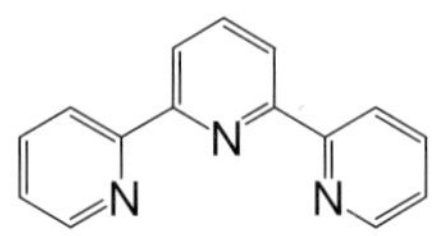

- tren = 2,2′,2″-triaminotriethylamine =

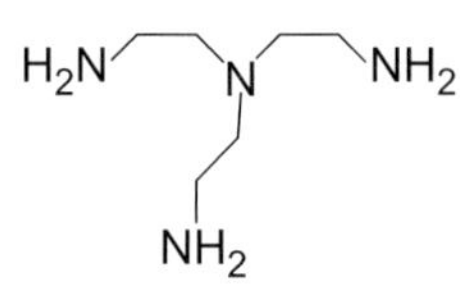

- TREN-bis(6-Me-HOPO)-(TAM-TRI) =

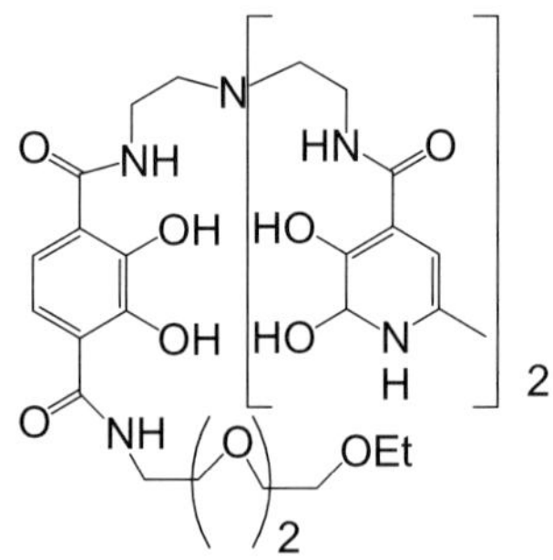

- trien = triethylentetramine =

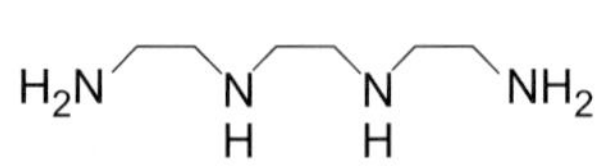

- TRITA^{4-} = 1,4,7,10-tetraazacyclotridecane-1,4,7,10-tetraacetate =

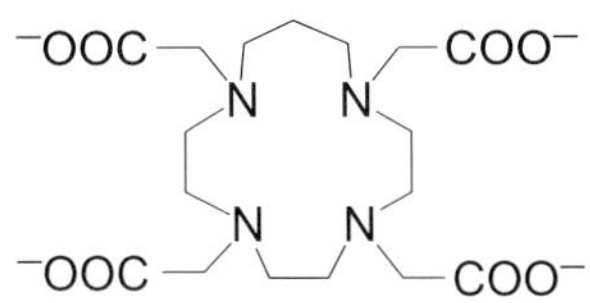

Acknowledgments

The authors gratefully acknowledge financial support from the Swiss National Science Foundation and the Swiss Office for Education and Science (COST Program). Furthermore we would like to thank the large number of co-workers who have contributed to the work performed in Lausanne.

References

1. Richens, D. "*The Chemistry of Aqua Ions*", John Wiley & Sons: Chichester, 1997
2. Cusanelli, A.; Frey, U.; Richens, D. T.; Merbach, A. E. *J. Am. Chem. Soc.* **1996**, *118*, 5265–5271.
3. Caravan, P.; Tóth, É; Rockenbauer, A.; Merbach, A. E. *J. Am. Chem. Soc.* **1999**, *121*, 10403–10409.
4. Moreau, G.; Helm, L.; Purans, J.; Merbach, A. E. *J. Phys. Chem. A* **2002**, *106*, 3034–3043.
5. Lincoln, S. F.; Merbach, A. E. *Adv. Inorg. Chem.* **1995**, *42*, 1–88.
6. Dunand, F. A.; Helm, L.; Merbach, A. E. *Adv. Inorg. Chem.* **2003**, *54*, 1–69.
7. Radnai, T. *J. Mol. Liq.* **1995**, *65–66*, 229–236.
8. Smirnov, P.; Wakita, H.; Yamaguchi, T. *J. Phys. Chem. B* **1998**, *102*, 4802–4808.
9. Pasquarello, A.; Petri, I.; Salmon, P. S.; Parisel, O.; Car, R.; Tóth, É; Powell, D. H.; Fisher, H. E.; Helm, L.; Merbach, A. E. *Science* **2001**, *291*, 856–859.
10. Grundler, P. V.; Laurenczy, G.; Merbach, A. E. *Helv. Chim. Acta* **2001**, *84*, 2854–2867.
11. Lippard, S. J.; Berg, J. M. "*Principles of Bioinorganic Chemistry*", University Science Books: Mill Valley, 1994.
12. Merbach, A. E.; Tóth, É. "*The Chemistry of Contrast Agents in Medical Magnetic Resonance Imaging*", 1st edn; John Wiley & Sons: Chichester, 2001.
13. Langford, C. H.; Gray, H. B. "*Ligand Substitution Processes*", W.A. Benjamin, Inc.: New York, 1965.
14. Merbach, A. E. *Pure Appl. Chem.* **1982**, *54*, 1479–1493.
15. Merbach, A. E. *Pure Appl. Chem.* **1987**, *59*, 161–172.
16. Pittet, P.-A.; Elbaze, G.; Helm, L.; Merbach, A. E. *Inorg. Chem.* **1990**, *29*, 1936–1942.
17. van Eldik, R. "*Inorganic High Pressure Chemistry: Kinetics and Mechanisms*", vol. 7; Elsevier: Amsterdam, 1986.
18. van Eldik, R.; Klärner, F.-G "*High Pressure Chemistry*", Wiley-VCH: Weinheim, 2002.
19. Eckert, C. A. *Ann. Rev. Phys. Chem.* **1972**, *23*, 239–264.
20. Stranks, D. R. *Pure Appl. Chem.* **1974**, *38*, 303–323.
21. van Eldik, R. "*High Pressure Kinetics: Fundamental and Experimental Aspects*", Ed. van Eldik, R.; Elsevier: Amsterdam, 1986, pp. 1–68.
22. Swaddle, T. W.; Mak, M. K. S. *Can. J. Chem.* **1983**, *61*, 472–480.
23. Swaddle, T. W. *Adv. Inorg. Bioinorg. Mech.* **1983**, *2*, 97–138.
24. Shannon, R. D. *Acta Crystallogr., Sect. A: Found. Crystallogr.* **1976**, *32*, 751–767.
25. Sandström, J. "*Dynamic NMR Spectroscopy*", Academic Press: London, 1982.

26. Laurenczy, G.; Rapaport, I.; Zbinden, D.; Merbach, A.E. *Magn. Reson. Chem.* **1991**, S45–S51.
27. Bernhard, P.; Helm, L.; Ludi, A.; Merbach, A. E. *J. Am. Chem. Soc.* **1985**, *107*, 312–317.
28. Hugi Cleary, D.; Helm, L.; Merbach, A. E. *Helv. Chim. Acta* **1985**, *68*, 545–554.
29. Helm, L.; Elding, L. I.; Merbach, A. E. *Inorg. Chem.* **1985**, *24*, 1719–1721.
30. Neely, J.; Connick, R. E. *J. Am. Chem. Soc.* **1970**, *92*, 3476–3478.
31. McConnell, H. M. *J. Chem. Phys.* **1958**, *28*, 430–431.
32. Leigh, J. S. *J. Magn. Reson.* **1971**, *4*, 308–311.
33. McLaughlin, A.; Leigh, J. S. *J. Magn. Reson.* **1973**, *9*, 296–304.
34. Swift, T. J.; Connick, R. E. *J. Chem. Phys.* **1962**, *37*, 307–320.
35. Swift, T. J.; Connick, R. E. *J. Chem. Phys.* **1964**, *41*, 2553–2554.
36. Meiboom, S.; Gill, D. *Rev. Sci. Instrum.* **1958**, *29*, 688–691.
37. McConnell, J. R. "*The Theory of Nuclear Magnetic Relaxation in Liquids*", Cambridge University Press: Cambridge, 1987.
38. Micskei, K.; Powell, D. H.; Helm, L.; Brücher, E.; Merbach, A. E. *Magn. Reson. Chem.* **1993**, *31*, 1011–1020.
39. Hugi Cleary, D.; Helm, L.; Merbach, A. E. *J. Am. Chem. Soc.* **1987**, *109*, 4444–4450.
40. Bleuzen, A.; Pittet, P.-A.; Helm, L.; Merbach, A. E. *Magn. Reson. Chem.* **1997**, *35*, 765–773.
41. Cossy, C.; Helm, L.; Merbach, A. E. *Inorg. Chem.* **1988**, *27*, 1973–1979.
42. Alsaadi, B. M.; Rossotti, F. J. C.; Williams, R. J. P. *J. Chem. Soc., Dalton Trans.* **1980**, 2147–2150.
43. Bertini, I.; Capozzi, F.; Luchinat, C.; Nicastro, G.; Xia, Z. *J. Phys. Chem.* **1993**, *97*, 6351–6354.
44. Farkas, I.; Bányai, I.; Szabó, Z.; Wahlgren, U.; Grenthe, I. *Inorg. Chem.* **2000**, *39*, 799–805.
45. Alei, M.; Jackson, J. A. *J. Chem. Phys.* **1964**, *41*, 3402–3404.
46. Frahm, J.; Füldner, H.-H. *Ber. Bunsenges. Phys. Chem.* **1980**, *84*, 173–176.
47. Yamaguchi, T.; Ohtaki, H.; Spohr, E.; Palinkas, G.; Heinzinger, K. Z. *Naturforsch.* **1986**, *41A*, 1175–1185.
48. Marx, D.; Sprik, M.; Parrinello, M. *Chem. Phys. Lett.* **1997**, *273*, 360–366.
49. Swaddle, T. W. *Inorg. Chem.* **1983**, *22*, 2663–2665.
50. Phillips, B. L.; Casey, W. H.; Crawford, S. N. *Geochim. Cosmochim. Acta* **1997**, *61*, 3041–3049.
51. Kowall, T.; Caravan, P.; Bourgeois, H.; Helm, L.; Rotzinger, F. P.; Merbach, A. E. *J. Am. Chem. Soc.* **1998**, *120*, 6569–6577.
52. Nordin, J. P.; Sullivan, D. J.; Phillips, B. L.; Casey, W. H. *Inorg. Chem.* **1998**, *37*, 4760–4763.
53. Glass, G. E.; Schwabacher, W. B.; Tobias, S. R. *Inorg. Chem.* **1968**, *7*, 2471–2478.
54. Eigen, M.; Tamm, K. *Z. Elektrochemie Angew. Phys. Chem.* **1962**, *66*, 107–121.
55. Bol, W.; Welzen, T. *Chem. Phys. Lett.* **1977**, *49*, 189–192.
56. Caminiti, R.; Licheri, G.; Piccaluga, G.; Pinna, G. *J. Chem. Phys.* **1979**, *71*, 2473–2476.
57. Caminiti, R.; Radnai, T. *Z. Naturforsch.* **1980**, *35A*, 1368–1372.
58. Maeda, M.; Ohtaki, H. *Bull. Chem. Soc. Jpn.* **1977**, *50*, 1893–1894.
59. Caminiti, R.; Paschina, G. *Chem. Phys. Lett.* **1981**, *82*, 487–491.
60. Connick, R. E.; Fiat, D. *J. Chem. Phys.* **1963**, *39*, 1349–1351.
61. Fratiello, A.; Lee, R. E.; Nishida, V. M.; Schuster, R. E. *J. Chem. Phys.* **1968**, *48*, 3705–3711.
62. Swift, T. J.; Fritz, O. G.; Stephenson, T. A. *J. Chem. Phys.* **1967**, *46*, 406.
63. Helm, L.; Merbach, A. E. "*Water Exchange on Metal Ions: The Effect of Pressure*", Eds. Eldik, R. v.; Klärner, F.-G.; Wiley-VCH: Weinheim, 2002, pp. 131–160.
64. Spees, S. T.; Perumareddi, J. R.; Adamson, A. W. *J. Phys. Chem.* **1968**, *72*, 1822–1825.
65. Rotzinger, F. P. *J. Am. Chem. Soc.* **1996**, *118*, 6760–6766.
66. Rotzinger, F. P. *Chimia* **1997**, *51*, 97–99.
67. Rotzinger, F. P. *J. Am. Chem. Soc.* **1997**, *119*, 5230–5238.
68. Rotzinger, F. P. *Helv. Chim. Acta* **2000**, *83*, 3006–3020.
69. Ducommun, Y.; Zbinden, D.; Merbach, A. E. *Helv. Chim. Acta* **1982**, *65*, 1385–1390.
70. Ducommun, Y.; Newman, K. E.; Merbach, A. E. *Inorg. Chem.* **1980**, *19*, 3696–3703.
71. Powell, D. H.; Helm, L.; Merbach, A. E. *J. Chem. Phys.* **1991**, *95*, 9258–9265.
72. Powell, D. H.; Furrer, P.; Pittet, P.-A.; Merbach, A. E. *J. Phys. Chem.* **1995**, *99*, 16622–16629.
73. Hartmann, M.; Clark, T.; van Eldik, R. *J. Am. Chem. Soc.* **1997**, *119*, 7843–7850.
74. Magini, M.; Licheri, G.; Paschina, G.; Piccaluga, G.; Pinna, G. "*X-Ray Diffraction of Ions in Aqueous Solutions: Hydration and Complex Formation*", CRC Press: Boca Raton, 1988.

75. Ohtaki, H.; Radnai, T. *Chem. Rev.* **1993**, *93*, 1157–1204.
76. Magini, M. *Inorg. Chem.* **1982**, *21*, 1535–1538.
77. Hugi, A. D.; Helm, L.; Merbach, A. E. *Inorg. Chem.* **1987**, *26*, 1763–1768.
78. Hugi, A. D.; Helm, L.; Merbach, A. E. *Helv. Chim. Acta* **1985**, *68*, 508–521.
79. Xu, F. C.; Krouse, H. R.; Swaddle, T. W. *Inorg. Chem.* **1985**, *24*, 267–270.
80. Grant, M. W.; Jordan, R. B. *Inorg. Chem.* **1981**, *20*, 55–60.
81. Swaddle, T. W.; Merbach, A. E. *Inorg. Chem.* **1981**, *20*, 4212–4216.
82. Hartmann, M.; Clark, T.; van Eldik, R. *J. Phys. Chem. A* **1999**, *103*, 9899–9905.
83. Rapaport, I.; Helm, L.; Merbach, A. E.; Bernhard, P.; Ludi, A. *Inorg. Chem.* **1988**, *27*, 873–879.
84. Helm, L.; Elding, L. I.; Merbach, A. E. *Helv. Chim. Acta* **1984**, *67*, 1453–1460.
85. Kallen, T. W.; Earley, F. *Inorg. Chem.* **1971**, *10*, 1149–1151.
86. Aebischer, N.; Laurenczy, G.; Ludi, A.; Merbach, A. E. *Inorg. Chem.* **1993**, *32*, 2810–2814.
87. De Vito, D.; Sidorenkova, E.; Rotzinger, F. P.; Weber, J.; Merbach, A. E. *Inorg. Chem.* **2000**, *39*, 5547–5552.
88. Cross, R. J. *Adv. Inorg. Chem.* **1989**, *34*, 219–292.
89. Frey, U.; Merbach, A. E.; Powell, D. H. "*Solvent Exchange on Metal Metal: A Variable Pressure NMR Approach*", vol. 6; Ed. Delpuech, J.-J.; John Wiley & Sons: Chichester, 1995, pp. 264–307.
90. Deeth, R. J.; Elding, L. I. *Inorg. Chem.* **1996**, *35*, 5019–5026.
91. Palmer, D. A.; Wesolowski, D. J. *Geochim. Cosmochim. Acta* **1993**, *57*, 2929–2938.
92. Wesolowski, D. J.; Palmer, D. A. *Geochim. Cosmochim. Acta* **1994**, *58*, 2947–2969.
93. Biryuk, E. A.; Nazarenko, V. A. *Russ. J. Inorg. Chem.* **1973**, *18*, 1576.
94. Yamada, S.; Iwanaga, A.; Funahashi, S.; Tanaka, M. *Inorg. Chem.* **1984**, *23*, 3528–3532.
95. Campisi, A.; Tregloan, P. A. *Inorg. Chim. Acta* **1985**, *100*, 251–259.
96. Wang, K.; Jordan, R. B. *Inorg. Chem.* **1993**, *32*, 895–897.
97. Bányai, I.; Glaser, J.; Read, M. C.; Sandström, M. *Inorg. Chem.* **1995**, *34*, 2423–2429.
98. Fong, D.-W.; Grunwald, E. *J. Am. Chem. Soc.* **1969**, *91*, 2413–2422.
99. Swift, T. J.; Stephenson, T. A. *Inorg. Chem.* **1966**, *5*, 1100–1105.
100. Bertini, I.; Fragai, M.; Luchinat, C.; Parigi, G. *Inorg. Chem.* **2001**, *40*, 4030–4035.
101. Stephenson, T. A.; Swift, T. J.; Spencer, J. B. *J. Am. Chem. Soc.* **1968**, *90*, 4291–4296.
102. Zetter, M. S.; Grant, M. W.; Wood, E. J.; Dodgen, H. W.; Hunt, J. P. *Inorg. Chem.* **1972**, *11*, 2701–2706.
103. Aime, S.; Anelli, P. L.; Botta, M.; Brocchetta, M.; Canton, S.; Fedeli, F.; Gianolio, E.; Terreno, E. *J. Biol. Inorg. Chem.* **2002**, *7*, 58–67.
104. Liu, G.; Dodgen, H. W.; Hunt, J. P. *Inorg. Chem.* **1977**, *16*, 2652–2653.
105. Desai, A.; Dodgen, H. W.; Hunt, J. P. *J. Am. Chem. Soc.* **1969**, *91*, 5001–5004.
106. Hunt, J. P. *Coord. Chem. Rev.* **1971**, *7*, 1–10.
107. Rablen, D. P.; Dodgen, H. W.; Hunt, J. P. *J. Am. Chem. Soc.* **1972**, *94*, 1771–1772.
108. Coates, J.; Hadi, D.; Lincoln, S. F.; Dodgen, H. W.; Hunt, J. P. *Inorg. Chem.* **1981**, *20*, 707–711.
109. Moore, P. *Pure Appl. Chem.* **1985**, *57*, 347–354.
110. Grant, M.; Dodgen, H. W.; Hunt, J. P. *J. Am. Chem. Soc.* **1970**, *92*, 2321–2323.
111. Rablen, D. P.; Gordon, G. *Inorg. Chem.* **1969**, *8*, 395–397.
112. Hunt, J. P.; Grant, M.; Dodgen, H. W. *J. Chem. Soc., Chem. Commun.* **1970**, 1446–1447.
113. Neubrand, A.; Thaler, F.; Koerner, M.; Zahl, A.; Hubbard, C. D.; van Eldik, R. *J. Chem. Soc., Dalton Trans.* **2002**, 957–961.
114. Powell, D. H.; Merbach, A. E.; Fábián, I.; Schindler, S.; van Eldik, R. *Inorg. Chem.* **1994**, *33*, 4468–4473.
115. Priimov, G. U.; Moore, P.; Helm, L.; Merbach, A. E. *Inorg. Reaction Mechanisms* **2001**, *3*, 1–23.
116. Persson, I.; Persson, P.; Sandström, M.; Ullström, A.-S. *J. Chem. Soc., Dalton Trans.* **2002**, 1256–1265.
117. Benfatto, M.; D'Angelo, P.; Della Longa, S.; Pavel, N. V. *Phys. Rev.* **2002**, *65*, 174205/174201–174205/174205.
118. Schneppensieper, T.; Zahl, A.; van Eldik, R. *Angew. Chem., Int. Ed.* **2001**, *40*, 1678–1680.

119. Schneppensieper, T.; Seibig, S.; Zahl, A.; Tregloan, P.; van Eldik, R. *Inorg. Chem.* **2001**, *40*, 3670–3676.
120. Laverman, L. E.; Hoshino, M.; Ford, P. C. *J. Am. Chem. Soc.* **1997**, *119*, 12663–12664.
121. Cossy, C.; Barnes, A. C.; Enderby, J. E.; Merbach, A. E. *J. Chem. Phys.* **1989**, *90*, 3254–3260.
122. Helm, L.; Merbach, A. E. *Eur. J. Solid State Inorg. Chem.* **1991**, *28*, 245–250.
123. Habenschuss, A.; Spedding, F. H. *J. Chem. Phys.* **1979**, *70*, 3758–3763.
124. Habenschuss, A.; Spedding, F. H. *J. Chem. Phys.* **1979**, *70*, 2797–2806.
125. Habenschuss, A.; Spedding, F. H. *J. Chem. Phys.* **1980**, *73*, 442–450.
126. Yamaguchi, T.; Nomura, M.; Wakita, H.; Ohtaki, H. *J. Chem. Phys.* **1988**, *89*, 5153–5159.
127. Spedding, F. H.; Cullen, P. F.; Habenschuss, A. *J. Phys. Chem.* **1974**, *78*, 1106–1110.
128. Spedding, F. H.; Shiers, L. E.; Brown, M. A.; Derer, J. L.; Swanson, D. L.; Habenschuss, A. *J. Chem. Eng. Data* **1975**, *20*, 81–88.
129. Miyakawa, K.; Kaizu, Y.; Kobayashi, H. *J. Chem. Soc., Faraday Trans. I* **1988**, *84*, 1517–1529.
130. Cossy, C.; Helm, L.; Merbach, A. E. *Inorg. Chem.* **1989**, *28*, 2699–2703.
131. Laurenczy, G.; Merbach, A. E. *Helv. Chim. Acta* **1988**, *71*, 1971–1973.
132. Fay, D. P.; Litchinsky, D.; Purdie, N. *J. Phys. Chem.* **1969**, *73*, 544–552.
133. Powell, D. H.; Merbach, A. E. *Magn. Reson. Chem.* **1994**, *32*, 739–745.
134. Salmon, P. S.; Howells, W. S.; Mills, R. *J. Phys. C Solid State Phys.* **1987**, *20*, 5727–5747.
135. Caravan, P.; Merbach, A. E. *J. Chem. Soc., Chem. Commun.* **1997**, 2147–2148.
136. Kowall, T.; Foglia, F.; Helm, L.; Merbach, A. E. *Chem. Eur. J.* **1996**, *2*, 285–294.
137. Tóth, É.; Helm, L.; Merbach, A. E. "*Relaxivity of Gadolinium(III) Complexes: Theory and Mechanism*", 1st edn; Eds. Merbach, A. E.; Tóth, E.; John Wiley & Sons: Chichester, 2001, pp. 45–119.
138. Caravan, P.; Ellison, J. J.; McMurry, T. J.; Lauffer, R. B. *Chem. Rev.* **1999**, *99*, 2293–2352.
139. Nicolle, G. M.; Yerly, F.; Imbert, D.; Böttger, U. A.; Bünzli, J.-C.; Merbach, A. E. *Chem. Eur. J.* **2003**, *9*, 5453–5467.
140. Laus, S.; Ruloff, R.; Tóth, É.; Merbach, A. E. *Chem. Eur. J.* **2003**, *9*, 3555–3566.
141. Zhang, S.; Matthew, M.; Woessner, D. E.; Lenkisnki, R. E.; Sherry, A. D. *Acc. Chem. Res.* **2003**.
142. Botteman, F.; Nicolle, G. M.; Vander Elst, L.; Laurent, S.; Merbach, A. E.; Muller, R. N. *Eur. J. Inorg. Chem.* **2002**, 2686–2693.
143. Micskei, K.; Helm, L.; Brücher, E.; Merbach, A. E. *Inorg. Chem.* **1993**, *32*, 3844–3850.
144. Gonzalez, G.; Powell, D. H.; Tissieres, V.; Merbach, A. E. *J. Phys. Chem.* **1994**, *98*, 53–59.
145. Pubanz, D.; Gonzalez, G.; Powell, D. H.; Merbach, A. E. *Inorg. Chem.* **1995**, *34*, 4447–4453.
146. Zhang, S.; Wu, K.; Sherry, A. D. *J. Am. Chem. Soc.* **2002**, *124*, 4226–4227.
147. Benetollo, F.; Bombieri, G.; Calabi, L.; Aime, S.; Botta, M. *Inorg. Chem.* **2003**, *42*, 148–157.
148. Yerly, F.; Borel, A.; Helm, L.; Merbach, A. E. *Chem. Eur. J.* **2003**, *9*, 5468–5480.
149. Aime, S.; Barge, A.; Bruce, J. I.; Botta, M.; Howard, J. A. K.; Moloney, J. M.; Parker, D.; De Sousa, A. S.; Woods, M. *J. Am. Chem. Soc.* **1999**, *121*, 5762–5771.
150. Dunand, F. A.; Aime, S.; Merbach, A. E. *J. Am. Chem. Soc.* **2000**, *122*, 1506–1512.
151. Woods, J.; Aime, S.; Botta, M.; Howard, J. A. K.; Moloney, J. M.; Navet, M.; Parker, D.; Port, M.; Rousseau, O. *J. Am. Chem. Soc.* **2000**, *122*, 9781–9792.
152. Zhang, S.; Kovacs, Z.; Burgess, S.; Aime, S.; Terreno, E.; Sherry, A. D. *Chem. Eur. J.* **2001**, *7*, 288–296.
153. Dunand, F. A.; Dickins, R. S.; Parker, D.; Merbach, A. E. *Chem. Eur. J.* **2001**, *7*, 5160–5167.
154. Dunand, F. A.; Borel, A.; Helm, L. *Inorg. Chem. Commun.* **2002**, *5*, 811–815.
155. Aime, S.; Barge, A.; Borel, A.; Botta, M.; Chemerisov, S.; Merbach, A. E.; Müller, U.; Pubanz, D. *Inorg. Chem.* **1997**, *36*, 5104–5112.
156. Toth, E.; Ni Dhubhghaill, O. M.; Besson, G.; Helm, L.; Merbach, A. E. *Magn. Reson. Chem.* **1999**, *37*, 701–708.
157. Yerly, F.; Dunand, F. A.; Tóth, E.; Figueirinha, A.; Kovacs, Z.; Sherry, A. D.; Geraldes, C. F. G. C.; Merbach, A. E.. *Eur. J. Inorg. Chem.* **2000**, 1001–1006.
158. Aime, S.; Barge, A.; Bruce, J. I.; Botta, M.; Howard, J. A. K.; Moloney, J. M.; Parker, D.; De Sousa, A. S.; Woods, M. *J. Am. Chem. Soc.* **1999**, *121*, 5762–5771.
159. Hajela, S.; Botta, M.; Giraudo, S.; Xu, J.; Raymond, K. N.; Aime, S. *J. Am. Chem. Soc.* **2000**, *122*, 11228–11229.

160. Nicolle, G. M.; Yerly, F.; Imbert, D.; Böttger, U.; Bünzli, J.-C.; Merbach, A. E. *Chem. Eur. J.* **2003**, *9*, 5453–5467.
161. Tóth, É.; Helm, L.; Merbach, A. E.; Hedinger, R.; Hegetschweiler, K.; Jánossy, A. *Inorg. Chem.* **1998**, *37*, 4104–4113.
162. Burai, L.; Tóth, É.; Seibig, S.; Scopelliti, R.; Merbach, A. E. *Chem. Eur. J.* **2000**, *6*, 3761–3770.
163. Seibig, S.; Tóth, E.; Merbach, A. E. *J. Am. Chem. Soc.* **2000**, *122*, 5822–5830.
164. Burai, L. Private Communication.
165. Burai, L.; Scopelliti, R.; Tóth, É. *J. Chem. Soc., Chem. Commun.* **2002**, 2366–2367.
166. Tóth, É.; Ni Dhubhghaill, O. M.; Besson, G.; Helm, L.; Merbach, A. E. *Magn. Reson. Chem.* **1999**, *37*, 701–708.
167. Yerly, F.; Dunand, F. A.; Tóth, É.; Figueirinha, A.; Kovacs, Z.; Sherry, A. D.; Geraldes, C. F. G. C.; Merbach, A. E. *Eur. J. Inorg. Chem.* **2000**, *5*, 1001–1006.
168. Bretonnière, Y.; Mazzanti, M.; Pecaut, J.; Dunand, F. A.; Merbach, A. E. *Inorg. Chem.* **2001**, *40*, 6737–6745.
169. Thompson, M. K.; Botta, M.; Nicolle, G. M.; Helm, L.; Aime, S.; Merbach, A. E.; Raymond, K. N. *J. Am. Chem. Soc.* **2003**, *125*, 14274–14275.
170. Aime, S.; Barge, A.; Batsanov, A. S.; Botta, M.; Castelli, D. D.; Fedeli, F.; Mortillaro, A.; Parker, D.; Puschmann, H. *J. Chem. Soc., Chem. Commun.* **2002**, 1120–1121.
171. Barge, A.; Botta, M.; Parker, D.; Puschmann, H. *J. Chem. Soc., Chem. Commun.* **2003**, 1386–1387.
172. Aime, S.; Barge, A.; Botta, M.; Parker, D.; De Sousa, A. S. *J. Am. Chem. Soc.* **1997**, *119*, 4767–4768.
173. Aime, S.; Crich, S. G.; Gianolo, E.; Terreno, E.; Beltrami, A.; Uggeri, F.; *Eur. J. Inorg. Chem.* **1998**, 1283–1289.
174. Aime, S.; Botta, M.; Fasano, M.; Paoletti, S.; Terreno, E. *Chem. Eur. J.* **1997**, *3*, 1499–1504
175. Aime, S.; Gianolio, E.; Barge, A.; Kostakis, D.; Plakatouras, I. C.; Hadjiliadis, N. *Eur. J. Inorg. Chem.* **2003**, 2045–2048.
176. Tóth, É.; Burai, L.; Merbach, A. E. *Coord. Chem. Rev.* **2001**, *216–217*, 363–382.
177. Caravan, P.; Merbach, A. E. *Chem. Commun.* **1997**, 2147–2148.
178. Yee, E. L.; Gansow, O. A.; Weaver, M. J. *J. Am. Chem. Soc.* **1980**, *102*, 2278–2285.
179. Seibig, S.; Tóth, É.; Merbach, A. E. *J. Am. Chem. Soc.* **2000**, *122*, 5822–5830.
180. Burai, L.; Tóth, É.; Moreau, G.; Sour, A.; Scopelliti, R.; Merbach, A. E. *Chem. Eur. J.* **2003**, *9*, 1394–1494.
181. Burai, L.; Tóth, É.; Seibig, S.; Scopelliti, R.; Merbach, A. E. *Chem. Eur. J.* **2000**, *6*, 3761–3770.
182. Moreau, G.; Scopelliti, R.; Helm, L.; Purans, J.; Merbach, A. E. *J. Phys. Chem. A* **2002**, *106*, 9612–9622.
183. Vallet, V.; Wahlgren, U.; Schimmelpfennig, B.; Szabo, Z.; Grenthe, I. *J. Am. Chem. Soc.* **2001**, *123*, 11999–12008.
184. Bardin, N.; Rubini, P.; Madic, C. *Radiochim. Acta* **1998**, *83*, 189–194.
185. Farkas, I.; Grenthe, I.; Bányai, I. *J. Phys. Chem. A* **2000**, *104*, 1201–1206.
186. Yang, T.; Tsushima, S.; Suzuki, H. *J. Phys. Chem. A* **2001**, *105*, 10439–10445.
187. Moll, H.; Denecke, M. A.; Jalilehvand, F.; Sandström, M.; Grenthe, I. *Inorg. Chem.* **1999**, *38*, 1795–1799.

NUCLEAR MAGNETIC RELAXATION STUDIES ON ACTINIDE IONS AND MODELS OF ACTINIDE COMPLEXES

JEAN F. DESREUX

Coordination and Radiochemistry, University of Liége, Sart Tilman B6, B-4000 Liége, Belgium

I. Introduction

There is continuing interest in the study of actinide elements for over six decades because of their fundamental role in the nuclear industry. Despite the extensive chemistry reported for these elements, little is known about their relaxivity properties. The purpose of this chapter is to arouse the interest of the reader for unusual relaxation behaviors in an often overlooked family of elements. Potential applications of relaxivity to practical problems involving actinides will also be stressed upon. The many experimental difficulties encountered when working with radioactive nuclides is of course at the origin of the dearth of nuclear magnetic resonance studies on the actinides and on their compounds. Relaxometry has received even less attention than NMR spectroscopy. There is, however, a wealth of information on the electronic structure of the actinide ions that has been collected by electronic paramagnetic resonance, magnetic susceptibility, fluorescence, and electronic spectroscopy. Relaxometric studies on solutions of actinide ions should become an interesting complement to these investigations.

The actinide elements result from the filling up of the 5*f* orbitals and are the analogs of the lanthanides (*1*). There is a close similarity between the trivalent 4*f* and 5*f* ions. However, the actinides differ from the lanthanides in

ADVANCES IN INORGANIC CHEMISTRY
VOLUME 57 ISSN 0898-8838 / DOI 10.1016/S0898-8838(05)57008-9

many respects as follows:

- Relativistic effects are more pronounced for the actinides because of their higher nuclear charge. As a result, the *s* and *p* orbitals screen the charge of the nucleus better and the *d* and *f* orbitals expand, and are destabilized (*2,3*). The shielding of the 5*f* orbitals by filled outer *s* and *p* orbitals is thus not as effective, and actinide ions form more covalent bonds and are found in higher oxidation states, at least at the beginning of the 5*f* series.
- The spin–orbit coupling constants of the actinides are about twice as large as those of the corresponding lanthanides and the interelectronic repulsion parameters are approximately 30% smaller. Covalency effects can cause a reduction in the orbital momentum but no quenching unlike in the case of transition metal ions. There is a considerable departure from the Russell–Saunders coupling scheme with a tendency towards a distribution of electronic levels that is expected for *jj* coupling. However, neither schemes allows an accurate computation of the electronic levels, and models must be based on intermediate couplings (*4*). The calculation of energy levels is thus much more complicated than for lanthanides (*5*).
- The ground state of the lighter actinides could vary depending on the nature of the ligands. Although often mentioned in text books (*6,7*), there are not many clear examples of a change of ground state with the crystal field. One interesting case has been reported for an organometallic derivative of trivalent thorium that was shown to be a $6d^1$ rather than a $5f^1$ compound as expected (*8,9*).

Other differences between the lanthanides and the actinides are of particular relevance for relaxivity studies.

- All actinides are radioactive and many can only be handled in glove-boxes located in specially designed laboratories. At present, the only field-cycling relaxometer installed in a "hot" environment is at the University of Liége (*10*). Also, one is rarely welcome with a highly radioactive solution in a NMR spectroscopy laboratory, and it is thus difficult to measure relaxation time at high fields. This problem may be alleviated thanks to ACTINET-6, a Network of Excellence in the Sixth Framework Program of the European Atomic Energy Community (EURATOM) that will give access to high-field NMR spectrometers in governmental nuclear facilities.
- Relaxivity measurements are more easily performed with long-lived actinide nuclides that do not cause too much radiation damage (*11*). This is particularly true for curium whose most common nuclide ^{244}Cm ($t_{1/2}$ = 18.11 y) causes extensive radiolysis. It would be much more convenient to work with ^{248}Cm ($t_{1/2}$ = 3.4 10^5 y) for instance but only minute quantities of this nuclide are available in Europe. The situation is of course even more difficult with heavy actinides (Bk–Lr) which are available only at the trace level and there is thus no hope of being able to study the relaxivity of the whole actinide family as has been done for the lanthanides (*12*).

– Many oxidation states of the actinides are poorly stable or stable only under certain conditions. Great care must thus be taken in preparing samples for relaxometry studies. Working under the same chemical conditions with different actinides in the same oxidation state is sometimes impossible. Plutonium is particularly noteworthy because it is the only element in the Mendeleev table that can exist simultaneously in solution in four different oxidation states. This unusual situation stems from the fact that the ions Pu^{4+} and PuO_2^+ have a tendency to undergo dismutation according to:

$$2Pu^{4+} + 2H_2O \rightleftharpoons Pu^{3+} + PuO_2^+ + 4H^+$$

$$2PuO_2^+ + 4H^+ \rightleftharpoons Pu^{4+} + PuO_2^{2+} + 2H_2O$$

while the reactions involving the formation and the rupture of Pu–O bonds are much slower than simple electron transfers such as:

$$Pu^{3+} + PuO_2^{2+} \rightleftharpoons Pu^{4+} + PuO_2^+$$

Plutonium solutions are also slowly reduced under the action of the α radiations produced by the isotopes ^{239}Pu or ^{240}Pu (*11*). Finally, PuO_2^{2+} is reduced by some organic ligands (*13*).

Since this involves a lot of difficulties, the reader will not be surprised that there are only a limited number of studies devoted to the relaxation properties of the actinides.

The present chapter starts with a survey of the relaxometric measurements performed so far on actinide ions. These ions are divided into two groups depending on whether they are of the simple Me^{n+} type or if they are combined with oxygen atoms in the form of MeO_2^{n+} species. A second part of this chapter is devoted to studies of practical interest in nuclear chemistry that have been carried out by relaxometry using Gd^{3+} as a model of the trivalent actinides. The processing of nuclear wastes before disposal relies on solvent extraction procedures and the NMRD technique yields interesting information on the dynamic behavior of metal complexes with extracting agents. The distribution of actinides in nature is another field of great interest in radiochemistry. Humic acids form complexes with the actinide ions and play an active role in the migration of these ions. NMRD throws some light on the speciation of metal humates.

II. Actinide Ions

A. Me^{n+} Actinide Ions

A few simple Me^{n+} actinide ions will be considered here, the most interesting one being Cm^{3+}, the analog of Gd^{3+}. These two ions feature a half-filled *f* orbital and their ground state is $^8S_{7/2}$ in Russell–Saunders term. The crystal field should not split the orbitally non-degenerate *S* state and crystal field splittings of the order of 0.2 cm^{-1} have indeed been reported for Gd^{3+}. The electronic relaxation time for this ion is thus much longer than for the other lanthanides, and Gd^{3+} is the ion of choice in magnetic resonance imaging because of its high relaxivity. The same behavior can be expected in the case of Cm^{3+} but this ion is a much less effective relaxing agent as illustrated in Fig. 1. Electron paramagnetic resonance studies (*14,15*) and theoretical interpretations of the electronic spectra (*16,17*) of Cm^{3+} compounds yield information on the reasons why the relaxivities of Gd^{3+} and Cm^{3+} differ so much. The spin-orbit coupling constant ζ of the Cm^{3+} ion is about 3235 cm^{-1} (*18*) and is about twice the value found for the lanthanides. Intermediate coupling effects mix the $^8S_{7/2}$ ground state of Cm^{3+} with excited non-*S* states of the same *J* value at 16,000–20,000 cm^{-1} (*16*). The leading factors in the eigenvector of the ground state of Cm^{3+} in a $LaCl_3$ host lattice has been given as:

$$0.891|^8S_{7/2}> +0.414|^6P_{7/2}> -0.090|^6D_{7/2}>$$

while the ground state eigenvector of Gd^{3+} in an ethyl sulfate host lattice is:

$$0.987|^8S_{7/2}> +0.162|^6P_{7/2}> -0.012|^6D_{7/2}>$$

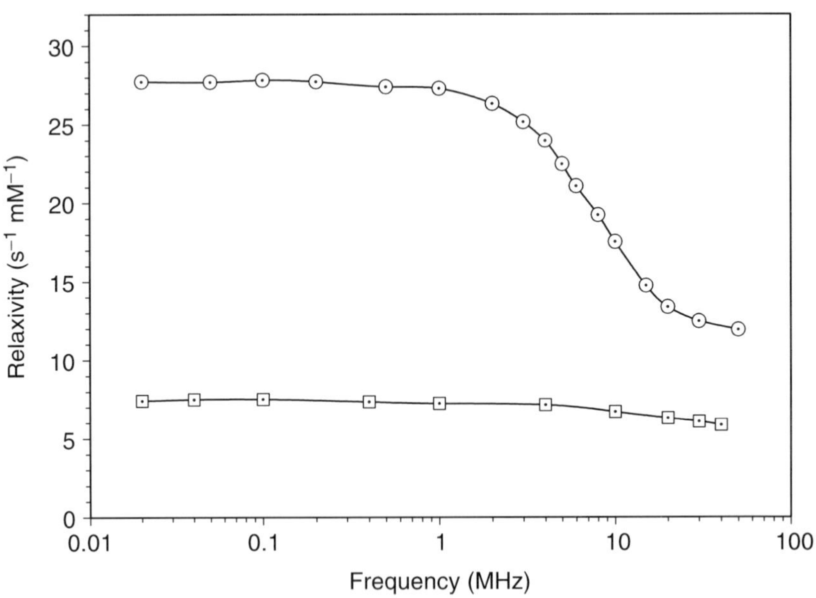

FIG. 1. NMRD curves of acidic aqueous solutions of Gd^{3+} (⊙) and $^{244}Cm^{3+}$ (⊡) (*10*) at 25°C.

The Gd^{3+} ion is thus approximately in a 98% pure $^8S_{7/2}$ state while Cm^{3+} is only 79–87% pure $^8S_{7/2}$, and is only nominally a S state ion (*4,16,19*). The zero field splittings of Cm^{3+} in different hosts varies between 2 and 20 cm^{-1}. Figure 2 is a comparison of the energy levels of the two ions. Particularly noteworthy is the very strong effect of a small departure from an S state on relaxivity. NMRD is a relatively simple experiment that clearly shows the effect of small changes in electronic structures.

The larger splitting of the ground state of Cm^{3+} brings about a shortening of the electronic relaxation time and the relaxivity is thus decreased as shown in Fig. 1. The NMRD curve of Cm^{3+} was interpreted with the help of Solomon–Bloembergen equations given later in this chapter with $S = 7/2$, even though it is not rigorous. The computations were performed with a hydration number $q_{\text{water}} = 9$ as found by time resolved fluorescence spectroscopy (*20*), a Cm^{3+}–H(water) distance $r = 3.0$ Å and a rotational correlation time $\tau_r = 41$ ps as reported for the aqua Gd^{3+} ion (*21*). A value of the electronic relaxation time τ_s (300 K) $= 1 \times 10^{-11}$ s at 20 MHz was found and must be compared with τ_s (300 K) $= 2 \times 10^{-10}$ s for Gd^{3+} (*21*). The ^1H NMR spectra of Gd^{3+} complexes feature only large unresolved peaks because of the strong shortening of the proton relaxation time T_{1M} brought about by the slow electronic relaxation rate of this ion. With a shorter electronic relaxation rate, Cm^{3+} should not cause as much broadening of the NMR resonances and this is exactly what has been observed by the author of this review (*22*). However, no NMR spectrum of a Cm^{3+} complex is presented here because the γ and α radiolysis due to the ^{244}Cm nuclide is so intense that a ligand such as DTPA (diethylenetriaminepentaacetate) is rapidly destroyed (*23*).

Figure 3 presents the NMRD curves of Pu^{3+} ($5f^5$, $^6H_{5/2}$) and Np^{4+} ($5f^3$, $^4I_{9/2}$). These two ions have non-spherically symmetric distribution of their unpaired electronic spins. Also included is the NMRD curve of Pr^{3+} which is the lanthanide analog of Pu^{3+} (*12*). All the ions are very poor relaxation agents

FIG. 2. A comparison of the zero-field splittings of Cm^{3+} in CaF_2 (left) (*19*) and of Gd^{3+}.

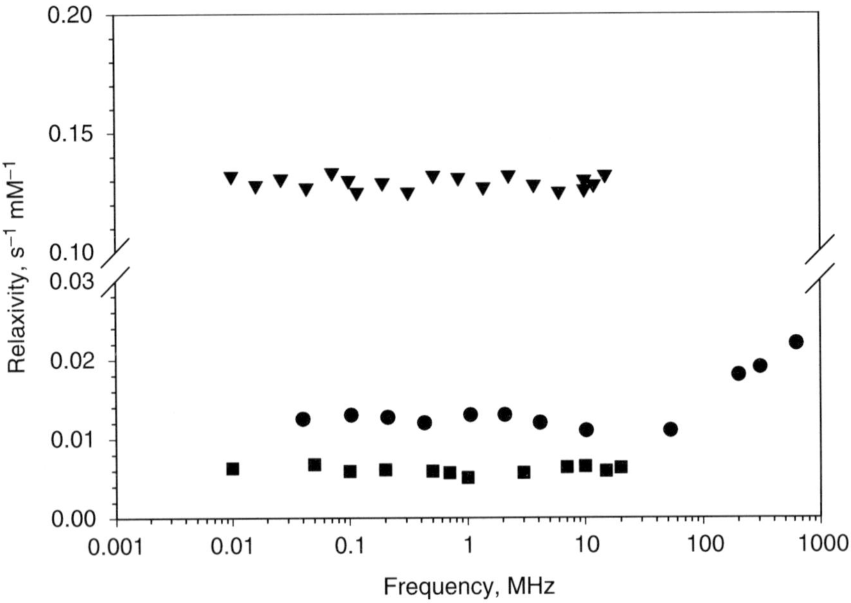

FIG. 3. NMRD curves of acidic aqueous solutions of $^{237}Np^{4+}$ (▼), Pr^{3+} (●) *(12)*, and $^{244}Pu^{3+}$ (■) *(10)* at 25°C.

as expected from the nature of their ground state *(12,24)*. The relaxivity of Pu^{3+} is lower than that of Pr^{3+} presumably because of its larger spin–orbit coupling constant (2550 cm^{-1} vs. 710 cm^{-1} *(18)*). No relaxation data have yet been recorded at frequencies higher than 80 MHz and it is thus impossible to ascertain whether there is a Curie contribution at high fields as already observed for the lanthanides *(12)*.

B. ACTINYL DIOXO CATIONS MeO_2^+ AND MeO_2^{2+}

Unlike the lanthanides, the actinides U, Np, Pu, and Am have a tendency to form linear actinyl dioxo cations with formula MeO_2^{2+} and/or MeO_2^{2+}. All these ions are paramagnetic except UO_2^{2+} and they all have a non-spherical distribution of their unpaired electronic spins. Hence their electronic relaxation rates are expected to be very fast and their relaxivities, quite low. However, two ions, namely NpO_2^+ and PuO_2^{2+}, stand out because of their unusual relaxation properties. This chapter will be essentially devoted to these ions that are both $5f^2$. Some comments will be included later about UO_2^{2+} ($5f^0$) and NpO_2^{2+} ($5f^1$). One should note here that there is some confusion in the literature about the nomenclature of the actinyl cations. The "yl" ending of plutonyl is often used indiscriminately for PuO_2^+ and PuO_2^{2+} and the name neptunyl is applied to both NpO_2^+ and NpO_2^{2+}. For instance, SciFinder® Scholar™ makes no difference between "yl" compounds in different oxidation states. Here, the names neptunyl and plutonyl designate two ions of the same $5f^2$ electronic structure but of different electric charge and containing metal ions in a different oxidation state, namely Np(V) and Pu(VI).

As mentioned in the Introduction, the actinyl ions are not stable under all chemical conditions. Plutonium can coexist in solution in several oxidation states, the stability of which often depends strongly on acidity (*26*). As a result, great care must be taken to obtain pure solutions of PuO_2^{2+} (*27*). On the other hand, the neptunyl ion NpO_2^+ is the most stable form of neptunium in aqueous solution. It is noteworthy that the exchange between the oxygen atoms of PuO_2^{2+} and H_2O^{18} is very slow ($t_{1/2} > 10^4$ h) (*25*), whereas it is quite fast ($t_{1/2} \sim 2.2$ s) in the case of NpO_2^+.

Only a handful of experimental data have yielded information on the ground state of NpO_2^+ and PuO_2^{2+} and many of these data were published in the 1950s. This is not surprising in view of the experimental difficulties encountered when preparing single crystals doped with radioactive actinyl compounds for EPR studies. Comparing relaxivity data with other magnetic measurements is thus not straightforward because of the paucity of experimental analyses, and also because of some discrepancy between theoretical studies.

One can envision two ground states with parallel electronic spins for the $5f^2$ ions in the strong electrostatic field of the Me–O bonds (*28*):

(a) in accordance with Hund's rule, the ground state has the maximum value of the axial component of the total orbital momentum $L_z = l_z(\pm 3) + l_z(\pm 2) = \pm 5$ (this approach is not rigorous as mentioned in the introduction). With $S_z = \pm 1$, this leads to a ^{4}H state that is divided by spin–orbit coupling into states of $J_z = \pm 6$, 5, and 4, the ground level being obtained for a doublet with $J_z = \pm 4$ (the standard notation of molecular spectroscopy is used here, see reference *34*).

(b) electrostatic repulsion due to bonding electrons along the O-Me-O axis leads to states with $l_z = +3$ and $l_z = -3$ and thus to $L_z = 0$. With $S_z = \pm 1$, $J_z = \pm 1$ and 0 would be obtained, the ground state being then $^3\Sigma$ with $J_z = 0$.

Magnetic susceptibility measurements do not allow choosing between these two possibilities (*28*). The ground state $^3H_4$ is characterized by $g_{\|} = 6, g\perp = 0$ and the corresponding magnetic moment is $\mu_{eff} \leq 3.0\ \mu_B$ (in Bohr magneton), while a spin-only value $2\sqrt{S(S+1)} = 2.83\mu_B$ is obtained for configuration (b). It is noteworthy that the theoretical value in the *L–S* coupling scheme applicable to the lanthanide $4f^2$ ion Pr^{3+} is $\mu_{eff} = 3.58\ \mu_B$ (*29*). Gruen *et al.* (*30*) reported a value $\mu_{eff} = 3.17\ \mu_B$ for the mono-protonated oxalate complex $NpO_2(OOC–COOH) \cdot 2H_2O$ and recently, Nakamoto *et al.* (*31*) published values of $\mu_{eff} = 2.83–3.32\ \mu_B$ for formate salts of NpO_2^+ in polymeric chains or sheets. Very few magnetic susceptibility measurements have been carried out so far on PuO_2^{2+} compounds. A magnetic moment $\mu_{eff} = 2.6$ μ_B was obtained by the Gouy method for a bipyridine complex of PuO_2Cl_2 (*32*). A value of $\mu_{eff} = 2.80-2.83\ \mu_B$ was reported (*33*) for a powder of $NaPuO_2(acetate)_3$. However, no distinction between configurations (a) and (b) can be made because of the cubic symmetry of this compound (*28*).

Measurements on single crystals of lower symmetry doped with NpO_2^+ or PuO_2^{2+} should remove this uncertainty but no such experiment has been performed so far. However, the distinction between configurations (a) and (b) has been possible in an electron paramagnetic resonance study at 20 K on a single uniaxial crystal of $RbUO_2(NO_3)_3$ doped with plutonyl ions. This study conclusively showed that the ground state is $^3H_4$ with $g_{\|} = 5.4$ and $g_{\perp} = 0$ (*28*).

Molecular orbital models of actinyl ions have been proposed to interpret the available magnetic susceptibility and EPR measurements. In a model of UO_2^{2+}, Eisenstein and Pryce (*34*) suggested that strong σ bonds are formed by the overlap of 5*f* and 6*d* orbitals (for instance $5f_{z^3}$ and $6d_{z^2}$ denoted $5f_\sigma$ and $6d_\sigma$) with one hybrid sp_z (or sp_σ) orbital from each oxygen atom. In addition, the $2p_\pi$ orbitals of the oxygen atoms form weaker π bonds with the metal $5f_\pi$ and $6d_\pi$ orbitals. The bonding orbitals $\sigma_g^2\sigma_u^2\pi_g^4\pi_u^4$ essentially arise from the oxygen 2*p* orbitals and are completely filled in all actinyl ions. The seven *f* orbitals give four nearly degenerate non-bonding δ_u and ϕ_u followed at higher energies by π_u and σ_u orbitals that are essentially antibonding. This model of UO_2^{2+} has recently been extended to the case of NpO_2^+ and PuO_2^{2+}. It was shown that the degenerate pair of δ_u orbitals is somewhat lower in energy than the ϕ_u orbitals. The two unpaired electrons of the f^2 ions can occupy the δ_u orbitals in a δ_u^2 configuration corresponding to $^3\Sigma_g$. This is the configuration suggested by Craw *et al.* (*35*). The same configuration was found by Hay *et al.* (*36*) using density functional theory. However, these authors also performed configuration interaction calculations that take into account the spin–orbit effects and they obtain a $\delta^1\phi^1$ ground state for PuO_2^{2+} that corresponds to $^3H_{4g}$. In these calculations, the $^3\Sigma_{0g}$ level is much less stable. Similar results have been reported by Matsika *et al.* (*37,38*) with a difference of 40.2 kJ/mol between the $^3H_{4g}$ and $^3\Sigma_g$ states. The same conclusion was reached by Maron *et al.* (*39*) with an energy difference of 51.4 kJ/mol and by Ismail *et al.* (*40*) with a difference of 46.15 kJ/mol. The spin–orbit coupling was not taken into account in this last study but the ground state is lowered by 69.66 kJ/mol according to Maron *et al.* if corrections are made for the *L–S* coupling (*39*). Moreover, some of these calculations were made in the gas phase ("naked ion") while others were performed on hydrated actinyl ions. For instance, Matsika *et al.* (*41*) reported a detailed analysis of the electronic spectrum of an aqueous solution of NpO_2^+ that was based on a model of this actinyl ion surrounded by five water molecules in the axial plane in a D_{5h} symmetry. These authors found an energy difference of 15.7 kJ/mol between the $^3H_{4g}$ ground state and the $^3\Sigma_{0g}$ level, a value that is significantly lower than in the absence of water molecules. Hay *et al.* (*36*) calculations were also performed on a hydrated ion, $PuO_2(H_2O)_5^{2+}$, but it led to a higher energy separation. Figure 4 is a schematic presentation of the energy levels in the actinyl ions (*36*).

The crystal field interacts directly only with the orbital motion of the unpaired electrons and it has an effect on the electronic spins only through the spin–orbit coupling. The strongest spin–lattice interaction will therefore occur for ions with ground states having an appreciable orbital character.

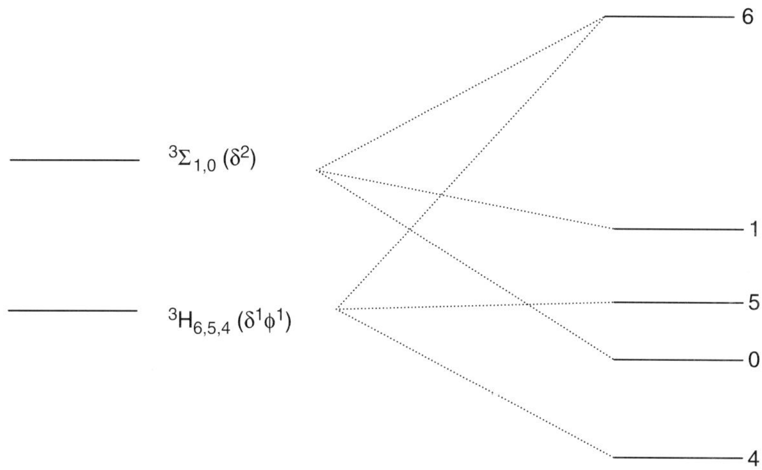

FIG. 4. Schematic presentation of the low-lying states of PuO_2^{2+} according to Hay *et al.* (*36*).

It is thus expected that the electronic relaxation time for the $5f^2$ actinyl ions in the $^3H_{4g}(\delta^1\,\phi^1)$ ground state will have a short electronic relaxation time T_{1e} because of its high L_z value. On the contrary, the T_{1e} relaxation time of a $5f^2$ actinyl ion in the $^3\Sigma_{0g}(\delta^2)$ state should be long because of the null value of its L_z parameter. The dependence of the relaxivities of solutions of NpO_2^+ and PuO_2^{2+} upon the exact nature of the ground states of these ions was investigated by Glebov *et al.* (*42*) in a pioneering work. A ground state $^3\Sigma_{0g}(\delta^2)$ was selected because these ions induce a sizeable reduction of the proton relaxation time of water. There is thus a discrepancy between the ordering of the energy levels deduced from relaxivity measurements, and inferred from EPR data and recent theoretical computations. Glebov *et al.* (*42*) made approximate calculations of the eigenfunctions of the $5f^2$ dioxo cations and suggested that the ordering of the energy levels depends on the relative magnitude of (a) the electrostatic field due to the two bonded oxygen atoms and (b) the Coulomb repulsion between the 5*f* electrons. The ^3H_g level would be the ground state as long as the Coulomb repulsion remains larger than the O–M–O electrostatic field but if the reverse is true, then the ground state would be the $^3\Sigma_g$ level. The crossover would take place for an electrostatic field of about 11,000 cm^{-1}. Glebov *et al.* used the Solomon–Bloembergen–Morgan theory to interpret the dependence of the relaxivity of aqueous solutions of NpO_2^+ (*43*) and PuO_2^{2+} (*44*) on temperature, at 20 MHz. The water exchange correlation time τ_m was assumed to be much larger than the rotational correlation time τ_r and than the electronic relaxation time τ_s and its influence on τ_c was neglected. The authors applied the classical Solomon–Bloembergen Eqs. (1)–(4) given below for a 1 mM solution in the form used by most authors (*21,45–47*) (see Chapters 2–4):

$$\frac{1}{T_{1\,\mathrm{is}}} = \frac{q_{\mathrm{solvent}} \times 10^{-3}}{55.5} \frac{1}{(T_{1\mathrm{M}} + \tau_{\mathrm{m}})} \tag{1}$$

$$\frac{1}{T_{1M}} = \frac{2}{15}\left(\frac{\mu_0}{4\pi}\right)^2 \frac{\gamma_H^2 \mu_B^2 g_e^2 S(S+1)}{r^6}\left[\frac{3\tau_c}{1+\omega_I^2\tau_c^2} + \frac{7\tau_c}{1+\omega_S^2\tau_c^2}\right] \quad (2)$$

$$\frac{1}{\tau_c} = \frac{1}{\tau_r} + \frac{1}{\tau_s} + \frac{1}{\tau_m} \quad (3)$$

$$\frac{1}{\tau_s} = \frac{1}{25}\Delta^2\tau_v\{4S(S+1)-3\}\left(\frac{1}{1+\omega_s^2\tau_v^2} + \frac{4}{1+\omega_s^2\tau_v^2}\right) \quad (4)$$

where $q_{solvent}$ is the number of solvent molecules in the first coordination sphere of the metal ions, μ_0 is the permeability of vacuum, γ_H is the proton magnetogyric ratio, μ_B is the electron Bohr magneton, g_e is the free spin electron g factor, r is the unpaired electronic spins-proton distance, ω_I and ω_S are the proton and electron resonance frequencies at the applied magnetic field. The electronic relaxation time τ_s is a function of the mean-square zero field splitting energy Δ^2 with a modulation correlation time τ_v. Glebov *et al.* (*42,43*) considered the term before the square brackets in Eq. (2) as a variable and thus did not make any assumption on the quantum number to be used in Eq. (2). These authors also assumed that both τ_s and τ_r depend on temperature according to $\tau = \tau_0 \exp(E/RT)$. Values of the parameters τ_s (300 K) $= 2\text{–}5 \times 10^{-10}$ s and $\tau_r = 80-90$ ps were obtained.

Figure 5 shows recent measurements of the magnetic relaxation dispersion of NpO_2^+ and PuO_2^{2+} between 0.01 and 80 MHz and the dependence of

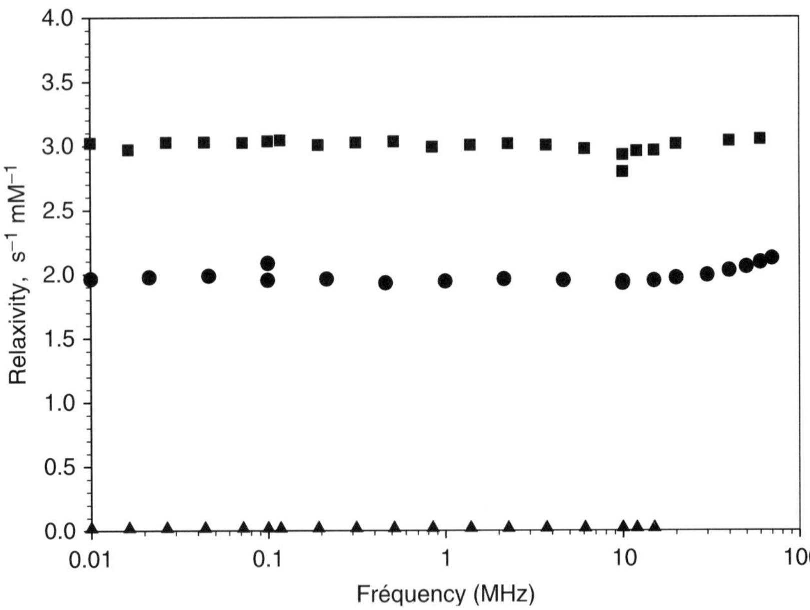

FIG. 5. NMRD curves of acidic aqueous solutions of NpO_2^+ (■), NpO_2^{2+} (▲), and PuO_2^{2+} (●) at 25°C (*10*).

the relaxivity of these ions on temperature at 20 MHz is presented in Figure 6 (*10*). As expected, the measured relaxivities are much lower than in the case of Gd^{3+} (*48*) but they are still higher than for the non-*S* state lanthanides. No dispersion is observed as already noted for all lanthanides except Gd^{3+} (*12*). The electronic relaxation time τ_s obviously dominates the correlation time τ_c. Furthermore, the relaxivity decreases when the temperature increases as expected if the rate of exchange of the water molecules has little or no effect on τ_c. However, the dependence of the relaxivity upon temperature presented in Fig. 6 cannot be interpreted if the influence of the rotational correlation time is not taken into account as already reported by Glebov *et al.* (*42,43*). The influence of temperature was accounted for by

$$\tau_v = \tau_v^{298} \exp\left\{\frac{E_v}{R}\left(\frac{1}{T} - \frac{1}{298.15}\right)\right\} \tag{5}$$

and

$$\tau_r = \tau_r^{298} \exp\left\{\frac{E_r}{R}\left(\frac{1}{T} - \frac{1}{298.15}\right)\right\} \tag{6}$$

as suggested by several authors (*21,45*). In addition, the value of $q_{solvent}$ was set to 5 as it is reasonably well established that five water molecules are coordinated in the axial plane of the metal ions. This hydration number is

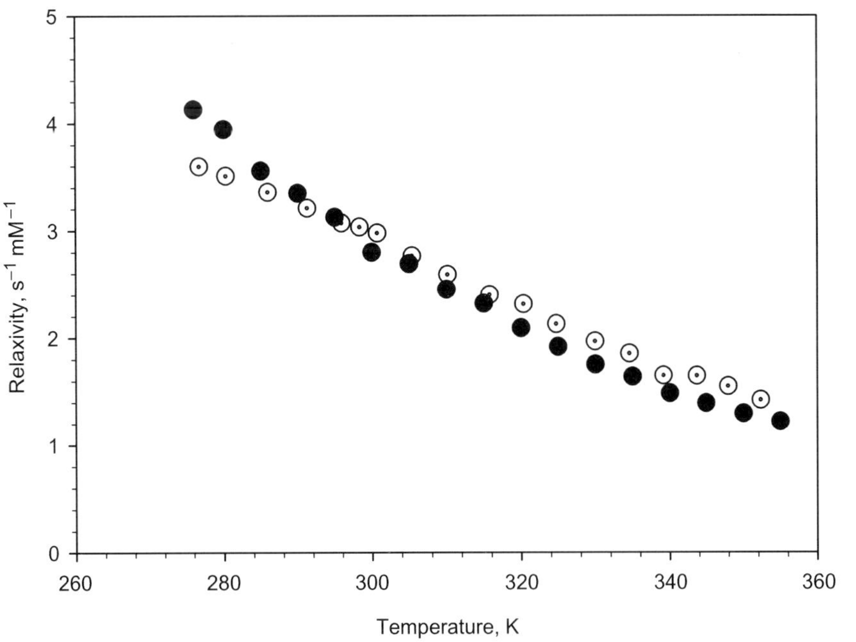

FIG. 6. Relaxivity of NpO_2^+ (●) and PuO_2^{2+} (⊙) as a function of temperature (*10*).

supported by a density functional theory (DFT) study in which the energies of tetra-, penta-, and hexahydrates were compared (*36*). The value of the distance factor *r* can be estimated from a few crystallographic structures of NpO_2^+ complexes (*49*), but there are no known crystallographic structures of a hydrated PuO_2^{2+} salt. It should also be noted that the partial covalency of the bonds formed by the actinides could cause some delocalization of unpaired spin densities into the coordinated water molecules and the *r* distance in Eq. (2) could thus be different from the value deduced from solid state studies in contrast with what has been assumed for the lanthanides (*12*). The interpretation of the variable temperature data presented in Fig. 6 on the basis of Eqs. (1)–(6) leads to τ_r (300 K) ~ 85 ps, E_r ~ 35 kJ/mol, τ_s (300 K, 20 MHz) ~ 7×10^{-11} s. These values are in the expected range for small ions in rapid rotation. The electronic relaxation times are intermediate between the values found for the lanthanide ions and for Gd^{3+}. The calculations were performed assuming that the spin quantum number *S* is a valid quantum number, an obvious oversimplification, and are thus only indicative. No clear indication of a Curie contribution (*12*) could be found as no measurement of the relaxation time T_1 at high fields has yet been carried out for security reasons.

The question then remains as to why the relaxivity is relatively high for ions whose ground state does not have an *S* configuration. An inversion between the $^3H_{4g}$ and $^3\Sigma_{0g}$ states as suggested by Glebov *et al.* (*42*) appears unlikely in view of recent theoretical investigations of the absorption frequencies and intensities in the electronic spectrum of NpO_2^+ (*38,41*). An admixture of states is more likely as Matsika *et al.* (*38*) found that the ground state of NpO_2^+ is only 84% $^3H_{4g}$ ($\delta_u \phi_u$) and as Ismail *et al.* (*40*) reported that the ^3H_g state of PuO_2^{2+} is not purely ^3H_g but a combination of this state and of other triplets. Further work will be needed to completely unravel the relationship between the relaxivity of the actinyl ions and their electronic structure, but one should note here that small departures from a pure state seem to have drastic effects on relaxivity as already observed for Cm^{3+}.

It should be noted here that the variable temperature data presented for NpO_2^+ in Fig. 6 are close to those reported by Glebov *et al.* (*43*) but that the relaxivities presented in this figure for PuO_2^{2+} are different from those published by these authors (*44*) (the concentration of the PuO_2^{2+} is uncertain in this reference). It has already been mentioned that special care must be taken in the preparation of solutions of plutonyl salts to avoid the presence of lower oxidation states. UV-visible spectroscopy (*27*) and liquid scintillation detectors (*50*) are particularly useful to assess the concentration and the nature of Pu species in solution. The purity of the PuO_2^{2+} can also be tested by liquid–liquid extraction with thenoyltrifluoroacetone, an agent known to be able to extract Pu^{4+} from acidic aqueous phases but unable to extract PuO_2^{2+} (*51*).

Two other actinyl ions, UO_2^{2+} ($5f^0$) and NpO_2^{2+} ($5f^1$), are worth mentioning here. Although the uranyl ion contains no unpaired *f* electrons, its complexes may display a weak temperature-independent paramagnetism because of second-order Zeeman effects (*29,34*). However, the diamagnetic contribution of the ligand atoms often exceeds this paramagnetism and UO_2^{2+} is thus of little interest in relaxivity. The NpO_2^{2+} features only one unpaired 5*f* electron and the theoretical treatment of its ground state is thus easier than for the other dioxo cations as one no longer has to take into account the interelectronic repulsion, as the spin–orbit coupling scheme is much simpler. Despite these simplifications, the exact nature of the ground state $^2F_{5/2}$ of NpO_2^{2+} is still debated upon. Matsika *et al.* (*38*) showed that the configuration δ_u^1 and ϕ_u^1 must be taken into account with the major contribution from the ϕ_u^1 orbital. According to these authors the ground state is only 70 % $^2\Phi_{5/2u}$. The relaxivity is expected to be very low as found experimentally (*10*) (see Fig. 5).

The paramagnetism of the actinyl ions can be exploited with more practical applications. For instance, Fig. 7 shows the change in relaxivity of a solution of NpO_2^+ brought about by the addition of the strongly complexing agent DTPA. The stability constant of the NpO_2^+–DTPA complex deduced from this relaxation titration curve is in good agreement with literature data (*52,53*). The method is easily implemented and it requires only limited amounts of radionuclide. Such an experiment would not be possible in the case of PuO_2^{2+} because aminopolycarboxylic acids easily reduce this ion in acidic media (*13,54*).

Relaxivity has also been applied to attempts at better understanding an unusual behavior of the dioxo actinyls. In 1961 itself, Sullivan *et al.* (*55*) showed that the relaxivity of solutions of NpO_2^+ was increased by the

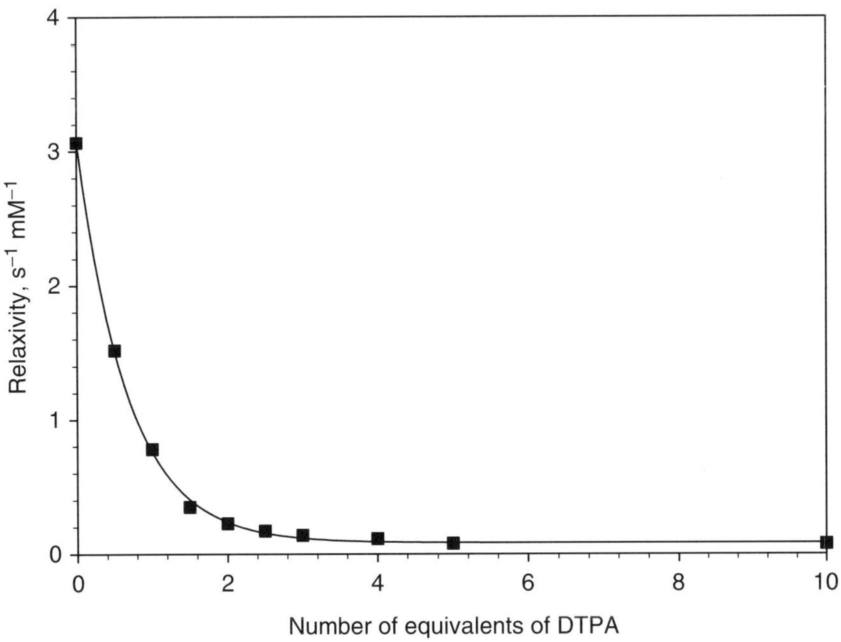

FIG. 7. Relaxivity of NpO_2^+ as a function of the number of added equivalents of diethylentriaminepentaacetate, DTPA (*10*).

addition of large amounts of UO_2^{2+} and assumed that a cation–cation complex was formed. A dimer could also be formed by NpO_2^+ itself in concentrated solutions. One oxygen atom of an NpO_2^+ unit could be coordinated to the central metal atom of another cation instead of a water molecule or bonds could be formed by solvent molecules between two cations. Recent EXAFS studies did not allow a clear distinction between these two binding modes (*56*) but actinyl ions are known to easily interact with oxygen donor molecules (*57,58*). Glebov and Tikhonov (*59*) deduced from variable-temperature relaxivity measurements in concentrated nitrate media that the interaction between NpO_2^+ and UO_2^{2+} increases the rotational correlation time of the former by a factor of 4. A reduction of the electronic relaxation time by a factor of 4.5 was assigned to a weakening of the O–Np–O bonds.

III. Questions of Nuclear Interest

A. Reprocessing of Nuclear Fuels by Solvent Extraction

The separation of radionuclides is a major problem in radiochemistry and in the nuclear industry. The selective extraction of metal ions from an aqueous phase into an organic phase with the help of an extracting agent is both a simple and a very effective method for achieving the isolation of radionuclides for reprocessing, or before storing as wastes. Lanthanides and actinides are major components of nuclear fuels and have to be separated before advanced reprocessing techniques such as transmutation can be used (*60,61*). The efficiency of a liquid–liquid extraction process depends on a large number of factors that are not yet fully understood. For instance, the aggregation of the metal complexes extracted in the organic phase seems to have a direct influence on the extraction efficiency. NMRD is a very useful technique to better characterize the aggregation processes taking place in organic phases. It is noteworthy that so far, NMRD has nearly always been applied to investigations of aqueous solutions but this technique is not limited to water alone as a solvent. Many organic solutions can be analyzed by NMRD provided the solvent molecules feature only one type of protons and provided, the metal ions interact with the solvent.

NMRD has been applied to detailed studies of the dynamic behavior of calixarene Gd^{3+} complexes in anhydrous acetonitrile. Here, Gd^{3+} is considered as a model of the trivalent actinides, which is more easily handled. As shown in Fig. 8, calixarenes are polyaromatic macrocyclic structures such as **1** that can be substituted with a variety of complexing units either on the phenolic "narrow" rim or on the "wide" rim after removal of the *tert*-butyl groups. Calix[4]arene ligands substituted with carbamoylmethylphosphine oxide functions (CMPO) proved to be very selective extracting agents for the lanthanides and the actinides (*62*). Distribution coefficients are several orders of magnitude higher than for CMPO itself because of the pre-organization of the coordinating groups on the cyclic core.

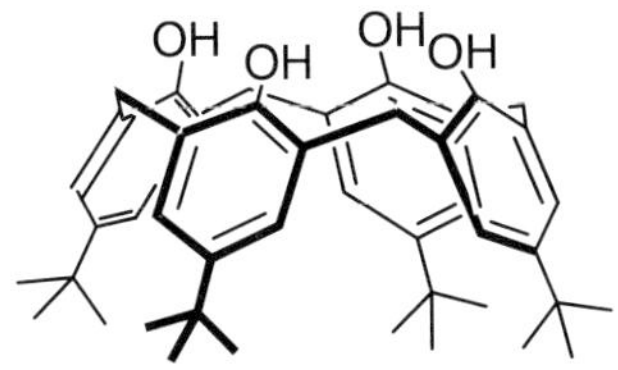

1

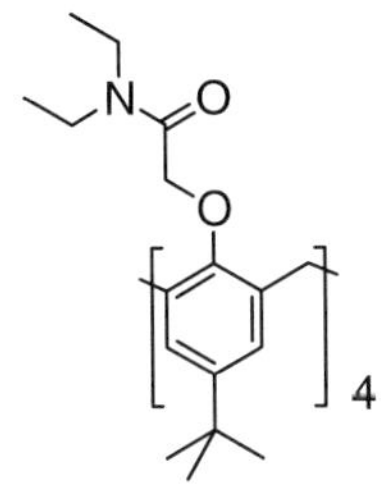

2

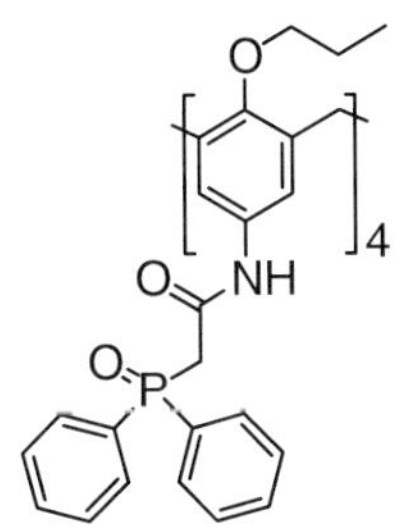

3

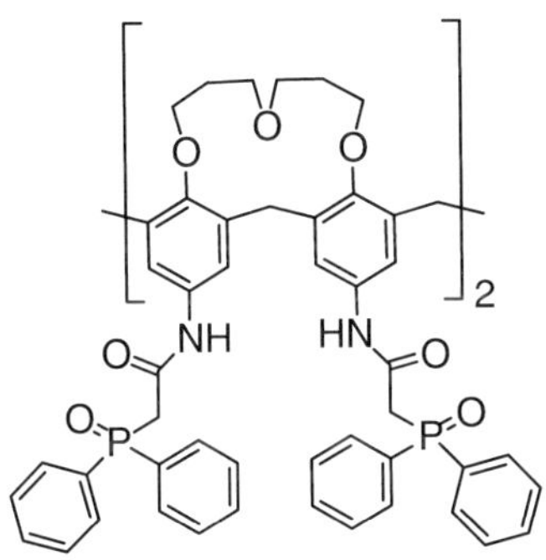

4

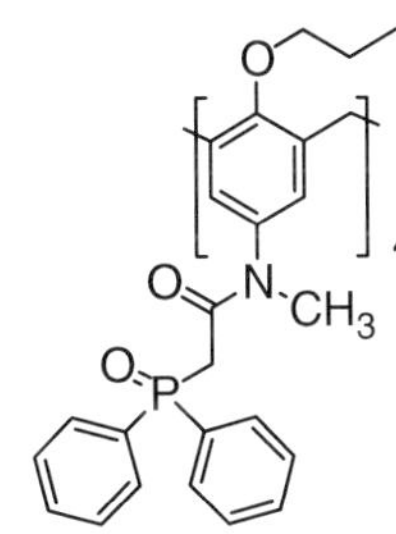

5

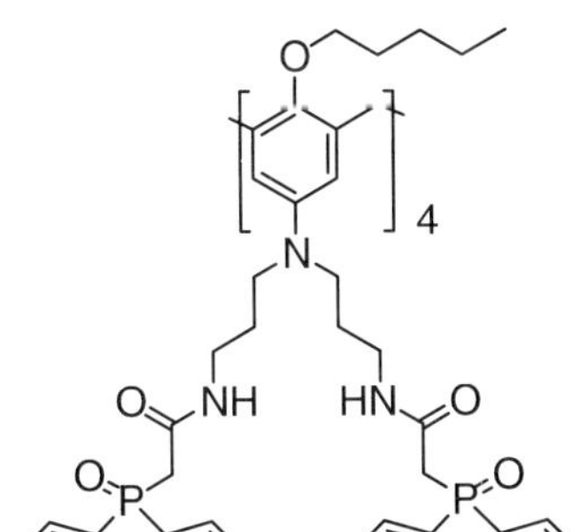

6

FIG. 8. Structures of the calix[4]arenes whose Gd^{3+} complexes were investigated by relaxivity in acetonitrile (*63,66,67*).

Relaxivities were measured either as a function of the applied field or in titration experiments at 20 MHz in anhydrous acetonitrile solutions of $Gd(ClO_4)_3$, i.e., in the simplest possible conditions with no competition from water or coordinating anions for the complexation of the metal ions.

Some relaxivity titration curves are collectively shown in Fig. 9 (*63*). The expected behavior is observed when Gd^{3+} is complexed by a calixarene substituted at the narrow rim by four amide groups (compound **2** in Fig. 8). The relaxivity of the solution decreases until a plateau is reached for a 1:1 concentration ratio. Solvent molecules are thus progressively removed from the first coordination sphere of the metal ions and a highly stable Gd^{3+} complex is fully formed with a 1:1 stoichiometry. On the contrary, the relaxivity of a Gd^{3+} solution in acetonitrile increases when a wide-rim substituted calix[4]arene$(CMPO)_4$ (**3** in Fig. 8) is added until a maximum is reached for a ligand to Gd^{3+} ratio of about 1. The relaxivity then slowly decreases until a plateau is obtained for ratios higher than 2.5. The NMRD curves presented in Fig. 10 also illustrate the difference between ligands **2** and **3**. The dispersion curve for free Gd^{3+} is a simple *S*-shape curve similar to the one obtained in water but with relaxivities that are about 10 times lower as expected because of the longer distance *r* between the metal ions and the acetonitrile protons ($Gd^{3+} \cdots N \equiv C\text{–}CH_3 = 5.3$ Å, see Eq. 2). Upon complexation by **2**, a simple dispersion curve at lower relaxivity is obtained while the coordination of Gd^{3+} by the wide-rim substituted CMPO ligand **3** yields a NMRD curve with a maximum at about 30 MHz. This maximum is assigned to the formation of slowly tumbling oligomers. A quantitative interpretation of these NMRD curves is thwarted by the lack of information on the solvent

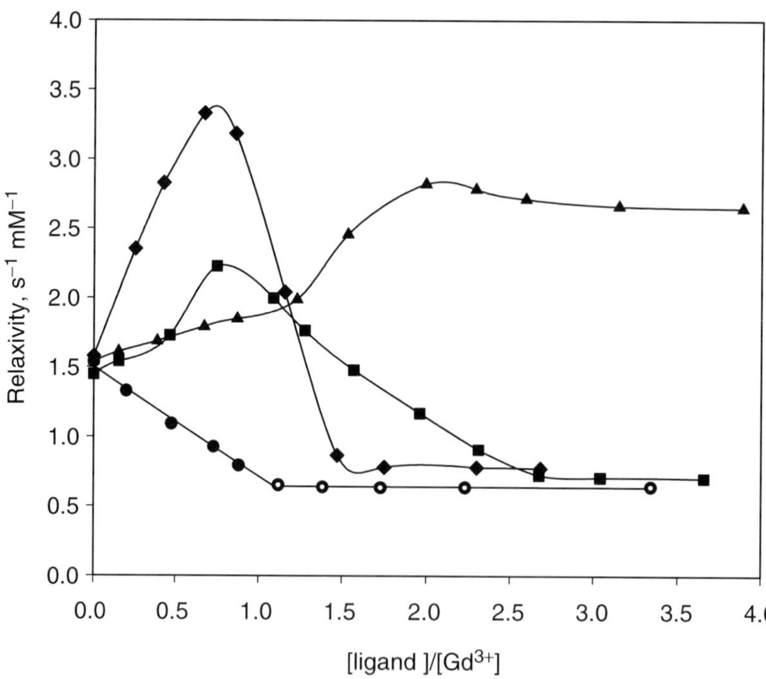

FIG. 9. Relaxivity titration curves of Gd^{3+} by calix[4]arenes **2** (●), **3** (■), **4** (▲), and **5** (◆) in anhydrous acetonitrile at 25°C (*63,66,67*).

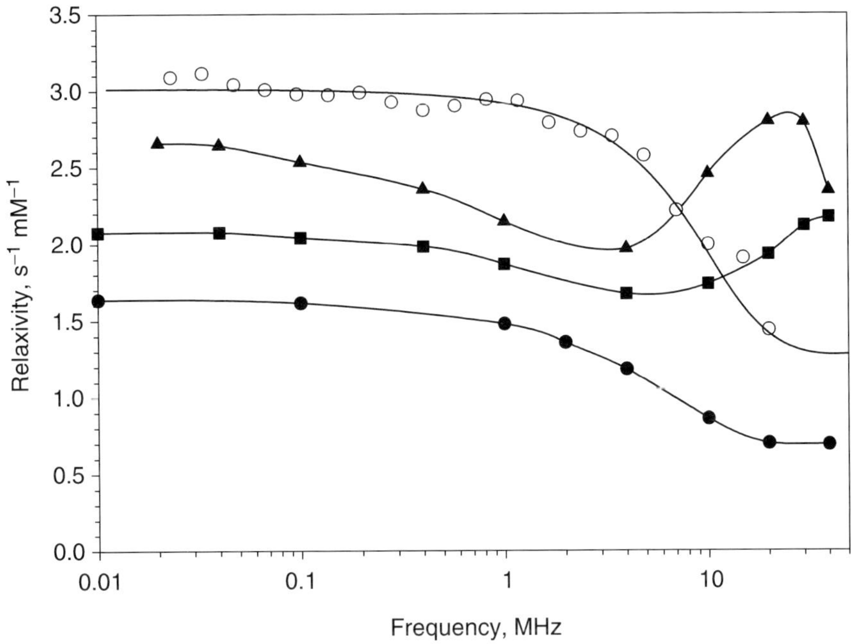

FIG. 10. NMRD curves of free Gd^{3+} (○) and its complexes with calix[4]arenes **2**(●) (at the maximum of the relaxivity titration curve), **3** (■) (at the maximum of the relaxivity titration curve), **4** (▲) (ligand to metal ratio = 2) in anhydrous acetonitrile at 25°C (*63,66,67*).

exchange correlation time τ_m, but information is available on the solvation number of uncomplexed Gd^{3+} ($q_{solvent} = 8$) (*64*). The best fit of the experimental NMRD data calculated by Bertini *et al.* computer program (*65*) was obtained for $\tau_r = 51$, 193, and 649 ps, for free Gd^{3+} and its complexes with **2** and **3**, respectively, with D = 0.016–0.060 cm^{-1}. The relaxivity maximum in Fig. 10 is thus assigned to the formation of oligomers that are tumbling sufficiently slowly so that the longer relaxation time of the solvent molecules released to the bulk of the solution is more than compensated by the shorter relaxation times of solvent molecules contained in oligomeric assemblies. The calculated rotational correlation times should be taken as approximate values because of the assumptions that had to be made in the calculations, but NMRD clearly indicate that calixarene extractants behave differently depending on the location of their coordinating groups. The formation of aggregates is systematically observed for all the calix[4]arenes substituted on the wide rim and a simple monomeric behavior is noted for the ligands featuring donor groups on their narrow rim. One is thus led to assume that an oligomerization takes place because of the greater flexibility of the ligating functions once they are located on the wide rim (*63,66,67*). The formation of these aggregates is in agreement with a crystallographic analysis (*68*) and with nOe measurements on a diamagnetic complex (*63*). An even more pronounced aggregation and a larger range of ligand/metal ratios was noted for CMPO calixarene **4** that is conformationally rigidified by polyether chains (see Figs. 9 and 10) (*66*). Also noteworthy is that small structural changes can bring about unexpected differences in the aggregation behavior. For instance, the only difference between ligands **3** and **5** is that the latter features methylated amide groups. However, the oligomerization is more pronounced as shown in Fig. 9. The rotational correlation time τ_r is about 1350 ps for $q_{solvent} = 2$ (*67*) and dynamic light scattering studies support the formation of dimeric and trimeric structures. The addition of nitrate ions to acetonitrile solutions of the Gd^{3+}-**3** complex leads to an even greater aggregation likely due to the bridging NO_3^- ions. It is noteworthy that differences between calixarenes **3** and **5** are also observed in extraction experiments. However, a direct correlation between NMRD measurements and extraction experiments is not yet firmly established. More work is required to take into account all the parameters that can influence the extraction processes. For instance, it was expected that the dendritic ligand **6** would also form oligomers as it is substituted on the wide rim. Unexpectedly, this dendritic calixarene forms only a stable monomeric Gd^{3+} complex of 2:1 metal:ligand stoichiometry.

NMRD has several added advantages in the analysis of the complexation of ions in organic solvents. The complexation processes can be analyzed step by step starting from an anhydrous solution of a Gd^{3+} salt with a poorly coordinating anion and adding a ligand, a complexing anion, water, changing the composition of a contacted aqueous phase, etc. In addition, only a few milligrams of painstakingly synthesized ligands are needed to collect the relaxivity data.

B. Migration of Actinides in Nature

Deep geological disposal is the most favored solution for the permanent disposal of nuclear wastes with long half-lives. Although the locations of the burial places are selected with outmost care to avoid migration of the wastes in nature over a very long period of time, no barrier can be safe forever, so, numerous studies are in progress to determine the main factors that could cause leaks of radioactive nuclides. Soluble compounds in ground water are likely to play a major role in the release of actinides.

Many organic species dissolved in natural water belong to the family of the humic and fulvic acids (*69*). These compounds are ubiquitous in nature and are found whenever there is decay of organic matter. They are polymeric polyelectrolytes of ill-defined composition known to contain carboxylic and phenolic groups in addition to many other functional groups. Humic acids are good complexing agents of metal ions including actinides in different oxidation states but the speciation of the metal complexes remains extremely difficult and can only be carried out by a combination of spectroscopic techniques (*70*). Relaxivity is one new approach that could yield interesting information about the nature of the complexing sites in humic acids. Once again, Gd^{3+} is used as a model of the trivalent actinides. Figure 11 shows the relaxivity increases obtained at 20 MHz upon the addition of three fractions of a sample of Aldrich humic acid that was fractionated by ultrafiltration (*10*). It is not the highest molecular weight fraction that brings about the highest relaxivity increase. Presumably, the complexation sites are more rigid in the middle-weight fraction. The corresponding NMRD curves are reproduced in Fig. 12. Very high relaxivities are achieved with the middle molecular weight fraction. A best fit treatment of the NMRD curves yields a mean hydration number of 6–7 and τ_r values of 510, 2110, and 1630 ps for the low, middle, and high molecular weight fractions respectively. Similar experiments have been performed with polyacryclic acids of low dispersity (*10*). The same NMRD curves were obtained for fractions of molecular weights 85,000 and 1,300,000 D. Relaxivity is also suitable for competition experiments in which Gd^{3+} ions are extracted from natural or synthetic polymers by a complexing agent. An unexpected application of relaxivity is presented in Fig. 13 that shows the change in relaxation rate with time for a solution of polyacrylic acid highly loaded in Gd^{3+} (one metal ion per carboxylic group). The metal complex slowly sediments in a glass tube and the relaxivity decreases accordingly until equilibrium is reached. NMRD could be an interesting supplement for the investigation of the hydrodynamic properties of metal-containing macromolecular complexes.

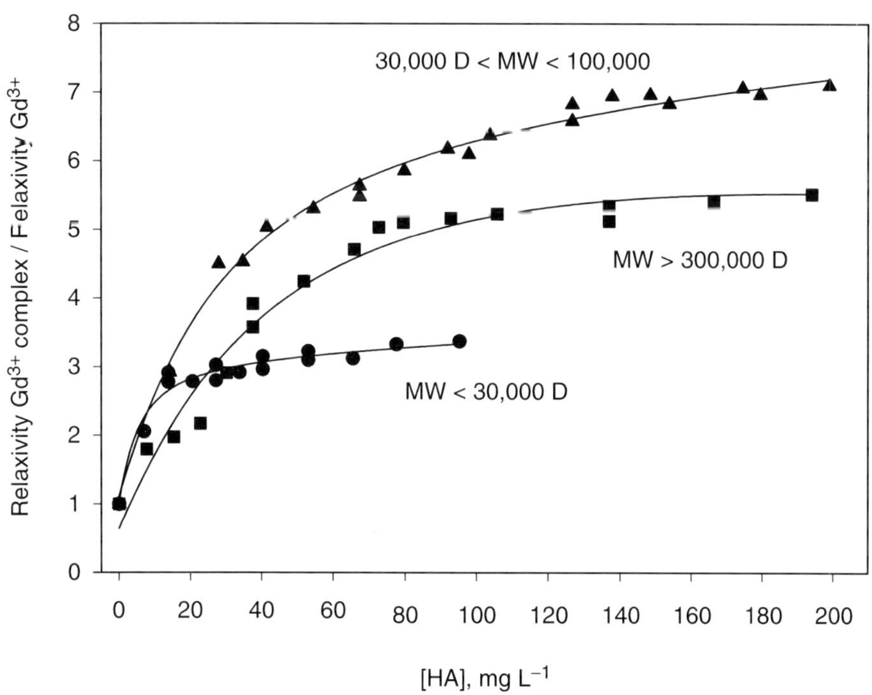

FIG. 11. Relaxivity increases of Gd^{3+} aqueous solutions (pH = 5) upon the addition of fractions of Aldrich humic acid of different molecular weights (25°C, 20 MHz, $[Gd^{3+}] = 10^{-5}$ M) (*10*).

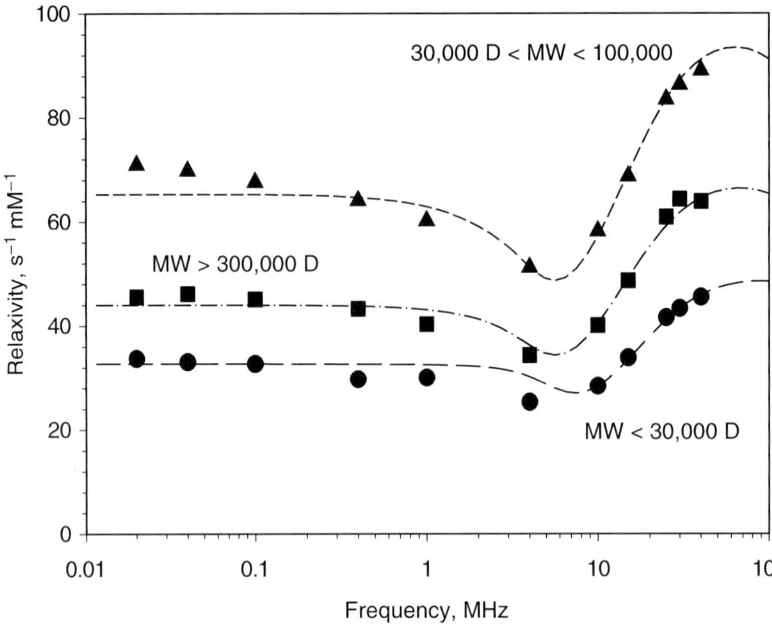

FIG. 12. NMRD curves of aqueous solutions of Gd^{3+} complexed by Aldrich humic acid of different molecular weights (25°C, 20 MHz, $[Gd^{3+}] = 10^{-5}$ M) (*10*).

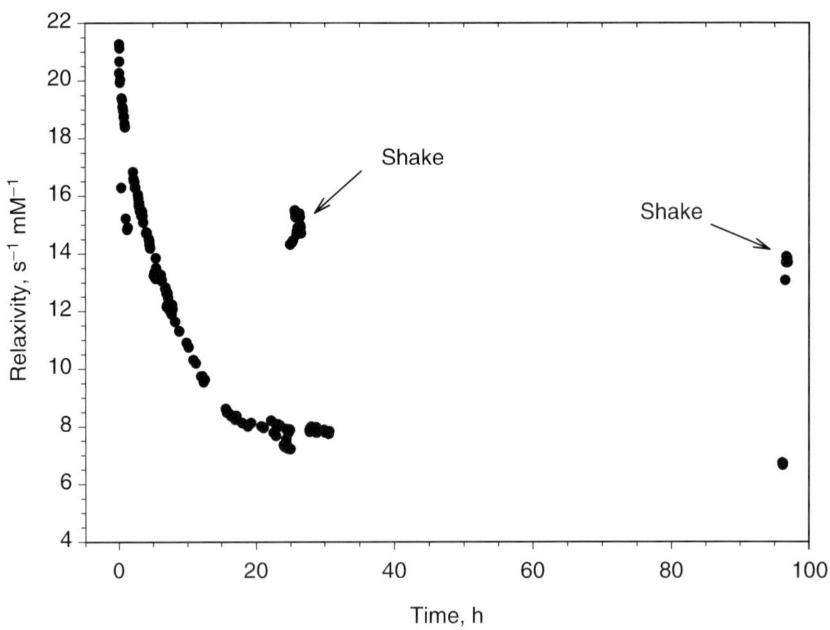

FIG. 13. Sedimentation of a Gd^{3+} polyacrylic acid complex (25°C, molecular weight 85,000 D, one metal ion per carboxylic function). Higher values of the relaxivity are obtained after shaking the solution (*10*).

IV. Conclusions

This chapter is an overview of what magnetic nuclear relaxation can bring to our knowledge of the actinide ions and their use in the nuclear industry. Most results are quite recent and the field is wide open. The author hopes this chapter will attract the attention of NMR specialists who should not be distraught by the experimental difficulties that always accompany the handling of radioactive materials. On the other hand, nuclear chemists and physicists will hopefully discover that NMRD is an interesting supplement to their favorite spectroscopic techniques.

Acknowledgments

The author sincerely thanks Dr. B. Lambert, Mr. A. Joassin and G. Gridelet for their invaluable assistance. The author also thanks the Institut Interuniversitaire des Sciences Nucléaires of Belgium and the European Community for financial support.

References

1. "*The Chemistry of the Actinide Elements*"; Eds. Katz, J.J.; Seaborg G.T.; Morss, L.R.; Chapman & Hall: London, 1986.
2. Lewis, W. B.; Mann, J. B.; Liberman, D. A.; Cromer, D. T. *J. Chem. Phys.* **1970**, *53*, 809–820.
3. Söderlind, P.; Nash, C. S. Electronic structure, "*Advances in Plutonium Chemistry 1967–2000*"; Ed. Hoffman, D. C.; American Nuclear Society: La Grange Park, Illinois, USA, 2002, pp. 6–23.
4. Carnall, W. T.; Wybourne, B. G. *J. Chem. Phys.* **1964**, *40*, 3428–3433.
5. Pepper, M.; Bursten, B. E. *Chem. Rev.* **1991**, *91*, 719–741
6. Cotton, F. A.; Wilkinson, G.; Murillo, C. A.; Bochmann, M. "*Advanced Inorganic Chemistry*"; Wiley: New York, 1999, pp. 1130–1164.
7. Greenwood, N. N.; Earnshaw, A. "*Chemistry of the Elements*"; Pergamon Press: Oxford, 1984, pp. 1450–1486.
8. Kot, W. K.; Shalimoff, G. V.; Edelstein, N. *J. Am. Chem. Soc.* **1988**, *110*, 986–987.
9. Parry, J. S.; Cloke, F. G. N.; Coles, S. J.; Hursthouse, M. B. *J. Am. Chem. Soc.* **1999**, *121*, 6867–6871.
10. Desreux J. F. *Proceedings of the Migration 03 Conference*, Gyeongju, Korea, September 2003.
11. Pikaev, A. K.; Shilov, V. P.; Gogolev, A. V. *Russ. Chem. Rev.* **1997**, *66*, 763–788.
12. Bertini, I.; Capozzi, F.; Luchinat, C.; Nicastro, G.; Xia, Z. C. *J. Phys. Chem.* **1993**, *97*, 6351–6354.
13. Cauchetier, P.; Guichard, C. *J. Inorg. Nucl. Chem.* **1975**, *37*, 1771–1778.
14. Kolbe, W.; Edelstein, N. M.; Finch, C. B.; Abraham, M. M. *J. Chem. Phys.* **1973**, *58*, 820–821.
15. Abraham, M. M.; Boatner, L. A. *Phys. Rev., Sect. B* **1982**, *26*, 1434–1437.
16. Systma, J.; Murdoch, K. M.; Edelstein, N. M.; Boatner, L. A.; Abraham, M. M. *Phys. Rev., Sect. B* **1995**, *52*, 12668–12676.
17. Murdoch, K. M.; Cavellec, R.; Simoni, E.; Karbowiak, M.; Hubert, S.; Illemassene, M.; Edelstein, N. M. *J. Chem. Phys.* **1998**, *108*, 6353–6361.
18. Figgis, B. N.; Hitchman, M. A. "*Ligand Field Theory and its Applications*"; Wiley: New York, 2000, pp. 311–324.
19. (a) Edelstein, N.; Easley, W. *J. Chem. Phys.* **1968**, *48*, 2110–2115 (b) Sytsma, J.; Murdoch, K. M.; Edelstein, N. M.; Boatner, L. A.; Abraham, M. M. *Phys. Rev. B* **1995**, *52*, 12668–12676.
20. Kimura, T.; Choppin, G. R.; Kato, Y.; Yoshida, Z. *Radiochim. Acta* **1996**, *72*, 61–64.
21. Powell, D. H.; Ni Dhubhghaill, O. N.; Pubanz, D.; Helm, L.; Lebedev, Y. S.; Schlaepfer, W.; Merbach, A. E. *J. Am. Chem. Soc.* **1996**, *118*, 9333–9346.

22. Thouin, E.; Lubell, W. D. *Tetrahedron Lett.* **2000**, *41*, 457–460.
23. Bibler, N. E. *J. Inorg. Nucl. Chem.* **1972**, *34*, 1417–1425.
24. Kot, W. K.; Edelstein, N. M.; Abraham, M. M.; Boatner, L. A. *Phys. Rev., Sect. B* **1993**, *47*, 3412–3414.
25. Masters, B. J.; Rabideau, S. W. *Inorg. Chem.* **1963**, *2*, 1–5.
26. "*Advances in Plutonium Chemistry 1967–2000*"; Ed. Hoffman, D.; American Nuclear Society: La grange Park, Illinois, USA, 2002.
27. Newton, T. W.; Baker, F. B. *J. Phys. Chem.* **1957**, *61*, 934–938.
28. Bleaney, B. *Disc. Faraday Soc.* **1955**, *19*, 112–118.
29. Boudreaux, E. A.; Mulay, L. N. "*Theory and Applications of Molecular Paramagnetism*"; Wiley: New York, 1976, pp. 1–510.
30. Gruen, D. M.; Hutchison, C. A. Jr. , *J. Chem. Phys.* **1954**, *22*, 386–393.
31. Nakamoto, T.; Nakada, M.; Nakamura, A. *J. Nucl. Sci. Technol.* **2002**, *Suppl. 3*, 102–105.
32. Balakrishnan, P. V.; Patil, S. K.; Sharma, H. D.; Venkatasetty, H. V. *Can. J. Chem.* **1965**, *43*, 2052–2058.
33. Dawson, J.K. *J. Chem. Soc.* **1952**, 2705–2707.
34. Eisenstein, J. C.; Pryce, M. H. L. *Proc. R. Soc. Lond. A* **1955**, *229*, 20–38.
35. Craw, J. S.; Vincent, M. A.; Hillier, I. H.; Wallwork, A. L. *J. Chem. Phys.* **1995**, *99*, 10181–10185.
36. Hay, P. J.; Martin, R. L.; Schreckenbach, G. *J. Phys. Chem. A* **2000**, *104*, 6259–6270.
37. Matsika, S.; Zhang, Z.; Brozell, S. R.; Blaudeau, J. P.; Wang, Q.; Pitzer, R. M. *J. Phys. Chem. A* **2001**, *105*, 3825–3828.
38. Matsika, S.; Pitzer, R. M. *J. Phys. Chem. A* **2000**, *104*, 4064–4068.
39. Maron, L.; Leininger, T.; Schimmelpfennig, B.; Vallet, V.; Heully, J. L.; Teichteil, CV.; Gropen, O.; Wahlgren, U. *Chem. Phys.* **1999**, *244*, 195–201.
40. Ismail, N.; Heully, J. L.; Saue, T.; Daudey, J. P.; Marsden, C. J. *Chem. Phys. Lett.* **1999**, *300*, 296–302.
41. Matsika, S.; Pitzer, R. M.; Reed, D. T. *J. Phys. Chem. A* **2000**, *104*, 11983–11992.
42. Glebov, V. A. *Radiokhimiya* **1979**, *21*, 793–801.
43. Glebov, V. A.; Nikitina, T. M.; Tikhonov, M. F. *Koord. Khim.* **1976**, *2*, 1271–1280.
44. Glebov, V. A.; Tikhonov, M. F. *Koord. Khim.* **1981**, *7*, 1859–1865.
45. "*The Chemistry of Contrast Agents in Medical Magnetic Resonance Imaging*"; Wiley: Chichester, 2001.
46. Banci, L.; Bertini, I.; Luchinat, C. "*Nuclear and Electron Relaxation*"; VCH: Weinheim, 1991, pp. 1–208.
47. Bertini, I.; Luchinat, C. *Coord. Chem. Rev.* **1996**, *150*, 1–296.
48. Koenig, S. H.; Epstein, M. *J. Chem. Phys.* **1975**, *63*, 2279–2284.
49. Andreev, G. B.; Fedosseev, A. M.; Antipin, M.Y. *Mendeleev Commun.* **2001**, 58–59.
50. Choppin, G. R.; Liljenzin, J. O.; Rydberg, J. "*Radiochemistry and Nuclear Chemistry*"; Butterworth-Heinemann: Wonurn, MA, 2002, pp. 220–222.
51. Bajo, S.; Eikenberg, J. *Radiochim. Acta* **2003**, *91*, 495–497.
52. Sevost'yanova, E. P. *Radiokhimiya* **1980**, *22*, 720–726.
53. Moskvin, A. I. *Radiokhimiya* **1971**, *13*, 641–643.
54. Reed, D. T.; Wygmans, D. G.; Aase, S. B.; Banaszak, J. E. *Radiochim. Acta* **1998**, *82*, 109–114.
55. Sullivan, J. C.; Hindman, J. C.; Zielen, A. J. *J. Am. Chem. Soc.* **1961**, *83*, 3373–3378.
56. DenAuwer, C.; Grégoire-Kappenstein, A. C.; Moisy, P. *Radiochim. Acta* **2003**, *91*, 773–776.
57. Den Auwer, C.; Simoni, E.; Conradson, S.; Madic, C. *Eur. J. Inorg. Chem.* **2003**, 3843–3859.
58. Madic, C.; Guillaume, B.; Morisseau, J. C.; Moulin, J. P. *J. Inorg. Nucl. Chem.* **1979**, *41*, 1027–1031.
59. Glebov, V. A.; Tikhonov, M. F. *Koord. Khim.* **1982**, *8*, 48–54.
60. Schenkel, R., Magill, J.; Glatz, J.-P. Mayer, K. Partitioning and Transmutation - Technical Feasibility, Proliferation Resistance, and Safeguardability. 2-21-2003. Ref Type: Electronic Citation.
61. Wiles, D. R. "*The Chemistry of Nuclear Fuel Waste Disposal*"; Polytechnic International Press: Montreal, Canada, 2001.

62. Delmau, L. H.; Simon, N.; Schwing-Weill, M. J.; Arnaud-Neu, F.; Dozol, J. F.; Eymard, S.; Tournois, B.; Böhmer, V.; Grüttner, C.; Musigmann, C.; Tunayar, A. *J. Chem. Soc., Chem. Commun.* **1998**, 1627–1628.
63. Lambert, B.; Jacques, V.; Shivanyuk, A.; Matthews, S. E.; Tunayar, A.; Baaden, M.; Wipff, G.; Böhmer, V.; Desreux, J. F. *Inorg. Chem.* **2000**, *39*, 2033–2041.
64. Bünzli, J.-C. G.; Mabillard, C. *Inorg. Chem.* **1986**, *25*, 2750–2754.
65. Bertini, I.; Galas, O.; Luchinat, C.; Parigi, G. *J. Magn. Reson.* **1995**, *113*, 151–158.
66. Arduini, A.; Böhmer, V.; Delmau, L. H.; Desreux, J. F.; Dozol, J. F.; Garcia Carrera, A.; Lambert, B.; Musigmann, A.; Pochini, A.; Shivanyuk, A.; Ugozzoli, F. *Chem. Eur. J.* **2000**, *6*, 2135–2144.
67. Schmidt, C.; Saadioui, M.; Böhmer, V. Host, V.; Spirlet, M. R.; Desreux, J. F.; Brisach, F.; Arnaud-Neu, F.; Dozol, J.-F. *Org. Biol. Chem.* **2003**, *1*, 4089–4096.
68. Cherfa, S. unknown. **1998**. University of Paris Sud.
69. Choppin, G. R.; Allard, B. Complexes of actinides with naturally occurring organic compounds, "*Handbook on the Physics and Chemistry of the Actinides*"; Eds. Freeman, A. J.; Keller, C.; North-Holland: Amsterdam, 1985, pp. 407–429.
70. "*Humic Substances. Structures, models and functions*"; Royal Society of Chemistry: Cambridge, UK, 2001.

TECHNICAL ASPECTS OF FAST FIELD CYCLING

GIANNI FERRANTE and STANISLAV SYKORA

Stelar s.r.l., Via E. Fermi, 4 I-27035 Mede (PV), Italy

ADVANCES IN INORGANIC CHEMISTRY
VOLUME 57 ISSN 0898-8838 / DOI 10.1016/S0898-8838(05)57009-0

I. Introduction

Since the earliest days of Nuclear Magnetic Resonance (NMR), it has been clear (*1–10*) that relaxation mechanisms were to play an important role in all its applications. In fact, even the first detection of an NMR signal (*11–12*) was delayed (*12–13*) by several years because the chosen compounds had, unfortunately, excessively long relaxation times.

From the pioneering BPP (Bloembergen–Purcell–Pound) formula (*7*), published in 1948 itself, it became immediately evident that, qualitatively speaking,

- NMR relaxation times, particularly the longitudinal ones, depend through the Larmor frequency on the magnetic field induction B_0.
- The relaxation mechanisms require some kind of nuclear interaction subject to stochastic fluctuations, typically due to random molecular motions.
- The most pronounced relaxation phenomena (in terms of field dependence) were to be expected at relatively low fields where low-frequency molecular motions can have a very large impact on the longitudinal relaxation times T_1.

The dependence of $T_1(B)$ on the field B was soon nicknamed as the T_1 *dispersion curve* or, more recently, as the Nuclear Magnetic Relaxation Dispersion (*NMRD*) *profile*. The first experimental curve of this type (Fig. 1) was published in 1950 by Ramsey and Pound (*15,16*).

Many more such curves were measured in subsequent years, some of which were reported by Abragam (*17*). When Abragam's work was published it was already quite clear that the dispersion curves could become a valid tool for the study of *molecular dynamics*, thus laying down the foundation for *variable field NMR relaxometry*. In principle, the dispersion curves are potentially powerful tools to discriminate between various molecular dynamics models.

The development of this branch of NMR, however, has been quite slow compared to the explosive progress of NMR spectroscopy and, later on, NMR imaging. There are many reasons for this slow start, the most obvious ones being:

- Complexity of NMR relaxation theories (*1,7–9,18–48*).
- Lack and/or complexity of molecular dynamics models.
- Practical difficulties inherent in measuring the dispersion curves.

Since this is essentially an engineering chapter, we shall dwell only on the last point. From the BPP formula, it was already qualitatively clear that, in order to become efficient and useful tools, the dispersion curves must extend over a wide interval of relaxation field values (preferably several orders of magnitude).

Achieving this goal using the traditional, fixed-field approach was almost impossible. One can, of course, use an electromagnet and a broad-band NMR console and, re-tuning the system at every point of the measured profile,

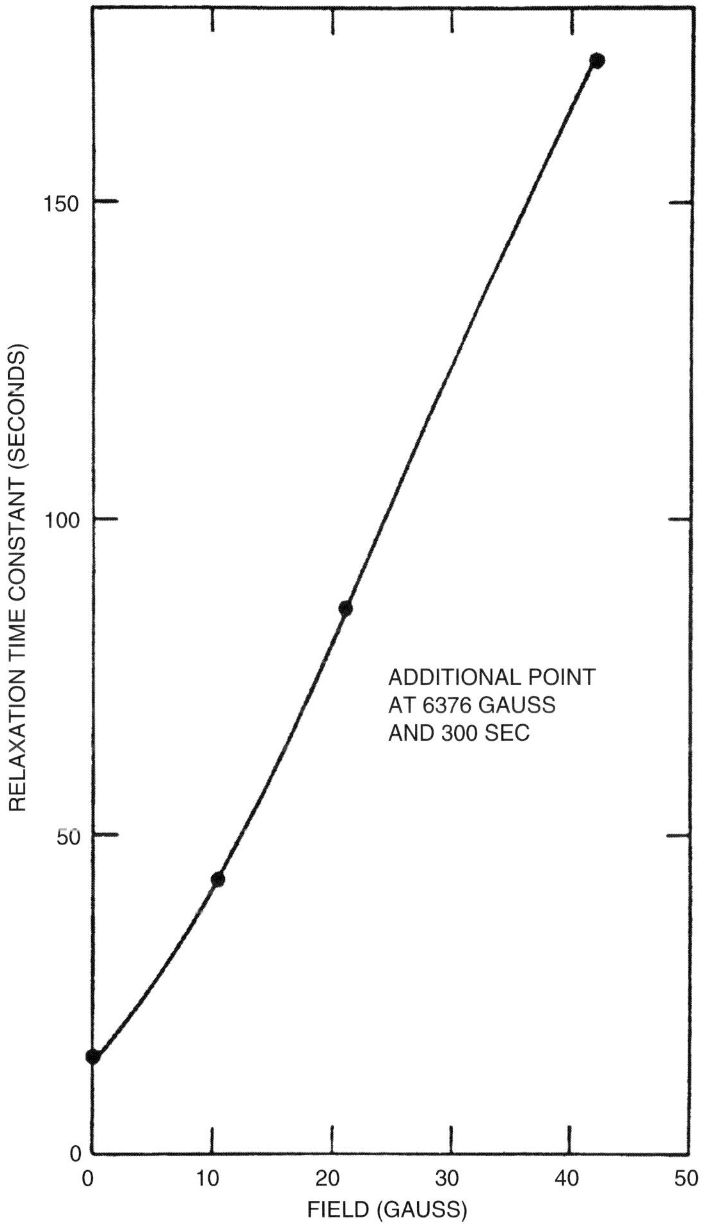

FIG. 1. Relaxation time constant as a function of magnetic field for Li^7 in LiF. This very first NMRD profile is reprinted from the article published by Ramsey and Pound (*15*) in 1950.

carry out a conventional relaxation time measurement at many field values. Apart from being painfully slow, however, such an approach is limited to at best one decade of rather high field values. At fields corresponding to less than about 1 MHz of the ^1H Larmor frequency, the signal excitation and detection technologies in fact change too much to use the same type of instrument and, in addition, the signal becomes often too weak to be detected.

Limited fixed-field, traditional relaxation measurements at very low fields, including the Earth field, were of course carried out (*49*) using specially built NMR systems. Such measurements confirmed the general tendency of relaxation times to be more "discriminating" at low fields than at high fields. The fact has been used even to produce a medical low-field NMR system capable of diagnosing particular fetal pathologies by means of in vivo measurements of the longitudinal relaxation time of the amniotic liquid. Even such systems, however, were limited to quite a narrow relaxation field interval.

It soon became clear that in order to cover comfortably a wide range of relaxation field values, one had to use an excitation/detection assembly operating at some fixed field and, during the relaxation periods of a relaxation measurement NMR sequence, subject the sample to another, easily variable field. One rather obvious solution for such an arrangement was to combine a fixed field, conventional NMR relaxometer operating at a high field with an auxiliary variable electromagnet and, during the sequence, *mechanically shuttle* the sample between the two magnetic fields.

For almost three decades, many T_1 dispersion curves (including the first one shown in Fig. 1) were actually measured by moving the sample manually from one magnet to the other. Quite soon, however, mechanical devices (*50–64*) were developed to achieve the task, some of which were quite sophisticated.

Since, during an actual measurement, the shuttling process is repeated many times in a cyclic manner, the technique has been named *field-cycling (FC) NMR relaxometry*, a term which underlines the fact that it is the *magnetic field variation* that matters and not the manner in which it is achieved.

The main drawback of mechanical shuttling consists in the relatively long time needed to move the sample physically between the two fields. Even with the best devices, there are severe limits on the maximum acceleration/deceleration, dictated by the mechanical stability of the sample transport system as well as of the sample. Consequently, relaxation times much shorter than 100 ms (R_1 above 10) are almost impossible to measure, ruling out an enormous segment of potentially interesting applications. On the other hand, mechanical shuttling systems had been successfully combined with high-field, high-resolution NMR spectroscopy (*64*), a feature which is still quite unique.

A new approach in the early 1970s consisted in keeping the sample fixed while the field, produced by an air-core electromagnet, is switched between different field values. This approach, named *fast field-cycling (FFC) NMR relaxometry*, explored primarily by Redfield (*35,75*) Noack (*61,65,67,77*) Koenig (*58,66,78*) and Kimmich (*61,73,76,77,94*) has the potential of handling much faster relaxing samples (the current upper limit of manageable R_1 is between 1000 and 10000, depending upon the shape of the NMRD profile). It requires novel type of magnets and power supplies, the development of which is still in progress.

The advent of the FFC instruments has opened a number of important application areas (molecular dynamics of liquid crystals, paramagnetic contrast MRI agents, proteins, polymers, etc.) and has thus provided a powerful impulse for further development of variable-field NMR relaxometry. Since 1996, Stelar entered the field and, building on the Noack–Schweikert technology (*67*), started producing the first commercial FFC NMR relaxometers. The availability of such instruments has further enhanced the drive towards new applications, apart from confirming the enormous potential of the technique as a primary tool for the study of molecular dynamics of even quite complex systems.

In what follows, we wish to describe the most important technical aspects of FFC NMR relaxometry, including both the required special hardware (magnet, power supply, etc.) and the measurement methodology (data acquisition sequences and, to some extent, the subsequent data evaluation). Naturally, the description is based primarily on our own experience which has not yet been described in detail elsewhere.

II. Premises of FFC NMR Relaxometry

In a conventional NMR instrument the resonant magnetic induction of a nuclide immersed in a field with magnetic induction B occurs at the Larmor frequency $\omega_L = \gamma B$, where γ is the gyromagnetic ratio of the nuclide. Since the NMRD profiles cover several orders of magnitude of the field value B, they necessarily include sections with very low B values.

Unfortunately, the detection of the signals induced by a given nuclide becomes difficult when B becomes small because of the progressive decrease of the signal to noise ratio (S/N). It is often quoted that the sensitivity decreases with the square of the field or, according to more realistic estimates (*65,68,69*), with factors such as $B^{3/2}$ *or* $B^{7/4}$. The use of field-cycling measurement methods (for details, see Section VIII) is a truly neat way of avoiding all the problems of low-field NMR and allows one to measure relaxation rates over something like five decades of the relaxation field magnitudes without ever changing the RF excitation and detection hardware. In principle, it makes it even possible to measure relaxation rates at zero-field which would be completely impossible to do in any other way.

A field-cycling experiment requires that the sample be subject at different times, and for variable durations, to different values of the magnetic field B. From the instrumental point of view, there are basically just two ways to achieve this goal.

The first method consists in mechanical translation of the sample between areas with different field intensities (*15,16,50–64*). However, such mechanical "shuttling" methods are inherently slow. It follows that they are applicable only to samples with long relaxation times, limited essentially by the shortest possible time it takes to move the sample from one position to another which is typically about 50 ms.

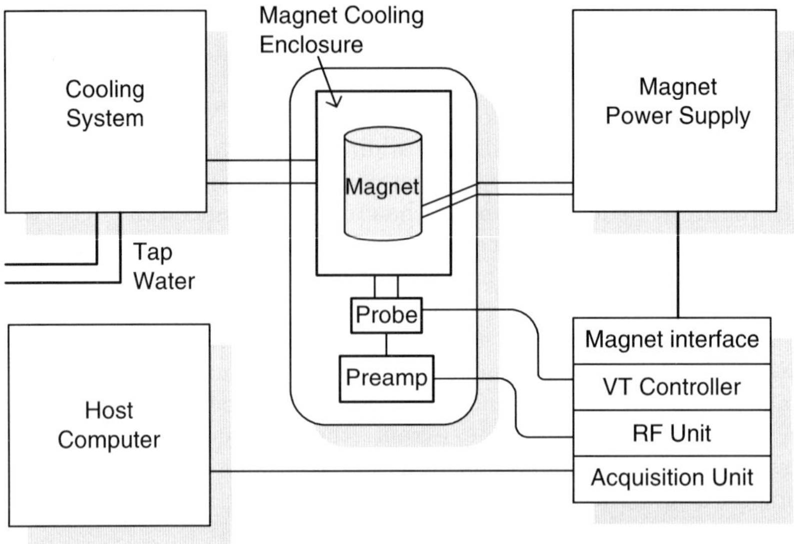

FIG. 2. Basic block diagram of a FFC NMR relaxometer.

The second field-cycling method (*65–67*) uses electronic modulation of the current flowing through the coil of an electromagnet. This technique, commonly called Fast Field Cycling (FFC) NMR relaxometry, permits much faster variations of the field induction and thus extends the applicability of the field cycling approach to very short relaxation times T_1, at present, down to fractions of a millisecond.

The instrumental aspects which will be discussed in the following sections concern primarily the fundamental characteristics and the functional behavior of those parts of an FFC NMR relaxometer which are characteristic of the FFC technique (Fig. 2). Subsequently, some of the methodological aspects of FFC NMR relaxometry, such as different data acquisition sequence and data accumulation and evaluation methods will be discussed.

Above all, the authors wish to share the experience they have acquired during the development of a commercial series of routine and research FFC NMR relaxometers.

III. The FFC Magnet

A. General Considerations

Any NMR field-cycling (FC) relaxometry experiment presumes that the sample is subject to a magnetic field of various intensities for time intervals of varying durations. More specifically, between the various intervals of a relaxation-time measurement, the external *magnetic field induction*

B (the Zeeman field) must be made to commute rapidly between several predefined values, such as the polarization field B_p, the relaxation field B_r, and the acquisition field B_a.

These values belong to the most important parameters of the experiment and their total range is given by the electrical characteristics of the apparatus. The time interval required to commute, or *cycle* between the different field levels is usually called *switching time*. The minimum switching time values depend on the desired size of the field jump ΔB, defined by the experiment, as well as on the electrical characteristics of the magnet (primarily its inductance and resistance), and of the power supply (primarily its maximum voltage). The latter factors, in fact, define the maximum field-variation rate (i.e., the *slewing rate* $s = dB/dt$) of an electromagnet, an aspect which shall be discussed in detail in Section IV.C. Evidently, one of the prerogatives of FC relaxometry is to keep the switching times of magnetic fields as short as possible, in principle much shorter than the relaxation rates T_1 of the sample we want to measure. For the sake of completeness, one should mention also the fact that very special situations exist in which extremely high slewing rates may be undesirable. In particular, this regards relaxation time measurements in extremely low fields where the transversal components of Earth and environmental fields may interfere in a manner which depends upon whether the field switching is adiabatic (Larmor periods much shorter than switching times) or non-adiabatic (Larmor periods comparable to switching times). Detailed analysis of such phenomena, however, is beyond the scope of this exposition (*70*).

From the above it follows that the geometry of an FFC electromagnet must be carefully studied taking into account not just the maximum field $B_{\max}$ but also the maximum slewing rate $(dB/dt)_{\max}$ achievable with a given power supply. Optimization of both $B_{\max}$ and $(dB/dt)_{\max}$ leads to the necessity of a compromise, whose resolution defines the final specifications of an FFC magnet (naturally, the compromise must include another aspect of the problem also, such as the maximum electrical power, and all technological limits and manufacturing constraints).

B. Optimal Configuration of an FFC Magnet

A traditional electromagnet with a ferromagnetic yoke would be totally inadequate for FFC experiments since its design aims at the maximization of the field, while no attention is being paid to the slewing rates which turn out to be extremely low. The high inductance values of such magnets are achieved exploiting the very high magnetic permeability μ of ferromagnetic materials. At room temperature, the relative permeability μ/μ_0 is, for example, about 1200 for pure iron, 1300 for cobalt, and 400 for nickel (the values for special alloys can be substantially higher). Since field induction $B = \mu H$ is directly proportional to μ, while the magnetic intensity H is given only by the total energization current, high values of μ make it possible to efficiently minimize the electric power needed to produce a given field. On the other hand, the same fact implies an extremely high magnet inductance L. As a result, the typical inductance values of iron-based electromagnets (1–10 Henry) make them unsuitable for field-cycling experiments. To drive such inductive loads at speeds characteristic of FFC experiments, in fact, the produced over-voltages $V = -L(dI/dt)$ would be enormous, implying extremely costly and complex high-voltage power supplies.

In general, in order to be able to achieve high slewing rates, one has to keep the magnet inductance L reasonably low which, on the other hand, means that the current required to achieve a desired value of B is going to be rather large (hence the necessity to find a suitable compromise).

Despite the above objections to iron-based magnets, a limited use of materials with high magnetic permeability μ might still be advantageous in the design of low-inductance FFC magnets. It might permit, for a given value of L, to construct magnets with large useful volume and/or lower the required power. Unfortunately, the presence of such materials leads to a number of design and employment problems. Ferrites, for example, exhibit a rather large hysteresis which complicates the management of measurement cycles. Metallic ferromagnets, such as soft iron or Alnico, have a rather limited frequency response since their magnetization involves a reorientation (rather than re-magnetization) of elementary domains. In both cases, moreover, the permeability has a strong temperature dependence which constitutes an additional source of magnetic field instabilities.

Several special electromagnet configurations described in the literature (*79–83*) have characteristics which, theoretically, might be compatible with FFC requirements. However, considering the above objections, it is safe to claim that, at present, the configuration most suitable for FFC NMR relaxometry is a cylindrical magnet composed of one or several coaxial air-core solenoids.

Such a configuration permits to reach elevated maximum field values B_{max} with acceptable slewing rates dB/dt. On the other hand, it presents also a number of design problems and some practical disadvantages.

One of the disadvantages concerns the fact that in axial magnets it is rather difficult to use probes with solenoid RF coils. The difficulties are related to sample insertion/removal complications and to numerous spatial constraints, exacerbated by the presence of a glass dewar for sample-temperature control (see Section VI). This is unfortunate because the alternative saddle coils are substantially less efficient, especially at the relatively low excitation/detection frequencies used in FFC NMR.

The magnet design problems are mostly related to the fact that in any real solenoid the field inhomogeneity $\Delta B/B$ at its center is completely insufficient for NMR experiments as, to reach the required homogeneity in a simple solenoid, its length-to-diameter ratio would have to be extremely large. Consequently, in a cylindrical magnet, it is always necessary to adopt special designs which optimize field homogeneity over the whole sample volume at the center of the magnet.

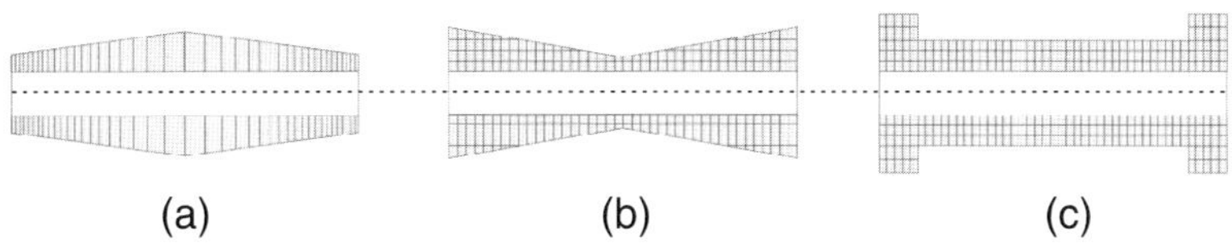

FIG. 3. Modulation of magnetic flux density B along the axis of an solenoid. The figures show three examples of how one can modulate the density of the current I along the axis of an solenoid and thus indirectly modulate the magnetic field B along its axis (a) Graduated current density (b) Linear varied winding density (c) Outer notch.

One of the simplest and most direct approaches is to modulate the current density distribution along the solenoid (*85–97*) (Fig. 3). A suitable longitudinal current density distribution can be obtained by varying the gaps between the loops or by adding special correction coils at the extremes of the solenoid or, as shall be described later, by varying the cross-section of the individual loops. All these methods aim at reducing the length of the solenoid without an excessive degradation of field homogeneity at its center.

C. Electric and Geometric Parameters Determining Maximum B and dB/dt

Consider a cylindrical magnet formed by a simple solenoid with dimensions defined by the inner radius r_0 (half the bore), outer radius r_1, and length $2l$ as the one shown in Fig. 4. We shall assume that the magnet has ohmic resistance R and an inductance L, and when we pass a variable current I through it, it leads to a corresponding variation of the magnetic field induction B at its center. In order to do that, we shall use a power supply with an output voltage V, controlled so as to guarantee the desired current I (we shall see later the conditions under which this can be done).

According to Ampere's law, for any magnet which does not contain ferromagnetic materials, there is a linear dependence between the current I and

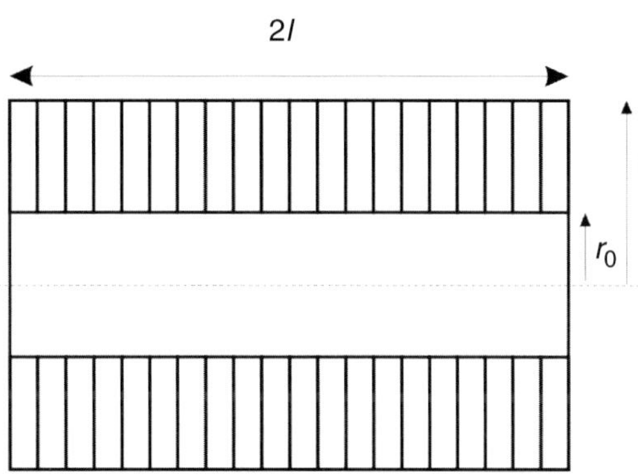

FIG. 4. Cross section of a simple solenoid.

the central field B. In the case of a thin layer solenoid with $r_0 = r_1 = r$, the relationship is:

$$B = \mu H = \mu \frac{n}{l}\left[1 + \left(\frac{r}{l}\right)2\right]^{-1/2} I, \tag{1}$$

where μ is the magnetic permeability of the medium (in our case $\mu_{air} \approx \mu_0 = 4\pi \times 10^{-7}$ Vs/Am), l and n are length and number of turns of the solenoid.

For a real coil of the type shown in Fig. 4, the magnetic field B can also be expressed by the following equation:

$$B = G\sqrt{\frac{Pf}{\rho_0 r_0}} \tag{2}$$

which highlights the relationship between electrical power and, geometric parameters and geometric shape for a cylindrical coil (*71*).

In Eq. (2) P is the electrical power applied to the magnet while G is a coefficient, known as the Fabry factor, which depends exclusively upon the normalized geometry of the magnet (for example, in the case represented by Fig. 4, G depends only on the ratios r_1/r_0 and l/r_0).

For such a cylindrical solenoid, the typical value of the G factor ranges from a minimum of 0.18 (for an uniform distribution of currents I), up to 0.22 depending on the shape of current distribution along the axis of the solenoid. Notice that the non-linear dependence between the field B and the power P required to generate it, in fact, grows with B^2.

The coefficient ρ_0 in Eq.(2) defines the resistivity of the solenoid conductor. It is evident that by decreasing the resistivity, one linearly decreases the power required to generate a given field. This is the quantitative advantage of using metals with the lowest possible resistivity. The solenoid bore r_0 also affects very strongly the generated field inductance B. Though it is not directly evident from Eq. (2), other *relative* dimensions of the magnet (i.e., its form as opposed to its dimension) affect B only through the Fabry factor and that dependence is relatively modest.

The factor f, known as the *packing factor*, expresses the ratio between the total conducting volume and the total magnet volume. It should be kept as high as possible, provided that the magnet design minimizes efficiently the isolating gaps between individual winding loops.

Equations (1) and (2) describe the relation between the current (or power) and the generated field. This direct relationship is inherently independent of any variation of the current. When the current I is made to vary at a rate dI/dt, it induces across the magnet an over-voltage $L(dI/dt)$ which needs to be compensated using part of the applied power supply voltage V. In other words, the voltage V across the magnet is composed of two terms, one used to overcome the ohmic resistance (RI) and one to overcome the magnet inductance ($L = dI/dt$). For future convenience, we express the situation by means of the following linear differential equation:

$$\frac{dB_0}{dt} \approx \frac{dI}{dt} = \frac{V - IR}{L}. \tag{3}$$

One must also bear in mind that the inductance L is, in general, a complex function of the magnet geometry. In the case of a thin-layer solenoid, for example,

$$L = \mu \frac{n^2 \pi r^2}{l} \tag{4}$$

Equations (1)–(4) are used here essentially to illustrate the intricate interdependence between the maximum achievable field, the employed electric power, the maximum slewing rate dB/dt and the geometric parameters of the solenoid. A detailed, quantitative treatment (which must necessarily be carried out when designing an actual magnet) is beyond the scope of this review since, for example, the calculation of G and L for a real magnet is quite complex and requires numerical methods.

Analyzing Equations (2) and (3), one can show that:

- When B and dB/dt are optimized independently, their maxima are not reached with the same design parameters of the magnet.
- The design parameters therefore always represent a compromise between the maximum achievable field and satisfactory slewing rates. The compromise can be resolved once one has defined the maximum available power, the basic geometry of the solenoid (in particular its volume), and any optimization constraints (see the next point).
- The optimization procedure must take into account constraints such as maximum admissible local power dissipation and maximum tolerable magnetic field inhomogeneity. In this way, such constraints influence, often quite seriously, the final performance of the magnet.

D. Example of a Real 1 Tesla FFC Magnet

We have seen that magnets for FFC NMR experiments should be designed in a way to obtain, for an *a priori* given maximum electric power P, the highest possible field B, a satisfactory homogeneity $B/\Delta B$ over a defined volume, and high slewing rates dB/dt (this implies low values of the magnet inductance L), under a number of constraints of which the most important ones are the effective sample volume, an upper limit to maximum local power dissipation (the hot-spot limit) and, last but not least, reasonably easy manufacturing. All these requirements are strongly interrelated and the systematic optimization of coil geometry is a difficult problem.

A numerical optimization procedure which keeps track of all these factors has been published by Schweikert, Krieg and Noack (*67*). The algorithm used in this procedure is almost unique; as far as we know, there is only one alternative published by Kimmich *et al.* (*73*), that achieves a similar degree of completeness.

We have found the Noack–Schweikert approach very advantageous. Once we had mastered the rather complex cutting and assembly technology, we used it with excellent results to develop commercially available magnets for FFC NMR relaxometers with maximum field from 0.5 T up to 1 T (*72*)

The cylindrical configuration of a Noack–Schweikert type foresees the use of several coaxial solenoids connected electrically so that their individual fields are summed together. The winding of each solenoid is realized by cutting a variable-step spiral in a thick-walled metal cylinder (Fig. 5). The slope of the spiral along the magnet length is continuously varied according to a pre-calculated, optimized function. In this way the thickness of the individual winding loops varies in a way which optimizes the current distribution.

The mathematical formalism used to calculate the best current distribution consists of a numerical inversion of the Biot–Savart differential relations between field and current for a given geometric configuration of the conductive loops. At the same time, all imposed constraints are being taken care of by means of the standard Lagrange formalism.

Having fixed the basic geometry and the required power, the algorithm iteratively identifies the current density distribution which maximizes the generated field while maintaining the necessary field homogeneity over a pre-defined volume.

In a refined form, this design method has been used by Stelar to develop and manufacture FFC magnets, commercially available since 1997. In what follows, we shall discuss in detail some of the practical aspects of this design approach when applied to the development of a 1 T magnet optimized for the highest Fabry factor and the highest slewing rates compatible with current technology.

The first task in a Noack–Schweikert magnet design is to define the best initial set of parameters and constraints, such as the number and dimensions of individual solenoids (layers), the maximum electrical power to be used, and the best materials. To define these parameters, one must consider numerous limitations arising from the manufacturing technology as well as the performance characteristics of the cooling system, and of the power

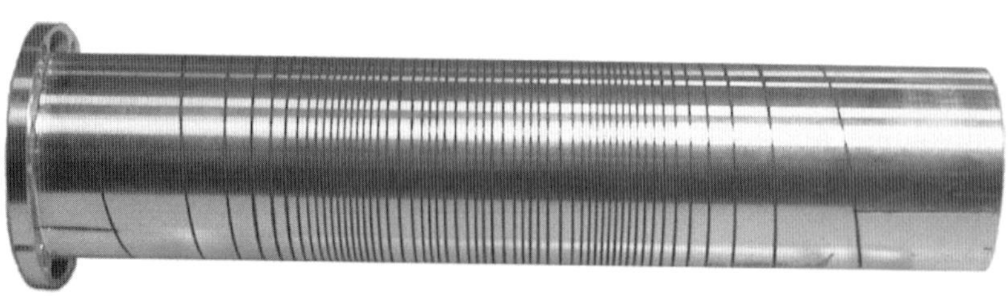

FIG. 5. Example of a solenoid of the Noack–Schweikert type.

supply. In fact, magnet cooling efficiency leads to one of the most stringent limits in using higher electric power. Moreover, because of finite manufacturing tolerances and other technological problems, the magnetic field homogeneity obtained in practice is quite far from the calculated theoretical values.

In principle, one could consider a number of metals and alloys to be used for the construction of the magnet but, considering their physical and electrical characteristics, copper and silver are undoubtedly the best choices. This assertion sounds obvious but the use of other metals with higher resistivity, such as aluminum alloys, is sometimes justified because of their negligible cost and mechanical properties which simplify the manufacturing process. The most important physical characteristics of the best conductors such as OF copper (Oxygen Free) and silver, are shown in Table I.

Apart from its higher costs and difficulties in obtaining suitably dimensioned silver tubes or cylinders from commercial sources, silver is certainly the most favorable choice since it is superior to copper in all comparison parameters. It is also worth mentioning at this point that gold, while better than aluminum, is substantially worse than both copper and silver.

On the basis of Eq. (2), it is evident that the lower silver resistivity proportionally reduces the electrical power required to produce a given field. At the same time, it reduces the time constant R/L of the magnet which is an important factor in minimizing the final field-switching times. Section IV C discusses how the magnet time constant R/L and the power supply output voltage affect the maximum achievable slewing rate dB/dt.

The lower temperature coefficient and the higher thermal conductivity of silver contribute to improve the final field stability of the system. In fact, the geometrical dimensions of the magnet layers change with the temperature of the metal, because of its thermal dilatation, and thus modify its geometry, causing undesired thermal drifts of the field and, to a lesser extent, even some field homogeneity degradation. The variations of the field B with magnet temperature are not negligible and, when not corrected by suitable countermeasures discussed in Sections IV.D and XI.A, can affect the precision of relaxation rate measurements.

Our choice of the metal for the 1 T magnet was silver. Cutting of the magnet layers from silver tubes was done by means of a specially designed, numerically controlled tooling machine (Fig. 6). Traditional tooling machines

TABLE I

MAIN PHYSICAL PARAMETERS OF COPPER AND SILVER IN SI UNITS

	Resistivity ρ	Temperature coefficient α	Thermal conductivity σ	Mechanical cutting
Copper OF	1.78×10^{-8}	0.0068	383	Very difficult
Silver	1.58×10^{-8}	0.0061	419	Difficult

Notes: Resistivity is in [Ω.m], its relative temperature coefficient $\alpha = (d\rho/dT)/\rho$ is in K^{-1} and thermal conductivity σ is in $Wm^{-1}K^{-1}$.

FIG. 6. The magnet layer cutting system.

in fact do not have all the degrees of freedom required to cut a variable-slope spiral in a thick-walled cylindrical tube.

The specifications of the cutting system give the designer complete freedom in defining the magnet geometry. The only restriction (if it can be considered as such) is that the cut thickness cannot be smaller than 0.16 mm. This defines the minimum thickness of isolation between adjacent layers which, in order to minimize the packing factor f in Eq. (2), should be as small as possible.

Table II lists the initial electrical and mechanical constraints used to calculate the magnet. The values were chosen on the basis of not only the target parameters of the magnet but also all of the considerations mentioned above.

TABLE II

STARTING ELECTRICAL AND MECHANICAL CONSTRAINTS

Electrical		Mechanical	
B_{max}	1 Tesla	Number of layers	4
Max. available power	12–14 kW	Magnet length	100 mm
Max. power density	1.5 W/mm^2	Isolation thickness	0.16 mm
Metal resistivity	1.55×10^{-8} Ω.m	Internal radius (layer 1)	11.8 mm

TABLE III

PARAMETERS OF THE CALCULATED MAGNET

B_{max}	1.143 T	Inductance	330 μH
V_{max}	30.1 V	Number of layers	4
I_{max}	400 A	Magnet volume	220 cm^3
Max. power density	1.1 W/mm^2	Internal bore	23.6 mm
Resistance	0.0747 Ω	Magnet length	100 mm

FIG. 7. The four silver layers of a 1 T FFC magnet. The picture shows also a detail of the inner wall of the smallest, most internal layer. The thickness of the cuts is 0.16 mm. At 400 A, the hottest spot power density dissipation reaches 1.1 W/mm^2. One of the most difficult engineering challenges is to cool these hot spots of the innermost layer efficiently enough to extract all the heat. If the cooling is inefficient, the layer can be seriously damaged or even destroyed.

After subjecting these starting data to the above described algorithm, and optimizing the values of B and dB/dt, one obtains the results listed in Table III.

Apart from the calculated magnet parameters, the optimization software generates also a complete data set for cutting each of the four layers. Using these, one produces the real layers shown in Fig. 7.

Subsequently, the package of four layers is mounted inside a special glass and Plexiglas container (Fig. 8) which is designed to make the assembly mechanically rigid and, above all, to admit high throughput for the cooling liquid, which it directs to flow flush with the surfaces of all the magnet layers.

The cooling of a magnet of the Noack–Schweikert type would not represent a problem if the power were dissipated uniformly on all of its surfaces. What considerably complicates the task is the fact that the distribution is very non-uniform and, moreover, the highest power dissipation densities occur on loops with the smallest thickness, and therefore the smallest surface area exposed to the cooling liquid (Fig. 9). In addition, in order not to compromise the packing factor and thus the maximum achievable field, the gap between the layers must be kept as small as possible, leaving just the minimum clearance which can still guarantee a good cooling efficiency.

FIG. 8. The 1 T magnet assembled inside its enclosure.

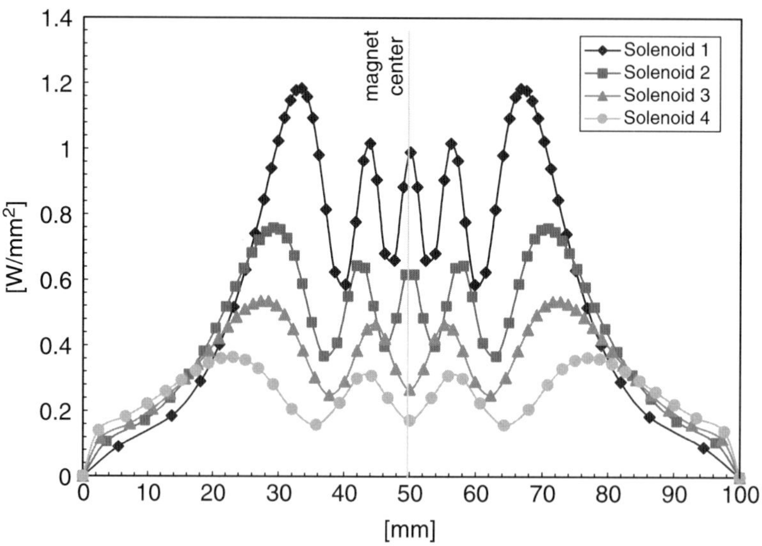

FIG. 9. Distribution of power density along the four solenoids of the magnet.

IV. Magnet Power Supply

A. General Considerations and Peculiarities of an FFC Power Supply

Besides all the requirements dictated by any NMR application, a power supply designed to drive an FFC NMR electromagnet, whose peculiarities have been discussed in Section III. D, should satisfy a number of specific additional features.

The FFC magnet power supply should be designed as a *fast bipolar current source*. In order to match the low impedance of the magnet (fractions of Ohm), the output impedance of the power supply should be extremely low. We will discuss that, in fact, the most convenient way to control the magnetic field B in an air-cored magnet is to control the current I flowing through the magnet to exploit the proportionality between B and I.

When designing a power supply for an FFC electromagnet system, one of the main challenges is to meet the extremely high field-switching slewing rates required in some FFC experiments, especially considering that the shortest measurable T_1 depends on the minimum switching time achievable by the magnet and the power supply combination.

From the electrical point of view, the theoretical minimum switching time is limited by the maximum power supply voltage and by the time constant L/R of the magnet with resistance R and inductance L. In a basic circuit like that shown in Fig. 10 with a fixed power supply voltage V, when the switch trips on, the current evolves according to the equation:

$$I(t) = \frac{V}{R}\left(1 - e^{-\frac{R}{L}t}\right) \tag{5}$$

and the corresponding magnetic field $B(t)$ increases proportionally to $I(t)$.

It is evident that, for a given magnet, it is the maximum power supply *voltage* which determines the maximum field-slewing rates and switching times while, according to Eq. (2), the maximum available *power* determines the maximum achievable field.

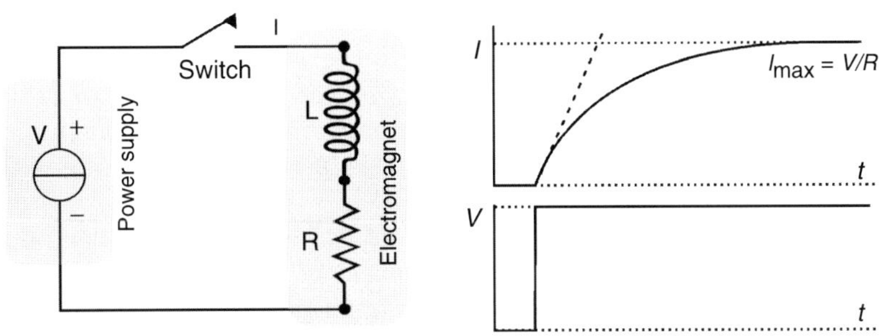

Fig. 10. Evolution of the current in a switched R, L circuit. When the switch is turned On, the voltage (bottom right) across the magnet jumps from 0 to V, while the current (top right) evolves according to Eq. (5). The starting slope of the $I(t)$ curve (the dotted line), corresponding to the maximum field-slewing rate, is given by $V/L = (R/L)I_{max}$.

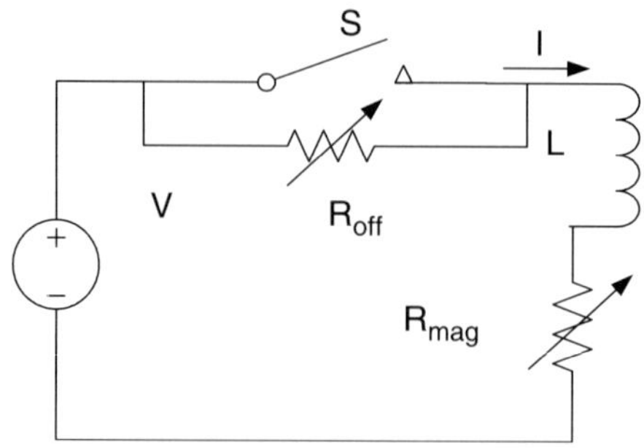

FIG. 11. An early switching FFC circuit (Packard and Varian).

B. HISTORIC SOLUTIONS AND A MODERN APPROACH

In 1954, Packard and Varian (*74*) realized the first simple electronic circuit for switching between two levels of field defined by two values of current, R_{off} and R_{mag}, as shown in Fig. 11.

The electronic switching methods used in the early days of the FFC technique look very simple nowadays. They were severely limited by the lack of suitable electronic power devices which would allow fast high-power switching on inductive loads. This problem was not limited to FFC and was eventually overcome by the development of modern power devices which allowed the introduction of switching methods with considerably better performance.

In 1968, A. Redfield *et al.* (*75*) proposed an FFC power supply in which they introduced an energy storage circuit, whose specific purpose was to overcome the magnet inductance and reach the desired value of B in a shorter time while minimizing the required power. The most obvious method to rapidly switch the current in an inductive load without a substantial increase in the mean power consumption consists in applying a rather high voltage, but doing so only during the relatively brief switching intervals. Redfield's idea actually went a step beyond this simple principle. When a magnet is energized, there is a considerable amount of energy stored in its magnetic field. Upon switching off the magnet, this energy, rather then being wasted, can be used to charge a large storage capacitor to a high voltage. In a subsequent cycle, the energy stored in the capacitor is used to help energize the magnet by inserting the capacitor into the charging circuit and using it as a voltage booster. The result is faster field switching with only a marginal mean power increase.

In subsequent years, the above principle was used and further improved by Koenig and Brown (*58,66,78*), Kimmich *et al.* (*61,76,77*) and Noack *et al.* (*65,75*), leading to the realization of magnet/power supply systems with elevated slewing rates.

At present, modern power components such as GTO (Gated Transistor On/Off device), IGBT (Isolated Gate Bipolar Transistor), Power Mosfet (Metal Oxide Field Effect Transistor), and high voltage capacitors are easily commercially available and perfectly adequate to realize the energy storage switching system. Nevertheless, the realization of a complete power supply based on this method turns out to be costly and problematic when high reliability and routine usage are required. For this reason, starting in late 1990s, Stelar introduced yet another approach to the problem which is equally efficient but easier to implement. This approach, whose characteristics shall be discussed in detail in the next two sections, consists of:

(i) controlling the magnet current rather than voltage and
(ii) during the switching interval, making the current to vary along suitable and well-defined waveforms.

The field-switching therefore becomes a rigidly controlled process rather than an on-off action, thus permitting to minimize field-switching times without having to use excessively high power-supply voltages and powers. The goal is to make sure that the system operates always in the linear region without reaching saturation conditions. The saturation of the power supply occurs when the sum of the induced voltage $L(dI/dt)$ across the magnet terminals and the ohmic voltage RI exceeds the maximum available power supply voltage. In such a case, the power supply control loop exits from the linear region and is no longer capable of operating as desired.

The functional block diagram of an FFC power supply based on these principles is shown in Fig. 12. The *Current Sensor* generates a signal V_{Imag} which is proportional to the current I_{mag} flowing through the magnet. The *Regulation and Control* unit generates a control signal V_{control} for the Power Mosfet Bank which, acting as a "variable resistance", guarantees that the actual current I_{mag} ends up to satisfy the condition $V_{\mathrm{ref}} = V_{\mathrm{Imag}}$ at the input of the control comparator. V_{ref} is a reference voltage whose time dependence is defined by the control software during the measurement sequence

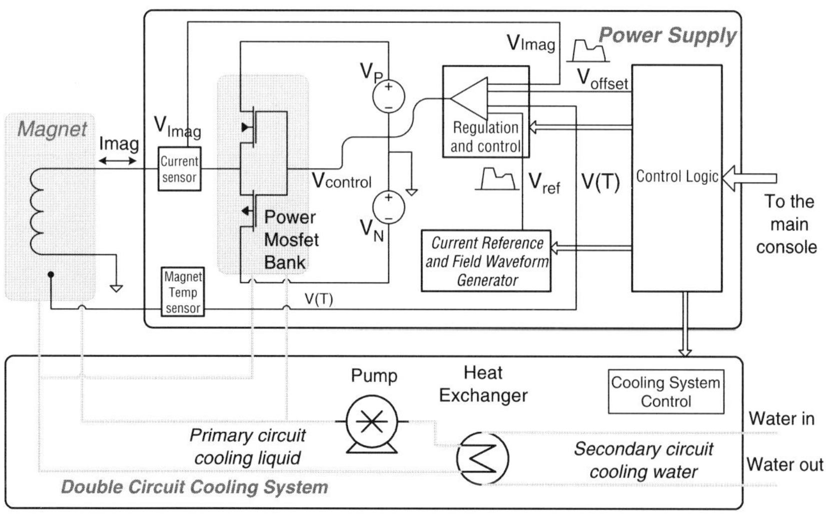

FIG. 12. Block diagram of an FFC magnet power supply. Interconnections with the magnet and the cooling system are also shown. For further details, see the text.

programming. As a final result, the control loop guarantees that, apart from a proportionality constant, the time dependence of the magnetic field B reproduces that of the reference voltage V_{ref}, provided that the desired field-switching waveforms never cause the system to reach the saturation region.

C. Switching Time Considerations and Limits

We have already anticipated that in a power supply of the type represented by Fig. 12, the current $I(t)$ and the field $B(t)$ are linearly related to the reference voltage V_{ref} only when the device operates in the linear region, clear of saturation. The power supply attains the saturation limit when:

$$L[dI(t)/dt] + RI(t) = V_P \quad \text{while switching up and} \tag{6}$$

$$L[dI(t)/dt] + RI(t) = V_N \quad \text{while switching down.} \tag{7}$$

As usual, L and R are the inductance and the resistance of the magnet, while V_P and V_N are the extreme positive and negative voltages available from the positive and negative sections of the power supply, respectively.

Equations (6) and (7) define the conditions at which the current $I(t)$ should be satisfied, in order to achieve the fastest possible transition between any two current values comprised between zero and the maximum admissible magnet current $I_{\max}$. When Eqs. (6) and (7) are satisfied, in fact, the power supply is exactly at the edge of saturation but does not exceed it.

During the switching periods, the maximum current-slewing rate $dI(t)/dt$ and, consequently, also the maximum field-slewing rate $dB(t)/dt$, are limited. For the current-slewing rate, the limits follow directly from Eqs. (6) and (7):

$$dI(t)/dt = [V_P - RI(t)]/L \quad \text{while switching up and} \tag{8}$$

$$dI(t)/dt = [V_N - RI(t)]/L \quad \text{while switching down.} \tag{9}$$

These equations can be easily integrated, giving

$$I(t) = (V_P/R)\{1 - \exp[-(R/L)t]\} \tag{10}$$

when switching from 0 to $I_{\max}$ and

$$I(t) = I_{\max}\exp[-(R/L)t] + (V_N/R)\{1 - \exp[-(R/L)t]\} \tag{11}$$

when switching from $I_{\max}$ to 0.

In order to achieve symmetry between the cases of switching-up and switching-down, one needs to impose the condition

$$-(V_N/R) = (V_P/R) - I_{\max} \tag{12}$$

under which Eq. (11) becomes

$$I(t) = (V_P/R)\exp[-(R/L)t] - [(V_P/R) - I_{\max}] \tag{13}$$

The resulting fastest-switching waveforms, given by Eqs. (10) and (13) and shown in Fig. 13, allow us to characterize the magnet–power supply system from the point of view of switching rates and times. In particular, they lead to the following expression for the minimum switching-up time $mSwt$ for a complete transition from 0 to $I_{\max}$:

$$mSwt = -(L/R) \cdot \ln[(V_P - RI_{\max})/V_P] \tag{14}$$

Thanks to the symmetry condition of Eq. (12), $mSwt$ is equal to the minimum switching-down time for the complete jump from $I_{\max}$ to 0.

More generally, one can ask what are the minimum switching-up and switching-down times SWT_{up} and SWT_{down} for a transition between some arbitrary current levels I_1 and I_2 comprised between 0 and $I_{\max}$. In this case, the resulting equations for the two times are slightly different:

$$SWT_{\mathrm{up}} = (L/R)\{\ln[(V_P - RI_2)/(V_P - RI_1)]\} \tag{15}$$

$$SWT_{\mathrm{down}} = (L/R)\{\ln[(V_P + R(I_1 - I_{\max}))/(VP + R(I_2 - I_{\max}))]\} \tag{16}$$

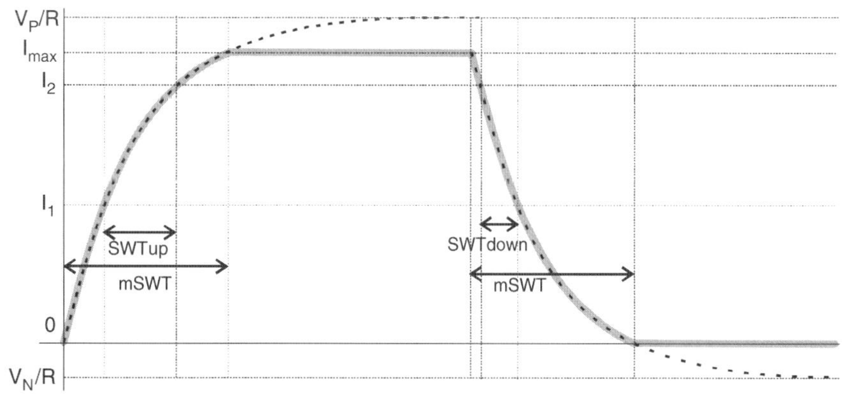

FIG. 13. The fastest current-switching waveform. The diagram shows the magnet current when switching between zero to the maximum admissible current $I_{\max}$. The horizontal line at V_P/R corresponds to the maximum positive current which a power supply with voltage V_P can deliver into a magnet with resistance R. The line V_N/R defines the extreme negative current which can be generated by the negative power supply section with its extreme negative voltage V_N. The thick curve defines the fastest up- and down-going waveforms, clipped by the imposed limits of 0 and $I_{\max}$. The two non-stationary sections have a generic exponential form with a time constant given by L/R, where L and R are the inductance and the resistance of the magnet. This defines the absolute minimum switching time $mSwt$ and the fastest-approach waveform. If one wants to actively control the approach, it is necessary to take into account that, at any given current level I, the slope of the actual $I(t)$ curve may not exceed that of the fastest-approach function.

Equations (*15*) and (*16*) express the shortest duration of the two switching intervals as a function of specific magnet and power supply parameters. The dependence of the SWT_{down} on V_N does not directly transpire from Eq. (16), but the importance of V_N in this context is evident from Eq. (12). We will discuss this point in more detail in the next Section.

D. Why the Bipolar Configuration

As shown in the block diagram of Fig. 12 and implicitly stressed by the discussion in the preceding section, our power supply is of bipolar design, even though the magnitude of its maximum negative voltage $|V_N|$ is substantially smaller than that of its maximum positive voltage V_P.

This makes it possible to program – within limits – the sign of the magnet current and thus the orientation of the magnetic field B along the magnet axis.

In experiments at relaxation fields close to zero, the precision, resolution and stability of the absolute value of B_r become critical and a bipolar configuration; though it makes the design and implementation of the hardware more complex, it improves the precision of the whole system and offers the following advantages:

- Reduced switching-down times, especially at low fields, as discussed in Section IV.C.
- Improved setting precision of the relaxation field for values close to zero.
- Possibility to null the current offset due to electronic control components.
- Possibility to null the axial component of the environmental magnetic field.

In order to better illustrate these points, consider, for example, a magnet such as the one described in Section III.D. It reaches the maximum field of 1.143 T, corresponding to 48.7 MHz of ^{1}H Larmor frequency, with a current of 400 A. It follows that for a very low magnetic field corresponding to, let us say, 1 kHz, one has to set a current of just 8 mA. In order to do that, one should be able to control the current with a precision and resolution of about 20 ppm of the maximum value. The required absolute precision is therefore of the same order of magnitude as the current offsets and thermal drifts of even the best analog electronic components.

Precise null calibration of current offsets (which can be positive as well as negative) is possible thanks to the bipolar configuration of the power supply whose range in the negative section, is much smaller than in the positive section, and can be set with a considerably higher resolution. This increases the absolute precision of the field values for very small relaxation field settings and, in practice, makes it possible to extend the NMRD profiles to fields as low as a few tens of μT.

In a laboratory environment, a measured sample is subject not just to the field generated by the magnet but also – unless one implements particular screening or compensation devices – to all kinds of dispersed environmental magnetic fields due to, for example, the Earth's magnetic field, electric power lines, other magnets, etc.

In a traditional magnet for NMR spectroscopy, the field B_0 of the magnet is much higher than the field components originating from outside sources. Moreover, devices such as efficient NMR field stabilizers are used to suppress all interfering external fields. Consequently, the presence of such field components can be usually ignored. On the contrary, during a FFC NMR measurement the sample may be subject to very low fields (ideally down to zero) which is practically impossible when the relaxation field value becomes comparable to the environmental fields. The amplitude of such fields, if not compensated, represents the lower relaxation field limit for a reliable NMRD profile.

Thanks to the bipolar configuration of the power supply, at least the static axial component ΔB_{env} of the environmental magnetic field can be easily compensated by means of a small current offset ΔI generating an axial field offset ΔB of the same value as ΔB_{env}, but of opposite sign.

For a complete suppression of the local magnetic field effects, however, one should also compensate its transversal component perpendicular to the magnet axis. This requires a pair of auxiliary coils, oriented in a correct direction in the magnet's azimuth plane, and capable of generating a transversal field component of the correct magnitude and sign (for more details, see Ref. (*70*)).

E. Power Regulation Banks

The power components used to drive the magnet current are an important part of the FFC power supply and can critically affect its performance. All together, these components constitute the power unit indicated in Fig. 12 as the Power Mosfet Bank.

The power bank contains an adequate number (generally several tens) of linear solid-state power devices, such as transistors or power Mosfets. Because of the availability of modern power Mosfets that combine high performance with easy control, classical transistors are nowadays no longer used in applications of this type (in our case, we use channel-N and channel-P Mosfets for the positive and negative power supply sections, respectively).

In order to control the large required currents, all the power devices are electrically connected into a parallel array and, in order to dissipate the large amounts of heat generated during particular phases of the FFC experiments, the power array is actively cooled.

Figure 16 in Section V shows the power dissipated by the magnet and the power supply as a function of the magnet current. The total dissipated power, as well as its distribution between the magnet and the power supply, depend upon the working point which varies, often quite sharply, during an FFC cycle. The dissipated electric power is converted to heat which must be eliminated by means of a suitable cooling system. The power dissipated on a single Mosfet device can easily reach peak values of the order of 200–300 W, while the peak power dissipated by the magnet can be as high as 10–12 kW.

This implies the use of an efficient cooling system employing, as we shall see later, a forced circulation of cooling liquid.

The positive section of the power banks shown in Fig. 14 uses 40 type-N power Mosfet devices which can drive currents of up to 400 A. The same banks mount also 4 type-P devices for the negative section. The number of devices in the negative section is much smaller since the negative side is subject to much smaller power requirements. All 44 devices are mounted, together with their electronic control boards, on four special liquid-cooled, copper heat sinks. These, thanks to the excellent thermal conductivity of copper, combined with a design which maximizes the contact area between the copper and the cooling liquid, makes it possible to reach the requested cooling efficiency.

From the electric point of view, each Mosfet device is equipped with its own driver circuit which, in a case of failure, disconnects it from the parallel array. This is very important since it permits the system to operate properly even when a few power components have burned out. It is also extremely

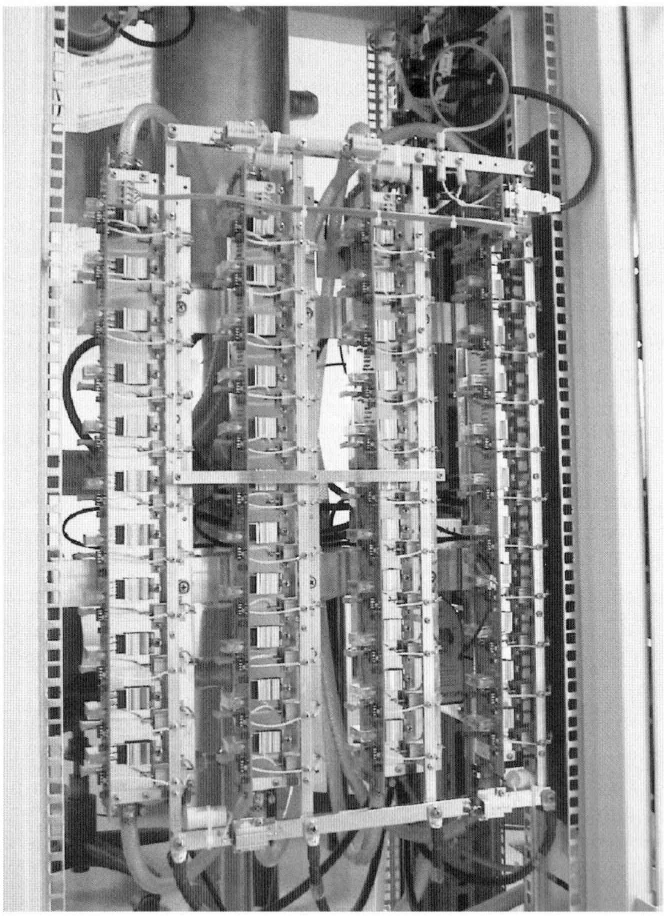

FIG. 14. The power mosfet bank.

useful since it incorporates a possibility to easily locate any of the broken components.

F. Active Compensation of the Temperature Dependence of B

We have mentioned earlier that, as shown in Fig. 15a, small but perceptible magnetic field variations occur when the magnet temperature changes. The magnet temperature variation ΔT is approximately proportional to the power dissipated by the magnet and therefore to the square of the magnet current I. The variation of ΔB is a function of ΔT which reflects the changes in magnet geometry due to thermal dilatations. The net result is

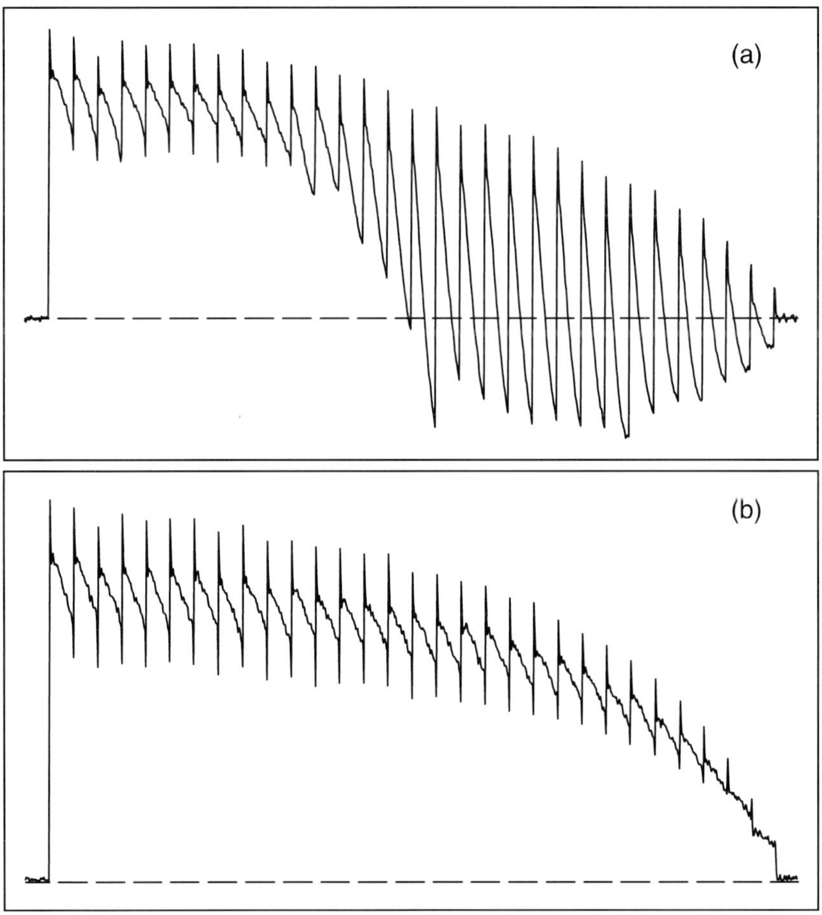

Fig. 15. Data acquired in a multi-block NP experiment (32 τ-values). (a) Experiment done with the thermal compensation turned Off. The FID offset is clearly not the same for all blocks, revealing magnetic field displacements due to magnet temperature fluctuations. The magnet temperature varies because the power dissipated on it in each block depends on the varying τ value. (b) The same experiment with the thermal compensation turned On.

not easy to determine since it depends on many factors such as the magnet's geometry and its mechanical rigidity, the thermal properties of the used materials, the efficiency of the cooling system, etc. In practice, it is best to consider the $\Delta B(\Delta T)$ dependence as an empirical characteristic of the particular magnet.

The knowledge of the relation $\Delta B(\Delta T)$, combined with the actual magnet temperature T_{mag}, makes it possible to correct, at least in part, the magnetic field variations due to the varying magnet temperature. The magnet temperature sensor shown in Fig. 12 is used primarily for this purpose. Its output, after a transformation based on the empirical knowledge of the function $\Delta B(\Delta T)$ and on the known $B(I)$ dependence, is fed-back to the control circuit of the power supply. Eventually, it generates a magnetic field variation equal to $\Delta B(\Delta T)$ but of opposite sign, and thus compensates the original field-drift effect.

V. Cooling System

The cooling of the magnet and its power supply constitutes in practice one of the major design challenges of an FFC relaxometer. The fact that one uses a resistive magnet and a power supply operating in the linear region implies a large power dissipation in at least some operating situations(Fig. 16). The generated heat must be removed from the system by means of a suitable cooling system, like the one shown in Fig. 17.

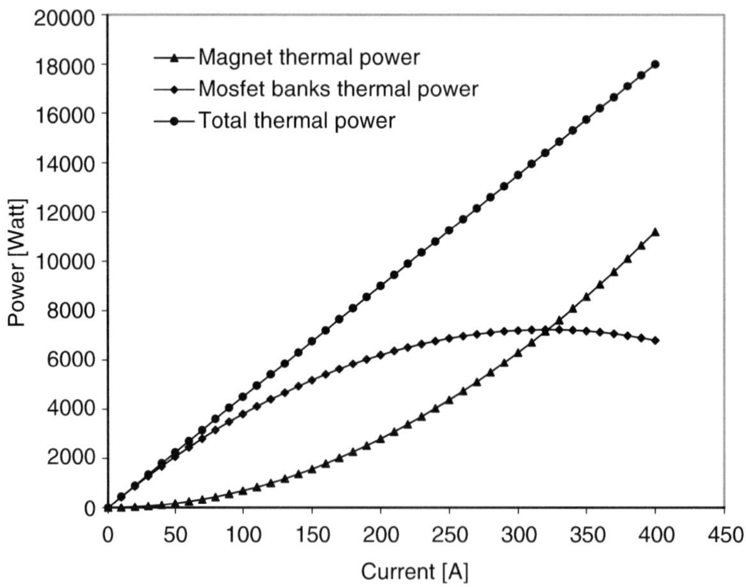

FIG. 16. Distribution of dissipated electrical power between the magnet and the Power Mosfet bank.

FIG. 17. Cooling system of a 1 Tesla FFC NMR relaxometer.

As shown in Fig. 12, the cooling system contains two thermally interconnected circuits.

The primary circuit is closed and filled with a special cooling fluid which, driven by a high-capacity pump, cools the magnet and the Power Mosfet banks. The cooling fluid is a commercially available product (Galden^{TM}) whose physical characteristics had been optimized for cooling electric devices wherever it is not possible to use water. It is chemically almost inert, absolutely non-toxic, and electrically non-conductive.

At first sight, it might appear preferable to use water as a cooling fluid since it has better thermal properties and lower viscosity. In practice, however, water had to be ruled out because of troublesome water electrolysis phenomena which can occur between windings of the solenoids when a voltage is applied accros the magnet.

The secondary circuit uses water from any water distribution network or, when possible, from a centralized cooled-water system. The heat dissipated from the magnet and from the power supply into the primary cooling circuit is transferred to the secondary circuit by means of a high-performance counter-current heat exchanger.

VI. Signal Detection Probe

As far as technological principles are concerned, an FFC NMR detection probe does not differ from the probes of any other NMR instrument. Nevertheless, there are a few design characteristics of an FFC probe which are dictated by the type of the magnet used and its geometrical constraints. Considering that and sensitivity permitting, one would like to measure NMRD profiles of any nuclide. The RF frequencies used on an instrument with a 1 T magnet like the one described in Section III.D range from about 3 MHz to over 40 MHz. One can hardly achieve a complete and efficient coverage of such a wide range with a single probe. In general, two or more tunable, broad-band probes are needed in order to be able to cover nuclides with low γ ratios such as deuterium, as well as high γ nuclides such as protons.

Considering the axial geometry of the FFC magnets, a number of practical problems makes it almost unavoidable to use a saddle-shaped detection coil with cylindrical Helmholtz geometry, even though it is well-known that such a configuration is less efficient than a plain solenoid. To some extent, the situation resembles the one encountered in NMR spectroscopy when using axial superconducting magnets. In our case, however, the problem is further exacerbated by the fact that saddle coils are very difficult to optimize at the relatively low RF frequencies characteristic of FFC NMR relaxometry.

An alternative to a saddle coil would be a solenoid coil which, however, would have to be oriented perpendicularly to the magnet bore and thus to the physical axis of the probe assembly. Due to spatial constraints, such an arrangement complicates considerably sample insertion, especially when the sample temperature has to be controlled and the assembly has to include an enveloping dewar for temperature control of the sample .

In our final realization (Fig. 18), the probes use the Helmholtz coil geometry, favoring ease of use and efficient sample temperature control over a wide range of temperature values. The tunable, broad-band probe is inserted into the magnet from below and fixed to the bottom part of the magnet assembly in a simple way reminiscent of most high-resolution NMR systems. Thanks to this design, it is possible to use standard 10 mm NMR sample tubes which are inserted comfortably from above without any need to manipulate the probe.

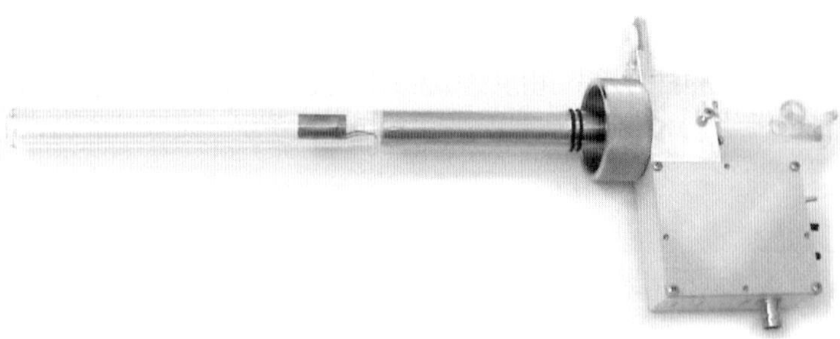

FIG. 18. A FFC signal detection probe.

VII. Control Console

By the term console we intend all the electronic functional blocks used in a traditional NMR spectrometer, excluding the field control. In particular, it includes the following subsystems:

- RF generation and gating
- Transmitter RF power booster
- RF receiver (gated quadrature phase detector)
- Low frequency signal handling (filters, amplifiers)
- Timing devices, including a pulser/sequencer unit
- Analog-digital converters
- Local CPU and firmware for Host-independent data acquisition management
- Sample temperature controller (VTC)
- Hardware interfaces
- Host data system
- Software

Actually, a console to be used with an FFC NMR relaxometer does not differ much from any conventional general-purpose NMR console. With the exception of the relatively simple interfaces controlling the magnet power supply and thus the field, all other hardware units are mostly the same as in any sufficiently versatile NMR spectrometer or relaxometer (what does differ quite a lot, of course, is the application software).

For this reason, a detailed description of the console in this paper is superfluous, except for a brief list of those features which, in our opinion, any research-grade FFC console should possess in order to guarantee maximum versatility of NMR dispersion measurements.

Firstly, all the RF units are broad-band and the console must be able to operate in the full range of frequencies compatible with the maximum magnet field $B_{\max}$. The first advantage which follows from such a choice is the possibility to observe, under optimal conditions, not only protons but also other interesting nuclides with smaller gyromagnetic ratios γ, such as ^{2}H, ^{31}P, ^{23}Na, ^{19}F, etc. In principle, a console operating at a fixed-frequency ω would permit the observation of all nuclides for which $\omega \leq \gamma B_{\max}$. However, in order to be able to measure nuclides with low γ values, one would have to choose quite low ω and thus penalize heavily the achievable signal-to-noise ratio (S/N) for nuclides with high γ values.

In order to understand this, consider that in an FFC experiment the amplitudes of the acquired signals are approximately proportional to the signal acquisition field B_a. For example, in the case of the basic pre-polarized sequence (described in Section VIII.C.), one can show (*65*) that

- When the relaxation period τ is zero, then $S \approx B_p B_a$, where B_p and B_a are respectively the polarization field and the acquisition field.
- When τ is much larger than T_1, $S \approx B_r B_a$, where B_r is the relaxation field.

While the first factor in these expressions varies, the proportionality with respect to the acquisition field is always present. In order to maximize S/N, it is therefore advantageous to use always the largest possible acquisition field B_a and make the acquisition frequency match the corresponding Larmor frequency γB_a of the measured nuclide. This, however, can be done only when using a broad-band console.

From the previous paragraph it is apparent that, ideally, one should keep $B_a = B_{\max}$. This is certainly true for slow-relaxing samples where the field-switching periods practically do not affect the measurements. In fast relaxing samples, however, one has to take into account the deleterious effects of the switching intervals when using very large field jumps. This matter shall be discussed in detail in Sections IX.B and E. What is relevant at this point is only the fact that, in order to maximize the measurement precision, keeping $B_a = B_{\max}$ is *not* the best choice for samples whose relaxation rates are comparable to the magnet switching time. It follows that in such cases, the versatility offered by a broad-band console comes handy.

Another important console feature regards the receiver detection method. This topic will be discussed from another point of view and in more detail in Section X. It is useful to mention, however, that the *only* type of receiver which offers *complete* information about the detected RF signal consists in a dual-channel quadrature phase detector. All other detector types (diode, envelope, power, etc.) provide only a *partial* description of the signal which may occasionally mask some instrumental artifacts and thus look inviting, but which can never really replace a complete quadrature detector. The signals acquired by a quadrature detector can be used to recover any conceivable signal feature, while this is not true for the signals generated by other types of detectors. In particular, the quadrature signals allows one to calculate the signal modulus and thus make data evaluation insensitive to frequency offsets and receiver phase settings while, at the same time, maintaining the possibility to estimate and automatically correct such parameters. A dual-channel quadrature phase detector is therefore a must.

Another important feature regarding the whole receiver chain (including the probe and the preamplifier) is the total dead time which must be kept as short as possible in order to allow measurements of NMRD profiles of solids with extremely fast decaying FIDs.

As mentioned above, the parts which truly distinguish an FFC console from the consoles of traditional NMR instruments are the hardware and software interfaces through which one controls the magnetic field (Fig. 19). They must permit the operator to not only set the different requested field levels (Off, Polarization, Relaxation, and Acquisition) but also to control the field switching from one level to another according to a precise time sequence which, moreover, must be rigorously synchronized with the concurrent RF pulse sequence. In addition, it must be possible to control the magnet slewing rate during the switching interval and thus, the optimal switching waveform according to the principles discussed in Section IV.C.

Finally, we should mention the sample temperature control. It is a direct consequence of all relaxation theories that, in any FFC NMRD application,

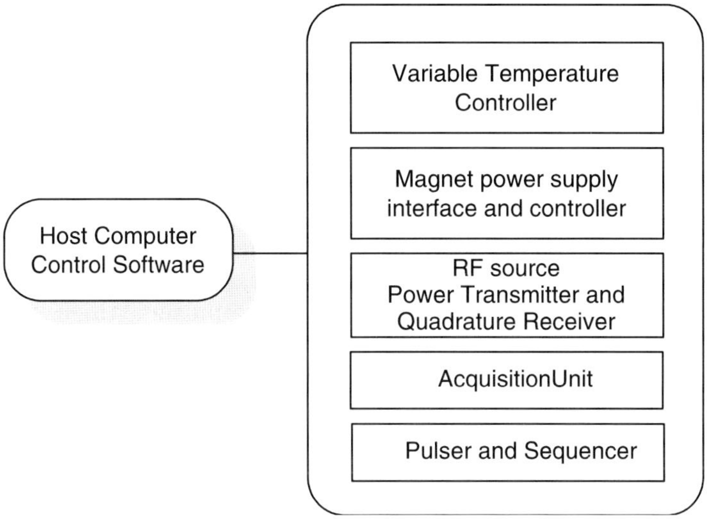

FIG. 19. The main functional blocks of a NMR Fast Field Cycling Console.

sample temperature is just as important as the dispersion curves themselves. In a sense, the temperature activates molecular motions while the field, by determining the Larmor frequency of the nuclide, defines a frequency "window" for their observation. There is no way how the two effects could be separated or even just considered independent of each other. A precise sample temperature controller usable over a large range of temperatures is therefore to be considered not an accessory but an essential part of a FFC relaxometer.

VIII. FFC Data Acquisition Sequences

A. Special FFC Features of Elementary Sequence Intervals

Users of any NMR instrument are well aware of the extensive employment of what is known as *pulse sequences*. The roots of the term go back to the early days of pulsed NMR when multiple, precisely spaced RF excitation pulses had been invented (*17,98–110*) and employed to overcome instrumental imperfections such as magnetic field inhomogeneity (*Hahn echo*) or receiver dead time (*solid echo*), monitor relaxation phenomena (*saturation-recovery, inversion recovery, CPMG*), excite and/or isolate specific components of NMR signals (*stimulated echo, quadrupole echo*), etc. Later on, employment of pulse sequences of increasing complexity, combined with the so-called *phase-cycling* technique, has revolutionized FT-NMR spectroscopy, a field where hundreds of useful excitation and detection sequences (*111,112*) are at present routinely used to acquire qualitatively distinct 1D, 2D, and 3D NMR spectra of chemical systems. A similar evolution occurred in the field of MRI (Magnetic Resonance Imaging) where specific pulse sequences (*113–116*) are used to enhance or modify the technique's spatial sensitivity and produce qualitatively distinct images of the investigated objects. It is therefore hardly surprising that pulse sequences play a crucial role in FFC relaxometry also.

Generation of precisely timed sequences of events requires the presence of a special device, known as the *pulser* (though *sequencer* would be a more fitting term). Considering that the development of new NMR techniques almost always involves new pulse sequences, research-grade instruments are necessarily equipped with general purpose pulsers which, using a suitable software, can be programmed by the operator to generate *any* desired pulse sequence. The pulser mounted on Stelar FFC-Spinmasters, for example, features twelve output *channels* which can be operated in a strictly synchronized manner to generate 12-channel control sequences of virtually any duration with absolute time resolution of 100 ns and a timing jitter lower than 1 ns.

During the evolution of NMR techniques, the terms *RF pulse sequence* and *RF phase cycling* have undergone a considerable modification in meaning. Originally, they were used to indicate just the gating and phase-switching timing of the principal RF transmitter. In NMR spectroscopy, this was soon complemented by the control of a second RF channel, the decoupler, leading to various *gated decoupling* sequences in which two physically distinct RF devices are operated in a synchronized manner. At present, pulse sequences also need to be synchronized with devices other than the RF generators. Consider, for example, the UV flashes in CIDNP (Chemically Induced Dynamic Nuclear Polarization) investigations, field-gradient pulses in PFGSE (Pulsed Field-Gradient Spin Echo) self-diffusion experiments (*117*), homogeneity spoiling pulses (*106*) used in high-resolution relaxometry and, last but not least, the rigorously synchronized magnetic field gradient pulses which are an essential part of all MRI techniques (*113–116*).

One thus arrives at the concept of *pulser sequences* (rather than *RF pulse sequence*) in which the programmable pulser generates a *sequence of synchronized events* and controls a number of distinct devices. Analogously, the term *RF phase-cycling* is no longer appropriate (in Stelar terminology, it is replaced by *X-device cycling*).

In FFC relaxometry, the most conspicuous pulser-controlled device (apart from the RF excitation channel) is the magnet system. In other words, we generate B_0 field pulses of considerable amplitude, often switching the magnet field between zero and a maximum value of over 1 T, and we rigorously synchronize such B_0 pulses with the RF signal-excitation and/or preparation pulses. This, moreover, does not exclude the possibility to control other devices as well.

Like any sequence of events, an FFC experiment can be intended as a sequence of elementary intervals during each of which all system *control lines* maintain constant values. One needs to keep in mind, however, that while a control line transition is always very fast (settling times of the order of 1ns), the controlled device/parameter may require a much longer time to complete the requested state transition. In particular, for technical reasons explained in Section IV.C., the main magnetic field intensity cannot change instantaneously and requires switching times of the order of a few ms.

In order to control the impact of the field-transient periods on the measured data, it is necessary to devise FFC sequences in such a way that each elementary interval falls into one of two possible categories:

- *fixed-field intervals*, in which B_0 is constant within the desired NMR precision, and
- *switching intervals*, during which the field varies in a controlled manner between a pre-programmed starting value and a pre-programmed final value.

In general, an FFC relaxation rate measurement requires a series of elementary experiments in which the duration τ of just one of the fixed-field intervals varies, while each of the switching intervals has always the same duration. Only in this way can one guarantee that the measured relaxation rate is correct and that it corresponds to the *relaxation field* present during the variable-duration interval (to be discussed later).

The performance of the field-switching circuitry (power supply and magnet dynamics) affects the minimum duration of the switching intervals. Figure 20 illustrates the principal characteristics of the field-switching

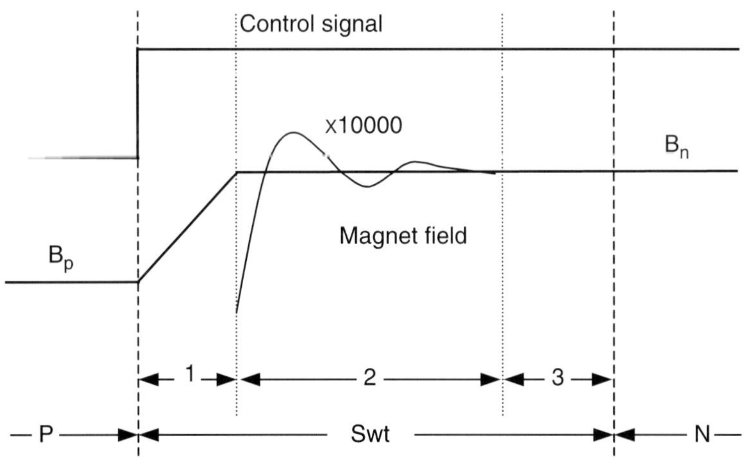

FIG. 20. Schematic illustration of a main-field switching interval. The single pulser interval *Swt* (switching time), nested between a previous interval *P* and a next interval *N*, is actually composed of three distinct phases (1,2,3). During phase (1) the magnetic field *B* is actively driven along a linear ramp from its previous value B_p to a close vicinity of its next value B_n. During the subsequent phase (2) it settles to its final value. The settling waveform is schematically represented by the thin line using a considerably expanded-scale. The switching margin phase (3) is a filler which simply completes the *Swt* interval. The operator controls the total duration of *Swt* and, within hardware-defined limits, the slope of the linear ramp but has no direct control of the settling waveform. One must make sure that the phases (1) and (2) fit within the *Swt*.

waveform. Essentially, even though we talk about a single interval, it is composed of three physically distinct phases:

(1) The *slewing phase* where the field is driven, following a linear ramp, from a starting value to a close vicinity of the desired final value. Following the arguments introduced in Section IV.C, one can show that, when a linear switching waveform is used, the maximum achievable slewing rate is given by

$$s_u \equiv dB/dt = (\kappa/L)V_p - (R/\kappa L)B = S_u - \alpha B \tag{17}$$

when switching up ($s_u > 0$), and

$$s_d \equiv dB/dt = (\kappa/L)V_N - (R/\kappa L)B = S_d - \alpha B \tag{18}$$

when switching down ($s_d < 0$).

Here κ is the proportionality constant between field and current ($B = \kappa I$), V_P and V_N are the maximum positive and maximum negative power supply voltage, L and R are the inductance and resistance of the magnet, B is the instantaneous main field value, $S_u = (\kappa/L)V_P$ and $S_d = (\kappa/L)V_N$ are the up-going and down-going slewing rates when $B = 0$, and $\alpha = R/\kappa L$.

Consequently, when switching from a field value B_a to a field value B_b, the minimum duration t_s of the linear switching ramp is

$$t_s = (B_b - B_a)/[S_u - \alpha B_b] \tag{19}$$

when switching up, and

$$t_s = (B_b - B_a)/[S_d - \alpha B_b] \tag{20}$$

when switching down.

Using reasonably dimensioned power supplies, the minimum achievable t_s values are typically of the order of 1 ms.

Notice the presence of the field-dependent terms in the above formulae. They indicate, for example, that when the target field value is very large, the fastest up-going slewing rate drops down. Similarly, when the target field value is very low, there is a deterioration in the fastest down-going slewing rate. Among other things, this implies that when maximum system performance is required, the operator should keep in mind the dependence between the target field values to be used in the experiment and the maximum achievable slewing rates.

(2) The *settling phase* where the field is already very close to its final value (with residual differences below 100 kHz of ^{1}H Larmor frequency), and later settles down to its final value whose required reproducibility and stability is of the order of 100 Hz. Since the linearly driven ramp is no

longer active, the dynamic characteristics of settling waveforms are different from those of switching waveforms. In general, there is a less conspicuous dependence on the field value (in some cases, the fine settling may actually take longer when the final field value is low). As far as the operator is concerned, it is usually sufficient to consider the settling phase as having an approximately constant duration.

It is actually not always necessary to wait until the field has completely settled. There are only two situations where such a precision is really required:

(a) When switching to the *acquisition field* at which the NMR signal is collected and
(b) When switching to a low field value (below 100 kHz) where the settling waveform amplitudes represent an unacceptable field-setting error.
(3) A *switching margin* which is treated as a part of the switching interval rather than a part of the subsequent fixed-field interval, even though the field is already stable. The usefulness of such switching margins stems from practical considerations. For example, they represent an important simplification for the operator. Since the duration of the switching margin does not affect the measured relaxation rates, the operator can often *uniformly specify the same duration for all switching times* appearing in a pulser sequence, paying attention only to the requirement that the combined phases (1) and (2) must always fit within the switching time setting. The fact that such an approach leads, for some of the switching intervals, to somewhat excessive switching margins is often of little importance. Counter indications to this approach regard only very fast relaxing samples where even a very small switching margin affects the precision of the measurements (we shall return to this point later).

B. Basic Structure of any FFC Sequence

Since all FFC experiments (except for a few set-up and/or diagnostic sequences) are a special variety of relaxation measurements, they are necessarily composed of three chronologically ordered sections or sub-sequences:

(1) *Preparatory section sub-sequence*, during which the sample is subject to a sequence of field and/or RF pulses inducing therein a specific nuclear magnetization M_0. In general, M_0 can be characterized as a zero-quantum coherence state which, in most cases means simple I_z magnetization aligned along the direction of the main magnetic field. In some types of experiments, the starting M_0 state may be simply a zero magnetization.
(2) *Relaxation sub-sequence section*, during which the sample is kept for a time τ at a constant *relaxation field* B_r. The magnetization then evolves towards an equilibrium value $M(B_r)$, following a relaxation evolution curve $M(B_r,\tau)$, such that $M(B_r,0) = M_0$ and $M(B_r,\infty) = M(B_r)$. The primary goal of an FFC relaxometer is the acquisition of the

$M(B_r, \tau)$ curves for a number of B_r values spread over many orders of magnitudes.

(3) *Detection sub-sequence section*, during which the sample magnetization is sampled by means of a more or less complex signal detection RF sequence. This occurs always at the signal *acquisition field* B_a determined by the current operating frequency (probe tuning etc.) and the gyromagnetic ratio of the measured nucleus.

Using various detection sub-sequences, it is possible to select distinct nuclear magnetization components and thus discriminate between various sample components and/or distinct relaxation mechanisms. In combination with various possibilities of initial state preparation, this represents a powerful NMR relaxometry tool which, at present, is far from being completely exploited.

C. Elementary FFC Sequences

As a starting example, let us now very briefly discuss the two simplest, text-book FFC experiments embodied in the PP (pre-polarized) and the NP (non-polarized) sequences.

The **non-polarized sequence** (**NP**, Fig. 21) is suitable for measurements of T_1 relaxation curves at relatively high relaxation fields (typically above a few MHz). The starting longitudinal magnetization M_0 is in this case zero and is prepared by letting the sample relax in a zero field for a time RD (recycle delay) which starts the sequence. The magnet is then switched to the desired relaxation field B_r where, after the rigorously constant switching period SWT_1 is over, it is kept for a variable time τ during which the sample magnetization grows towards the equilibrium value M_r corresponding to B_r. The magnet is then switched again, this time to the acquisition field value B_a and, once the constant switching period SWT_2 is over, a 90° RF pulse is applied and the resulting FID is acquired. Immediately after, the field is switched off and the whole sequence is repeated (notice that the duration of the last switching interval is completely irrelevant and it does not need to be explicitly programmed as a specific pulser interval).

The **pre-polarized sequence** (**PP**, Fig. 21) is suitable for T_1 relaxation measurements at low fields, theoretically down to zero value (in the present context we may neglect the phenomena (*70*) which make life a bit complicated at very low fields). The starting nuclear magnetization M_0 is in this case prepared by applying a strong *polarization field* B_p for a sufficiently long *polarization time* T_p. The result is a strong magnetization M_0 aligned with the field B_p. The magnet is then switched to the desired relaxation field B_r where, after the constant switching period SWT_1 is over, it is kept for a variable time τ during which the magnetization decays from M_0 towards M_r. Actually, starting from the end of the preparatory *sub-sequence* section, the two sequences are perfectly identical so that we can skip the details of switching to the acquisition field (SWT_2) as well as the description of signal acquisition.

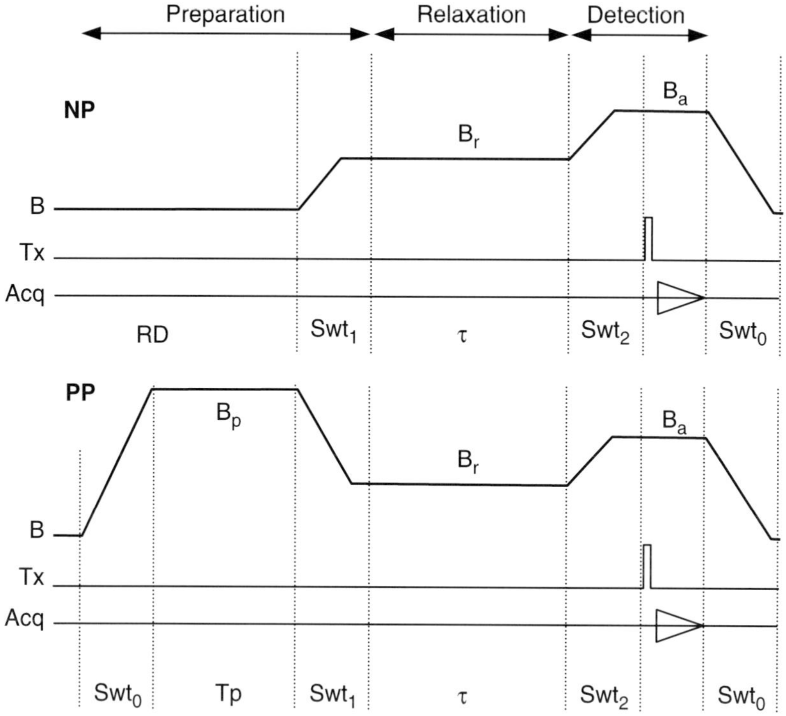

FIG. 21. The basic non-polarized (NP) and pre-polarized (PP) sequences. The thick line illustrates the behavior of the main magnetic field. The switching between the four preset field values (off, polarization B_p, relaxation B_r, and acquisition B_a) is achieved by means of two control lines driven by the pulser. The pulser control lines for the RF transmitter gate (*Tx*) and data acquisition trigger (*Acq*) are also shown. Notice that from the end of the polarization sequence on, the two sequences are identical what differs is only the sample magnetization at the beginning of the relaxation period. The various switching intervals Swt_i can be individually optimized, though this is rarely done. The start-up switching period Swt_0 and the final switching-off period Swt_0 are normally not even programmed as explicit parts of the sequence (they have no effect on the acquired data) and the switching periods Swt_1 and Swt_2 are usually set equal to a common value *Swt*. The slopes of the linear switching ramps are all identical and their value can be preset by the operator prior to executing the sequence. For further details, see the text.

IX. Acquisition and Evaluation of Complete Relaxation Curves

A. Arrayed T_1 Measurements

The acquisition of a complete relaxation curve $\mathbf{M}(\tau)$ consists in repeating an elementary NP or PP experiment while stepping τ through an array of values distributed according to a pre-defined *strategy* (linear, logarithmic, multi-range, etc., depending upon sample relaxation characteristics). All other instrument and sequence settings are kept rigorously constant so that any observed signal variation is due exclusively to the varying τ.

As shown in Fig. 22, the resulting procedure, referred to as a multi-block experiment, produces a two-dimensional data set, such as an array of FIDs (its exact nature depends upon the signal acquisition method). The data of each τ-block are then reduced to a single quantity, $S(\tau)$ which should be proportional either to the total sample magnetization $M_a(\tau)$ or to one of its components. Since the vertical scale of the relaxation curve is irrelevant, we can identify $S(\tau)$ with $M_a(\tau)$ at the exact time of detection (usually just after the first excitation pulse).

Regardless of what sequence one is using, reproducibility requires that the starting magnetization M_0 must be rigorously the same for all the τ-blocks, unaffected by whatever happened before each block started. One way to guarantee this is by imposing the following conditions:

(1) For the NP sequence, we need $M_0 = 0$ which means that we must keep

$$RD \geq f \times T_{1\max}(0), \tag{21}$$

where RD is the recycle delay between the individual τ-blocks where the field is off and $T_{1\max}(0)$ is the relaxation time of the sample at zero field and f is a factor to be discussed.

(2) For the PP sequence, we need $M_0 = M_{\rm p}$ (equilibrium magnetization at the field B_p) so that we must keep

$$T_p \geq f \times T_{1\max}(B_p), \tag{22}$$

where T_p is the duration of the polarization interval during which the sample is held at the polarization field B_p. In the PP-type sequences, the recycle delay RD is almost totally irrelevant and can be kept at 0 (magnet heating permitting).

In both cases, f is a numeric factor and $T_{1\max}(B)$ denotes the estimated relaxation time of the slowest-relaxing component of the sample magnetization in field B. One usually sets $f = 4$ which, assuming exponential curves, guarantees a relative precision of at least e^{-4} (about 1.8%).

An important feature of the above equations is the fact that in the NP sequence we refer to relaxation times at zero-field while in the PP sequence they are referred to as the polarization field B_p. This does often make a big difference since, in many samples, $T_{1\max}(0)$ may be much shorter than $T_{1\max}(B_p)$.

Notice also that proper setting of the parameters RD and T_p requires an approximate knowledge of relaxation times, which are in their own term, the objects of the investigation. This circular problem makes it often necessary to carry out a few preliminary experiments before starting an actual relaxation curve acquisition. Fortunately, the values of RD and T_p do not need to be precise (a 20% tolerance is quite reasonable) so that a simple recurrent estimation process converges to acceptable values in just one or two cycles.

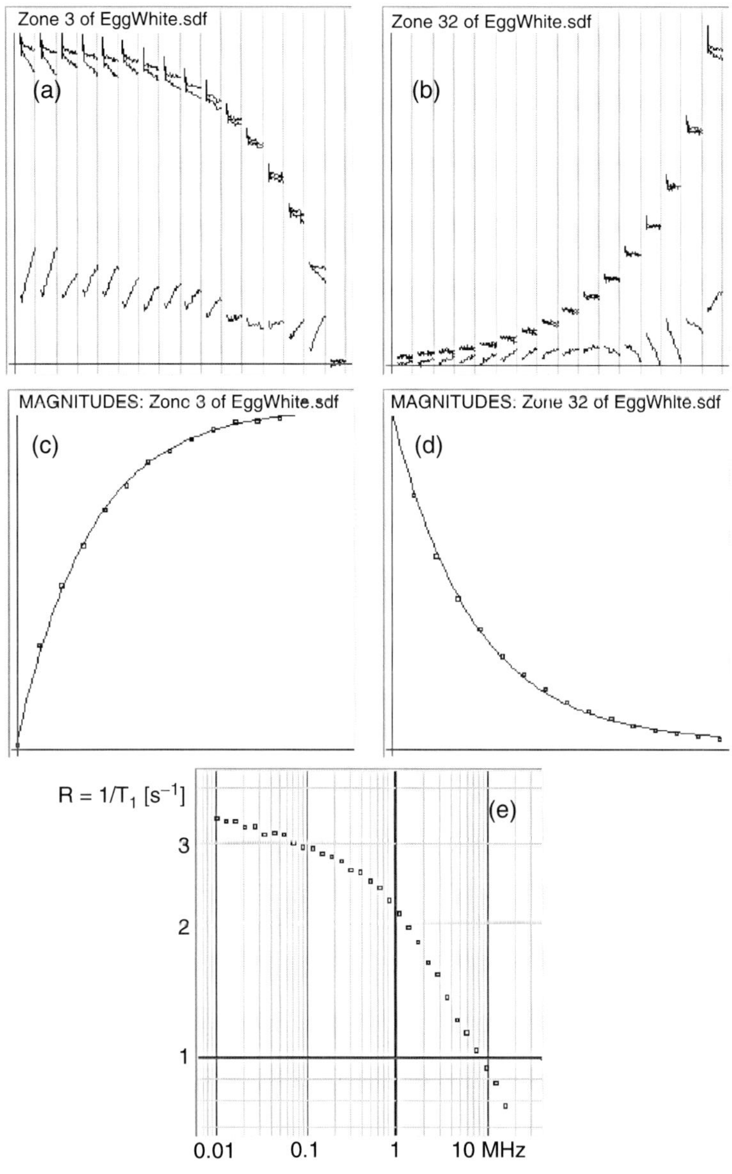

FIG. 22. Example of various types of experimental FFC data. These screenshots are shown exactly as they appear at the instrument. Graphs (a) and (b) show raw ^{1}H multi-block data (starting portions of FIDs), while (c) and (d) show the corresponding reduced data (1 point per block, obtained by averaging signal magnitudes over an FID window) and their mono-exponential fit, in this case the starting sections of simple FIDs. The data in (a) were obtained using the NP sequence and for two values of B_r (10 kHz and = 5 MHz, while those in (b) were obtained employing the PP sequence and $B_r = 10$ kHz respectively). To optimize magnet heating effects, the τ-values decrease from left to right. For each block, the real and imaginary components and the magnitudes of the signal are shown (originally colored). (c) and (d) show the reduced data (1 point per block, obtained by averaging signal magnitudes over an FID window) and their mono-exponential fit. Graph (e) shows the complete whole profile in which every point corresponds to a complete multi-block experiment.

B. Effects of Field-Switching Intervals

Though evaluation and interpretation of the relaxation curves do not directly relate to the instrument, there is an instrumentation-related feature which affects evaluation and needs to be discussed.

If one could neglect the effect of switching intervals, the intensity of the acquired signals would correspond to the sample magnetization value $M(\tau)$ at time τ (the end of the relaxation *sub-sequence* section) so that, by varying τ, it would be easy to trace the complete $M(\tau)$ curve. Because of the switching periods, however, the situation is a little more complex. Denoting M_0 as $= 0$, M_p, M_r, and M_a the equilibrium magnetizations corresponding to the null field, polarization field, relaxation field, and acquisition field, respectively, one has the following real scenarios (Fig. 23):

In the NP sequence, a certain amount of magnetization builds up during the field-switching interval Swt_1 so that the $M(\tau)$ curve never starts from exactly zero but from a value which is intermediate between M_0 and M_r. It then follows its relaxation curve in the field B_r, tending towards M_r. The acquisition period starts with a field-switching interval from B_r to B_a during which the magnetization again varies so that what we really detect is not the magnetization $M(\tau)$ at the end of the relaxation *sub-sequence* section but rather a value intermediate between $M(\tau)$ and M_a.

A similar analysis for the PP sequence shows that the experimental $M_{ex}(\tau)$ curve also differs from the theoretical one $M_{th}(\tau)$ in two respects: (i) the starting value is not M_p but something intermediate between M_p and M_r and (ii) the final value is again intermediate between $M(\tau)$ and M_a.

One consequence of this situation is that the FFC relaxation curves cannot be analyzed assuming any fixed starting or ending value. For example, under the mono-exponential hypothesis, the relaxation rate R (inverse of the relaxation time) must be estimated by fitting the three-parameter formula

$$M_{ex}(\tau) = c + w[1 - \exp(-R\tau)], \tag{23}$$

with no *a priori* assumptions about the values of the constants $c = M_{ex}(0)$ and $w = M_{ex}(\infty) - M_{ex}(0)$.

The starting and final discrepancies between $M_{ex}(\tau)$ and $M_{th}(\tau)$ can be quite large (Fig. 24 c,d). In fast relaxing samples, close to the high relaxation rate limit of an instrument, they may even exceed the total variation w of the relaxation curve $M_{ex}(\tau)$. It is therefore necessary to ask whether the $M_{ex}(\tau)$ dependence on τ is still "correct" in the sense that if its generic type is still given by Eq. (23) with a relaxation rate R identical to that of the theoretical curve $M_{th}(\tau)$. Fortunately, the answer is in the affirmative (see Appendix A for a proof), provided that the field-switching waveforms are rigorously reproducible and independent of τ (no memory effects within the magnet control system). On the other hand, as far as the problem at hand is concerned, the waveforms do not need to conform to any particular shape (such as a linear ramp or other shapes).

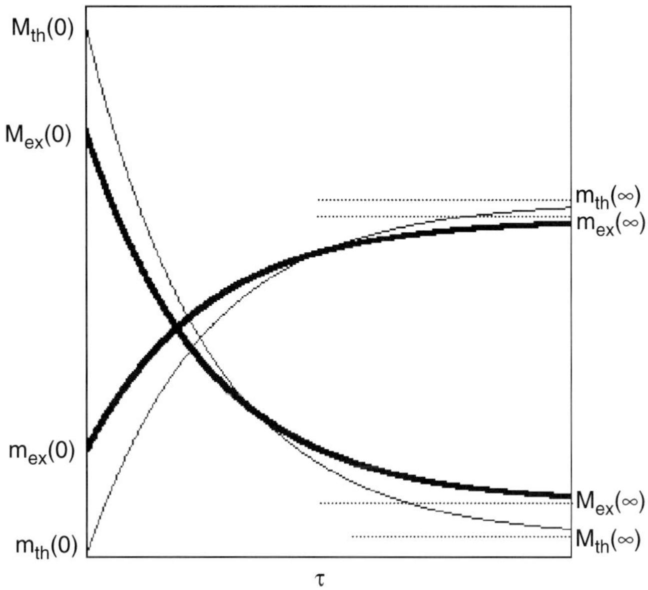

FIG. 23. Effect of switching intervals on experimental M(τ) curves. The two theoretical (thin) and experimental (thick) lines represent two ad-hoc examples of the effect of switching intervals on the magnetization relaxation curves. In the case of an NP-type experiment (increasing curves with small-letter labels), the theoretical $m_{\text{th}}(\tau)$ function starts at $m_{\text{th}}(0) \equiv 0$ and increases toward the limit value of $m_{\text{th}}(\infty) \equiv m_r$, corresponding to the relaxation field b_r. The experimental curve $m_{\text{ex}}(\tau)$ starts at $m_{\text{ex}}(0) > 0$ because of the first switching interval during which the field varies from zero to b_r. It then grows toward $m_{\text{ex}}(\infty) \neq m_r$ because of the second switching interval from b_r to the acquisition field B_a (in this example we assume $B_a < b_r$ so that the discrepancy has a negative sign). In the case of a PP-type experiment (decreasing curves with capital-letter labels), the theoretical $M_{\text{th}}(\tau)$ function starts at $M_{\text{th}}(0) \equiv M_p$ and decreases toward the limit value of $M_{\text{th}}(\infty) = M_r$, corresponding to the relaxation field B_r (assumed here to be small but not-zero). The experimental curve $M_{\text{ex}}(\tau)$ starts at $M_{\text{ex}}(0) < M_p$ because of the first switching interval during which the field varies from B_p to B_r. It then falls toward $M_{\text{ex}}(\infty) > M_r$ because of the second switching interval from B_r to the (higher) acquisition field B_a. Despite the apparent discrepancies due to relaxation during the switching intervals, the relaxation rate constants characterizing the evolution of the experimental curves with respect to τ are the same as those of the theoretical functions (all the curves plotted in this Figure have the same relaxation rate R).

The only consequences therefore consist in the need for an extra parameter (c) in the fit and in a reduced dynamic range of the decay curves.

C. DATA ACCUMULATION METHODS

Like always in NMR (*118*), when the single-scan signal-to-noise ratio (S/N) is insufficient for a precise evaluation of the acquired data, it can be enhanced by averaging or, more frequently, accumulating, of data sets obtained in a number of repeated scans. Multiple-scan data accumulation also opens a number of novel experimental possibilities extending beyond

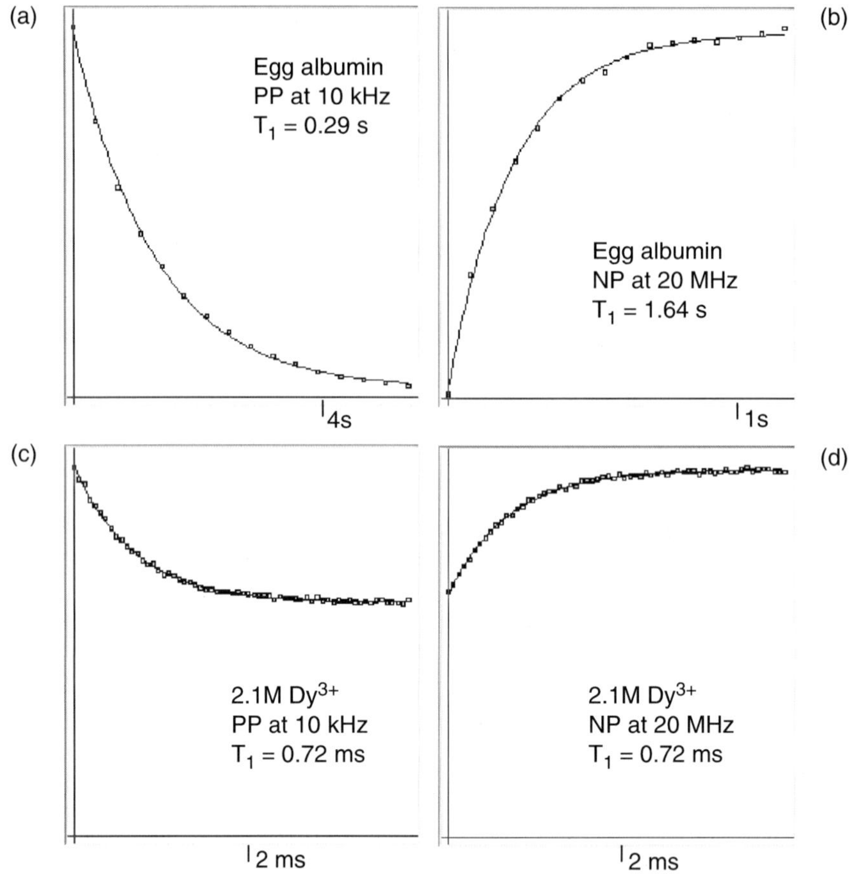

FIG. 24. Examples of experimental relaxation curves The vertical scales are arbitrary, horizontal scales (τ-values) are linear with a single label. In all cases, the axes cross at the origin (0,0). The upper two plots regard a slow-relaxing sample in which the switching effects described in the text are negligible. The bottom two plots regard a very fast relaxing sample in which switching-interval effects are very pronounced.

the original goal of S/N ratio enhancement. By adding extra sophistication, in fact, data accumulation can be used to do any of the following:

(a) Enhance S/N ratio,
(b) Suppress the effect of a number of instrumental imperfections, and
(c) Isolate particular components of the NMR signal.

The S/N ratio increases with the square-root of the number of scans, provided that the NMR signals are perfectly reproducible (except for the receiver noise). In FFC, this is not always true due to the main field instabilities, which often exceed the natural resonance line-width of the sample. In such cases, the rule applies well to the starting portion of the FIDs but not necessarily to the FIDs as a whole. This is one of the reasons why any free-evolution signals are rarely acquired for times longer than about 100 μs, even if the studied system would allow it. In order to extend this interval, one must resort to driven-evolution signal detection methods, such as the CPMG sequence with its spin-locking properties (to be discussed later).

Suppression of instrumental imperfections and/or selection of particular signal components are both based on the technique of phase cycling which exploits the dependence of NMR signals on the variations of the RF phases of the transmitter pulse(s) S (since phase-cycling is used in every branch of NMR, we assume that the reader is acquainted with the technique) (we will provide more information later, while discussing signal detection methods). At this point we just wish to point out that phase-cycling is extensively used in FFC and has to be supported by the console hardware – a requirement which implies pulser control of RF phases.

In practice, a particular "phase cycle" is defined by means of an array of RF pulse settings (to be used cyclically during consecutive scans) and an associated array of "receiver phases". The "receiver phase", however, does not correspond to any hardware device setting. Rather, it is an interlocution for the various modes of how each single-scan signal should be handled by the data accumulation procedure (add, subtract, quad add, quad subtract, etc.).

In multi-block experiments we are stepping through the values of the arrayed parameter (such as τ). If, in addition, we also want to accumulate N scans, the following alternative arises:

(a) Select the first arrayed-parameter value and carry out an N-scans accumulation (with phase-cycling). Then select the next arrayed-parameter value and repeat the whole process, stopping after all the arrayed-parameter values have been handled.
(b) Select the first phase-cycle settings and carry out a single elementary experiment for each arrayed-parameter value, storing the data in a global multi-block accumulation buffer. Then select the next phase-cycle settings and repeat the passage, terminating after N such complete passages have been completed.

We have here two possible cycles, one over the N scans with their nested phase-cycling, and the other one over the arrayed-parameter values. In the first case, the inner cycle is the one over N scans, while in the second case, usually referred-to as *interleaving*, it is the one over arrayed parameter values. Even though the two approaches are theoretically equivalent, the interleaved accumulation is preferred over the non-interleaved one. The reason is that, in a lengthy accumulation, any systematic long-term sample variations in the whole system (for example, sample deterioration) affect all the arrayed-parameter data blocks in approximately the same way. This is important, for example, when measuring samples subject to internal evolution on the time scale between several minutes and several days. Such systems are suprisingly common and include biological samples (for example, excised tissues) and materials subject to interesting internal-dynamics phenomena (phase transitions, polymerization, etc).

D. Evaluation of the Relaxation Curves

There are a number of ways by which the data of an elementary FFC experiment (such as NP or PP) can be reduced to a single point of the $M_{ex}(\tau)$ curve. Since these data-reduction procedures depend upon the signal detection technique, we shall postpone their discussion. It is useful, however, to provide a few comments on how the $M_{ex}(\tau)$ curves should be evaluated once we have a set of experimental points $\{\tau_k, y_k\}$, where $y \approx M_{ex}(\tau)$ and $k = 1, 2, \ldots, n$, and n is the number of blocks in the multi-block experiment.

In the case of bi- or multi-exponential relaxation curves the treatment involved can be rather complex (*119–123*). It becomes even more problematic. Needles to say, the same is true for systems with suspected continuous distributions of relaxation rates, whose evaluation by numerical analysis of the decay curves (*124–128*) represents one of the most arduous mathematical problems (*124–128*). In general, evaluation tasks of this kind need to be treated off-line, using specific programs and algorithms.

However, a fast and simple mono-exponential on-line evaluation procedure included in the control software of an FFC relaxometer is not only possible but, in reality, is a must since it provides the operator with relaxation-rate data estimates essential for correct setting of acquisition parameters. The fact that the mono-exponential hypothesis may be inaccurate does not really change the fact that some kind of a preliminary estimate is essential for correct data acquisition.

In the following we review the on-line mono-exponential evaluation procedure we have chosen for on-line used on Stelar instruments. We believe that the qualitative features of this algorithm, such as the method used to estimate the probable error of the relaxation rate, represent a good example of how data should be handled in more complex cases also.

As explained above, under the mono-exponential hypothesis the data $\{\tau_k, y_k\}$ must be fitted by the three-parameter theoretical formula

$$y = c + w[1 - \exp(-r\tau)], \tag{24}$$

where c, w, and r are as yet unknown parameters. This requires a non-linear least-squares fit in which one minimizes the total quadratic deviation

$$Q(c,w,r) = \Sigma_k [y_k - \{c + w(1 - \exp(-r\tau))\}]^2 \tag{25}$$

with respect to c, w, and r. It is convenient to split the task into two distinct parts.

(1) Assuming the value of r to be fixed, Eq. (24) is linear with respect to c and w. The constrained optimal values of these two parameters, denoted as c_1 and w_1, are therefore easily determined using standard linear correlation formulae. Since c_1 and w_1 depend on r, this procedure defines the non-linear function $Q_1(r) = Q(c_1(r), w_1(r), r)$.

(2) The minimum value of the function $Q_1(r)$ can be determined numerically using a standard algorithm such as Brendt interval bisection. Assuming that the minimum occurs at $r = r_2$, the value $Q_2 = Q_1(r_2) = Q(c_1(r_2), w_1(r_2), r_2)$ coincides with the absolute minimum of $Q(c, w, r)$.

In addition to finding the optimal fit and thus the optimal relaxation rate $R = r_2$, this two-step procedure provides us with the possibility of properly evaluating the *confidence interval* for R (Fig. 25).

We first notice that (i) it is easy to evaluate the function $Q_1(r)$ for any r and (ii) along the curve $Q_1(r) = Q(c_1(r), w_1(r), r)$, the parameters c and w vary so as to remain optimal for every value of r. The latter fact is essential since otherwise the confidence-interval estimates for r would be grossly over-optimistic. Numeric values of the *confidence interval* for R (regardless of the less interesting values of c and w) can now be based on the least significant increment of $Q_1(r)$. Assuming that the optimum value Q_2 of the total quadratic deviation $Q(c, w, r)$ is due entirely to random experimental errors, its least significant increment $\Delta_\alpha Q \equiv Q_{si}$ can be determined for any given significance level α by means of the *Fisher statistic* with both degrees of freedom set to $(n-1)$. The confidence interval $\Delta_\alpha R$ for R then comprises all r values for which $Q_1(r) - Q_2 \leq Q_{si}$ and its *probable error* $e = (\Delta_\alpha R)/2$ is obtained, as usual, by setting $\alpha = 0.69$. When, as expected, $Q_1(r)$ is approximately quadratic in the vicinity of the optimum, it turns out that the result can be excellently approximated by the simple formula:

$$e = \sqrt{\frac{1}{(n-1)}\frac{Q_2}{Q_1''(r_2)}} \tag{26}$$

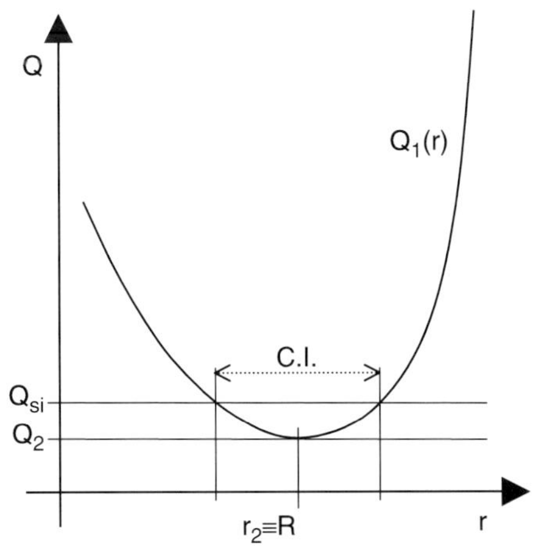

FIG. 25. Determination of the R_1 confidence interval. Q_1(r) is the total quadratic deviation assuming a relaxation rate value r and fitting all other parameters in Eq.(24). Its absolute minimum at r_2 defines the most probable relaxation rate R. Q_2 is the minimum of $Q_1(r)$ and Q_{si} equals Q_2 plus the least significant increment determined by statistical methods. This defines the confidence interval C.I. comprised between the two vertical lines. For more details, see the text.

where $Q_1''(r)$ is the second derivative of $Q_1(r)$ which can be easily estimated by standard numeric methods.

When using the above algorithm as an on-line help, the operator must pay attention to the fact that, in general, the mono-exponential hypotheses need not hold. If, for example, the estimated fitting error is too large, the usual course of action would be to increase either the number of scans or the number of blocks. When, however, the apparent fitting errors are due to the fact that the relaxation curves are non-exponential, rather than insufficient data quality, improving the precision of the $M_{ex}(\tau)$ curve will not be to be of much help.

E. Factors Influencing the Precision of Relaxation Rate Estimates

Whatever is the fitting hypotheses and the corresponding evaluation algorithm, the evaluation results are burdened by experimental errors which depend primarily upon the following principal factors:

- Inherent signal-to-noise ratio $\rho = S/N$.
- Maximum relative magnetization $\mu = M_{max}/M_a$.
- Relative magnetization variation $\nu = (M_{max} - M_{min})/M_{max}$.
- Distribution and number n_b and of τ-values.
- Number of scans N.

A detailed discussion of the exact impact of each of these factors is beyond the scope of this chapter. Nevertheless, a few guidelines born out by experience are appropriate.

For a given number of nuclides placed in the sample coil volume, the inherent S/N ratio ρ is a parameter depending only on the probe and pre-amplifier assembly. It is usually measured in terms of the maximum FID amplitude after a 90° pulse applied after the sample has reached its equilibrium magnetization M_a in the acquisition field B_a. Defined in this way, it is independent of the details of any FFC sequence.

The impact of ρ ratio on the relative precision of relaxation rate/time estimates is close to linear, meaning that doubling ρ reduces the relative errors by a factor of two.

The importance of the factor $\mu = M_{max}/M_a$ stems from the fact that in an actual FFC sequence, the maximum measured magnetization M_{max} is not M_a. For example, when switching-time effects are negligible, we have $\mu_{NP} = M_r/M_a = B_r/B_a$ for the basic NP sequence and $\mu_{PP} = M_p/M_a = B_p/B_a$ for the basic pre-polarized sequence.

The relative magnetization-variation factor $\nu = (M_{max} - M_{min})/M_{max}$ is related to the fact that if the magnetization did not vary with respect to τ, we could never determine R, no matter how large ρ and μ might be. The reason why we consider this factor independently of ρ is that, theoretically, it depends only on the pulser sequence type and, in pre-polarized sequences, on the ratio $\chi = M_r/M_p = B_r/B_p$. For example, one can easily verify that for the basic non-polarized and pre-polarized sequences $\nu_{NP} = 1$ and

$\nu_{PP} = (M_p - M_r)/M_p = 1 - \chi$, respectively, provided that switching effects are negligible.

As in the case of the S/N ratio, the relative precision of relaxation rate/time estimates is linearly proportional to both the μ-factor as well as to the ν-factor.

The effect of switching times on the two factors can be inferred from Fig. 23. For non-polarized sequences, we have $\nu_{NP} = [m_{ex}(\infty) - m_{ex}(0)]/m_{ex}(\infty)$ which is always smaller than the theoretical value of 1. This is also true for pre-polarized sequences, since it can be easily seen that the experimental value $\nu_{PP} = [M_{ex}(0) - M_{ex}(\infty)]/M_{ex}(0)$ is again smaller than the theoretical value.

The effect of switching intervals on ν-factor is that there is always a decrease in measurement precision – a problem which becomes dramatic in samples relaxing on a scale comparable to, or faster than, the switching times. In the case of the μ-factor, the switching interval effects are less dramatic and there is no universally valid rule, though a tendency toward degradation prevails (in any case, whenever there is a marginal enhancement of the μ-factor, it is always accompanied by a marked degradation of the ν-factor).

The type of the distribution of τ-values is a much discussed topic. Experience shows that, in a mono-exponential case, the values should spread over an interval of more than $3 \times T_1$ but not much over $4 \times T_1$, and a linear distribution appears to be slightly better than a logarithmic one. This is probably due to the fact that in a three-parameter exponential fit, the points with large τ-values play as crucial a role in determining the relaxation rate as the slope at small τ-values, and one needs both to determine R. On the other hand, it is evident that in multi-exponential cases, logarithmic distribution is often better suited for the task, especially when the relaxation rates of different sample components differ by an order of magnitude or more.

To describe the impact of the τ-values distribution type on the relative precision of relaxation rate estimates, we shall use a phenomenological factor f_d. We expect it to be independent of all the other factors, but dependent upon the type of relaxation rate quantity to be determined (for example, the fastest- or the slowest-relaxing component in a multi-component mixture).

For a given type of τ-values distribution, the size of the τ-values array (number of blocks n_b) plays approximately the same role as the number of scans N. Theoretically, the relative precision of any relaxation rate estimate is proportional to the square root of both n_b and N. This, of course, presumes that n_b is anyway large enough to carry out the analysis. For example, values as small as 4 may be sufficient in mono-exponential cases, while continuous distributions spreading over several orders of magnitude require a logarithmic distribution of τ-values and n_b values of over 100.

If we denote as ε the relative error of a particular relaxation rate estimate (or, for that matter, of any quantity related to relaxation-rates), the above discussion can be summed up by the following formula

$$\varepsilon^{-1} \approx \rho \cdot \mu \cdot \nu \cdot f_d \sqrt{n_b} \cdot \sqrt{N} \tag{27}$$

It is necessary to point out that the above discussion has been centered completely on the intrinsic experimental errors playing a role in the evaluation of a *single* relaxation curve. This is not the same as the *reproducibility* of the results (scatter) when the whole multi-block measurement is repeated. When comparing single-fit errors with the overall scatter, two situations arise:

(a) In a mono-exponential case the scatter is typically about twice as large as the fitting errors and the two quantities are correlated.

The discrepancy is due to sources of errors not considered in the above discussion, such as field noise and reproducibility, thermal effects, etc. In particular, thermal effects on the magnet are important since from experience, it shows that the scatter increases when the relaxation field and/or the polarization field are close to the upper limit of the magnet. Since all such contributions are random, prolonged data accumulation reduces both the fitting errors and the scatter.

(b) When the mono-exponential hypotheses does not hold, the fitting error reflects the discrepancy between the hypotheses and the data, rather than any random characteristics of the experimental $M_{ex}(\tau)$ curve. Consequently, situations may arise when the fitting errors are much larger than the scatter. Prolonged data accumulation, in particular, reduces the scatter but, beyond a certain point, has little or no effect on the fitting errors.

F. Optimization of Relaxation Rate Measurements

When is an experiment, or a series of experiments, optimal? The answer to this often asked question (*129*) is not unambiguous because, as in most optimization problems, it involves multiple and mutually contradictory criteria, such as

- Minimum measurement time
- Minimum scatter (best reproducibility) achievable in a given time.
- Capability to falsify/confirm specific application hypotheses.

Here, of course, we can only concentrate on the first two points. There is always the necessity to find a compromise between the maximum affordable measurement scatter and the data accumulation time necessary to reach it. To a considerable extent, the operator can influence the outcome of this compromise by placing a premium either on the precision or on the speed which is achieved primarily by adjusting the number of blocks in the multi-block experiments and the total number of scans.

A little less obvious is the setting of the recycle delay *RD* (for NP-type sequences) or T_p (for PP-type sequences) which is linked through the factor f in Eqs. (3) and (4) to the estimated relaxation time T_{1max} of the slowest-decaying component of sample magnetization at a specific field. One cannot influence the sample relaxation times, of course. On the other hand, the relaxation times usually dominate the overall duration of a single multi-block scan so that, except for fast relaxing samples, substantial measurement-time savings can be achieved only by acting on the parameters linked to the relaxation.

All this points to the factor f, which guarantees that the sample magnetization at the beginning of each block is the same with a relative precision of e^{-f}. However, the actual reproducibility is much better than this, since we do not really need M_0 to be exactly zero (in NP) or M_p (in PP) but only that they be the same for all τ-blocks of a multi-block experiment. Theoretically, considering that the acquisition period of the previous block normally destroys the longitudinal magnetization and the subsequent sequence of events until the start of the next relaxation period is the same for every τ-block, any value of f should be theoretically acceptable. Further investigation of these aspects is currently under way.

In very fast relaxing samples ($R \gg 10\ \mathrm{s}^{-1}$), the contribution of the switching intervals to the total measurement time becomes appreciable and is therefore susceptible to possible optimization. The common practice is to set the duration of all switching intervals in a sequence to the same value Swt which has a safety margin to include the ramp phases as well as the settling phases of all switching periods (see Fig. 20). This is quite fine for samples in which all relaxation times are consistently much longer than Swt. Otherwise, individual optimization of the switching times is advisable for two reasons:

- it reduces total measurement time (though this has usually a rather modest effect), and
- it enhances the relative magnetization variation $\nu = (M_{\max} - \mathrm{M}_{\min})/M_{\max}$ of the experimental $M_{\mathrm{ex}}(\tau)$ curve and thus, through Eq. (27), directly affects the measurement precision, often in a marked way.

The optimization of individual switching intervals exploits two principles:

- Extremely precise field settling (well below 0.1%) is required only when switching to the acquisition field where RF pulses are to be applied and/or the NMR signal is to be collected. In all other switching periods (for example, switching from the polarization field to the relaxation field), field-settling precision of the order of 0.1% is quite sufficient.
- The duration of the ramp phase of a switching interval depends upon the field levels between which the switching actually occurs and can be therefore individually adjusted see Eq. (2).

Before concluding this section, we should mention another, more basic choice the operator (or the instrument's software) has to make, namely the selection of the type of sequence to be used. Whether one should use a non-polarized or a pre-polarized sequence depends essentially upon the relaxation field value B_r. The choice affects the precision of the measurements basically through the product of the factors μ and ν in Eq. (27). For the ideal NP and PP sequences, these products are $\mu_{\mathrm{NP}}\nu_{\mathrm{NP}} = (B_r/B_a)$ and $\mu_{\mathrm{PP}}\nu_{\mathrm{PP}} = (B_p/B_a)(1 - B_r/B_p)$, and it is elementary to see that $\mu_{\mathrm{NP}}\nu_{\mathrm{NP}} > \mu_{\mathrm{PP}}\nu_{\mathrm{PP}}$ when $B_r > B_p/2$ and vice versa. Theoretically, therefore, one should use the

PP sequence for relaxation fields B_r smaller than half the polarization field value and the NP sequence for relaxation fields higher than that.

In practice, this is the best choice for all samples, except those which relax very fast and, in addition, relax much faster at low fields than at high fields (large overall dispersion). In such cases, the degradation of the ν-factor due to the effect of switching periods is substantially higher for the NP sequence (which starts always from zero field) than for the PP sequence (which never descends below B_r) and therefore the NP sequences should be disparaged by moving the B_r switchover level slightly above $B_p/2$ (typically 60–65% of B_p). Again, exact formulae describing this kind of optimization are not yet available and further research is in progress.

X. Signal Detection and Analysis

In FFC relaxometry, one is concerned with the time evolution of the parallel component M of the nuclear magnetization of a sample or, in more complex cases, of one or more of its constituents. The primary goal of the signal detection is to estimate M and not, like in NMR spectroscopy, to analyze the FIDs in any detail beyond a simple solid/liquid phase distinction.

In principle, therefore, FFC relaxometry could employ – and often does so – a number of detector types such as phase detectors, diode detectors, square detectors, modulus detectors, envelope detectors, SQUID detectors, etc.

In any case, however, the NMR signal can be acquired only after $M_{\parallel}$ has been brought to the XY detection plane which, of course, is achieved by means of a suitable NMR excitation sequence. There is little conceptual difference between the FFC excitation sequences and those used in other NMR techniques. The NMR types of signals used in FFC include all the "classical" ones, such as free induction decays (FID), spin-echoes, envelopes of CPMG spin-echo trains, etc. In principle, any NMR signal is acceptable, as long as the acquired signal reflects M at the end of the relaxation period. The Classical NMR signal excitation and preparation methods can also be used to select or/ enhance particular components of nuclear magnetization.

In general, the signal acquisition process provides us with more data than needed. To extract relaxation parameters, we need a single value $M(\tau)$ for each τ setting, but we generally acquire a whole array of values. Clearly, some kind of data-reduction process must be implemented before the acquired signals can be used in the way we intend it to be. Like the NMR excitation techniques, the data reduction process can be exploited to enhance or suppress particular signal components.

This delimits three aspects of FFC signal detection (detector hardware, excitation & detection method, and data-reduction algorithm). The following paragraphs explain briefly the most popular choices we have made to handle these aspects.

A. Hardware Detection

From the practical point of view, dual-channel phase detectors operated in quadrature appear to be the best hardware-detection choice. Unlike other techniques, phase detection is sensitive to RF frequency offset from resonance and to the RF phase which, on the one hand, makes the signals more complex to use (as well as and more sensitive to instrument instabilities) but, on the other hand, leads to a number of important advantages, such as:

- *Linearity.* Phase detection is truly linear, in the sense that a multicomponent sample magnetization leads to a signal which is guaranteed to be a simple linear superposition of contributions from the individual components. All other detection methods lead to signals containing cross terms between different components as well as non-linear coupling cross terms between the signal and the noise.
- Applicability of standard NMR *phase-cycling* data accumulation methods.
- Precise *definition of receiver bandwidth* by means of audio filters (improves sensitivity).
- *Sensitivity enhancement* due to the fact that we are using two uncorrelated detectors.
- Improved *control of experimental conditions.* Phase detection prevents the operator from straying from resonance, makes possible reliable offset estimates and receiver phase estimates and even completely automated maintenance of optimal offset settings.
- Possibility of *complete spectral analysis* of the acquired signals. Though FFC is still a low resolution technique, this is sometimes useful and it appears to be an important potential advantage for future developments.

Modern instruments usually offer an on board choice between a quadrature phase detector and some kind of diode or square detector. The latter, however is mostly used just for instrument setup (probe tuning, etc.) while signal acquisition is done almost exclusively by the phase detector.

As usual, we shall from now on consider the output of the dual-channel phase detector as a unique complex signal, with the outputs of the two channels identified by its Cartesian components.

B. Post-Detection Signal Handling

By revealing all aspects of the signal, the phase detector makes evident all instrumental artifacts which would not be observable with another type of detection. On an FFC instrument, this typically includes thermal field drifts (see Section IV.D) and field instabilities associated with the large dynamic bandwidth of the switching magnet system.

In a certain sense, the detector provides us with more details than that bargained for. The goal of the primary post-detection signal handling is to get rid of those features which are irrelevant in a given context and enhancing those which are essential. In many cases, for example, it makes no sense to analyze the shape of an acquired FID, and the only desired quantity is the total amount of the signal (a single number rather than a dual data array).

There are various approaches to the data-reduction task. An often used one consists of computing the modulus of the complex phase-detector signal. This removes all offset imperfections as well as any receiver phase misadjustment, bringing us theoretically to what we would have by summing the outputs of two independent, ideal diode detectors. In this case, however, the original signals are still available and can be used to check various aspects of data quality, carry out additional corrections (such as removal of noise-rectification artifacts), or submitted to alternative evaluation algorithms.

The modulus data are then used to estimate the total signal amount by means of various algorithms. The simplest one uses the average of the modulus over a pre-defined data-array window.

For evident reasons, this topic overlaps with the much broader field of application-specific data evaluation methods which we will occasionally mention but whose comprehensive exposition exceeds the scope of this chapter.

C. NMR Signal Excitation and Detection Sequences

In this section we shall list and briefly discuss the pros and cons of some classical NMR techniques used for signal excitation and acquisition.

(1) Simple FIDs (Fig. 22)

The simplest method of detecting longitudinal magnetization consists in applying a 90° RF pulse and acquiring the resulting FID. Though at first sight this appears as the most obvious approach, it is not void of drawbacks.

One problem is the dead time of the probe-preamplifier subsystem (the combined effect of probe RF ringing and of preamplifier recovery from saturation). While irrelevant in samples with long enough FID (above 0.1 ms or so), it may become a major limitation with fast-decaying FIDs (solid samples and/or samples with very short T_2s) because it can obscure a substantial portion of the FID signal.

Another problem is the relatively low field stability of present FFC systems which causes considerable fluctuations in the tailing portions of FIDs, far from the beginning position, and thus affects negatively the data for samples with long FIDs. To eliminate this problem, one normally uses only the starting FID portions of the FIDs (typically covering the first 0.1–0.2 ms).

The shape of a simple, low-resolution FID is usually not suitable for discriminating between various sample magnetization components except, perhaps, in the co-presence of a very fast-decaying component and a very slow-decaying one.

The simplest data reduction algorithm for FIDs consists in averaging the magnitudes of the complex FID signal over a data window positioned within its starting portion (Fig. 26). The window can be freely positioned in a way to cut out any dead-time distortions and, at the same time, minimize field-fluctuation effects.

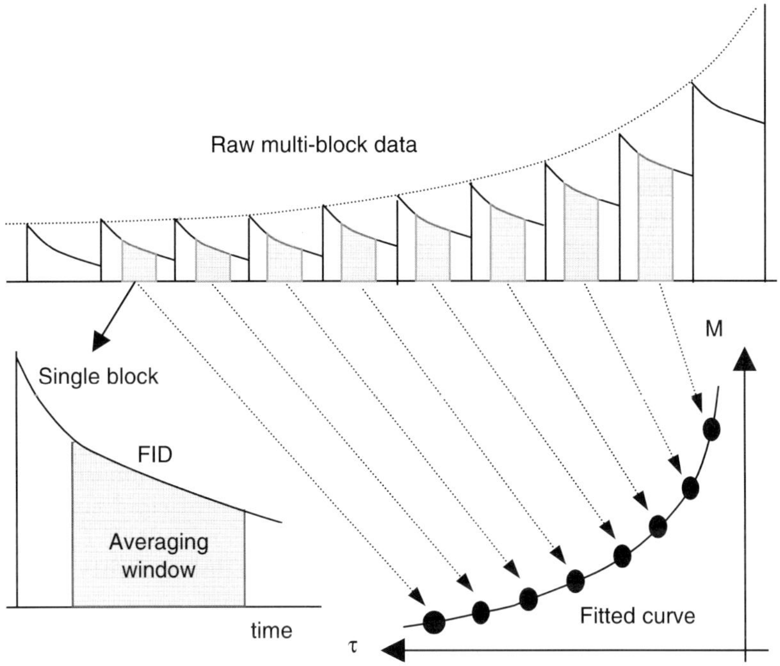

FIG. 26. Example of the data reduction process. Each data block of a multi-block sequence (in this case simple FIDs) is "reduced" to a single value by means of averaging over a predefined data window and plotted against the block's τ-value. The resulting relaxation curve is then fitted to estimate its decay rate(s). The algorithm gives a lot of freedom in setting the data window and including/excluding any number of initial or final blocks. Notice that in the PP case shown here, the τ-value decreases from left to right. This helps to minimize thermal variations of the magnet.

One of the advantages of window averaging is of course an additional S/N enhancement, which is roughly proportional to the square root of the number of data points present in the window (provided they are well-reproducible). When too long portions of the FIDs are used, however, the S/N gain is eventually invalidated by FID signal fluctuations due to offset instabilities.

(2) Spin echo

The classical Hahn-echo technique employs the basic sub-sequence $90_x - \delta - 180_y - \delta' -$Acq. It is applicable primarily to samples with long T_2 values (over 0.2 ms) in which it eliminates all dead-time problems, refocuses magnet inhomogeneity and enlarges by a factor of almost 2 the applicable data-reduction window which, in this case, should be centered over the top of the echo. When δ is longer than typical field fluctuation times, spin echoes become sensitive to field instabilities. Consequently, the echo delay parameter δ must be kept small enough to avoid any field instability effects while, on the other hand, it should be long enough to make the top of the echo fall within the central part of the data evaluation window (typical values are 0.05–0.1 ms). The second delay δ' needs to be long enough to cover the dead-time effects due to the second RF pulse but not necessarily longer (typical values are 0.01–0.02 ms).

It must be stressed that the spin-echo sequence is applied only during the detection period and its unique purpose is to estimate the signal amplitude (in a sense, it is a replacement for the simple 90° pulse). Consequently, in an arrayed multi-block experiment whose purpose is to measure $T_1(B_r)$, only the τ value is varied, while the delays δ and δ' are kept constant in order to make sure that no $T_2(B_A)$ effects leak into the experimental relaxation curves. Moreover, to avoid contamination of the echo by FID residues due to imprecise settings of RF pulses and to B_1 inhomogeneity, proper phase cycling is highly recommended.

Spin-echo detection suppresses the sample magnetization components with short T_2 values (comparable to, or shorter than $\delta + \delta'$). Depending upon the context goal of the measurement, this can be either a disadvantage or an advantage. It makes sequences with spin-echo detection unsuitable for samples with very fast-decaying FIDs, such as those of rigid solids. On the other hand, it allows one to isolate the components with long $T_2(B_A)$ values (high mobility) from those with short $T_2(B_A)$ values (low mobility) in samples where such distinct components exist.

One should mention also the possibility of using the solid-echo subsequence of the type $90_x - \delta - 90_x - \delta' - \text{Acq}$ (also known as quadrature echo). The principal purpose of this detection method is a reduction in the loss of signal due to the dead time. This makes it interesting for solid samples with strong dipolar interactions. One sets δ' just about long enough to suppress the dead time due to the second pulse. The value of δ, theoretically identical to δ', is adjusted so as to maximize the amplitude of the resulting solid echo. Depending upon the structure of the sample, the echo can be substantially higher than a plain FID signal.

(3) Multiple spin echoes

It is possible to use a whole series of 180° pulses to repeatedly refocus the sample magnetization using the CPMG-like detection sub-sequence $90_x–[\delta - 180_y - \delta]_n$ with n anywhere between one and a few tens. Data acquisition can in this case proceed either all the time, starting almost immediately after the first RF pulse, or – more efficiently – in short segments centered around the top of each echo. When δ is smaller than typical field-fluctuation times, the train of RF pulses refocuses all field inhomogeneites and, due to its spin-locking properties, compensates quite efficiently any field instability.

The result is an enhancement of all the advantages of spin-echo detection. The number of usable data points in each acquired data array can in this case exceed in an FID by a factor much larger than two. Since T_2 relaxation is going on during the detection, it is again important to keep δ rigorously constant during the whole multi-block experiment.

As in the case of the single spin-echo detection, the multiple spin-echoes method attenuates magnetization components with short $T_2(B_A)$ values and the magnitude of the attenuation increases with increasing echo number – a fact which can be exploited for partial separation of sample components based on their $T_2(B_A)$ values.

(4) CPMG echo train

An even more dramatic increase in S/N enhancement is achieved using a variety of the CPMG detection sub-sequence $90_x–[\delta - 180_y - \delta - \mathrm{AcqS}]_n$ called LR-CPMG. This time, however, n is very large (for example, 1000) and each echo is sampled just once at the moment when it attains its maximum amplitude (this is the acquisition strobe AcqS). The result is an array of points describing the echo-tops envelope. When δ is kept very small (typically, one uses 0.05 ms), such envelope is completely insensitive to field inhomogeneity as well as – due to its excellent spin-locking performance – to field instabilities. One can, in fact, use the envelope to determine the value of $T_2(B_A)$, even though this is not the primary purpose of this detection sub-sequence in the present context.

Due to its excellent S/N enhancement properties, LR-CPMG detection is probably the only possible choice when the primary S/N ratio is very small. This regards in particular measurements of low-abundance nuclides and nuclides with low gyromagnetic ratio.

As in the previous cases, the LR-CPMG sequence is applied exclusively during the detection period and its main purpose is again to estimate the signal amplitude at the end of the relaxation period. However, the LR-CPMG envelope, whose data points are all quite insensitive to instrumental artifacts, can be analyzed by any of the standard methods in order to detect distinct components of sample magnetization on the basis of their $T_2(B_A)$ values. Having large arrays of data points, even the quasi-continuous Laplace inversion methods are easily applicable which paves the way for FFC investigations in complex multi-component systems.

A disadvantage of the LR-CPMG detection method is its total insensitivity to field/frequency offset which must be adjusted before a profile measurement and cannot be corrected by means of a simple procedure during an automatic profile measurement. This requires a higher degree of long-term field stability (including any thermal effects) than the other methods. Despite the insensitivity of the technique, in fact, the field may not be allowed to drift too far from resonance where the RF pulses would lose their efficiency (excursions up to about 5 kHz are, however, quite tolerable).

(5) Special detection techniques suppressing specific imperfections and artifacts

Once the acquisition field has been reached, the detection *sub-sequence* section of an FFC sequence is actually quite the same as in any other branch of NMR. This implies that the same categories of problems crop up and the same methods to solve them can be employed. We shall briefly mention two such cases.

(i) *Composite*: the use of *composite pulses to suppress B_1 inhomogeneity effects.* (*130*)

The effects of B_1 inhomogeneity in various NMR sequences are well-known and there are a number of ways to combat them. In FFC relaxometry, B_1 inhomogeneity is actually not much of a problem since it does not directly affect T_1 measurements. Whatever effect it has consists essentially in a loss of signal due to imperfect sample excitation and/or imperfect refocusing (in sequences using spin echoes and/or magnetization inversion).

The classical cure (*130*), involving the use of various composite pulses, is perfectly applicable in FFC relaxometry also.

The use of composite pulses is subject to several counter-indications, the principle of which is the fact that they last much longer than their simple prototypes. This makes their employment problematic in the case of rigid solids, as well as in the detection of sub-sequences which rely on extremely closely-spaced echoes.

In general, there is not much need for composite pulses in FFC relaxometry where the only aim is to encounter the inherent instrumental B_1 inhomogeneity. One usually delimits the sample height so as to make it fit inside the measurement coil, which is a less controversial way of reducing B_1 inhomogeneity than composite pulses. However, the employment of composite pulses is indicated in two cases:

- when the sample height simply cannot be reduced to fit inside the measurement coil, and
- when the sample itself induces a strong B_1 inhomogeneity.

The latter case is of interest, for example, when the sample contains metal particles (such as in some types of MRI relaxation contrast agents).

(ii) *and Signal: signal detection sequences suppressing acoustic ringing.* (*131,132*)

Acoustic ringing of the probe assembly after an RF pulse is a pesky problem which often limits the measurements of nuclides with low gyromagnetic ratios (it can also strongly interfere with measurements of samples containing piezoelectric components). The disturbance is often misinterpreted as a particularly long dead-time disturbance, until one notices that, unlike normal dead-time components, it disappears when B_a is set to zero. It is difficult to remove because it follows the phase of the RF pulse and thus cannot be eliminated by any simple RF phase-cycling.

The classical cure (*131,132*), apart from special probe construction precautions, is a pulse sequence using a *phase & device* detection cycle in which one exploits the fact that acoustic ringing increases linearly with pulse width while NMR signal follows the sinusoidal nutation-angle curve. In its most elementary form, the cycle is composed of four steps (ideally with null δ):

Step (1)	$0_x - \delta - 90_x -$ Acq (add)	introduces 90_x ringing
Step (2)	$180_x - \delta - 90_x -$ Acq (subtract)	compensates 90_x ringing, introduces 180_x ringing
Step (3)	$0_x - \delta - 90_{-x} -$ Acq (subtract)	introduces 90_{-x} ringing
Step (4)	$180_x - \delta - 90_{-x} -$ Acq (add)	compensates 90_{-x} ringing, compensates 180_x ringing

Full anti-ringing quadrature cycle is a bit more complex, while extension of the anti-ringing pulse technique to signal detection sub-sequences other than the simple FID is quite simple.

XI. Advanced FFC Sequences

So far we have discussed two different magnetization preparation methods (NP and PP) and several signal detection methods. The two aspects of an FFC sequence were so far independent of each other, thus giving rise to the full set of possible cross-combinations.

The original NP- and PP-type preparatory sub-sequences can be refined to partially compensate instrumental problems such as magnet heating during the measurements (see Section IV.D). The results are the so-called balanced NP and balanced PP preparatory sub-sequences.

There are important FFC sequences in which the preparatory *sub-sequence* sections includes RF pulses. A typical example is the FFC version of the classical inversion recovery sequence (IR).

In some sequences, such as the FFC variety of the Jeener–Broekaert sequence, the RF pulses applied during the preparatory sub-sequence need to be coordinated in phase with those applied during the detection sub-sequence. In such cases the preparatory and detection *sub-sequence* sections of the FFC sequence are no longer mutually independent.

A. Thermally Balanced Sequences

In a multi-block measurement sequence with interleaved phase cycle, there is a systematic, τ-dependent variation of the average magnet heating. In PP sequences with high B_p and small B_r, for example, the power dissipated on the magnet during each block increases with decreasing τ. In the case of NP sequences with high B_r, the situation is reversed. Whichever is the case, the magnet temperature variations induce signal offset variations which, being correlated with the τ values, could affect the relaxation curves.

The problem can be partially mitigated by hardware compensation devices (see Section IV.D). A complementary approach consists in setting up preparatory sequences which balance the average per-block thermal dissipation making it independent of τ. Though this does not remove the differences between the magnet temperature cycle within each block, it at least removes the systematic τ-dependent thermal drift.

The timing diagrams of the thermally balanced NP and PP sequences are shown in Fig. 27. Comparing them with those of the corresponding unbalanced sequences, one notices immediately the price to be paid for the improved thermal performance consisting in an increased duration of each cycle (particularly those with small τ values) which makes the experiments longer by a factor of about 2.

Whether to use the balanced sequences or not thus becomes a matter of compromise. When magnet cooling is sufficiently efficient, all the used fields are far from their maxima (keep in mind that heat dissipation on the magnet

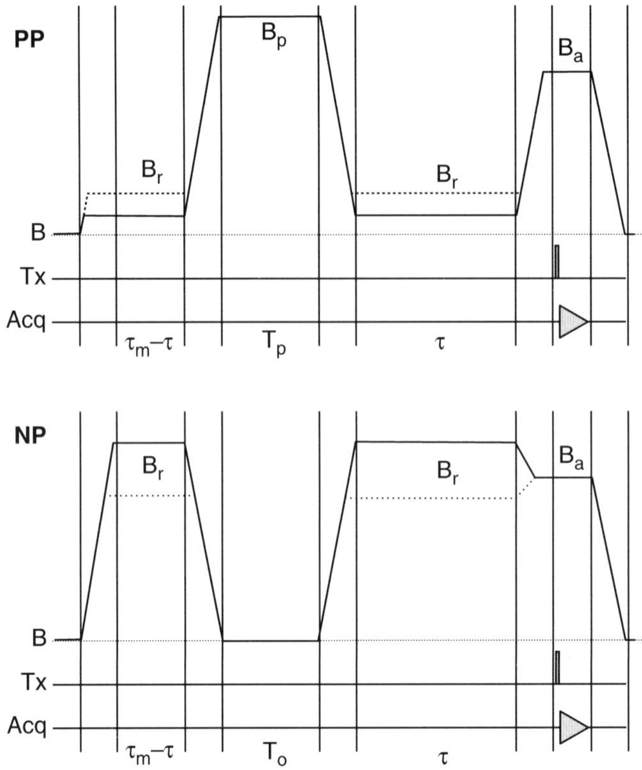

FIG. 27. Thermally balanced PP and NP sequences. (PP) In the balanced PP sequence, the sample is first kept at the relaxation field B_r for a time $\tau_m - \tau$ and, then pre-polarized at the polarization field B_p for a time T_p, and finally allowed to relax for time τ before the start of the detection period. The time T_p should be set to about $4T_1(B_p)$. As τ varies during a multi-block sequence, the polarization interval position moves horizontally but the total block duration and the mean power dissipation remain constant. (NP) The balanced non-polarized sequence is conceptually similar, except for the fact that the polarization interval is replaced by a magnetization annihilation interval in which the field is zero and whose duration should be about $4T_1(0)$. In both cases, the time τ_m should be about or more than $4T_1(B_r)$. The concept can be combined with any detection mode, not just the simple FID detection shown here.

grows with the square of the field), and since sensitivity is an issue, it may be advisable to use the unbalanced sequences and rely only on the hardware temperature compensation. Whenever the magnet is under a considerable thermal stress, however, it is advisable to use both the hardware compensation and the balanced sequences.

B. INVERSION RECOVERY

Figure 28 shows the diagram of the FFC version of the classical IR sequence. Notice that the first RF pulse has to be applied at the acquisition field B_a because the probe is tuned to the Larmor frequency at that field. This implies the presence of an extra field-switching interval unless, of

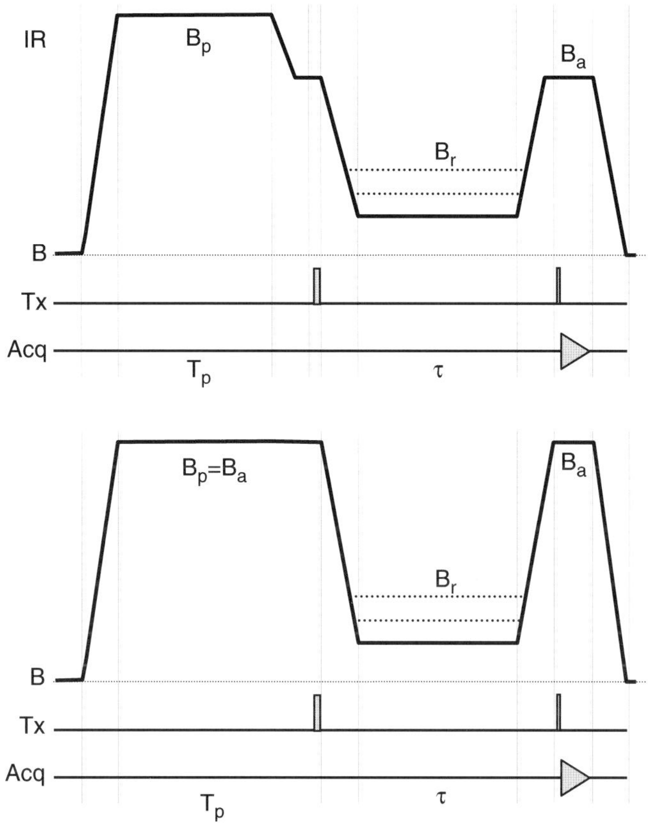

FIG. 28. FFC Inversion Recovery sequence. In the upper case the sample is first pre-polarized in a filed B_p, then switched to the "acquisition" field B_a where the first RF pulse of 180° is applied and the sample magnetization is inverted. The field is then switched to B_r and the sample is allowed to relax for the variable time τ. Finally, the field is switched again to the acquisition value and the magnetization is sampled by any of the sample-detection methods (here, a simple FID following a 90° RF pulse). Notice that, as shown in the lower diagram, in the special case when $B_p = B_a$ it is possible to neatly avoid the extra switching interval prior to the inversion pulse.

course, B_a equals B_p. For the IR preparatory sub-sequence, the latter condition represents an optimum (this goes hand-in-hand with the fact that a high acquisition frequency tends to improve the inherent S/N ratio). When $B_a = B_p$, the magnetization inversion is achieved by an RF pulse applied at the very end of the polarization interval.

In order to optimize the inversion, it is a good idea to make this pulse a composite one (except, maybe, in the case of rigid solid samples). As far as signal detection is concerned, all methods are acceptable so that, for example, IR preparation can be combined with a simple FID detection just as well as with the CPMG detection. Likewise, it is easy to combine IR with the balanced PP preparatory sub-sequence.

The IR-type sequences have a number of advantages, of which the two most important ones are:

(a) Assuming ideal inversion and negligible switching-time effects, their relative magnetization variation turns out to be $\nu = (M_{max} - M_{min})/M_{max} = 1 + (B_r/B_p)$ which is always greater than 1 and, for $B_r \geq B_p$, reaches values $\nu \geq 2$. With respect to both NP- and PP-types of sequences, this amounts to better final precision due to an increased magnetization-variation range.
(b) Since, unlike in the PP sequences, the factor ν never crosses zero, the IR sequences can be used throughout the full range of relaxation fields with no necessity of switching the sequence type at some particular relaxation field value. This makes the measurements of whole profiles internally more coherent.

C. Jeener–Broekaert Dipolar Order Relaxation Sequence

The classical Jeener–Broekaert sequence (*133*) is used to determine the dipolar-order relaxation time T_{1D} (in systems of spin 1/2 nuclides) and the T_{1Q} relaxation time (in systems with spin 1 nuclides) of spin 1 nuclides with quadrupolar contributions to T_1. Its FFC version is similar to the Inversion Recovery, except that the first 180° pulse is replaced by the sequence $90_y - \delta - 45_x$, the detection pulse becomes 45_x and a special phase cycle is required. We shall not dwell on the details and purpose of the sequence since they go beyond the scope of this chapter. We wish to underline, however, the fact that sequences of this type require a close coordination of the preparatory sub-sequence with the signal-detection sub-sequence in order to isolate not just a particular magnetization component but a particular relaxation pathway.

XII. Conclusions and Perspectives

The instruments currently produced by Stelar prove that Fast-field-cycling NMR relaxometry instrumentation is industrially viable in the sense that (i) its specifications are reasonably reproducible and (ii) once installed, the reliability of the instruments is comparable to that of any other NMR equipment. The overall performance, unthinkable-of only a few years ago, is best witnessed by the examples shown in Fig. 29.

This, by itself, might be considered as a nice accomplishment. However, the instrumentation is at present still very far from the underlying utopistic ideal represented by a hypothetical system with a maximum field in excess of 10 T, high-resolution-grade field homogeneity and stability, and switching times of the order of just a few microseconds.

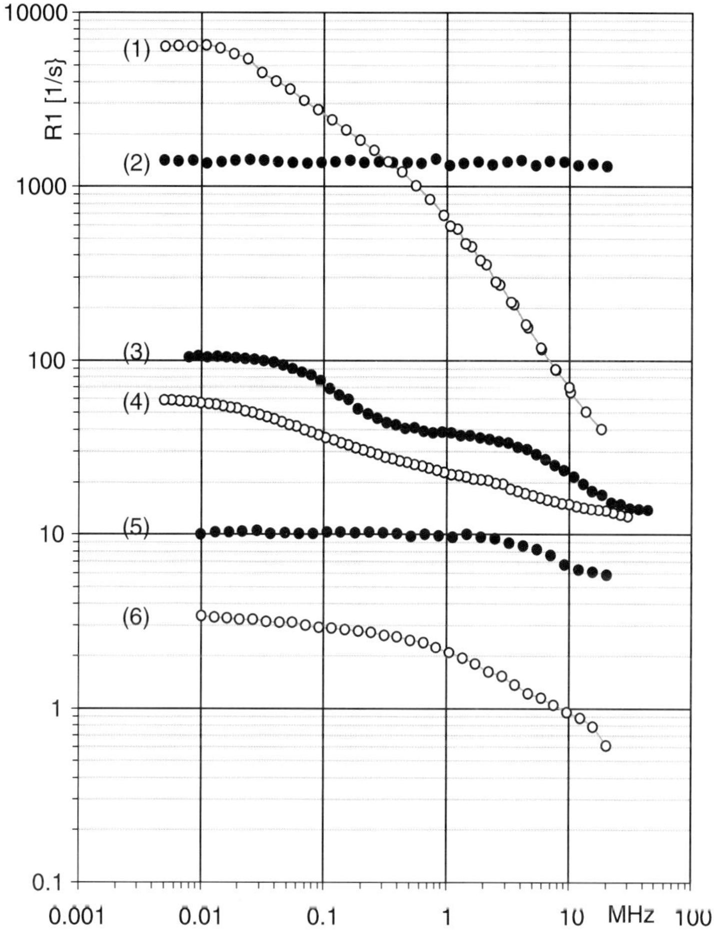

FIG. 29. Examples of FFC NMRD profiles. All NMRD profiles shown here were measured with the Stelar FFC NMR Relaxometer at 25°C using the automated profile acquisition Wizard. The observed nuclide was ^1H in all cases except (5) where it was ^2D. The individual curves show: (1) Parafilm M (American National Can Co.), (2) 2.1M $Dy(ClO_4)_3$ in H_2O (courtesy Dr.L.Holm), (3) 2mM $MnCl_2$ in H_2O, (4) egg yolk, (5) 10 mM Gd^{3+} in D_2O, (6) egg albumin.

A more realistic near-future evolution shall certainly include the following steps, some of which are already under development:

- Further increase of the maximum field beyond the present limit of about 1 T.
- Implementation of efficient devices for the compensation of environmental magnetic fields, both stationary and variable/alternating and a push toward more reliable measurements in the relaxation-field region of 100 Hz–10 kHz.
- Progressive improvements of the system stability and reproducibility.

- Refinements of magnet technology in order to achieve better field homogeneity.
- Use of double-irradiation methods for the exploration of cross-relaxation phenomena.
- Novel approaches to the study of relaxation dynamics in complex systems (sequences, etc).
- A drive towards *high-resolution variable-field FFC NMR relaxometry* (HR-FFC-NMRD), possibly combining the respective advantages of the FFC and the sample-shuttling methods.

A lot depends upon the development of the many potential application fields. Though, from the chemical point of view, FFC is still a low-resolution NMR technique, it has already proved itself as an excellent research-tool in many application areas such as the study of contrast agents (*134–141*), dynamics of prosteins (*142–153*), polymers (*154–162*) and liquid crystals (*163–168*), dynamics of water in rocks and cements (*169–176*), etc.

The increasing interest in FFC NMR relaxometry is also evidenced by the fact that there have been several international conferences in recent years specifically dedicated to the method. Even then, there are still many untapped application areas and additional ones will come to the front with further developments in technology. Opening such novel fields will undoubtedly provide a strong feedback, posing newer challenges to FFC NMR engineering.

References

1. Broer, L. J. F. *Physica* **1943**, *10*, 801.
2. Torrey, H. C. *Phys. Rev.* **1946**, *69*, 680.
3. Pound, R. V.; Purcell, E. M.; Torrey, H. C. *Phys. Rev.* **1946**, *69*, 681.
4. Bloch, F. *Phys. Rev.* **1946**, *70*, 460.
5. Bloch, F.; Hansen, W. W.; Packard, M. E. *Phys. Rev.* **1946**, *70*, 474.
6. Bloembergen, N. "*Nuclear Magnetic Relaxation*"; Martinus Nijhoff: The Hague, 1948, reprinted by Benjamin: New York, 1961.
7. Bloembergen, N.; Purcell, E. M.; Pound, R. V. *Phys. Rev.* **1948**, *73*, 679.
8. Drain, L. E. *Proc. Roy. Soc. (London)* **1949**, *A62*, 301.
9. Torrey, H. C. *Phys. Rev.* **1949**, *76*, 1059.
10. Wright, A. *Phys. Rev.* **1949**, *76*, 1826.
11. Bloch, F.; Hansen, W. W.; Packard, M. E. *Phys. Rev.* **1946**, *69*, 680.
12. Purcell, E. M.; Torrey, H. C.; Pound, R. V. *Phys. Rev.* **1946**, *69*, 36.
13. Gorter, C. J. *Physica* **1936**, *3*, 995.
14. Gorter, C. J.; Broer, L. J. F. *Physica* **1942**, *9*, 591.
15. Ramsey, N. F.; Pound, R. V. *Phys. Rev.* **1950**, *77*, 278.
16. Pound, R. V. *Phys. Rev.* **1951**, *81*, 156.
17. Abragam, A. "*The Principles of Nuclear Magnetism*"; Clarendon Press: Oxford, 1961.
18. Kubo, R.; Tomita, K. *J. Phys. Soc. Japan* **1954**, *9*, 888.
19. Kubo, R. *J. Phys. Soc. Japan* **1954**, *9*, 935.
20. Torrey, H. C. *Phys. Rev.* **1953**, *92*, 962.
21. Koster, G. F.; Slater, J. C. *Phys. Rev.* **1954**, *95*, 1167.
22. Redfield, A. G. *IBM J. Res. Develop.* **1957**, *1*, 19.
23. Bloch, F. *Phys. Rev.* **1957**, *105*, 1206.
24. Argyres, P. N.; Kelley, P. L. *Phys. Rev.* **1964**, *134*, A98.

25. Solomon, I. *Phys. Rev.* **1955**, *99*, 559.
26. Hubbard, P. S. *Phys. Rev.* **1958**, *109*, 1153.
27. Bloembergen, N.; Shapiro, S.; Pershan, P. S.; Artman, J. O. *Phys. Rev.* **1959**, *114*, 445.
28. Gutowsky, H. S.; Lawrenson, I. J.; Shimomura, K. *Phys. Rev.* **1961**, *6*, 349.
29. Shimizu, H. *J. Chem. Phys.* **1962**, *37*, 765.
30. Ter Haar, Ed. "*Fluctuation, Relaxation and Resonance in Magnetic Systems*"; Oliver and Boyd: Edinburgh, 1962.
31. Hubbard, P. S. *Phys. Rev.* **1962**, *128*, 650 and **1963**, *131*, 275.
32. Tomita, K.; Tanaka, M. *Progr. Theor. Phys. (Kyoto)* **1963**, *29*, 528 and **1963**, 29, 651.
33. Runnels, L. K. *Phys. Rev.* **1964**, *134A*, 28.
34. Hilt, R. L.; Hubbard, P. S. *Phys. Rev.* **1964**, *134A*, 392.
35. Redfield, A.G. "*Advances in Magnetic Resonance*", Ed. Waugh, J.; Academic Press, San Diego, 1965, Vol. 1, p. 1.
36. Bonera, G.; Rigamonti, A. *J. Chem. Phys.* **1965**, *42*, 171 and **1965**, *42*, 175.
37. Noggle, J. H. *J. Chem. Phys.* **1965**, *43*, 3304.
38. Gordon, R. G. *J. Chem. Phys.* **1966**, *44*, 228; **1966**, *44*, 1184 and **1966**, *45,* 1635.
39. Blicharski, J. S. *Phys. Letters* **1967**, *24A*, 608.
40. Huntress, W.T.Jr. "*Advances in Magnetic Resonance*", Ed. Waugh, J.; Academic Press, 1967, Vol. 3, pp. 1–42.
41. Gordon, R.G. "*Advances in Magnetic Resonance*", Ed. Waugh, J.; Academic Press, 1968, Vol. 4, p. 1.
42. Blicharski, J. S. *Acta Phys. Polonica* **1969**, *36*, 211; **1970**, *A38*, 19 and **1970**, *A38*, 25.
43. Hubbard, P. S. *J. Chem. Phys.* **1969**, *51*, 1647.
44. McBrierty, V. J.; Douglas, D. C. *J. Magn. Reson.* **1970**, *2*, 352.
45. Ivanov, E. N. *Phys. Status Solidi (Germany)* **1970**, *42*, 453.
46. Sykora, S. *J. Chem. Phys.* **1970**, *52*, 4818 and **1971**, *54*, 2469.
47. Hubbard, P. S. *J. Chem. Phys.* **1970**, *52*, 563.
48. Noack, F. "*NMR-Basic Principles and Progress*", vol. 3; Eds. Diehl, P.; Fluck, E.; Kosfeld, R; Springer Verlag, 1971.
49. Béné, G. J. *Physics Reports* **1980**, *58*, 213.
50. Abragam, A.; Proctor, W. G. *Phys. Rev.* **1958**, *109*, 1441.
51. Schumacher, R. T. *Phys. Rev.* **1958**, *112*, 837.
52. Pershan, P. S. *Phys. Rev.* **1960**, *117*, 109.
53. Pfeifer, H. *Z. Naturforsh.* **1962**, *17a*, 279.
54. Hauser, R.; Noack, F. *Z. Phys.* **1964**, *182*, 93.
55. Johnson, B. C.; Goldburg, W. I. *Phys. Rev.* **1966**, *145*, 380.
56. Sprinz, H. *Ann. Phys.* **1967**, *20*, 168.
57. Hauser, R.; Kolb, H.; Siegle, G. *Z. Angew. Phys.* **1967**, *22*, 375.
58. Koenig, S. H.; Shillinger, W. E. *J. Biol. Chem.* **1968**, *244*, 3283 and **1969**, *244*, 6520.
59. Jones, G. P.; Daycock, J. T.; Roberts, T. T. *J. Phys. E (Sci. Instr.)* **1968**, *2*, 630.
60. Florkowski, Z.; Hennel, J. W.; Blicharska, B. *Nucleonica* **1969**, *14*, 563.
61. Kimmich, R.; Noack, F. *A. Angew. Phys.* **1970**, *29*, 248.
62. Blinc, R.; Luzar, M.; Mali, M.; Osredkar, R.; Seliger, J.; Vilfan, M. *J. Physique Colloque* **1976**, *37*, C3.
63. Edmonts, D. T. *Phys. Rep.* **1977**, *29*, 233.
64. Thayer, A. M.; Pines, A. *Acc. Chem. Res.* **1987**, *20*, 47.
65. Noack, F. "*Progress in NMR Spectroscopy*", vol. 18; Eds. Emsley, J. W.; Feeney, J.; Sutcliffe, L. H.; Pergamon Press: Oxford, **1986**.
66. Koenig, S. H.; Brown, R. D. "*Progress in NMR Spectroscopy*", vol. 22; Eds. Emsley, J. W.; Feeney, J.; Sutcliffe, L. H.; Pergamon Press: Oxford, **1990**.
67. Schweikert, K. H.; Krieg, R.; Noack, F. *J. Magn. Reson.* **1988**, *78*, 77.
68. Hoult, D. I.; Richards, R. E. *J. Magn. Reson.* **1976**, *24*, 71.
69. Sykora, S. *Magn. Reson. Imaging* **1991**, *9*, 833.
70. Anoardo, E.; Ferrante, G. M. *Appl. Magn. Reson.* **2003**, *24*, 85–96.
71. Kroon, D. J. "*Laboratory Magnets*" *Philips technical library* –1968.
72. Anoardo, E.; Galli, G.; Ferrante, G. M. *Appl. Magn. Reson.* **2001**, *20*, 365–404.

73. Grössl, G.; Winter, F.; Kimmich, R. *J. Phys.* **1985**, *E 18*, 358.
74. Packard, M.; Varian, R. *Phys. Rev.* **1954**, *93*, 941.
75. Redfield, A. G.; Fite, W.; Bleich, H. E. *Rev. Sci. Instrum.* **1968**, *39*, 710.
76. Kimmich, R. *Bull. Magn. Reson.* **1980**, *1*, 195.
77. Seitter, R.-O.; Kimmich, R. "*Encyclopedia of Spectroscopy and Spectrometry*", Eds. Lindon, J. C.; Tranter, G. E.; Holmes, J. L.; Academic Press: London, **1999** 2000–2008.
78. Brown, R.D.; Koenig, S.H., **1977**, IBM Research Rep. RC6712, Yorktown Heights.
79. Westphal, W. *Handbuch der Physik Bd. XV (Magnetismus/Elektromagnetisches Feld)* 1927.
80. Hart, P. J. *Universal Tables for Magnetic Fields of Filamentary and Distributed Circular Currents*, Elsevier, New York, 1967.
81. Kohlrausch, F. *Praktische Physik Bd. II* **1968**.
82. Eder, F. X. *Moderne Methoden der Physik Bd. II* **1972**.
83. Schnell, G. *Magnete, Grundlagen, Aufbau, Anwendungen.* **1973**.
84. Neuman, H. *Archiv für Technisches Messen* **1940**.
85. Garret, M. W. *J. Appl. Phys* **1951**, *22*, 1091; **1967**, 38, 2563; **1969**, 20, 3171.
86. Saint-Jalmes, H.; Taquin, J. *Rev. Sci. Instrum.* **1981**, *52*, 1501.
87. Ruark, A. E.; Peters, M. F. *J. Opt. Soc. Am.* **1926**, *131*, 205.
88. Fanselau, G. *Z. Phys.* **1936**, *54*, 260.
89. McKeehan, L. W. *Rev. Sci. Instrum.* **1936**, *7*, 150.
90. Hak, J. *Arch. Elektrotech.* **1936**, *30*, 736.
91. Montgomery, D. B. "*Solenoid Magnet Design*"; Wiley Interscience, **1969**.
92. Cesnak, L.; Kabat, D. *J. Phys.* **1972**, *5*, 944.
93. Reményi, G.; Kirchner, I.; Poiuesz, T. *Cryogenics* **1977**, *17*, 565.
94. Grössl, G.; Winter, F.; Kimmich, R. *J. Phys. E. (Sci. Instr.)* **1985**, *18*, 538.
95. Gaume, F. "*High Magnetic Fields*", Eds. Kolm, H.; Lax, B.; Bitter, F.; Mills, R; Wiley: New York, **1962**.
96. Hart, P. J."*Universal Tables for Magnetic Fields of Filamentary and Distributed Circular Currents*"; Elsevier: New York, 1967.
97. Schnell, G.; Thiemig, K. "*Magnets: Basic Relations, Engineering, Applications*"; Munich, 1973.
98. Pucell, E. M.; Pound, R. V. *Phys. Rev.* **1950**, *77*, 279.
99. Hahn, E. L. *Phys. Rev.* **1950**, *80*, 580.
100. Carr, H. Y.; Purcell, E. M. *Phys. Rev.* **1954**, *94*, 630.
101. Meiboom, S.; Gill, D. *Rev. Sci. Instrum.* **1958**, *29*, 688.
102. Powles, J. G.; Mansfield, P. *Phys. Letters* **1962**, *2*, 58.
103. Slichter, C. P. "*Principles of Magnetic Resonance*", 2nd edn.; Harper & Row: New York, 1963; Springer-Verlag: New York, 1978.
104. Freeman, R; Wittekoek, S. *Colloque Ampère* **1969**, *XV*, 205.
105. Farrar, T. C.; Becker, E. D. "*Pulse and Fourier Transform NMR*"; Academic Press: New York, 1971
106. Fukushima, E.; Roeder, S. B. W."*Experimental Pulse NMR. A Nuts and Bolts Approach*"; Addison-Wesley: New York, 1981
107. Gerstein, B. C.; Dybowski, C. R."*Transient Techniques in NMR of Solids*"; Academic Press: London, 1985
108. Freeman, R.; "*Nuclear Magnetic Resonance*"; Longman, 1987.
109. Homans, S. W. "*A Dictionary of NMR Concepts*"; Oxford University Press: Oxford, 1989.
110. Kimmich, R. "*NMR Tomography, Diffusometry, Relaxometry*"; Springer-Verlag: Berlin, 1997.
111. Ernst, R. R.; Bodenhausen, G.; Wokaun, A. "*Principles of Nuclear Magnetic Resonance in One and Two Dimensions*"; Clarendon Press: Oxford, 1987.
112. Atta-ur-Rahman "*One and Two Dimensional NMR Spectroscopy*"; Elsevier: Amsterdam, 1989.
113. Bonera, G.; Paolucci, G.; Ruffato, C.; Sykora, S. "*Risonanza Magnetica Nucleare in Medicina*"; Piccin Nuova Libreria: Padova, 1984.
114. Ernst, R. R. *Quart. Rev. Biophys.* **1987**, *19*, 183.

115. Haacke, E. M.; Brown, R. W.; Thomson, M. R.; Venkatesan, R. "*Magnetic Resonance Imaging. Physical Principles and Sequence Design*"; Wiley-Liss, 1999.
116. Zhi-Pei, L.; Lauterbur, P. C. "*Principles of Magnetic Resonance Imaging*"; IEEE Press, 2000.
117. Stilbs, P. "*Progress in NMR Spectroscopy*", vol. 19; Eds. Emsley, J. W.; Feeney, J.; Sutcliffe, L. H.; Pergamon Press: Oxford, **1987**.
118. Lindon, J. C.; Ferrige, A. G. "*Progress in NMR Spectroscopy*", vol. 14; Eds. Emsley, J. W.; Feeney, J.; Sutcliffe, L. H.; Pergamon Press: Oxford, **1980**.
119. Brown, R. J. S. *J. Magn. Reson.* **1989**, *82*, 539.
120. Clayden, N. J.; Hesler, B. D. *J. Magn. Reson.* **1992**, *98*, 271.
121. Lin, Y. Y.; Ge, N. H.; Hwang, L. P. *J. Magn. Reson.* **1993**, *A105*, 65.
122. Jakes, J. *Czech. J. Phys.* **1993**, *B43*, 1.
123. Lupu, M.; Todor, D. *Chemometrics and Intelligent Lab. Systems* **1995**, *29*, 11.
124. Borgia, G. C.; Brown, R. J. S.; Fantazzini, P. *J. Magn. Reson.* **1998**, *132*, 65.
125. Borgia, G. C.; Brown, R. J. S.; Fantazzini, P. *Magn. Reson. Imaging* **1998**, *16*, 549.
126. Borgia, G. C.; Brown, R. J. S.; Fantazzini, P. *J. Magn. Reson.* **2000**, *147*, 273.
127. Borgia, G. C.; Brown, R. J. S.; Fantazzini, P. *Magn. Reson. Imaging* **2001**, *19*, 473.
128. Song, Y. Q.; Venkataramanan, L.; Huerlimann, M. D.; Flaum, M.; Frulla, P.; Straley, C. *J. Magn. Reson.* **2002**, *154*, 261.
129. Weiss, G. H.; Ferretti, J. A. "*Progress in NMR Spectroscopy*", Eds. Emsley, J. W.; Feeney, J.; Sutcliffe, L. H.; Pergamon Press: Oxford, **1988**, Vol. 20, p. 317.
130. Levitt, M. H. "*Progress in NMR Spectroscopy*", Eds. Emsley, J. W.; Feeney, J.; Sutcliffe, L. H.; Pergamon Press: Oxford, **1986**, Vol. 18, p. 61.
131. Patt, S. L. *J. Magn. Reson.* **1982**, *49*, 161.
132. Gerothanassis, I. P. "*Progress in NMR Spectroscopy*", vol. 19; Eds. Emsley, J. W.; Feeney, J.; Sutcliffe, L. H.; Pergamon Press: Oxford, **1987**.
133. Jeener, J.; Broekaert, P. *Phys. Rev.* **1967**, *157*, 232.
134. Aime, S.; Botta, M.; Fasano, M.; Terreno, M. *Chem. Soc. Rev.* **1998**, *27*, 19.
135. Aime, S. *Angew. Chem. Int. Ed.* **2000**, *39*, 747.
136. Muller, R.N.; Raduchel, B.; Laurent, S.; Platzek, J.; Pierart, C.; Mareski, P.; Vander Elst, L. *Eur. J. Inorg. Chem.* **1999**, 1949–1955.
137. Toth, E.; Connac, F.; Helm, L.; Adzamli, K.; Merbach A. *Eur. J. Inorg. Chem.* **1988,** 2017.
138. Gossuin, Y.; Roch, A.; Lo Bue, F.; Muller, R.; Gillis, P. *Magn. Reson. Med.* **2001**, *46*, 476–481.
139. Roch, A.; Muller, R.; Gillis, P. *J. Magn. Reson. Imaging.* **2001**, *14*, 94–96.
140. Çorsi, DM.; Vander Elst, V.; Muller, R. N.; van Bekkum, H.; Peters, J. A. *Chemistry* **2001**, *7*, 1383–1389.
141. Comblin, V.; Gilsoul, D.; Hermann, M.; Humblet, V.; Jacques, V.; Mesbahi, M.; Sauvage, C.; Desreux, J. F. *Coord. Chem. Rev.* **1999**, *185–186*, 451.
142. Bertini, I.; Luchinat, C. *Coord. Chem. Rev.* **1996**, *150*, 1.
143. Bertini, I.; Luchinat, C.; Brown, R. D.; Koenig, S. H. *J. Am. Chem. Soc.* **1989**, *111*, 3532–3536.
144. Koenig, S. H.; Schillinger, W. E. *J. Biol. Chem* **1969**, *244*, 3283–3289.
145. Banci, L.; Bertini, I.; Hallewell, R. A.; Luchinat, C.; Viezzoli, M. S. *Eur. J. Biochem.* **1989**, *184*, 125–129.
146. Kimmich, R.; Noack, F. *Z. Naturforsch.* **1970**, *25a*, 299–301.
147. Kimmich, R.; Noack, F. *Z. Naturforsch.* **1970**, *25a*, 1680–1684.
148. Kimmich, R. *Z. Naturforsch.* **1971**, *26b*, 1168–1170.
149. Kimmich, R. *Makromol. Chem., Macromol. Symp.* **1990**, *34*, 237–248.
150. Korb, J. P.; Bryant, R. G. *Magn. Reson. Med.* **2002**, *48*, 21.
151. Korb, J. P.; Bryant, R. G. *J. Chem. Phys.* **2001**, *115*, 10964.
152. Korb, J. P.; Van-Quynh, A.; Bryant, R. G. *Chem. Phys. Lett.* **2001**, *339*, 77.
153. Kimmich, R. "*Encyclopedia of Nuclear Magnetic Resonance*", vol. 5; John Wiley Ltd: Chichester, **1996**, pp. 3083–3088.
154. Bryant, R. G.; Brown, R. D.; Koenig, S. H. *Biophys. Chem.* **1982**, *16*, 133–137.
155. Halle, B.; Denisov, V. P.; Venu, K. "*Biological Magnetic Resonance*"; Kluwer-Plenum: New York, 1990.
156. Kimmich, R.; Winter, F. *Progr. Colloid Polymer Sci.* **1985**, *71*, 66–70.
157. Kimmich, R.; Winter, F.; Nusser, W.; Spohn, K.-H. *J. Magn. Res.* **1986**, *68*, 263–282.

158. Zeumer, U.; Dippel, T.; Noack, F.; Muller, K.; Mayer, C.; Heaton, N.; Kothe, G. *J. Chem. Phys.* **1992**, *97*, 3794.
159. Voigt, G.; Kimmich, R. *Polymer* **1980**, *21*, 1001.
160. Koch, H.; Bachus, R.; Kimmich, R. *Polymer* **1980**, *21*, 1009.
161. Stapf, S.; Kimmich, R. *Macromolecules* **1996**, *29*, 1638–1641.
162. Kimmich, R. *Polymer* **1975**, *16*, 851–852.
163. Noack, F.; Schweikert, K. H. "*Molecular Dynamics in Liquid Crystals*", Eds. Luckhurst, G. R.; Veracini, C. A.; Kluwer Academic Publications: London, **1994**, p. 233.
164. Struppe, J.; Noack, F.; Klose, G. *Z. Naturforsch* **1997**, *A 52*, 681–694.
165. Ribeiro, A. C.; Sebastiao, P. J.; Vilfan, M. *Liq. Cryst.* **1988**, *3*, 937–945.
166. Anoardo, E.; Pusiol, D. *Phys. Rev. Lett.* **1996**, *76*, 3983–3986.
167. Burnell, E. E.; Capitani, D.; Casieri, C.; Segre, A. L. *J. Phys. Chem.* **2000**, *B104*, 8782.
168. Stapf, S.; Kimmich, R. *Mol. Physics* **1997**, *92*, 1051–1060.
169. Korb, J. P.; Hodges, M. W.; Bryant, R. G. *Phys. Rev. E.* **1997**, *56*, 1934–1945.
170. Stapf, S.; Kimmich, R.; Nieß, J. *J. Appl. Phys.* **1994**, *75*, 529–537.
171. Stapf, S.; Kimmich, R.; Seitter, R. O. *Phys. Rev. Lett.* **1995**, *75*, 2855.
172. Korb, J. P.; Hodges, M. W.; Gobron, T.; Bryant, R. G. *Phys. Rev. E.* **1999**, *60*, 3097–3106.
173. Levitz, P.; Korb, J. P.; Bryant, R. G. *J. Chem. Phys.* **1999**, *96*, 1494–1505.
174. Korb, J. P.; Whaley, M.; Bryant, R. G. *Phys. Rev.* **1997**, *E 56*, 1934–1945.
175. Barberon; Korb, J.-P.; Petit, D.; Morin, V.; Bermenjo, E. *Phys. Rev. Lett.* **2003**, *90*, 116103.
176. Porteneuve, C.; Korb, J.-P.; Petit, D.; Zanni, H. *Cement Concrete Res.* **2002**, *32*, 97.

INDEX

CONTENTS OF PREVIOUS VOLUMES

VOLUME 52

VOLUME 53

VOLUME 54

VOLUME 55